MOLD: The War Within

Win the Battle Against Mold & Chemical Exposures
Arm Yourself with Knowledge from
Experts in Mold, Health, & Nutrition

By: Kurt and Lee Ann Billings

Partners Publishing LLC
Kodak, Tennessee

Partners Publishing LLC
Kodak, Tennessee

MOLD: The War Within

Copyright © 2007, 2010, 2013, 2018 Kurt and Lee Ann Billings

DISCLAIMER:
The information and research presented in this book is for educational purposes only. It is not intended to be medical advice or replace the services of a licensed health care professional. The supplements referenced in this book have not been evaluated by the Food and Drug Administration and are not approved drug products. The information and/or products presented in this book are not intended to diagnose, treat, cure, or prevent any disease. All physical and mental conditions should be diagnosed, treated, and monitored by a licensed, knowledgeable health care provider. The publisher and authors are not liable and/or responsible for any loss, injury, and/or damage allegedly arising from the information and/or use of any products mentioned in this book.

ISBN 978-0-9721016-0-8

Published by:
Partners Publishing LLC
P.O. 441, Kodak, Tennessee 37764
www.MoldTheWarWithin.com

Printed in the United States of America
First Printing June 2007; Second Edition Printing August 2010; Third Edition Printing June 2013, Fourth Edition Printing 2018,

Logo design by Mariah Billings
Logo copyright © Mariah and Matthew Billings

Cover photo of Stachybotrys provided by Dennis Kunkel: www.denniskunkel.com. Front cover image copyright © Dennis Kunkel Microscopy, Inc.

What Are Professionals Saying about the Book

MOLD: The War Within?

Mold: The War Within provides an excellent guidebook to patients and others who suspect they are suffering from mold poisoning.
—Raymond Singer, Ph.D., FACPN, FNAN, FAPS

A Must-read for Anyone Exposed to Toxic Mold. The Billings should be commended on their extensive research and tireless work on this book. Their contribution to the field of toxic mold poisoning is monumental. This book is very well written and organized and loaded with important information, so it can be referenced time and again as those exposed to toxic mold struggle in their journey back to health. —Mary Short Ray, DO, Environmental Specialist

One excellent resource is the book *Mold: The War Within* by Kurt and Lee Ann Billings. The Billings learned the hard way about the damaging health effects of mold—and the level of ignorance about mold's effects by the medical profession as a whole. —Joseph Mercola, DO

Mold: The War Within, is an excellent, easy-to-read, eye-opening verification of the major public health problem we have in this country with mold and mycotoxin exposure—best illustrated after floods and hurricanes.
—Lisa Nagy, MD, Environmental and Integrative Medicine Specialist

Mold: The War Within provides one of the few available maps to guide patients through the science of mold-related illnesses and a view of both natural and pharmaceutical treatment options in the field. —Andrew Lange, N.D.

As a "victim" of the bureaucracies, ill-informed medical community and confusing, sometimes contradictory, regulations, Kurt and Lee Ann waded through the mire of information and emerged with a clear, comprehensive understanding of the problem and solution. They have unselfishly shared this information through their book and are probably unaware of the thousands of people they have helped through this effort.—Doug Hoffman, Executive Director of the National Organization of Remediators and Mold Inspectors (NORMI)

The Billings offer an educational process, albeit technical at some points (out of necessity), that is truly second to none.
—Doug Kaufmann, national TV host of *Know the Cause*

This book will be of value to the lay person who might have just experienced a mold infestation, because it contains information from scientists who are active in the field and provides some of the latest up-to-date scientific information in the field of mold. Of further value are explanations, for the novice, of some of the tecniques on the procedures that mold inspectors use to examine houses for mold.
—David C. Straus, PhD

FOREWORD BY:

Doris J. Rapp, MD, FAAP, FAEM

The book *Mold: The War Within* by Kurt and Lee Ann Billings is truly outstanding. They personally learned the pervasive nature of molds and chemicals when their family became sick from exposures to hurricane-size molds and toxins during and after Katrina. The Billings knew that their family was only one among many sick from Katrina exposures as cataclysmic levels of molds and toxins had devastated so many families. In addition, after months of trials and tribulations, they came to realize that all across our nation and even in other countries people are suffering health challenges from unavoidable exposures to molds and chemicals. *MOLD: The War Within* is not just a "Katrina" book, but far more. It is a source of valuable information for the growing ranks of mold and chemical survivors.

Each year countless numbers of mold and chemical victims exposed through hurricanes, flooding, water leaks, oil spills, and sick buildings are faced with similar unexpected monumental challenges. Until now, there was a paucity of places to turn for inclusive help or knowledge. In their search for answers to their family's health problems, the Billings have written the ultimate "Why, When, and How-to" book. *MOLD: The War Within* confirms the many frustrating and challenging changes that can occur in the health and lives of people who have been exposed to molds, formaldehyde (in trailers), and other harmful pollutants or chemicals. It gives people hope and living proof that recovery is possible; it enables people to make informed health decisions by providing information (in an understandable and practical format) with which people can educate and help themselves.

One of the most significant reasons their book is so valuable is that they made their readers aware that molds can cause much more than respiratory and skin problems. Many common, but less, well-acknowledged symptoms of mold-related illnesses (see pages 116 and 119) are all too often wholly attributed to emotions or deep-seated psychological issues. While stresses from mold symptoms, remediation factors, and sudden unanticipated expenses cannot be ignored, too few recognize the full devastating scope of the purely medical aspects of mold-related illnesses. The potential immense variety of symptoms that can be caused by such exposures are unfortunately not fully appreciated, or even believed, by many in the medical profession. These symptoms can include medical complaints, such as depression, fatigue, joint pain, twitching, heart symptoms, and headaches. Some notice inability to walk, write, speak or learn normally. Others have an inability to act and behave in an acceptable manner. Unfortunately, these types of complaints are not typically recognized or even considered as possible manifestations of mold and/or mold toxin sensitivities.

As a physician practicing for well over 40 years, I have dedicated my life to providing conventional medical care while keeping an open mind to complementary and alternative medicine. After the first twenty years in practice, I became aware, for the first time, of a process called provocation and neutralization allergy treatment. It is a newer version of what many of you know as "allergy shots," but this method can demonstrate exactly how mold, pollen, dust, and food exposures can cause specific symptoms and how they can then be treated. When I started trying this method of testing for molds, I was astounded to find that discrete areas of the brain, for example, could be affected so some people became tired and docile while others might become aggressive, angry, and hyperactive when tested with a drop of a mold allergy extract. Allergies and sensitivities to molds are especially problematic to health because, as the Billings so thoroughly explain in their book, toxigenic molds are a double-faceted foe. The mold itself can grow in and on human tissues and cavities, such as the lungs, and the mycotoxins they produce can cause chronic systemic poisoning.

I have witnessed and documented untold numbers of allergic reactions in children and adults over the years as an allergist. In my book, *Is This Your Child?* I show some examples of how allergies manifest themselves in children—red ears, puffy and wrinkled eyes, mottled tongues, and red cheeks—and provide pictures of hand-writing changes in children and adults before and after allergy treatments. Readers really identified with the allergy characteristics outlined in this book which became a New York Times Best Seller.

There are many paths to healing. If one doesn't work, try another, especially if it is safe and possibly effective. After conventional treatment methods failed the Billings, they embarked on a natural treatment plan that included diet modification and supplements known to detox the body and boost the immune system. I applaud the Billings for their initiative and ingenuity in seeking and identifying an effective, non-invasive form of treatment for them that was backed by sound scientific and medical logic. I am all for safer, easier, and less expensive forms of treatment that help, even if they would be considered nonconventional by most medical doctors. As you read about the various components of the Billings' treatment plan, you will be amazed at the simplicity and practicality of most of what they found helpful. For example, they supplemented with magnesium, which I have recommended to patients for years. It is more important than most people realize. Many patients have reported that it promotes strong bones, healthy muscles and nerves, keeps the digestive tract working properly, and sometimes even reduces heart pain. Check with your doctor to find out if it might help you.

As a pediatrician, I have grave concern about the toxic overload that is taking place in our bodies today—especially in the bodies of our children. Detoxification regimes, such as what the Billings followed and detailed in their book, are an integral part of maintaining and restoring health in today's toxic world. Parents especially need to be aware of the effects of toxic buildup in their children, which is the reason I wrote another book *Is This Your Child's World?* We must all care enough about our health and the health of our children to read books like *MOLD: The War Within* and *Is This Your Child's World?* to better equip ourselves with knowledge on how to retain our health while living in a dangerously toxic world.

Too many people continue to hope for our government to take appropriate and responsible action for what has turned their lives and health upside down, be it devastation from natural disasters, toxic exposures in the workplace, or mold-deteriorating schools. It took a while for Gulf Coast residents, rescue workers, and remediation crews who were harmed from Katrina exposures to find out that genuine meaningful help was not forthcoming. This book, *Mold: The War Within*, explains in detail why the Katrina fiasco and deception continues to remain an open, oozing sore and how the lessons learned from Katrina can benefit us all.

This book provides the kind of insight and knowledge you need to know to help resolve some of the health and other challenging issues many mold- and chemical-sensitive individuals face. It is all there, in this outstanding book. I hope everyone will not only read but study this book if they have mold and chemical issues, because so many who claim to be informed in the health field do not fully appreciate the immense scope of illness that molds or chemicals can cause. *MOLD: The War Within* offers practical ("I've been there") realistic help and suggestions so others who are stressed and sick from recent or long-term water or mold-related damage realize their health can be acutely and chronically harmed physically, emotionally and mentally. This book will share in detail what you can do and where you can go for the type of help and assistance so many people need when battling illnesses from mold and chemical exposures.

Doris J. Rapp, MD, FAAP, FAEM
Board Certified in Pediatrics, Allergy and Environmental Medicine
Clinical Ass't Prof. Pediatrics (emeritus) at SUNYAB
New York Times Best-selling Author of *Is This Your Child?*
www.drrapp.com for DVDs of reactions to molds

TABLE OF CONTENTS

reported physical and mental health effects from mold and mycotoxin exposures, and Sudden Infant Death Syndrome (SIDS).

SECTION TWO: HEALTH MATTERS
The chapters in this section contain potentially life-saving information on mold, mycotoxin, and other toxic exposures regarding health issues; avenues of re-exposure; risks of certain pharmaceutical treatments; effective and ineffective alternative treatments; mycotoxin contamination of food; and governmental regulation of mycotoxins in food and animal feed.

4. Andrew Puccetti, PhD, CIH (certified industrial hygienist), has been an independent consultant providing litigation support in the field of industrial hygiene for the past 25 years.   **372**

5. Jim Pearson, CMH (certified mechanical hygienist), is president and CEO of Americlean Corporation, a disaster restoration company that he helped start over 27 years ago. Mr. Pearson is coauthor of IICRC S520 Standard and Reference Guide for Professional Mold Remediation.   **382**

6. Richard L. Lipsey, PhD, is a forensic toxicologist and president of Lipsey & Associates Inc. Dr. Lipsey has served as the liaison professor with and consulted for the USDA, the EPA, and the U.S. State Department, both nationally and internationally.   **391**

7. Cynthia Coulter Mulvihill, Attorney at Law, is the managing partner at the Law Offices of Cynthia Mulvihill. She specializes in mold-related legal cases and provides consultation services.   **399**

8. Melinda Ballard, MBA, founded Policyholders of America (POA), a nonprofit organization that helps policyholders receive all the benefits to which they are entitled after she legally battled an insurance giant over a water-related loss in a lawsuit that drew media attention worldwide.   **411**

9. Jia-Sheng Wang, MD, PhD, is a professor in and head of the Department of Environmental Sciences, College of Public Health at the University of Georgia, and has authored/coauthored over 80 peer-reviewed publications. Dr. Wang is one of several doctors who first identified the two major factors that cause liver cancer (hepatitis B and aflatoxin) while conducting research in Qidong and Guangxi in China nearly 30 years ago.   **425**

10. David L. Eaton, PhD, is a professor of environmental and occupational health sciences, Toxicology Program at the University of Washington (UW) in Seattle; associate vice provost of research, School of Public Health & Community Medicine at UW; and director of the Center for Ecogenetics and Environmental Health at UW.   **437**

11. Regina M. Santella, PhD, is a professor of environmental health sciences at the Mailman School of Public Health at Columbia University, director of the Epidemiology Program at the Herbert Irving Comprehensive Cancer Center (HICCC), director of the Biomarkers Core Facility at HICCC, and director of the Columbia Center for Environmental Health in Northern Manhattan.   **451**

ACKNOWLEDGMENTS

This book would not have come into being without the valuable information from leading professionals in the fields of science and medicine who have painstakingly and meticulously documented the health effects of mold and mycotoxin exposure in published, peer-reviewed reports and studies. We thank them for chronicling the history and health dangers of mold and mycotoxins and creating a base of literature from which others could draw. Furthermore, we thank the many professionals who so generously gave of their time, providing multiple interviews, written correspondences, references to published literature, referrals to colleagues and peers, and avenues of product research.

We thank David Backstrom, ND; Charles Bacon, PhD; Timothy Callaghan, MD, DC; Linda Casey; Ginger Chew, ScD; David Christopher, MH; Udo Erasmus, PhD; Michael Fagan; Becky Gillette; Thad Godish, PhD; Michael Gray, MD, MPH, CIME; Ellen Harrison, MS; Brian Issell, MD; Dennis Kunkel, PhD; Curt Kurtz, MD; Anthony Lupo; David Marcus; Scott Needle, MD; Katrine Stevens; and Garnett Wood, PhD. We especially thank Melinda Ballard, MBA; David Eaton, PhD; Richard Lipsey, PhD; Cynthia Mulvihill, JD; Jim Pearson, CMH; Andrew Puccetti, PhD, CIH; Regina Santella, PhD; Gina Solomon, MD, MPH; David Straus, PhD; Jack Thrasher, PhD; and Jia-Sheng Wang, MD, PhD. Their contributions to this project, which have been invaluable, enabled us to provide readers with knowledge far beyond what would be expected to be found in the confines of one book.

We give many thanks to Doris J. Rapp, MD, for her enthusiasm and support of our efforts to educate the public about the health effects of mold and chemical exposures. She has been an inspiration to us since the early 90's when we first started reading her books. To get to speak with her personally and have her write a foreword to our book, has been a real honor.

Additionally, we thank all the individuals who provided support during this project including our dedicated transcriptionists, proofreaders, and test-market readers; book cover designer; logo designer; and those individuals who generously provided products, some of which became effective components of our treatment plan. We are especially grateful to Kelley Shoffner, without whose help and inspiration from God we may never have fully recovered; to our children who are a constant inspiration; and to God. We pray that others will be led in their own inspirational journey of healing as we share the wisdom, knowledge, and insight of those people who helped us understand the complex world of mold.

God Bless. May each day bring His knowledge and inspiration for a fruitful life.

PREFACE

The August 29, 2005, thrashing from Hurricane Katrina ravaged the landscape of hundreds of miles of southern coastal states and altered the lives of millions of people: those personally devastated by loss of life and stripped of generations of material heritage; and those who suffered illness, sometimes fatal, from exposure to the proliferating mold and toxic cocktails that permeated the air, ground, and water. The assault on health was not limited to just hurricane-hit area residents as those who came to help (first-response rescuers, medical teams, church volunteers, professional remediators, building specialists, and others) were also exposed to the health-degrading toxins in varying degrees. Likewise, the life of our family was affected, forever changed.

In *MOLD: The War Within* we share how our family's health was systemically "ambushed" by the toxic effects from airborne fungi, stemming from mold growth in our house that occurred from water leaks during the hurricane. Exposure to these indoor contaminants was compounded by the additional exposure to outside airborne contaminants: fungi, other biologicals, chemicals, and aerosolized debris that was prevalent in the aftermath of Hurricane Katrina. We share our story of exposure, the ensuing illnesses, and the challenges of medical misdiagnoses and medical mistreatments made by ignorant and misperception-filled medical professionals. This lack of knowledge in the medical community left us no choice but to research scientific and medical literature ourselves, seeking not only answers as to why we had gotten so sick, but also searching for a treatment plan that would effectively restore our health. This book originated in our quest for information, which led to months of extensive research, countless interviews, and the revelation of shocking facts that still astound us to this day. We detail our family's medical trials and errors, and share what worked medically and nutritionally for us—what didn't—and why.

Since the purpose of sharing our story is not to focus on the individual medical professionals in our case who were uninformed and/or refuted the documented scientific and medical findings that prove mold *can* make you sick, we felt it appropriate to withhold the names of all the doctors who medically treated our family. Instead, we devote our energy to the educational process and make available information that may not be readily accessible. For example, we provide answers straight from experts in both scientific and medical fields, who clarify and explain complex topics and data from cutting-edge research published in peer-reviewed scientific and medical journals and governmental documentation.

It is our deepest desire that this book, *MOLD: The War Within*, will provide the much-needed information to help those individuals sick from toxic exposures from mold and chemicals—from any source—and expose the inclination of medical professionals to diagnose the multi-systemic symptoms of mold and mycotoxin exposures as mental disorders. Consulting expert Timothy Callaghan, MD, DC, reports that quite often an incorrect psychiatric diagnosis is made by the examining physician in order to have something to treat. These types of misdiagnoses are perpetuated by barriers of ignorance created by lack of knowledge and, sadly, are not isolated incidents. We sincerely hope that *MOLD: The War Within* will help open the doors of public knowledge and understanding in regard to the health effects of mold, mycotoxin, and chemical exposures, enlighten those uninformed, and foster an environment of learning and healing.

As we journeyed through the medical mine field of mold treatments, we became increasingly empathetic of other mold victims as we experienced how challenging it was to restore health after becoming sick from mold and chemical exposures, and how difficult it was to find a doctor knowledgeable in this area who could provide an effective treatment plan. The more medical travails we endured, the more driven we became not only to research the cause of our deteriorating health and identify a solution that would effectively restore our health to pre-exposure vitality, but also to share the knowledge we had gained from making sense of reams of scientific and medical data with the help of experts. We wanted others struggling with the same life-threatening, systemic effects from toxic exposures to be able to learn from our research, draw their own conclusions, and potentially benefit as well. For these reasons, we have heavily documented *MOLD: The War Within* to ensure readers the veracity and integrity of our research.

It should be noted that although the book addresses many issues specific to Hurricane Katrina, New Orleans, and the other affected areas, the information herein is applicable to everyone—everywhere—as ill health effects can occur anytime enough structural moisture is present to support fungal growth, regardless of location. Mold exposure becomes most problematic when an indoor source of mold growth develops, concentrating the contaminants in the indoor air. The health effects of exposure to indoor sources of mold contaminants are addressed in *MOLD: The War Within*. Obviously, issues of toxic chemical exposures are not limited to just hurricane-devastated areas, and will be of interest to all.

To increase ease of clarity and readability, we wrote *MOLD: The War Within* in first person (Lee Ann), instead of third person or identifying one or the other of us throughout the book. Furthermore, we divided the book into three sections. Section One focuses on mold, mold-related illnesses, and toxic exposures; Section Two focuses on health issues and alternative treatment solutions; and Section Three contains full-length interviews with medical, scientific, legal, and industry professionals with whom we consulted. Because of the contributions of these professionals and the many others with whom we interviewed, some of which information is included in the text, *MOLD: The War Within* addresses a wide range of topics not normally found in between the pages of just one book.

It should be noted that the material in each chapter, which includes the latest research data from leading experts, builds on the information collectively presented in previous chapters. Thus, it will be helpful to most to read the book from front to back. Essentially, the book is designed to bring readers up to speed—scientifically, medically, and legally—by the end of Section Two so readers have the knowledge base to understand and enjoy the full-length, in-depth interviews with the professional experts that are presented in Section Three.

The professionals who provided information for this book have done so in order to share their expertise and foster a learning process. Inclusion in this book does not constitute support of or endorsement of the authors thoughts, conclusions, or the products and treatment plan that ultimately restored our family's health. Likewise, we are not recommending or endorsing treatment from or the treatment methods recommended by the professionals quoted in this book. Any consideration to use any of the treatment options mentioned in this book should be evaluated by each person's licensed physician. We provide this information solely for educational purposes.

The many levels of information in *MOLD: The War Within* will "arm" readers with knowledge to more successfully combat the health effects from mold and chemical invasions that may occur in their lives. We hope with all sincerity that this book ignites in readers a driving desire to restore, preserve, and protect their health from toxic exposures with as much intensity as has been awakened in us. Our family has not only completely recovered our pre-hurricane level of health, but we now enjoy an even higher quality of life.

God Bless!

SECTION ONE

MOLD & TOXINS

CHAPTER ONE:

MOLD CAN'T HURT YOU—OR CAN IT?

Most people who get exposed to mold and chemicals do so in the quiet of their own homes or workplaces. Oftentimes this silent attack on health occurs in the aftermath of a hurricane, flood, roof leak, or some other cause of wet building materials—not during the harrowing event itself. Our family's experience was no different. We had not hung from our rooftop nor had our house collapsed on top of us during Hurricane Katrina. In fact, our town, Lake Charles, Louisiana, was located on the western side of the designated disaster area and was hit by only the outer bands of the historic hurricane; yet wind-driven rains penetrated the roof of our rental home leaving streaks of moisture on portions of the walls, creating a perfect breeding ground for mold. Unknown to us at the time, the wrath of Hurricane Katrina had disrupted untold numbers of mold spores, fragments, and other debris into the air. As the outside spore-laden air floated inside, spores germinated on the moist sheetrock paper. The early stages of mold growth were evident as the streaks on the walls darkened. Calls to our landlord went unheeded. Soon thereafter, the air quality inside our home noticeably deteriorated. It hurt to breathe the contaminant-filled air inside our home, just as it hurt to breathe the particulate-laden air outside. The ductwork of our air conditioning system moist with seepage from the hurricane-driven rains acted as an incubator for the airborne microorganisms.

The initial discomfort from inhaling the air filled with mold spores, chemicals, and fine particulates increased. Our chests began to tighten, hurting and burning as we breathed; our skin burned and itched; and our eyes scratched with dryness. With our lack of knowledge and understanding of the invasive ability of mold, we did not know that we, and others, should have been wearing lung-protecting face masks. Regardless, adult-sized face masks would not have been effective forms of protection for our children or anyone else's. We did, however, limit our exposure to the outside air that was thick with larger, more visible particulates, which unfortunately increased our exposure to the less visible fungal contaminants inside our home. At the time, we did not understand the impact these various air contaminants would have on our long-term health, especially Kurt's.

When Hurricane Katrina roared ashore on August 29, 2005, no government agency notified the people in our town, or in towns even closer to the center of destruction, to wear respirators—face masks—to protect their lungs from the potential damaging effects from airborne mold and toxic contaminants. No public announcements alerted people that breathing the air in the intact outlying communities could pose health risks and wearing respirators were warranted because of harmful conditions. Sure, everyone saw pictures of some of the first responders wearing face masks (the kind we've all worn a time or two while painting or remodeling). However, no local, state, or federal government officials warned citizens outside the center of destruction about the associated health risks from inhalation of this never-seen-before combination of air contaminants. People, like us, thought government agencies would alert them if the post-hurricane air could cause significant health risks. Mold-naïve residents, like us, had no idea the air they were breathing was heavily laden with both toxigenic and non-toxigenic mold spores that could take hold in high-risk individuals and people genetically predisposed to mold susceptibilities. In the chaos of the aftermath of the hurricane, no local, state, or federal government officials notified disaster-area residents of the harmful health effects of mold exposure. This lapse of judgment and accountability regarding the health of citizens in the affected areas came with a price, one we and other residents would unfortunately later pay.

Our family faced the challenge, along with many, many families, of having lost our source of income due to the hurricane. We had relocated to Louisiana in May 2005 for employment and had accepted a new position in New Orleans just days before Katrina wreaked her destruction. Needless to say, we were in need of alternative employment. With thousands of people evacuated to our town, jobs were nonexistent; with our rental home becoming increasingly uninhabitable due to mold, we evacuated with the help of the Red Cross to Billings, Montana two weeks after Katrina. We thought fresh air and clean water would restore our health, which was not the case. The spores of illness, so to speak, had been planted in our bodies from the mold and chemical toxins to which we had been exposed post-Katrina.

Unknown to us at the time, the members of our family, who all had clean bills of health before the hurricane, fell into high-risk groups in regard to mold exposure. Kurt was at high risk because he had pre-existing allergies and borderline asthma, both of which the Centers for Disease Control and Prevention (CDC) documents as high-risk conditions.[1] Regular use of prescription steroids for allergy management, which weaken the immune system,[2] further compounded Kurt's already increased risk. Our children were at high risk because of their ages.[3] The CDC states that children ". . . may be affected to a greater extent than most healthy adults."[4] In fact, the CDC states that infants and children ". . . should avoid mold-contaminated environments entirely."[5] I was at high risk because I had recently undergone major surgery. According to the New York City Department of Health and Mental Hygiene, who published the first industry guidelines regarding mold, ". . . [P]ersons recovering from recent surgery . . . may be at greater risk for developing health problems associated with certain fungi."[6]

When we reached Billings, the Red Cross placed us in a shelter for two weeks, during which time we realized our health was not springing back to normal. We were sick from the contaminants we had inhaled in the aftermath of Hurricane Katrina. The extent of these toxins would later be revealed as various government agencies and environmental groups conducted testing.

We took Kurt to the doctor first, as his symptoms were the most severe. During the appointment we reviewed Kurt's myriad of symptoms, which we were all experiencing, albeit to a lesser degree. A physical exam and a test revealed Kurt had only about a quarter of his full lung capacity. We watched the doctor's face with anticipation, patiently waiting, expecting to hear the latest scientific and medical research relating to mold and chemical exposures. Instead, the doctor said, "Mold cannot make you sick." He diagnosed Kurt with bronchitis and acute sinusitis, handed him prescriptions for antibiotics and steroids, and sent him on his way. We did not know, at the time, Kurt had been misdiagnosed, improperly treated, and misinformed—he would not rebound to 100 percent in a week as the doctor had assured us.

We soon came to the realization that mold-related illnesses could very well be the most misunderstood medical condition of our decade. We found, as have many other people sick from mold exposure, that medical doctors (MDs) often don't keep abreast of the latest scientific and medical studies relating to mold exposure or other conditions for that matter. Without knowledge of the most current research, doctors cannot consider newly discovered treatment options and/or dangerous side effects of the pharmaceuticals they prescribe. Most physicians are not well-read or educated in the research that has been conducted and published over the past few decades on the health effects of mold exposure. Worse yet, as we found in our case, some doctors have no desire to read scientific medical research and studies and remain unwilling to do so, even when patients bring in easily accessible data from scientific and medical online data banks such as Medscape and PubMed. Although many doctors are caring and competent, some, unfortunately, appear to view patients as no more than their next house or car payment.

Prudence dictates that we, as patients, take the initiative to protect, preserve, and restore our health. We need to research physical symptoms, the disease itself (if already diagnosed), and all available treatment options including alternative solutions. Specifically, we must evaluate all possible side effects. Self-education also enables us to identify a doctor or specialist who is up-to-date on the latest scientific and medical research regarding our specific condition(s) and to discuss and evaluate all possible treatment options. However, finding a doctor both well-educated and open to less invasive and less toxic alternative treatments can require time, patience, and perseverance. Just because doctors have degrees after their names—be them MD, ND, or what have you—doesn't necessarily mean that they will provide effective and ultimately successful care. Our personal search left us dismayed and disgusted. Professionals in the field of industrial hygiene knew more about the latest research regarding the health effects of mold exposure than did most of the medical doctors with whom we consulted for our family's medical treatment—a sad and alarming reality.

For example, we found many doctors completely unaware of the results of a widely publicized study conducted by doctors at the Mayo Clinic that implicates fungus as the cause of most cases of chronic sinusitis.[7] A 1999 Mayo Clinic press release states, "Mayo Clinic researchers say they have found the cause of most chronic sinus infections—an immune system response to fungus. They say this discovery opens the medical door to the first effective treatment for this problem, the most common chronic disease in the United States."[8] Two peer-reviewed journals, the *Journal of Allergy and Clinical Immunology* and the *Mayo Clinic Proceedings*, published the findings the same year.

Many doctors, including those who treated Kurt, are either unaware of the Mayo Clinic study or do not understand the significance of the investigating doctors' following conclusion: "Antibiotics and over-the-counter decongestants are widely used to treat chronic sinusitis. In most cases, antibiotics are not effective for chronic sinusitis because they target bacteria, not fungi."[9] Many MDs do not realize that treatment with antibiotics and steroids can actually worsen fungal-related conditions by destroying the body's natural biological terrain that protects it from fungi (molds and yeasts). In fact, frequent improper treatments with antibiotics and steroids can create an internal incubating ground for fungi![10] Thus, the common practice of treating sinusitis with antibiotics and steroids may be contributing to the increase in frequency of fungal-related sinusitis. According to the Mayo Clinic, "An estimated 37 million people in the United States suffer from chronic sinusitis, an inflammation of the membranes of the nose and sinus cavity. Its incidence has been increasing steadily over the last decade."[11]

Past medical paradigms are to blame for the misguided antibiotic and steroid treatments. According to the Mayo Clinic, prior to its study, "'Fungus was thought to be involved in less than ten percent of cases,' says Dr. [David] Sherris [a then-Mayo Clinic ear, nose, and throat doctor]. 'Our studies indicate that, in fact, fungus is likely the cause of nearly all of these problems. And it is not an allergic reaction but an immune reaction.'"[12] These conclusions were evident when Mayo Clinic researchers ". . . found eosinophils (a type of white blood cell activated by the body's immune system) in the nasal tissue and mucus of 96 percent of the patients" in a 101-patient subset (of the 210-participant study group) that had surgery to remove nasal polyps.[13] According to Mayo Clinic researchers, the discovery that eosinophils migrated to the fungal-infested nasal tissue ". . . clearly portray[s] a disease process in which, in sensitive individuals, the body's immune system sends eosinophils to attack fungi and the eosinophils irritate the membranes in the nose. As long as fungi remain, so will the irritation."[14]

To reach these conclusions, Mayo Clinic researchers compared clinical profiles of affected patients to those of healthy controls. A discovery that warrants clarification is the revelation that extramucosal fungi were found in almost all patients with chronic rhinosinusitis as well as in all healthy controls.[15] In fact, a total of 40 different kinds of fungi were identified in the mucus of 96 percent of the entire study group, averaging 2.7 kinds of fungi per patient.[16] The discovery that fungi were prevalent in the mucus of all study participants does not undermine the study results but rather confirms the ubiquitous nature of fungi and its genetically determined effects on the health of individuals. The study findings clearly demonstrate that everyone has some fungi in his/her mucus, yet not everyone develops chronic rhinosinusitis; individuals whose bodies launch an immune response to fungal stimuli are most at risk to develop chronic rhinosinusitis.

Since fungal stimuli do not elicit the same reaction in all people, Mayo Clinic researchers were able to identify a scientifically provable differentiating component that separated the affected study participants from the controls, which enabled them to formulate their conclusions. For the sake of clarity, the Mayo Clinic researchers stress that the mere presence of fungal organisms is not the cause of disease but rather an individual's sensitized reaction to the fungi that activates the formation of eosinophilic mucin.[17] Findings from both the original 1999-released study and subsequent validation studies by the research team confirm that the presence of fungi in sensitized individuals act as ". . . a trigger to stimulate the immunologic (eosinophilic) response . . ."[18]

According to Mayo Clinic reports, the research group found that the eosinophils cluster in the nasal and sinus mucus and scatter a toxic protein onto the nasal and sinus membranes.[19] While the study revealed that this toxic protein did not invade the nasal and sinus tissue, the level of this protein in the mucus of study participants with chronic sinus infections ". . . far exceeded that needed to damage the nasal and sinus membranes and make them more susceptible to infections such as chronic sinus infection."[20] The Mayo Clinic doctors point out that this fungal-triggered eosinophil phenomenon was ". . . absent in healthy controls as well as in patients with allergic rhinitis."[21] Thus, study findings suggest that nearly all—96 percent—of the patients who suffer from chronic sinusitis are fungal sensitized and have immune responses triggered by inhaled fungal organisms.

Interestingly, chronic sinusitis as an immune response, not an allergic reaction, was studied and published over two decades before the Mayo Clinic research team released the results of its study. The earlier findings of the 1988 study also revealed the presence of extensive eosinophils in the paranasal tissue from patients suffering from chronic sinusitis.[22] The study also addressed similarities between the existence of eosinophils present in both chronic asthma and chronic sinusitis.[23] These studies have also documented the medical certainty that mold affects people differently, which has been and continues to be a reoccurring "theme" in scientific and medical mold literature.

Now that we've talked about chronic sinusitis, let's take a look at acute sinusitis. According to the University of Maryland Medical Center (UMMC), acute sinusitis lasts anywhere from two to eight weeks. When symptoms of acute sinusitis linger past the initial eight-week period, the sinusitis is considered a chronic condition.[24] In other words, the length of time a patient experiences symptoms determines whether a case of sinusitis is diagnosed as acute versus chronic. Essentially, if acute sinusitis cannot be effectively treated with antibiotics that target bacterial sources, symptoms linger; after the initial eight-week period, the sinusitis is considered a chronic condition.

UMMC reports that sinusitis is ". . . generally caused by a viral, bacterial, or fungal infection."[25] Therefore, we can logically deduce that the antibiotic-treated cases of sinusitis that linger and later become reclassified as chronic are caused by antibiotic-resistant bacteria, viruses, or most likely, fungi—as was proven by Mayo Clinic researchers. It is logical to consider the possibility that these cases of chronic sinusitis were initially misdiagnosed as bacterial infections and medically mistreated with antibiotics when in the acute stage. It is also reasonable to contemplate the probability that these chronic sinusitis cases were fungal-related—not bacterial—from their inception and, consequently, never should have been treated with antibiotics in the first place.

During the 17-plus-year period from the 1988 study to 2005 when Hurricane Katrina hit, medical paradigms in the treatment of fungal-related sinusitis did not shift and keep pace with scientific and medical studies, which led to no formal or continuing education of medical doctors regarding the prevalence of fungal-related sinusitis and the proven ineffectiveness of antibiotic treatments. It appears, based on our experiences and research, that much of the medical community is stuck in a time warp when it comes to fungal-induced illnesses—even in regard to the notably researched and highly publicized condition of fungal-induced sinusitis. Because of this apparent overall lack of understanding of mold-related illnesses by professionals in the medical community, it could be, more likely than not, that unless people pre-educate themselves by reading scientific and medical research such as is presented and referenced in this book, they may receive improper medical treatment, as we did, when it comes to treatment of fungal-related illnesses.

Health risks from mold exposure can escalate and acutely impact a person's health in conditions such as Katrina. Other times, exposure to mold and the accompanying biological compounds can slowly deteriorate a person's health until a degenerative illness, induced from exposure to the collective fungal components, ultimately causes death. A prolonged, yet unwitting, mold encounter in my own family may serve as a good example. My grandparents' house had a hidden source of black-colored mold growth for years. When I was in lower elementary school, a baby died in their basement rental. The cause of death was attributed to SIDS (sudden infant death syndrome) before SIDS was widely known. My grandmother succumbed to a form of cancer—leukemia—and my grandfather ultimately died from complications after lung cancer surgery. Only after the new owners of the house remodeled the inside stairwell was the black mold discovered and the severity of the situation explained by a remediation company. Was the black mold the cause of SIDS and two forms of cancer? The possibility exists, given scientific and medical research, but, frankly, we will never know.

We share this sad account as it could exemplify some of the possible health outcomes from mold exposure, which can snatch away anyone's health. It also illustrates how lack of information can prevent people from understanding the full impact of mold exposure; not until Kurt and I began searching for answers to our own health restoration some 30 years later did we understand the possible consequences from the mold growth in my grandparent's basement.

The preservation of other people's health through education was the catalyst to share the forthcoming information in this book on not only mold and mycotoxins but also on heavy metals and other contaminants. Even in times of chaos and pandemonium commensurable with post-disaster conditions, we need to take the time to educate ourselves for the benefit and longevity of our own health. By educating ourselves and others about the dangers of mold and mycotoxin exposure, we ready for battle with a strong defense should mold and mycotoxins launch an attack on our most valued territory—our bodies. Battle preparation is crucial when it comes to fighting fungus. We must create an internal biological terrain resistant to fungal invasion to ensure, at worst, we lose only a few battles, not *The War Within*.

Mold is an unwelcome guest in homes anytime enough moisture accumulates. It can grow inside the walls of homes from a leak or water vapor condensation due to energy efficient building methods. Tight construction doesn't allow houses to adequately "breathe," which can cause a buildup of moisture that doesn't evaporate. When enough moisture accumulates, from whatever source, a sick building situation can occur. So even if we don't live in the heart of New Orleans or a surrounding area, we all have a vested interest in gaining a mold education. We cannot rely on others to educate us—it could cost us our health.

The fungal-infested environment of post-Katrina provided a prime example of the importance of self-education. Richard Lipsey, PhD, a toxicologist who performed fungal and bacterial sampling in St. Bernard Parish on February 11, 2006, states, "I was shocked to find out that the Habitat for Humanity volunteers that I toured with and who were trained by FEMA [Federal Emergency Management Agency] had been told that mold cannot hurt you and you do not need any protective equipment."[26]

The statement "Mold cannot hurt you" told to the Habitat for Humanity volunteers echoes what we had heard repeatedly from medical doctors. This type of ignorance and propagation of misinformation perpetuates lack of knowledge and complacency. Without accurate information regarding associated health risks from mold exposure, people will not take the appropriate measures to protect themselves. If truth be told, fungal exposure can lead to mycosis, an infection or disease caused by a fungus,[27] and/or mycotoxicosis, the toxic effect of mycotoxins on animal and human health.[28]

To facilitate a full understanding of forthcoming discussions, let's review some mold terminology. Molds are identified by a genus name and a species name, somewhat similar to a last and first name. Molds are first identified by a genus name, such as *Aspergillus*. Within each genus, molds are further classified by species, for example, *Aspergillus flavus*. Molds breed only within the same species. The genus name is always capitalized and the species name is always in lowercase. Sometimes the genus name is represented by only the first letter capitalized, for instance, *A. flavus*. When the designation *sp.* (species) follows a genus name, it means that the species is unknown or unspecified in that particular discussion. The plural abbreviation *spp.* (species) is used to refer to all the species in a genus. Molds are classified as fungi. Fungi are a kingdom of organisms. They are not plants nor are they animals.[29] Scientists use two methodologies to measure airborne fungi: Viable (capable of reproducing), which is measured in colony forming units per cubic meter (cfu/m³), and nonviable, which is measured in spores per cubic meter (spores/m³). The viable method quantifies living spores, while the nonviable method quantifies both living and dead.

Andrew Puccetti, PhD, a certified industrial hygienist (CIH), explains, "In the viable methodology, the spores that are captured are identified very specifically as to the genus and species because they are cultured (or grown) into mold colonies that have significant differences in their morphology. This approach is analogous to cultivating lemon seeds and orange seeds and growing a lemon tree and an orange tree. Although the lemon and orange seeds cannot be differentiated, the orange tree and lemon tree can easily be."[30]

Dr. Puccetti continues, "In the nonviable method, the spores are trapped on what are normally referred to as a spore-trap device. There are several different types that are commonly used. The spores are not cultured; they are just examined microscopically. Under these examination conditions, microscopists can't often identify with very high specificity the type of spore that they see through the microscope. It's like looking at a lemon seed versus an orange seed. They're not that different and are difficult to differentiate. However, if the spores present are significantly morphologically different, then their identities can be specified. This situation is analogous to a lemon seed and an avocado seed. They are easily differentiated because of their significant morphological differences."[31]

"Ideally, you want to use both methods because there are advantages and disadvantages to both. Often what you can see using one method, you can't see using the other method," according to Dr. Puccetti. "You want to be able to see as much as you can so that you get as much information from your sampling as possible. For example, a lot of times it's very difficult to culture *Stachybotrys* spores. On the other hand, you can see them quite readily and identify them microscopically on spore traps."[32]

Other methodologies used to measure mold are bulk, swab, and contact-tape-lift

sampling, explains Jim Pearson, a certified mechanical hygienist (CMH). He and
several colleagues coauthored the publication *ANSI/IICRC S520 Standard and
Reference Guide for Professional Mold Remediation*, which is a reference for
industry procedural standards. The Occupational Safety and Health Administration
(OSHA) cites the manual on its website as one of three sources that ". . . may be
referenced by OSHA inspectors for informational purposes."[33] Mr. Pearson chaired
the committee that oversaw the completion of the second edition, which was
published in 2008.[34]

Mr. Pearson explains, "A bulk sample is simply a piece of moldy material placed in
a baggie and sent to the lab for analysis. Taking a swab sample is accomplished by
dipping the swab in a solution and rolling it over a specific area, say 2 inches by 2
inches. A tape lift is clear tape pressed lightly on a hard, flat surface and then placed
on a glass slide for viewing. When the lab gives me the tests results, they'll report
it in spores per square centimeter [spores/cm^2] or colony forming units per square
centimeter [cfu/cm^2]. So the bulk collection, tape lift, and swab sampling are all
measured in amount/cm^2, whereas air sampling is measured in amount/m^3."[35] Bulk,
tape, and swab samples can also be measured in amount per square inch (amount/
in^2).

Traditional testing methods quantify spores. They do not measure primary or
secondary fungal metabolites, which can also negatively impact health. Primary
metabolites ". . . are produced during processes essential . . ." to the growth of
the mold and are called volatile organic compounds (VOCs). The musty smell
we associate with mold comes from these VOCs. Some molds produce secondary
metabolites called mycotoxins that are not essential to the growth of the organism.[36]
Hundreds of different mycotoxins exist, which can differ in chemical structure and
other characteristics. Mycotoxins are classified as a group because they ". . . can
cause disease and death in human beings and other vertebrates."[37] Because of this
common, lethal trait, mycotoxins remain a continual focus of research.[38]

For people who are not microbiologists or well-versed in the sciences, the concept
of mycotoxins can seem daunting at first, but keep in mind it is not necessary to
understand every detail in order to grasp the major concepts. It is important to
understand that some, but not all, mold species produce poisonous mycotoxins; in
other words, only certain species of fungi produce mycotoxins.[39] Additionally, "It
is very important to recognize that spores retain their adverse health characteristics
regardless of their ability to reproduce. In other words, nonviable spores are still
allergens, contain toxins, etc.," according to the Department of Environmental
Health and Safety at Minnesota State University Moorhead (MSUM).[40] Hence,
exposure to nonviable (dead) mold spores can cause the same negative health
consequences as exposure to viable (live) mold spores.

Recent scientific studies address whether certain fungi constantly produce
mycotoxins or if they produce mycotoxins under only adverse environmental
conditions. A commonly held, albeit disproven, belief is that fungi do not constantly

produce mycotoxins. This theory is espoused by the CDC in its published report "Mold Prevention Strategies and Possible Health Effects in the Aftermath of Hurricanes and Major Floods." The CDC states, "Some molds are capable of producing toxins (sometimes called mycotoxins) under specific environmental conditions, such as competition from other organisms or changes in the moisture or available nutrient supply. Molds capable of producing toxins are popularly known as toxigenic molds; however, use of this term is discouraged because even molds known to produce toxins can grow without producing them, and many fungi are capable of toxin production."[41]

Groundbreaking studies have disproved this long-held theory, at least relative to *Stachybotrys* and the trichothecenes it produces. David Straus, PhD, a professor of microbiology and immunology at Texas Tech University Health Sciences Center, recently published studies involving *Stachybotrys* and trichothecenes.[42] Based on the research conducted by him and his colleagues, Dr. Straus reports that *Stachybotrys* is constantly producing trichothecenes.[43] These findings are corroborated by other published studies that also suggest metabolites of *Stachybotrys* are constitutively produced, which means they are produced in relatively constant amounts without regard to cell environmental conditions.[44] A 2003 published article by Gregory et al. states that stachylysin, a hemolytic agent, is constitutively produced by *Stachybotrys chartarum*.[45] Additional findings published in 2004 by Gregory et al. report that satratoxin-G, the major macrocyclic trichothecene mycotoxin produced by *Stachybotrys chartarum*, is also constitutively produced.[46]

This disparity in published literature demonstrates the evolving nature of the study of science. Prior to the Straus and Gregory studies, scientists reported that mycotoxins were produced under only adverse conditions. Based on these new scientific revelations, Dr. Straus states, "None of that's true."[47] He explains, "This is not what we see. We see mycotoxins being produced as soon as the organism begins to grow. As spores are being formed, so are the trichothecene mycotoxins."[48] He further explains, ". . . when spores are produced, the mycotoxins are packaged into the spores at the same time. We don't think the spores produce the mycotoxins. We think the colony produces them and packages them into the spores when the spores are formed. The mycotoxins appear to be on the outer plasmalemma [cell membrane] surface and the inner wall layer of spores."[49] The scientific proof that fungi constantly produce mycotoxins is limited to *Stachybotrys* and its trichothecenes, according to Dr. Straus. He points out, "We have not looked at this in any other organism."[50] Dr. Straus is not aware of anyone else in the scientific community who has studied mycotoxin production in regard to other fungi.[51]

Although scientists are trained and required to limit their conclusions to what can be scientifically proven in the laboratory and/or controlled field studies, other professionals can surmise "logical" conclusions based on scientific studies. For example, in regard to the studies conducted by Straus, Gregory, and fellow researchers, Michael Gray, MD, MPH, states, "There really is not any reason to assume that mycotoxins are not being produced all the time by all fungi capable of producing them."[52] In other words, Dr. Gray's assumption puts forth what scientific studies may indeed later correlate—that all fungi capable of producing mycotoxins begin to produce them as soon as the fungi begin to propagate.[53] In addition, Dr. Gray theorizes that levels of mycotoxin production may fluctuate in response to stress beginning to desiccate the organism.[54] Lack of water or nutrients can create conditions of stress, which can also cause increased sporulation.[55] Once fungal colonies begin to grow from these newly released spores, mycotoxins begin to form. While not "technically" scientifically sound, these types of speculations are necessary by medical professionals, such as Dr. Gray, to best assess the levels of toxins to which their patients may have been exposed and identify all possible causative factors of deteriorating health.

Another commonly held belief is that mycotoxins are ". . . products whose function seems to be to give molds a competitive advantage over other mold species and bacteria."[56] Dr. Gray concurs, "The biological function of mycotoxins is to enhance the probability of survival of the next generation of mold."[57] He explains that from a species specific survival perspective, mycotoxins give an organism the edge and advantage in competing with other organisms because mycotoxins are actual poisons that help fungi defeat the other microorganisms growing in the same ecologically balanced microenvironment.[58] He further points out that by killing their neighbors, it allows the fungi to continue to propagate and supply an additional food source (from both the substrate and the dead fungi) to the new colony.[59] "The mycotoxins really represent a true species specific survival mechanism," states Dr. Gray. "So, far from being 'secondary' in their real impact, they really and truly are extremely important to the propagation of these organisms."[60] However—scientifically speaking—from a microbiologist's point of view, Dr. Straus says, "This may be true. We essentially don't know why fungi produce mycotoxins."[61] Again, not everything has been "scientifically" proven as of yet.

Scientific studies have proven two fascinating facts: (1) certain genuses of fungi produce several different types of mycotoxins, and (2) certain mycotoxins are produced by multiple different fungal genuses. For example, the genus *Aspergillus* metabolizes the mycotoxins aflatoxin, citrinin, and ochratoxin while the mycotoxin trichothecene is ". . . produced by a number of fungal genera [genuses], including *Fusarium, Myrothecium, Phomopsis, Stachybotrys, Trichoderma, Trichothecium,* and others."[62]

Scientific and medical research prove that biological aerosols such as fungi and mycotoxins can penetrate into the human body through the nose, mouth, bronchi (the major air passages of the lungs) and alveoli (the lung's air sacs), and skin.[63] Once exposure occurs, mycotoxins can absorb into the body, circulate through the blood, and can affect the body systemically.[64] The complexity of successful treatment of mold-related illnesses is compounded because the microorganism is multifaceted. Not only can the mold spores and the VOCs make people sick, but the toxic mycotoxins that some molds produce are akin to biological warfare agents. Mycotoxins, even in small quantities, can negatively affect our health because they ". . . are lipid-soluble and are readily absorbed by the intestinal lining, airways, and skin."[65] Because of these inherent characteristics, mycotoxins are studied, researched, and tested via experiments as possible effective biowarfare agents.

As far back as the 1940s and 1950s, scientists have studied and experimented with mold as a biological warfare agent.[66] Public Broadcasting Service (PBS) reports, "According to the U.S. Department of Defense, more than ten countries have, or are developing, biological warfare programs. According to the Office of Technology Assessment and U.S. Senate committee hearings, the number is about 17 and includes: Russia, Israel, Egypt, China, Iran, Iraq, Libya, Syria, and North Korea."[67] Michael Gray, MD, MPH, who is a certified industrial medical examiner (CIME) that treats mold-exposed patients at the Progressive Healthcare Group in Benson, Arizona, states, "Several nations that have developed biological warfare capabilities have harvested mycotoxins, which are poisons produced by specific species of molds."[68] In 1997 the U.S. government published *Medical Aspects of Chemical and Biological Warfare*, a textbook of military medicine that discuses in scientific detail various aspects of mycotoxins as biological warfare agents.[69]

Aflatoxins are an example of a mycotoxin purported to have been used as a biological warfare agent. One published report states, "There is considerable evidence that Iraqi scientists developed aflatoxins as part of their bioweapons program during the 1980s. Toxigenic strains of *Aspergillus flavus* and *Aspergillus parasiticus* were cultured, and aflatoxins were extracted to produce over 2,300 liters of concentrated toxin. The majority of this aflatoxin was used to fill warheads; the remainder was stockpiled."[70] On the surface it would appear that we could have direct health ramifications from exposure to these concentrated levels of aflatoxins if warheads containing them exploded in our vicinity and indirect health repercussions from exposure via ingestion of contaminated water and food supplies. However, scientists point out that aflatoxin concentrates would not perform as an effective bioweapon in an outdoor environment.

Regina Santella, PhD, a professor of environmental health sciences at the Mailman School of Public Health at Columbia University and director of the Biomarkers Core Facility of the Cancer Center, explains, "Aflatoxin is very light sensitive. When we work with it in the laboratory, if we leave it out on the bench, it is gone. So, there are ways of getting rid of aflatoxin because it is relatively unstable as it is light sensitive when it is out in the air."[71]

Additionally, Dr. Straus states that although aflatoxins are extremely carcinogenic and cause liver cancer, "The way in which aflatoxins cause liver cancer in human beings is by ingestion. For example, in certain regions of the world people eat grain that has *Aspergillus flavus* growing on it. When the organism grows on the grain, it produces the aflatoxins, and then people eat them. There's a much higher incidence of liver cancer in those regions than, say, in the United States. That's the way aflatoxins cause liver cancer. To aerosolize the aflatoxins, in my opinion, especially outside, would be a total waste of effort. . . . I would worry much more about the explosives in the missile killing me than I would about the aflatoxin inhalation, because if you're close enough to inhale the aflatoxins, that missile's going to blow you up!"[72]

"Remember, when you put a toxin in the air outside, essentially what happens is it's diluted by the whole world. That stuff immediately begins to diffuse away from the point from which it was sent, and it gets away very, very quickly, especially if the wind currents carry it away. So, using mycotoxins outside is really not very effective. Using mycotoxins inside would be very effective because they are trapped inside the building and can't diffuse outside the building," explains Dr. Straus.[73]

Indoor exposure to aflatoxins can be fatal. A case in point occurred when aflatoxin (B.sub.1) was ". . . detected in the lung tissue of a chemical engineer who had worked for three months on a method for sterilizing Brazilian peanut meal contaminated with *Aspergillus flavus* and who died of alveolar cell carcinoma [lung cancer]."[74] (The Iraqis used the same species as a cultivation source for aflatoxins.) After only three months of aflatoxin exposure via the respiratory route, the man died. Did other factors contribute? Possibly, however, the presence of the form of aflatoxin reported as ". . . the most potent natural carcinogen known . . ."[75] in the man's lung tissue strongly indicates that aflatoxin was a possible causative factor of death.

The toxicity of mycotoxins was discussed in *Clinical Microbiology Reviews*, "Unlike the aflatoxins, trichothecenes can act immediately upon contact, and exposure of a few milligrams of T-2 [a trichothecene] is potentially lethal. In 1981, then Secretary of State Alexander Haig of the United States accused the Soviet Union of attacking Hmong tribesman in Laos and Kampuchea with a mysterious new chemical warfare agent, thereby violating the 1972 Biological Weapons Convention. The symptoms exhibited by purported victims included internal hemorrhaging, blistering of the skin, and other clinical responses that are caused by exposure to trichothecenes. Leaf samples from Kampuchea were analyzed by Chester Mirocha [a professor of plant pathology] of the University of Minnesota, who found nivalenon, deoxynivalenon, and T-2 [mycotoxins]."[76]

"The purported chemical warfare agent came to be known as yellow rain,"[77] because it ". . . consisted of a shower of sticky, yellow liquid that sounded like rain as it fell from the sky."[78] The controversial incident became the focus of books that discussed the origins of the mycotoxins. Some of the asserted theories of origination include biological warfare agents comprised of fungi-harvested trichothecenes, a yellow residue caused by showers and deposits of bee feces as a result of massive bee swarms,[79] and the production from indigenous trichothecene-producing fungi.[80] The U.S. government publication *Medical Aspects of Chemical and Biological Warfare* claims, "An article written by L.R. Ember, published in 1984 in *Chemical Engineering News*, is the most exhaustive and authoritative account of the controversy surrounding the use of trichothecene mycotoxins in Southeast Asia during the 1970s."[81] Although the use of trichothecenes as a biological warfare was not determined beyond question, the aforementioned U.S. government publication includes—at the top of a list of identifying factors of a biological warfare attack with trichothecene mycotoxins—the discovery of clinical findings that match the symptoms experienced by the victims of yellow rain in Southeast Asia.[82]

According to toxicologist Dr. Lipsey, the U.S. government has also explored mycotoxins as a biological warfare agent. He reports that the U.S. Army developed highly purified forms of T-2, which were never used and have since probably been destroyed.[83]

We all know that biological toxins derived from Mother Nature can be deadly. Take, for instance, the anthrax attacks that occurred in 2001. "Anthrax is an acute infectious disease caused by the spore-forming bacterium *Bacillus anthracis*. Anthrax most commonly occurs in wild and domestic lower vertebrates (cattle, sheep, goats, camels, antelopes, and other herbivores), but it can also occur in humans when they are exposed to infected animals or tissue from infected animals or when anthrax spores are used as a bioterrorist weapon. . . . [H]umans can become infected with anthrax by handling products from infected animals or by inhaling anthrax spores from contaminated animal products."[84] Anthrax spores can be inadvertently inhaled, just like mold spores, due to their microscopic size. This characteristic predisposes both organisms for the purposes of bioterrorism. For example, in 2001 *Bacillus anthracis* spores were intentionally distributed through the U.S. postal system, causing 22 cases of anthrax poisoning, including five deaths.[85]

Since anthrax is a biological agent with which most of us are familiar, let's ask ourselves a few questions. If we knew anthrax was going to be spewed throughout our town and inside our homes, would we leave before the attack? You bet we would. Would we go back to check our property if a thick blanket of anthrax covered our town and homes? Would we clean our property without the use of personal protective equipment? If an anthrax flurry hit another town, would we travel to the devastated area, pay for sightseeing tours, and partake in week-long parties in the streets? Worse yet, would we allow our teenagers to travel to the affected area as part of a volunteer cleanup crew? Most of us would answer "No" to these questions, understanding that to go to the frontline of a biological attack could be fatal.

We must remember: biological toxins are used as weapons because they have the potential to kill. Governments of multiple countries aren't spending their war chests of military research dollars investigating toxins that aren't potentially deadly. Mycotoxin exposure concentrated and contained in an indoor environment should be taken no less seriously than exposure to anthrax—both could be deadly. Unfortunately, the lack of citizen health campaigns publicizing the toxicity of mold has left the general public ignorant, which results in poor decisions and spiraling negative health effects. If the Habitat for Humanity volunteers were not informed by FEMA of the necessity of using proper personal protective equipment and were incorrectly told that mold—some of which produce mycotoxins—could not hurt them, how many other people received this same detrimental advice?

We all understand anthrax can kill. However, when biological toxins arrive via a natural disaster with the potential to create equally toxic effects in indoor environments, it doesn't command our attention with the same urgency as when deadly biologicals are delivered by an enemy. Molds producing biological warfare agents may sound like science fiction, especially to the nonmilitary, but it's not. The twilight-zone conditions created in the wake of Katrina can have far-reaching health implications for individuals exposed to the disrupted molds, mycotoxins, and chemicals. When Hurricane Katrina roared into the Gulf of Mexico billions of mold spores, both active and dormant, were disrupted and spewed everywhere— into the air, the water, and inside any structure still standing.[86] The spores settled, had plenty of water and heat, and now during the cleanup and rebuilding phase, residents and remediation workers are harvesting the largest bumper crop of mold in U.S. history.

Mold growth of this magnitude has never been seen before; within a week after the hurricane, a returning resident was quoted in the newspaper saying, "'We're just surrounded by mold.'"[87] The cleanup and remediation process will go on for years, if not decades, and spawn a mold-related health crisis due to the widespread lack of use of personal protective equipment. Insidious coughs, wheezes, rashes, and upper and lower respiratory irritations that have been dubbed by the media as the Katrina Cough and the Katrina Syndrome very likely may be forerunner symptoms of brewing diseases. Only after millions of dollars fund multi-year studies will we know if the post-Katrina era has been marred by epidemic levels of cancers, life-threatening and fatal fungal infections, mold- and chemical-induced degenerative diseases, and deaths—a time of human assault, not directly from Hurricane Katrina herself but from exposure to the mold and toxins stirred up and created in her wake.

Had fungal biologicals received the widespread media attention bestowed on anthrax, people would have more vigilantly protected themselves against exposure to the molds and the mycotoxins many produce. Instead, media reports have documented people cleaning and performing repairs, even months after the hurricane—without the safeguard of personal protective equipment—in buildings heavily infested with mold. Clearly evident is the breakdown in the dissemination of health- and life-preserving information regarding mold in a manner that effectively reaches the people.

So where is the ball being dropped? Dr. Lipsey, a veteran government consultant, states, "On all levels—from FEMA to the CDC. The CDC doesn't want to scare anybody and neither does NIOSH nor the EPA, but if you go on their websites, you'll see stories about how some molds can cause you to vomit blood, can cause you to die from pulmonary hemorrhaging, and can cause cancer. This is on the websites."[88] Dr. Lipsey has had offices in both the EPA (Environmental Protection Agency) and the USDA (United States Department of Agriculture) facilities in Washington, DC, while working as a consultant for each agency. He has served also with NIOSH (National Institute for Occupational Safety and Health) and the CDC in Atlanta on cases related to toxic issues. He points out, "Government regulatory agencies have regulators; they don't have educators. . . . These agencies don't have enough time, money, or staff to educate the public properly."[89]

Because of these limitations, a large portion of the public remains unaware of the health risks posed from exposure to biological aerosols such as fungi and mycotoxins. This ensuing lack of public knowledge has impaired people's abilities to make informed decisions and has caused many to put themselves and their families unwittingly—and unnecessarily—in harm's way.

(ENDNOTES)

1 THE CDC MOLD WORK GROUP: MARY BRAND ET AL., "MOLD PREVENTION STRATEGIES AND POSSIBLE HEALTH EFFECTS IN THE AFTERMATH OF HURRICANES AND MAJOR FLOODS," *MMWR*, JUNE 9, 2006/55(RR08); P. 17. AVAILABLE AT HTTP://WWW.CDC.GOV/MMWR/PREVIEW/MMWRHTML/RR5508A1.HTM (ACCESSED JUNE 2006). ORIGINALLY published under the title "Mold Prevention Strategies and Possible Health Effects in the Aftermath of Hurricanes Katrina and Rita."

2 PBS, "Is It Safe to Return?" September 19, 2005. Available at http://www.pbs.org/newshour/bb/weather/july-dec05/return_9-19.html (accessed August 2006).

3 The CDC Mold Work Group, "Mold Prevention Strategies," p. 17 and 37.

4 Ibid., p. 17.

5 IBID., P. 37. (ITALICS ADDED.)

6 THE NEW YORK CITY DEPARTMENT OF HEALTH AND MENTAL HYGIENE, "GUIDELINES ON ASSESSMENT AND REMEDIATION OF FUNGI IN INDOOR ENVIRONMENTS." AVAILABLE AT HTTP://WWW.NYC.GOV/HTML/DOH/HTML/EPI/MOLDRPT1.SHTML (ACCESSED APRIL 2006).

7 PONIKAU JU ET AL., "THE DIAGNOSIS AND INCIDENCE OF ALLERGIC FUNGAL SINUSITIS," *MAYO CLIN PROC*, 1999; 74: 877-884. AVAILABLE AT HTTP://WWW.NCBI.NLM.NIH.GOV/ENTREZ/QUERY.FCGI?CMD=SEARCH& DB=PUBMED (ACCESSED JUNE 2006).

8 MAYO CLINIC PRESS RELEASE, "MAYO CLINIC STUDY IMPLICATES FUNGUS AS CAUSE OF CHRONIC SINUSITIS," SEPTEMBER 9, 1999. AVAILABLE AT HTTP://WWW.WEINPRODUCTS.COM/12.PDF#SEARCH=%22MAYO% 20CLINIC%20ROCHESTER%20NEWS.%20MAYO%20CLINIC%20STUDY%20IMPLICATES%20FUNGUS%20AS%2 0CAUSE%20OF%20CHRONIC%20SINUSITIS%20%5BNEWS%20RELEASE%5D%2C%20SEPTEMBER%209%2C% 201999.%22 (ACCESSED JULY 2006).

9 IBID.

10 GOLDBERG, *ALTERNATIVE MEDICINE*, P. 36 AND 685.

11 MAYO CLINIC PRESS RELEASE, "MAYO CLINIC STUDY IMPLICATES FUNGUS AS CAUSE OF CHRONIC SINUSITIS."

12 Ibid.

13 Ibid.

14 Ibid.

15 Mayo Clinic, "Allergic Fungal Sinusitis—Reply," *Mayo Clinic Proceedings*, 2000; 75:122-123. Available at http://www.mayoclinicproceedings.com/inside.asp?AID=1715 (accessed July 2006).

16 Mayo Clinic Press Release, "Mayo Clinic Study Implicates Fungus as Cause of Chronic Sinusitis."

17 Mayo Clinic, "Allergic Fungal Sinusitis—Reply."

18 Ibid.

19 Mayo Clinic, "Chronic Sinus Infection: Study Finds Cause with Implications for Treatment." Available at http://www.mayoclinic.org/spotlight/chronic-sinus-infection.html (accessed July 2006).

20 Ibid.

21 Mayo Clinic, "Allergic Fungal Sinusitis—Reply."

22 Harlin SL et al., "A Clinical and Pathologic Study of Chronic Sinusitis: The Role of the Eosinophil," *J Allergy Clin Immunol*, May 1988, 81(5 Pt 1): 867-75. Available at http://www.ncbi.nlm.nih.gov/entrez/query.fcgi?cmd=Retrieve&db=PubMed&list_uids=3286721&dopt=Citation (accessed August 2006).

23 Ibid.

24 UMMC, "Sinusitis." Available at http://www.umm.edu/ency/article/000647.htm (accessed February 2007).

25 Ibid.

26 Lipsey, "The Mold and Bacterial Results from the Sampling Dr. Lipsey did in St. Bernard Parish," March 9, 2006. Provided by Joe Wall, Times Picayune reporter, May 2006.

27 Medline Plus. Available at http://www.nlm.nih.gov/medlineplus/mplusdictionary.html (accessed June 2006).

28 Maja Peraica et al., "Toxic Effects of Mycotoxins in Humans." *Bulletin of the World Health Organization*, 1999; 77: 754-766. Available at http://www.moldacrossamerica.org/published2.htm (accessed November 2005).

29 Department of Environmental Health and Safety, USUM, "About Mold." Available at http://

www.mnstate.edu/ehs/Mold.htm (accessed April 2006).
30   Interview with Dr. Andrew Puccetti, April 2006.
31   Ibid.
32   Ibid.
33   OSHA, "Molds and Fungi Standards." Available at http://www.osha.gov/SLTC/molds/
standards.html (accessed February 2007).
34   International Sanitary Supply Association (ISSA), "IICRC Appoints New S520 Standard
Chairman," February 23, 2007. Available at http://www.issa.com/news/news_detail.jsp?typeId=23&ne
wsid=1488&page=startPage (accessed March 2007).
35   Interview with Jim Pearson, April 2006.
36   Department of Environmental Health and Safety, MSUM, "About Mold."
37   J.W. Bennett et al., "Mycotoxins," *Clinical Microbiology Reviews*, July 20, 2003, 497-516,
vol. 16, no. 3. Available at http://cmr.asm.org/cgi/content/abstract/16/3/497 (accessed November 2005).
38   Curtis et al., "Adverse Health Effects of Indoor Molds," *Journal of Nutritional
& Environmental Medicine*, vol. 14, no. 3, September 2004, 261-74. Also available at http://
taylorandfrancis.metapress.com (accessed February 2006); Bennett et al., "Mycotoxins," p. 1-4; and
Harriet M. Ammann, "Is Indoor Mold Contamination a Threat to Health?" Previously available at
http://www.doh.wa.gov/EHP/OEHAS/MOLD.HTML. Newly posted at http://www.mold-survivor.
com/harrietammann.html (accessed February 2006).
39   The CDC Mold Work Group, "Mold Prevention Strategies," p. 1.
40   Department of Environmental Health and Safety, MSUM, "About Mold."
41   The CDC Mold Work Group, "Mold Prevention Strategies," p. 1.
42   Straus, DC and Wilson, SC, "Respirable trichothecene mycotoxins can be demonstrated
in the air of *Stachybotrys chartarum*-contaminated buildings," *J Allergy Clin Immunol*. 2006 Sep;
118(3):760.
43   Interview with Dr. David Straus, April 2006.
44   Medline Plus. Available at http://www2.meriam-webster.com/cgi-bin/mwmednlm?book=M
edical&va=constitutively (accessed May 2006).
45   L. Gregory et al., "Immunocytochemical localization of stachylysin in *Stachybotrys
chartarum*." *Mycopathologia*, vol. 156, no. 2/September 2003. Available at http://www.springerlink.
com/content/tg5407598t64n518/ (accessed May 2006).
46   Gregory et al., "Localization of Satratoxin-G in *Stachybotrys chartarum* Spores and Spore-
Impacted Mouse Lung Using Immunocytochemistry." *Toxicologic Pathology*, vol. 32, no. 1, January-
February 2004, pp. 26-34(9). Available at http://www.ingentaconnect.com/content/tandf/utxp/2004/000
00032/00000001/art00005;jsessionid=dq0fqfupc7n2.alice?format=print&token=004312edf775686f235
7275c277b422c31464c7a763b2570504a2fa72156117d34d (accessed May 2006).
47   Interview with Dr. David Straus, April 2006.
48   E-mail from Dr. David Straus, September 2006.
49   Ibid.
50   Ibid.
51   Ibid.
52   Interview with Dr. Michael Gray, March 2007.
53   Ibid.
54   Ibid.
55   Ibid.
56   Ammann, "Is Indoor Mold Contamination a Threat to Health?"
57   Michael Gray, "Molds, Mycotoxins, and Human Health." According to Dr. Gray, the body
of this text is an excerpt from an affidavit he presented in the Ballard vs Farmers case. Available at
http://www.moldacrossamerica.org/published3.htm (accessed November 2005).
58   Interview with Dr. Gray, March 2007.
59   Ibid.
60   Ibid. (Internal quotation marks added.)
61   E-mail from Dr. David Straus, September 2006.
62   Bennett et al., "Mycotoxins," p. 15.
63   Rafal L. Górny, "Filamentous Microorganisms and Their Fragments in Indoor Air-A
Review," *Ann Agric Environ Med* 2004, 11, 185-197. Available at http://www.aaem.pl/pdf/11185.htm

(accessed April 2006).
64   Interview with Dr. Jack Thrasher, April 2006.
65   Committee on Environmental Health, American Academy of Pediatrics (AAP), "Toxic Effects of Indoor Molds," *Pediatrics*, vol. 101 no. 4, April 1998, 712-714. Available at http://aappolicy. aappublications.org/cgi/content/full/pediatrics;101/4/712 (accessed April 2006).
66   Freedom of Information Files, George Washington University, Washington D.C. (accessed by authors in 2000).
67   PBS, "Plague War." Available at http://www.pbs.org/wgbh/pages/frontline/shows/plague/etc/ faqs.html (accessed March 2006).
68   Interview with Dr. Michael Gray, March 2007. Progressive Healthcare Group: 520-586-9111 or 866-586-9111.
69   *Medical Aspects of Chemical and Biological Warfare*, Edited by Frederick Sidell (United States Government Printing, 1997). Available at http://www.bordeninstitute.army.mil/cwbw/Ch34.pdf (accessed March 2007).
70   Bennett et al., "Mycotoxins," p. 19-20.
71   Interview with Dr. Regina Santella, July 2006.
72   Interview with Dr. David Straus, April 2006.
73   Ibid.
74   Peraica et al., "Toxic Effects of Mycotoxins in Humans."
75   Bennett et al., "Mycotoxins," p. 6.
76   Ibid., p. 20.
77   Ibid.
78   Robert Wannemacher et al., "Trichothecene Mycotoxins," p. 656 in *Medical Aspects of Chemical and Biological Warfare*, Edited by Frederick Sidell (United States Government Printing, 1997). Available at http://www.bordeninstitute.army.mil/cwbw/Ch34.pdf (accessed March 2007).
79   Wannemacher et al., "Trichothecene Mycotoxins," p. 657.
80   Bennett et al., "Mycotoxins," p. 20.
81   Wannemacher et al., "Trichothecene Mycotoxins," p. 657.
82   Ibid., p. 668.
83   Lipsey, "The Mold and Bacterial Results from the Sampling Dr. Lipsey did in St .Bernard Parish."
84   CDC, "Anthrax." Available at http://www.cdc.gov/ncidod/dbmd/diseaseinfo/anthrax/faq/ (accessed March 2006).
85   Ibid.
86   Beth Daley, "The Next Menace: Mold," *Boston Globe*, Sept. 12, 2005. Available at http:// www.boston.com/news/globe/health_science/articles/2005/09/12/the_next_menace_mold?mode=PF (accessed February 2006).
87   Ibid.
88   Interview with Dr. Richard Lipsey, April 2006.
89   Ibid.

CHAPTER TWO:

BACK TO THE BASICS OF FUNGI

In order to keep ourselves safe from fungi, we must learn about these microorganisms. Three points are essential to understand. First, mold is ubiquitous, which means it is everywhere. Second, we are all potential victims of mold and its byproducts. Third, each one of us has a vested interest in gaining at least a basic understanding of molds, mycotoxins, and the knowledge that has been determined by scientific and medical studies and clinical observations. As we proceed through the following information, keep in mind that—given certain circumstances—mold is the enemy. To get a strategic edge in the battlefield, we must identify the strengths and weaknesses of our enemies and the effects they can have on us.

Let's first look at the classification of fungi in general. Mycology is the study of fungi and is a field in which definitive answers do not always exist. Some scientific conclusions are proven by scientific correlations and are universally agreed upon in the scientific community. Other conclusions drawn from scientific studies are debated. The one constant factor in the field of science is the continual progression of scientific research and study. Newly discovered data can disprove existing scientific theories and create new scientifically sound paradigms. Even the most basic questions can have a variety of answers. For example, when searching the question, "How many species of fungi exist?" we readily located the following three answers: The CDC states, "It is estimated that there are between 50,000 and 250,000 species of fungi; however, many of these species have not been classified or named."[1] The Washington State Department of Health reports, "There are more than 100,000 species of mold in the world."[2] MSUM notes, "The study of fungi is far from complete. Mycologists have described an estimated 80,000 species of fungi, and many believe this estimate is only a fraction of what remains to be discovered. Estimates of 1.5 million species and variants have been suggested."[3]

The term *fungi* refers to organisms of several subcategories of fungi. "In lay terms, organisms are grouped based on their familiar descriptions including molds, yeasts, mildew, rusts, smuts, mushrooms, and puffballs."[4] People often incorrectly use these terms interchangeably, not understanding the difference between one category of fungi and another. For example, "Many people refer to 'mildew' as the commonly occurring fungi that grow on damp clothing or bath tiles as if it is a more friendly organism. In reality, such growth is typically a filamentous mold of the types likely found on water-damaged building materials."[5]

"Fungi are found in every ecological niche and are necessary for the recycling of organic building blocks that allow plants and animals to live."[6] During this breakdown of organic matter, unshackled nutrients provide nourishment to fungi.[7] Fungi, along with bacteria, benefit our ecosystem by breaking down and disposing of damp and decaying matter in our outdoor environment.[8] According to MSUM, these microorganisms ". . . decompose the majority of dead organic matter on our planet."[9] Essentially, fungi and bacteria are in charge of "Mother Nature's recycling center!"[10]

"Most molds reproduce by forming spores that disperse into the air in search of more food and moisture (a reproductive activity similar to seed dispersal from plants). Due to the diversity of mold in our environment, outdoor air normally always contains some level of these airborne mold spores," according to MSUM.[11] "When mold spores that are floating around in the air land on a food source, they sit there patiently waiting for water. If the item they land on should contain sufficient moisture or water comes from another source (leaks, etc.), the spore germinates, and hyphae grow. The hyphae branch out, secrete enzymes to break down the food, form the mycelium, and absorb nutrients to grow."[12] Hyphae are microscopic, branched, thread-like filaments; mycelium is a mass formed by continued hyphal growth. Mycelium growth often becomes visible within 24–48 hours.[13]

"Fungi need external or organic food sources and water to be able to grow."[14] In other words, a spore, a food source, and a moisture source must all be present in order for mold to proliferate. "Fungi cannot produce their own nutrients as plants do through photosynthesis."[15] In an outdoor environment, a food source can be rotting trees, leaves, or other natural debris. In an indoor environment, "Molds can grow on cloth, carpets, leather, wood, sheetrock, insulation (and on human foods) when moist conditions exist."[16]

MSUM explains, "In most cases, the substrate mold is growing on provides the nutrients; however, in some instances, the substrate is simply a foundation. Mold growing on glass, ceramic tile, metals, or other inorganic materials is not obtaining nutrients from these substrates. In these cases, mold is feeding on microscopic organic matter that is on the surface or trapped in tiny pores of the material. Bath-tile mold is an example: Mold is typically consuming organic dust, dirt, debris, skin flakes, body oils, soap scum, etc., and the ceramic tile is simply a foundation for the colony."[17]

In addition to coming from leaks, moisture can occur in the form of natural humidity or steam produced from baths or showers. "Most building materials including concrete, brick, mortar, grout, drywall, wood, etc., are porous and do allow water vapor to pass."[18] This accumulated vapor often supplies enough moisture for fungi to thrive, which explains why bathrooms are prone to mold growth. Building materials, once moist, often become "dinner" to molds; for example, ". . . wood, paper, natural fabrics, leather, and even the starch in wallpaper paste are common examples of dead organic matter preferred by filamentous molds."[19] During this process of structural degradation, "[f]ungi secrete enzymes that digest the material the fungi are imbedded in and absorb the released nutrients."[20] MSUM states, "As the organism grows, reproductive spores are formed and released as part of its continuing life cycle."[21]

This proliferous process along with the tenacious nature of fungi is the reason mold is a structural threat. For example, fungal spores are ". . . capable of resisting dry, adverse environmental conditions and hence capable of surviving a long time,"[22] Dormant mold spores are like land mines that silently wait for a source of detonation to come along, which for mold is often just adequate moisture. For example, if spores land on a medium that doesn't have the necessary level of moisture to sustain growth, they can lie dormant for decades.[23] However, when adequate moisture occurs from sources such as a freeze that breaks a water pipe or a storm that causes a roof leak, then "poof," the mold starts to proliferate.

MSUM points out that ". . . most filamentous mold spores are microscopic and, therefore, invisible to the naked eye. It is not uncommon to find hundreds or even thousands of mold spores per cubic foot of outdoor air. . . . Most fungal spores range from 1 to 100 microns in size with many types between 2 and 20 microns. People with good vision may see 80–100 micron particles unaided but below that range, magnification is generally necessary. . . ."[24] Not many of us can relate to measurements in micrometers or microns. "To put things in perspective, you could place over 20 million five micron spores on a postage stamp," according to MSUM.[25]

Certain quantities of mold spores are inevitable in our indoor air as ". . . mold spores are a natural component in air . . . Many people have witnessed proof there are natural airborne mold spores indoors. After inadvertently leaving a cup of coffee or food out for a few days, the resulting colony will be visible!"[26] This same speed of mold growth is inevitable when the three necessary components of mold germination—spore, food, and moisture—are present, even if it's inside our walls!

Mold spores can be carried indoors on clothing, pets, or parcels and can travel indoors through open windows, doorways, and heating and air conditioning systems. Unhealthy living conditions can occur when viable spores make their way into our homes and proliferate due to adequate sources of food and moisture. As mentioned earlier, when a substrate provides adequate sources of both food and moisture, hyphae form. "Hyphae can intertwine into the fibers of the substrate, penetrating the pores."[27] They burrow into the substrate and penetrate beneath the surface, similar to a root. "This is one of the reasons it is so difficult to kill and/or clean up mold on organic substrates. If you remove the surface growth, those bits of hyphae within the substrate are ready for re-growth upon the return of moisture."[28]

Mold growth contained in indoor environments can lead to deterioration of human health because of the increased concentration of spores and related toxins. "When these organisms are allowed to grow in a closed indoor environment, they can release millions of spores, causing indoor levels to reach concentrations that are hundreds of times higher than outdoors . . ."[29] When the concentration level of spores increases so does the level of related fungal byproducts, such as fragments and mycotoxins. Multiple pathways of exposure exist. "Because molds grow in moist or wet indoor environments, it is possible for people to become exposed to molds and their [by]products either by direct contact on surfaces or through the air if mold spores, fragments, or mold [by]products are aerosolized."[30] Adequate moisture not only supports fungal growth, but it also solubilizes the mycotoxins, which enables them to become airborne on dust and other particulate. Therefore, even if mold growth is hidden deep inside a wall, the elevated levels of mold spores, fungal fragments, and mycotoxins can still be present in the indoor air and thus be inhaled, which can potentially cause adverse health effects in even healthy people.[31]

"Most fungi feed on dead organic matter," reports MSUM.[32] These fungi are saprobes—organisms that derive nutrients from the remains of dead organisms or dead vegetable matter. Some types of molds, however, are ". . . parasitic and feed on living things. Pathogenic (disease causing) fungi are known to cause adverse health in animals and humans."[33] We must all understand the following: "Parasitic fungi attack living organisms, penetrate their outer defenses, invade them, and obtain nourishment from living cytoplasm, causing disease and sometimes the death of the host. Most pathogenic (disease-causing) fungi are parasites of plants, but several are known to cause diseases of humans and lower animals."[34] In other words, parasitic fungi can live on plants, animals, humans, and even other fungi, while pathogenic fungi can cause disease and even death in plants, animals, and humans.

According to the CDC, "Fewer than 200 fungal species have been described as human pathogens that can cause infections."[35] These molds are also referred to as ". . . opportunistic pathogens, meaning the organism is disease-causing given the opportunity to infect the host."[36] These parasitic, pathogenic mold species can assault the health of plants, animals, and humans.

"Fungi don't actually 'eat' their food but rather release enzymes to break down complex organic compounds and absorb nutrients through their cell surfaces."[37] During this ". . . process of enzyme release and nutrient absorption, molds also produce volatile organic compounds (VOCs)."[38] These VOCs release into the environment as the fungi metabolize the substrate on which they exist. The smell of these VOCs is the odor that we associate as the smell of mold, explains Dr. Puccetti. He emphasizes the following: Even if you don't see any mold growth, but you can smell that moldy, musty smell, it means nonvisible, active mold growth is present in the structure and enough moisture exists to support its growth. Dr. Puccetti emphasizes, "It's not dormant, and it's not dead. It's actively growing because the odor that you're smelling is due to the metabolic process that the mold is undergoing."[39] So if you smell these odors, you must take action. You must locate the source of moisture, eliminate it, and remove the contaminated building materials.[40] Otherwise, the risk to health remains.

VOCs are volatile, which means they evaporate into the air we breathe and contribute to the detrimental effects of mycotoxins produced by certain species.[41] "A mold-contaminated building may have a significant contribution [of VOCs] derived from its fungal contaminants that is added to those VOCs emitted by building materials, paints, plastics, and cleaners . . . At higher exposure levels, VOCs from any source are mucous membrane irritants and can have an effect on the central nervous system, producing such symptoms as headache, attention deficit, inability to concentrate, or dizziness," according to Harriet M. Ammann, PhD, a senior toxicologist for the Washington State Department of Ecology Air Quality Program.[42]

An important point to understand is that if you smell a moldy, musty odor, mycotoxins may be present. As covered earlier, mycotoxins are fungal metabolites produced by certain species of fungi. "The term mycotoxin was coined in 1962 in the aftermath of an unusual veterinary crisis near London, England, during which approximately 100,000 turkey poults [young fowls] died. When this mysterious turkey X disease was linked to a peanut (groundnut) meal contaminated with secondary metabolites from *Aspergillus flavus* (aflatoxins), it sensitized scientists to the possibility that other occult mold metabolites might be deadly."[43] Ever since, mycotoxins have been a focus of scientific research.

Because of ever-constant research in the field of science, the number of mycotoxins is estimated based on current, available scientific data. One report states, "Excluding mushroom toxins, approximately 350 to 400 fungal metabolites are considered to be toxic."[44] Another medical report estimates ". . . some 300 to 400 compounds are now recognized as mycotoxins, of which approximately a dozen groups regularly receive attention as threats to human and animal health."[45]

In summary, health risks can occur from exposure to fungal spores, mycotoxins, and VOCs confined in an indoor environment. Additional health risks can occur from exposure to the fine particulate matter that accompanies fungal growth, according to immunotoxicologist Jack D. Thrasher, PhD. He states the following has been determined by scientific studies: One, colonies of fungi shed different types of particulate matter into the air—spores, fragments of their cell bodies, and fragments of the colonies; two, the size of the fine particulate is much smaller than the size of the spores (less than 2 microns in diameter down to 0.2 microns in diameter); three, the size of the fine particulate is 300–400 times more concentrated in quantity than is the spore count; and four, the increased quantity of fine particulate means that fragments are shed at a much higher rate than whole spores are released. Dr. Thrasher points out that these fragments ". . . contain all of the toxic chemicals that are produced by the molds and the spores."[46] The retention of toxic properties means that when the colonies of fungi shed, they are shedding particulates that are highly toxic. In fact, Dr. Thrasher states that the effective particulate load of fine particulates plus that of the spores is 300 times more than the particulate load of the spore counts alone.

Therefore, fungal growth indoors creates an environment of not just spores, VOCs, and possibly mycotoxins but also includes the toxically potent fine particulates. Dr. Thrasher explains several important points: One, fine particulates will become airborne, settle, and re-emit into the air as do spores; two, the shedding of fine particulates occurs from frequencies created by normal living activities; three, the shedding of fine particulates is independent of spore release; four, whether mold is actively growing or not, it will shed these fine particulates into the air we breathe; and five, once airborne, these microscopic toxins are easily inhaled because of their small size, which enables them to travel deep into our airways and absorb into our blood, potentially leading to serum sickness, systemic symptoms, and/or systemic infections. Dr. Thrasher points out that these toxic air particulates are not measured by the traditional viable or nonviable air sampling methodologies, which greatly underestimate the actual toxicity of the air we breathe.[47]

According to Dr. Puccetti, "Usually, when molds are drying out is when they emit the most spores."[48] We experienced this type of increased sporulation activity in the high desert climate of Billings, Montana, after moisture from rain or snow began to evaporate. Similar incidences of increased sporulation activity have been documented in other desert locations as well. Dr. Gray reports the regular occurrence of massive sporulation in the Sonoran Desert in Tucson, Arizona, after each rain. He states, "With every rain, the molds grow. Within 24 to 48 hours when humidity drops significantly, the ambient environmental spore counts reported by local meteorologists rise dramatically, because the organisms desiccate and fracture into respirable particles."[49]

Dr. Gray clarifies that whenever fungi grow, they produce their toxins and digestive enzymes and sporulate. However, once molds desiccate, they die and ". . . release millions of spores and tiny micron and submicron size fragments of their cell walls, which contain antigenic materials and toxins."[50] He points out that people often incorrectly assume mold will not grow in a desert environment and believe mold grows only where ample moisture exists; in actuality, mold does grow in the desert on a routine basis. "We see the counts reported meteorologically in the weather reports on an almost daily basis, and if anything, the desert selects the most tenacious forms of funguses and molds."[51]

During periods of elevated indoor sporulation comes the increased risk of inhalation of spores, VOCs, and mycotoxins if the fungi are mycotoxin-producing. Dr. Straus explains the production of mycotoxins, "We believe that when spores are produced, the mycotoxins are packaged into the spores at the same time. We don't think the spores produce the mycotoxins. We think the colony produces them and packages them into the spores when the spores are formed."[52]

After mycotoxins are produced, they are not necessarily released into the air. Dr. Straus explains that it takes the presence of water to solubilize the mycotoxins, at which point the mycotoxins are released. Dr. Straus points out, "Spores cannot solubilize. They are particles and will never go into solution."[53] He provides a helpful analogy, "Think of the mycotoxins like sugar, which can go into solution and the spores like grains of sand, which will never go into solution."[54] Mycotoxins can solubilize from the moisture present in the substrate on which the fungi is growing. Once solubilized, the mycotoxins can adhere to dust, which can become airborne from air movement, at which point the mycotoxins can be inhaled.[55] In addition, mycotoxins exist on spores, fungal fragments, and fine particulate. All of these fungal components can also become airborne and thus be inhaled, delivering accompanying mycotoxins into the human body. Once the mycotoxins get absorbed into the blood stream, they begin to kill cells.[56]

As mentioned earlier, molds can lie dormant for years if they are not getting enough moisture from their environment to germinate.[57] "When a mold's environment goes dry, its spores enter a kind of hibernation, able to sometimes exist for decades in an inactive state. These microscopic dry spores are lightweight, and wind blows them virtually everywhere—into homes, businesses, and schools; onto furniture, countertops, and rugs. In dry conditions, they're mostly invisible but can still make some people with allergies sneeze, cough, and rub their itchy eyes."[58]

Understanding how fungi respond when their environments go dry helps us understand the toxic fungal component of Katrina's dried flood residue. Water from the hurricane woke up billions of dormant fungi,[59] which settled into the sediment, mud, debris, and any still-standing structure. When everything began to dry, the fungi had less moisture. As a part of their natural survival mechanisms, these fungi—some toxigenic—responded by producing more spores. Some of these spores became airborne, resettled in areas with adequate food and moisture, and germinated. Others remained part of the toxic dust, which residents were at risk of inhaling as the spores became aerosolized from winds and activities.

As forthcoming test results will show, the predominant molds found post-Katrina were *Stachybotrys*, *Aspergillus*, *Cladosporium*, and *Penicillium*. In order to better understand the potential health risks from exposure to these molds, let's review some case histories and medical documentation.

"In the late 1930s, stachybotryotoxicosis was reported in humans working on collective farms in Russia. People affected were those who handled hay or feed grain infested with *S. chartarum* or were exposed to the aerosols of dust and debris from the contaminated materials. Some of these individuals had burned the straw or even slept on straw-filled mattresses. The infested straw was often black from growth of the fungus. Common symptoms in humans were rash, especially in areas subject to perspiration, dermatitis, pain and inflammation of the mucous membranes of the mouth and throat, conjunctivitis, a burning sensation of the eyes and nasal passages, tightness of the chest, cough blood, rhinitis, fever, headache, and fatigue. Workers developed symptoms within two to three days of exposure to the fungus. Some members of the Russian teams investigating this disease rubbed the fungus onto their skin to determine its direct toxicity. The fungus induced local and systemic symptoms similar to those observed in naturally occurring cases."[60]

"As recently as 1977, there was an outbreak of stachybotryotoxicosis among farm workers handling infested straw in Hungary. The symptoms were similar to those described in Russia and began appearing about 24 hours after exposure to the fungus. . . . In 1996 workers at a horticultural facility in Germany developed very painful, inflamed lesions on their fingertips followed by scaling of the skin when they handled decomposable pots infested with *S. chartarum*."[61]

"*Stachybotrys* produces cottony, rapidly growing colonies, which mature in about four days. From both front and reverse, the color of the colony is white initially and turns to black by aging."[62] The resultant black color spurred national media sources to herald *Stachybotrys* as the *Black Mold*, which awakened the public to the deleterious health effects of toxigenic molds. If you have read in the newspaper about a toxic black mold, most likely you were reading about *Stachybotrys*.

"The genus *Stachybotrys* has a single well-known species, *Stachybotrys chartarum*."[63] *S. chartarum* produces trichothecene mycotoxins. Dr. Gray points out, "Several types of mycotoxins, including trichothecenes, ochratoxins, patulins, and aflatoxins, induce human illnesses, some of which resemble radiation sickness as the result of the effect of being either random or specific DNA 'adduct formers.' Adduct formers are compounds whose molecular size and configuration allow them to insert themselves randomly into DNA and RNA, thus resulting in the inhibition of protein synthesis, bone marrow suppression, coagulation defects, and bleeding disorders resulting in nasal, pulmonary, and gastrointestinal hemorrhaging, bleeding into the adrenal glands, uterus, vagina, and the brain."[64]

In addition to trichothecenes, *Stachybotrys* releases other biological compounds that are immunosuppressive agents. Studies suggest that ". . . the combination of trichothecenes and these immunosuppressive agents may be responsible for the observed high toxicity of this fungus."[65] Other disturbing research shows, "In experimental animals, trichothecenes are 40 times more toxic when inhaled than when given orally [in animal feed]."[66] Lightly put, *Stachybotrys* is hardly a mold we should be disrupting without the complete "armor" of adequate personal protective equipment.

Regarding the genus *Aspergillus*, six or seven species are potentially very dangerous, according to Dr. Thrasher.[67] Dr. Ammann further reports *Aspergillus* as a genus that contains ". . . several toxigenic species, among which the most important are *A. parasiticus*, *A. flavus*, and *A. fumigatus*. Aflatoxins produced by the first two species are among the most extensively studied mycotoxins. They are among the most toxic substances known, being acutely toxic to the liver, brain, kidneys, and heart, and with chronic exposure, [are] potent carcinogens of the liver. They are also teratogenic [can produce or induce malformations in a developing embryo or fetus]. Symptoms of acute aflatoxicosis are fever, vomiting, coma, and convulsions."[68]

While aflatoxicosis is poisoning—acute or chronic—from aflatoxin exposure, aspergillosis is an infection or disease caused by inhalation of *Aspergillus* spores. Therefore, when a mycotoxin-producing *Aspergillus* species is present in an indoor environment, the risk exists for both aspergillosis and aflatoxicosis. Aspergillosis is a serious medical condition, especially for people who are immunocompromised. Dr. Gray states, "Aspergillosis in those who are immunosuppressed carries a 65% mortality rate."[69] The majority of aflatoxin studies focus on the ingestion of these substances rather than the inhalation. Hence, most data links aflatoxicosis to the ingestion of aflatoxins.

According to Dennis Kunkel, PhD, a photomicrographer at the University of Hawaii at Manoa, "The genus *Cladosporium* includes over 30 species." He states, "The most common ones include *Cladosporium elatum*, *Cladosporium herbarum*, *Cladosporium sphaerospermum* (common indoors), and *Cladosporium cladosporioides*. *Cladosporium* can cause mycosis and extrinsic asthma (immediate-type hypersensitivity: type I). Acute symptoms include edema and bronchiospasms; chronic cases may develop pulmonary emphysema. Certain species cause systemic fungal infections and . . . [have] been documented in cases of blastomycosis, candidiasis, chromoblastomycosis, histoplasmosis, entomophthoramycocis, onychomycosis, sinusitis, pulmonary infections, phaeophphomycocis, and keratomycosis. *Cladosporium* spp. can also be associated with the spoilage of refrigerated meats."[70]

Dr. Kunkel, who has found *Cladosporium* on the inner surface of a building's air supply duct, reports, "This genus is the most common outdoor airborne mold but may occur in homes. It is an allergenic mold producing over ten different types of antigens. The spores are easily made airborne and are a common cause of respiratory problems. It is frequently found at elevated levels in water-damaged environments."[71]

Texas Tech University Health Science Center (TTUHSC) documents that the health effects of the *Cladosporium* genus include the following: allergenicity, pulmonary, (sub)cutaneous infection, lesions, phaeohyphomycosis (a group of superficial and deep infections caused by dark fungi), possible skin, nail, eye, sinus infections, and foot mycetoma (tumorous infection caused by fungi).[72] According to TTUHSC, two main species of *Cladosporium* are important in terms of medical mycology. They are *Cladosporium carrionii* and *Cladosporium bantianum*.[73] The *Cladosporium* genus produces the mycotoxins epi- and fagi-cladosporic acids.[74] *Cladosporium cladosporioides*, specifically, produces mycotoxins such as cladosporin and isocladosporin, which are biologically active isomers of cladosporin.[75]

The genus *Penicillium* is probably the mold with which we are most familiar. "*Penicillium notatum*, also known as *Penicillium chrysogenum*, is a mold that is commonly found in homes," according to Dr. Kunkel.[76] Along with being prevalent in wet building materials, *Penicillium* grows on bread, fruits, and nuts and is used in the production of green and blue mold cheeses.[77]

According to Dr. Thrasher, three or four species of *Penicillium* are potentially very dangerous.[78] Medical literature further reports that the inhalation of *Penicillium* spores can result in ". . . initial pulmonary infection, followed by fungemia [the presence of fungi or yeasts in the blood] and dissemination of the infection. The lymphatic system, liver, spleen, and bones are usually involved. Acne-like skin papules [pimples] on [the] face, trunk, and extremities are observed during the course of the disease."[79]

Interestingly, the first miracle drug, penicillin, is derived from the fungus *Penicillium*, which is another reason many of us are familiar with the genus. This first antibiotic was discovered by Alexander Fleming, who published his findings in 1929.[80] Research and development efforts heightened during World War II, which led to the mass production of penicillin and the saving of many lives from once-fatal bacterial infections.[81] Somewhat confusing, some scientific literature refers to antibiotics as mycotoxins, which they are, technically. According to Dr. Lipsey, penicillin is the mycotoxin produced by *Penicillium*.[82] So, "While all mycotoxins are of fungal origin, not all toxic compounds produced by fungi are called mycotoxins."[83] Some are called antibiotics.

According to Dr. Ammann, "Antibiotics are isolated from mold (and some bacterial) cultures, and some of their bacteriotoxic or bacteriostatic properties are exploited medicinally to combat infections."[84] Dr. Kunkel points out, "The target and the concentration of the metabolite are both important. Fungal products that are mainly toxic to bacteria (such as penicillin) are usually called antibiotics."[85] On the other hand, fungal-derived pharmaceuticals that target fungal cells are generally called antifungals. A specific example of a fungal-derived antifungal is griseofulvin. According to Dr. Gray, the Physicians' Desk Reference (PDR) states, "Griseofulvin is an oral fungistatic antibiotic for the treatment of superficial mycosis and is derived from a species of *Penicillium*."[86] A fungistatic antibiotic is a drug designed to inhibit the growth of fungi.

"Because of their pharmacological activity, some mycotoxins or mycotoxin derivatives have found use as antibiotics, growth promotants, and other kinds of drugs . . ."[87] Due to these advancements in Western medicine, metabolites of *Penicillium* and some other mycotoxin-producing fungal species have the potential to restore health when properly and selectively used to treat specific health conditions. However, do not forget the dual nature of mycotoxins. A published scientific report points out the following important fact: "Toxigenic mold activities produce metabolites that are either broad-spectrum antibiotics or mycotoxins that are cytotoxic."[88] The term *cytotoxic* means toxic to cells; thus, mycotoxins kill cells. Which cells—human, fungal, and/or bacterial—they kill depends on which organisms each mycotoxin innately targets. Therefore, mycotoxins are much like a double-edged sword—some can save lives; yet some can prove fatal.

(Endnotes)
1 The CDC Mold Work Group, "Mold Prevention Strategies," p. 1.
2   Washington State Department of Health, "Got Mold? Frequently Asked Questions About Mold." Available at http://www.doh.wa.gov/ehp/ts/IAQ/Got_Mold.html (accessed March 2006).
3   Department of Environmental Health and Safety, MSUM, "About Mold."
4   Ibid.
5   Ibid.
6   Ammann, "Is Indoor Mold Contamination a Threat to Health?"
7   Department of Environmental Health and Safety, MSUM, "About Mold."
8   W. G. Sorenson, "Fungal Spores: Hazardous to Health?" *Environmental Health Perspectives Supplements*, vol. 107, no. S3, June 1999. Available at http://ehp.niehs.nih.gov/ members/1999/suppl-3/469-472sorenson/sorenson-full.html (accessed August 2006).
9   Department of Environmental Health and Safety, MSUM, "About Mold."
10   Ibid.
11   Ibid.
12   Randy Penn, "Mold: Volatile Organic Compound's & Mycotoxins: A Primer for Homeowners." Available at http://www.toxicmoldusa.com/mycotoxins_voc.htm (accessed April 2006).
13   Department of Environmental Health and Safety, MSUM, "About Mold."
14   Ammann, "Is Indoor Mold Contamination a Threat to Health?"
15   The CDC Mold Work Group, "Mold Prevention Strategies," p. 1.
16   Ammann, "Is Indoor Mold Contamination a Threat to Health?"
17   Department of Environmental Health and Safety, MSUM, "About Mold."
18   Ibid.
19   Ibid
20   The CDC Mold Work Group, "Mold Prevention Strategies," p. 1.
21   Department of Environmental Health and Safety, MSUM, "About Mold."
22   The CDC Mold Work Group, "Mold Prevention Strategies," p. 1.
23   E-mail from Dr. David Straus, April 2007.
24   Department of Environmental Health and Safety, MSUM, "About Mold."
25   Ibid.
26   Ibid.
27   Penn, "Mold: Volatile Organic Compound's & Mycotoxins: A Primer for Homeowners."
28   Ibid.
29   Department of Environmental Health and Safety, MSUM, "About Mold."

30   Ammann, "Is Indoor Mold Contamination a Threat to Health?"
31   Department of Environmental Health and Safety, MSUM, "About Mold."
32   Ibid.
33   Ibid.
34   "Fungus," Encyclopedia Britannica, 2006. Encyclopedia Britannica Premium Service.
Available at http://www.britannica.com/eb/article-57969/fungus (accessed August 2006).
35   The CDC Mold Work Group, "Mold Prevention Strategies," p. 1.
36   Department of Environmental Health and Safety, MSUM, "About Mold."
37   Ibid.
38   Ibid.
39   Interview with Dr. Andrew Puccetti, April 2006.
40   Ibid.
41   Peraica et al., "Toxic Effects of Mycotoxins in Humans."
42   Ammann, "Is Indoor Mold Contamination a Threat to Health?"
43   Bennett et al., "Mycotoxins," p. 3. (Italics added.)
44   Sorenson, "Fungal Spores: Hazardous to Health?"
45   Bennett et al., "Mycotoxins," p. 3.
46   Interview with Dr. Jack Thrasher, April 2006.
47   Ibid.
48   Interview with Dr. Andrew Puccetti, April 2006.
49   Interview with Dr. Michael Gray, March 2007.
50   Ibid.
51   Ibid.
52   Interview with Dr. David Straus, April 2006.
53   E-mail from Dr. David Straus, September 2006.
54   Ibid.
55   Ibid.
56   Interview with Dr. David Straus, April 2006, and Anyanwu E et al., "The neurological
significance of abnormal natural killer cell activity in chronic toxigenic mold exposure," *Scientific World
Journal* 2003; 3: 1128-37. Available at http://www.medscape.com/medline/abstract/14625399 (accessed
August 2006).
57   E-mail from Dr. David Straus, April 2007.
58   Daley, "The Next Menace: Mold."
59   Ibid.
60   Berlin D. Nelson, Professor, Department of Plant Pathology, North Dakota State University,
Forego, "*Stachybotrys chartarum*: The Toxic Indoor Mold," APSnet, November 2001, p. 5. Available at
http://www.apsnet.org/online/feature/stachybotrys/ (accessed February 2006).
61   Ibid., p. 6.
62   Dr. Fungus, "*Stachybotrys* sp." Available at http://www.doctorfungus.org/thefungi/
stachybotrys.htm (accessed February 2006).
63   Ibid.
64   Interview with Dr. Michael Gray, March 2007, and Gray, "Molds, Mycotoxins, and Human
Health."
65   Nelson, "*Stachybotrys chartarum*: The Toxic Indoor Mold," p. 4.
66   Peraica et al., "Toxic Effects of Mycotoxins in Humans."
67   Interview with Dr. Jack Thrasher, April 2006.
68   Ammann, "Is Indoor Mold Contamination a Threat to Health?"
69   Interview with Dr. Michael Gray, March 2007.
70   Dennis Kunkel, "Image Number: 20508B." Available at http://www.denniskunkel.com/DK/
DK/Fungi_and_Slime_Molds/20508B.html (accessed August 2006).
71   Ibid.
72   TTUHSC, "*Cladosporium* sp." Available at http://www.ttuhsc.edu/SOM/Microbiology/
mainweb/aiaq/FungalReferenceGuide/withoutframes/Fungi%20organisms/Cladosporium_sp.htm
(accessed March 2006).
73   Ibid.
74   Ibid.

75   Ibid.
76   Dennis Kunkel, "Image Number: 24301D." Available at http://www.denniskunkel.com/DK/DK/Fungi_and_Slime_Molds/24301D.html (accessed August 2006).
77   Ibid.
78   Interview with Dr. Jack Thrasher, April 2006.
79   Dr. Fungus, "*Penicillium* spp." Available at http://www.doctorfungus.org/thefungi/ penicillium.htm (accessed March 2006).
80   Utah State University, "*Penicillin*: the first miracle drug." Available at http://www.herbarium.usu.edu/fungi/FunFacts/penicillin.htm (accessed March 2006).
81   Ibid.
82   Interview with Dr. Richard Lipsey, April 2006.
83   Bennett et al., "Mycotoxins," p. 3.
84   Ammann, "Is Indoor Mold Contamination a Threat to Health?"
85   Kunkel, "Image Number: 24301D."
86   Interview with Dr. Michael Gray, March 2007.
87   Bennett et al., "Mycotoxins," p. 3.
88   Anyanwu et al., "The neurological significance of abnormal natural killer cell activity in chronic toxigenic mold exposure."

CHAPTER THREE:

MOLD TESTING—WHAT'S IT ALL MEAN?

Events in history can teach us a great deal—if we let them. Hurricane Katrina is no different. The resulting mold and toxic chemicals from this epic storm were of a magnitude previously incomprehensible. Ultimately, lessons learned, both good and bad, were drawn from the experiences of the people who battled these cataclysmic levels of mold and toxins. Medical data was created from the many victims who grappled with ensuing health effects from the multi-toxic exposures. Sitting at the advantage point of post-Katrina, we must seize the opportunity to learn from this historical natural disaster; we must not ignore the wisdom we can gain by learning about the plight of other human beings in fear that we will sink in the muck and mire of their present or past overwhelming circumstances. By earnestly reading the stories of survivors and victims of the moldy, noxious post-Katrina environment and vicariously experiencing the trauma that befell them, we can gain beneficial knowledge of the dangers of mold and toxins. With increased awareness and insight, we will be better equipped to minimize the effect that mold and toxins may have one day in our own individual lives. Basically, Katrina was the disaster so highly documented due to her size and impact on human life that she gave us the ability to learn what we otherwise might not have without personally living through our own mold infestation or toxin-laced mudslide. Let's make good use of the opportunity.

Two essential facts to understand are as follows: One, negative health effects from exposure to mold and mycotoxins do not occur from just Katrina-size levels of mold; and two, mold growth is not limited to humid climates. Mold is prolific and pervasive, given conducive conditions. A structural fungal invasion can begin from just a few spores and grow anyplace in any geographic region in the North, South, East, or West where enough heat and moisture exists to sustain growth. So no matter where we live, a careful study of the mold-related health hazards brought to light by Katrina is beneficial.

To start with, let's take a look at eye-opening media coverage of post-Katrina issues by *Newsweek*, which points out the repetitive nature of history when it quotes a disaster-seasoned government official, "Like the 9/11 workers, many of those working in the Katrina rubble are being exposed to deadly toxins, says Hugh Kaufman, a senior policy analyst for the Environmental Protection Agency (EPA) in Washington. With more than 35 years of experience in the field, he particularly worries about workers and citizens being exposed to harmful contaminants like asbestos and mold."[1]

Mr. Kaufman continues, "My big concern is civilians who are going down there trying to do the right thing, helping provide labor to help restore that area, and the environmental risks they're being exposed to. You have students and other folks going down there wearing—for all intents and purposes—'dust masks,' not being trained in the handling of asbestos and mold properly, and [spending time] inside tens of thousands of people's homes and exposing themselves to cancer-causing contaminants. What that means is the risk of cancer to the people who want to help has gone up . . . And the people who are bringing them down there are no more knowledgeable [of this] than the volunteers."[2]

"The government has waived certain environmental and occupational rules for the cleanup to save money. That means worker protection. [Prior to these changes,] you could not do asbestos remediation or cleanup without trained people certificated and with environmentally protective equipment requirements. But people are [now] being issued what are essentially just dust masks and if you read the package, it says these [masks] are not appropriate for asbestos or toxic chemicals or anything—only just dust! It's right there on the package. Federal rules will not allow [government] workers to work in potentially asbestos-contaminated areas with that type of protection and yet tens of thousands of students are down there, and have been down there, remediating asbestos and mold-contaminated areas. It's a major problem," states Mr. Kaufman.[3]

"Initial reactions are difficulty breathing and increase in asthma," Mr. Kaufman explains. "But down the line, the concern is lung cancer and other cancer. . . . You've got all these students who are involved in voluntary remediation without adequate protective gear putting themselves at risk . . . so you could have many more people impacted from all over the country."[4] Dr. Thrasher concurs there will likely be an increase in cancer. He states, ". . . I'll say, within a reasonable scientific probability . . . we will see an increase but what types of cancers, I don't know."[5]

With health outcomes as devastating as cancer potentially lurking around mold-infested corners, it important to review and understand the available Katrina mold data to gain health-protecting knowledge. We have interspersed various test results with forthcoming explanations from professional experts to transform the raw data into meaningful information that is valuable and applicable on a personal level. Many of the figures are staggering, which demonstrates just how concentrated indoor levels of mold spores can become from wet building materials. Please understand that these levels of elevated mold concentrations are not unique to post-Katrina buildings. Similar levels would likely be found across the nation by testing water-damaged buildings, which have been drenched by annual floods or other ravaging hurricanes such as Gustav and Ike. To translate the prolific nature of these microscopic intruders into quantifiable spore counts provides us with a tangible "image" of an oftentimes invisible enemy. Should a structural invasion occur, the strategy must always be to evacuate the mold-infested building, especially during the remediation process, and to use adequate personal protection equipment.

Mold testing must be interpreted by trained, knowledgeable professionals. For this reason, let's review various test results of post-Katrina collected mold samples and see what the experts have to say. (Issues relating to other forms of toxic exposures will be discussed in subsequent chapters.) Mold samples were taken by various independent groups that used a variety of methodologies, which will be explained and discussed.

Toxicologist Dr. Lipsey states, "The samples I collected on February 11, 2006, while on tour of the devastation in St. Bernard Parish have been analyzed by an independent microbiological lab. Out of 45,000 homes in the parish, only 500 were habitable, and I sampled six of them." He explains, "The mold levels were in the millions of spores per square inch of wallpaper or on furniture, and the most common pathogenic mold appeared to be *Stachybotrys*, which was seen at over 8 million colony forming units per square inch. *Stachybotrys* can produce mycotoxins and is sometimes called the black mold, and [it] is the most toxic of all toxic molds, producing some of the most potent and deadly mycotoxins. It is ten times more toxic than the most pathogenic of the *Penicillium* or *Aspergillus* molds, which also can produce mycotoxins. *Stachybotrys* is uncommon in contaminated homes, and I rarely find it in even the most sick of the sick buildings. *Stachybotrys* produces trichothecenes [mycotoxins] . . . There is also the concern that some of the bacteria and molds produce a synergistic effect in combination to make them even more toxic."[6]

Dr. Lipsey continues, "I inspected many homes, and the stench of rotting materials was in every home . . . there was significant water damage and high levels of pathogenic molds and bacteria in every home since the homes had been under water for days and have been growing mold for months. There were snakes living in some of the homes and marsh grass in most of the homes, and many had marsh grass on top of the roof indicating how deep the water was in those areas of St. Bernard Parish. None of the homes were safe to occupy or even be inside for any length of time without personal protective equipment including a HEPA [high-efficiency particulate air] respirator, rubber gloves, goggles, and a Tyvek suit."[7]

"The parish has 45,000 homes, of which 40,000 will have to be bulldozed," said Dr. Lipsey. "I recommended that they bulldoze the homes in one-square-block areas toward the center of each block and set them on fire. If it catches a house afire across the street, it's no big deal simply because that house is going to have to be bulldozed and burned anyway. There are very few homes in St. Bernard Parish that can be salvaged."[8]

In order to properly interpret the results from mold sampling, we must take note of the methodology used as well as the parameters of the chosen testing protocol. Dr. Lipsey used swab-sampling methodology, which (when microscopically analyzed) tells us the population and quantity of molds actually growing on the walls from which samples were collected. Cultured swab samples identify the species within a genus, which tells us if mycotoxin-producing fungi are present. However, without test results from accompanying air samples, swab samples will not tell us how many spores have become airborne and will thus be inhaled during normal respiration. Of course, when millions of spores are present per every square inch, common sense will tell us a sizeable amount of spores and related contaminants are getting into the air.

To get an indication of the level of airborne mold spores post-Katrina and gain a broader geographic perspective, let's take a look at the test results from the air samples taken in October and November of 2005 by the National Resources Defense Council (NRDC). In order to evaluate the health risks that could affect the largest portion of the returning population, the NRDC testing team selected the nonviable testing methodology, which measures both living and dead spores.[9] Outdoor air samples as well as some indoor air samples were collected by neighborhood in New Orleans and surrounding communities. Areas included in the survey were Bywater, Chalmette, French Quarter, Gentilly, Lakeview, Lower Ninth Ward, Mandeville, Metairie, Mid-City, New Orleans East-Little Woods, New Orleans East-Read Blvd. East, and Uptown/Carrollton.[10]

The NRDC lead scientist, Gina Solomon, MD, MPH, who oversaw the survey, cites budget limitations as the major reason both viable and nonviable sampling methodologies were not used. "We realized that we had to pick one or the other. As a physician, it seemed a little silly to me to look just at viable spores because whether a spore is living or dead is irrelevant to how your body's immune system perceives it. In other words, if you're going to mount an allergic reaction or an immune response to the mold, your immune cells will not differentiate between a dead spore and a living spore. The immune cells will go nuts, whatever that spore is, if they're going to go nuts—if you're a person who is sensitive. So, from a clinical perspective, I just wanted to know how many spores were out there. . . . Obviously, if the total spore count is high, you know that a fair number of those spores are viable, so you know there is a risk of all of the health effects related to mold."[11]

"The only reason the living spores matter is in the case, perhaps, of people who have very impaired immune systems who actually might get an infection from *Aspergillus* or some other invasive fungus," reasons Dr. Solomon. "Although this is definitely a concern, the other health concerns [allergy and asthma] are so much more widespread, so they are a much greater concern to me."[12]

She explains, "We wanted to test in a lot of neighborhoods, both indoors and outdoors, and in areas that were flooded as well as comparison locations that were not flooded. We also wanted to test for hours rather than just minutes in order to gather good estimates of 24-hour exposures. With all of those priorities, a limited budget, and limited time with a team on the ground in New Orleans, we had to make some hard decisions very quickly. One of the things we sacrificed was viable sampling. It would still be useful to do that."[13]

Nonviable methodology does not speciate as does the viable methodology. It simply gives a total spore count of both live and dead mold spores. Without identifying the species within the genuses, we do not know the amount of mycotoxin-producing fungi present. Dr. Solomon reflects, "I guess when I started this I was more focused on allergy. I am aware of the mycotoxin problem and, frankly, had been skeptical about mycotoxins. Obviously, they cause health problems in animal studies. There's lots of conflicting stuff in human studies regarding mycotoxins. I wasn't quite convinced of the serious problems involved with mycotoxins; mostly, I was thinking of people with allergies and asthma. Now I'm becoming more concerned about the mycotoxin issue. I think that, over time, that issue will become more widely understood and will be taken with the seriousness that it deserves."[14]

The objective of the NRDC study was to assess potential human exposure to bioaerosols in the New Orleans area after the flooding of the city.[15] "A team of investigators performed continuous airborne sampling for mold spores and endotoxin outdoors (in flooded and nonflooded areas) and inside homes (that had undergone various levels of remediation) for periods of 5–24 hr[s] during the two months after the flooding."[16] (Endotoxins are toxins produced by gram-negative bacteria.) Although it may initially appear that the following New Orleans' mold data would be of interest to only residents of the tested areas, such is not the case at all. Accompanying discussions provided by scientific experts highlight valuable information we would all need to know should we have to understand results from our own mold testing one day. Again, we are vicariously learning by way of Katrina.

The longer duration of the NRDC sample collections is certainly a strong point of the NRDC testing protocol. It gives a broader picture of the mold levels to which people would be exposed over a given period of time versus the shorter snapshot in time from samples collected in just a few minutes. Most homeowners are limited due to financial constraints to samples collected over shorter periods of time, which can distort the accuracy of results. "Because fungi (including molds) release their spores at different times during the day, brief sampling times may miss potentially important episodes of spore release and are less accurate for estimating daily spore concentrations," states Solomon et al.[17]

Because of the varied levels of flooding and remediation of the tested structures in New Orleans, the NRDC specified, "Indoor sites were categorized according to the degree of [previous] flooding and the level of remediation. Two indoor sites were 'minimally flooded' with < 4 cm [less than 4 centimeters] of water in the living space; one was currently inhabited. The remaining six indoor sites were 'severely flooded' with a history of water more than 1 m [meter] deep in the living space;

none were currently inhabited. The severely flooded indoor sites were further categorized as 'unremediated,' 'partially remediated,' or 'fully remediated.'"[18]

"The NRDC collected continuous volumetric samples using a Burkard Continuous Recording Air Sampler with a flow rate of 10 liters per minute for 6 to 24 hours. The results were extrapolated to estimated average 24-hour mold spore concentrations expressed as spores per cubic meter of air. The instrument has greater than 90 percent removal efficiency of particles . . ."[19] The sampling protocol chosen varied in duration of sampling time per site as well as time of day sampled. Some parameters of the sampling protocol may have resulted in underestimates of mold spore counts. According to the NRDC report, "Concentrations of some molds are typically higher at night. Some calculations, based on 6-hour continuous volumetric measurement during daytime hours, may underestimate the true 24-hour concentration."[20]

Dr. Puccetti concurs, "Fungi all have different preferences. Various species will not produce spores at certain times of the day. They have cycles in which they produce spores off and on. It's very species specific and they all vary. I'm not aware of any specific type of species that only sporulates at night. I think that the concept to take away from those statements is that when you're conducting air sampling from airborne spores, you're dealing with a highly variable situation; you can have the same amount of mold growth in a building, go in there at various times of day, and get results that differ quite significantly. The reason for that is you're going to get variability in the spore-producing propensities of the different types of molds that are there. Sometimes they will be emitting spores, while other times they won't."[21] As noted earlier, Dr. Puccetti believes molds usually emit the most spores when they are drying out.[22]

It is important to note that the NRDC's chosen sampling protocol called for collection of air samples without disturbing existing dust. According to Dr. Solomon, "The way we did our sampling was to try not to disturb the sampling environment. We actually would try to tiptoe into a house, try to not move anything around, carefully set up the equipment, and tiptoe out. The places we sampled were uninhabited. During the period of sampling, it would just be still and quiet."[23] Unfortunately, without any air disturbance, mold spores—*Stachybotrys* especially—are less likely to become airborne, which would have created artificially low test results of all genuses.

The NRDC acknowledges the effect its choice of protocol had on test results, "Indoor mold spore concentrations are typically far higher when dust is disturbed in a house. These samples were taken when there was no disturbance going on and may therefore underestimate the true spore concentrations."[24]

Not speaking directly in regard to the NRDC, Dr. Lipsey points out, ". . . [I]n unoccupied homes it's unethical to do air samples. In other words, measuring the number of spores/cm³ of air in an unoccupied home, unless you agitate the air to mimic a family living there, won't accurately reflect air quality. The federal agencies, NIOSH and the EPA, recommend a 20-inch box fan . . . to get the spores into the air to mimic a heating and cooling system running, children running, and doors opening and closing because if it's an unoccupied home, the spores are going to land on the floor, ceiling, and walls and stay there until they are disturbed."[25]

According to Mr. Pearson, the NRDC mold spore counts would have been astronomically higher if the dust had been disturbed. He explains, "With activity, you will certainly see much higher levels."[26] A published study that examined the effects of human activity on air sampling concurs, "Human activity resulted in retrieval of significantly higher concentrations of airborne spores."[27] Therefore, to put test results into proper perspective, it is important to consider not only the testing methodologies selected but also other parameters of the testing protocol.

However, even with knowledge of specific testing protocols, it is not easy to translate test results into medical cause-and-effect outcomes. In fact, according to the EPA, "Indoor air regulations and mold standards or threshold limit values (TLVs) for airborne concentrations of mold or mold spores have not been set. Currently, there are no EPA regulations or standards [for indoor or outdoor] airborne mold contaminants."[28] For this reason, the NRDC designated outdoor testing areas as having either *high* or *very high* mold spore counts based on average outdoor spore counts according to the following classifications set forth by the National Allergy Bureau (NAB): mold spore counts from 1–6,499 spores/m³ are *low*, from 6,500–12,999 spores/m³ are *moderate*, from 13,000–49,000 spores/m³ are *high*, and over 50,000 spores/m³ are *very high*.[29] The levels of mold in indoor testing areas were not classified as *high* or *very high* because these classifications do not apply to indoor air levels. According to the NAB, the aforementioned ". . . mold levels were determined based on outdoor exposure to natural[ly] occurring spores in the environment and should not be applied to indoor exposure, which may represent an entirely different spectrum of spore types."[30] The NAB has not established indoor air classifications for mold levels except to document that mold counts over 1,300 spores/m³ indicate an indoor environment is *moldy*.[31]

The significant molds revealed in the NRDC testing were *Aspergillus/Penicillium, Stachybotrys,* and *Cladosporium,*[32] which is not surprising to anyone mold literate as three of these four fungi are commonly associated with excess moisture in buildings. In fact, Dr. Straus notes, "The main organisms that grow in water-damaged buildings are *Aspergillus, Penicillium,* and *Stachybotrys.*"[33] These fungi are documented over and over again in damp buildings caused from hurricanes, floods, and/or leaks. For example, the *Aspergillus* genus was prevalent in water-damaged buildings in Florida in 2004 when four of the six hurricanes that season hit central Florida.[34]

The NRDC survey team, which used the nonviable methodology, notes in its published report, "We were unable to confidently differentiate the *Aspergillus* spores from *Penicillium* spores due to their morphological similarity."[35] For this reason, the NRDC reported *Aspergillus* and *Penicillium* spores as *Aspergillus/Penicillium*.

Dr. Straus further explains the nonviable methodology, "In the nonviable way, we just collect spores from the air and deposit them on a glass slide. Then we look at them under the microscope. We don't grow them, so we can't tell if they're dead or alive. Because *Aspergillus* and *Penicillium* spores under the microscope look almost identical, we can't tell what they are in the nonviable technique. That's why the spores are classified as *Aspergillus/Penicillium* or *Penicillium/Aspergillus*. What we're saying is that we really don't know if they are *Penicillium* or *Aspergillus*, and truthfully they may be something else."[36] For example, a reported 82% of *Aspergillus/Penicillium* may be 82% *Aspergillus*, 82% *Penicillium*, or most likely, a combination of both with possibly some other morphologically similar genuses.

The presence of both *Aspergillus* and *Penicillium* was confirmed by the CDC in its *Morbidity and Mortality Weekly Report* of January 20, 2006. *Aspergillus* and *Penicillium* were reported as the predominant fungi out of the four detected in both indoor and outdoor air samples taken October 22–28 by survey team members from both the CDC and the Louisiana Department of Health and Hospitals (LDHH).[37] Air samples were collected and filters were analyzed for culturable fungi.[38]

The NRDC survey team points out an additional limitation of the nonviable testing methodology. It reports, "We were also unable to determine the specific species present within these genera."[39] Without results from viable testing to speciate, we cannot tell if the fungi present are toxigenic (mycotoxin-producing), pathogenic (infection-causing), or even the genus as to which they were identified. Thus, we cannot assess the potential associated health risks from mycotoxin exposure in the mold-laden post-hurricane buildings tested by the NRDC.

According to Dr. Thrasher, to test for the presence of fungal organisms without identifying the species gives you a meaningless report.[40] He explains, "When we look at a report, we are also concerned not only about the genus but also about the species within the genus of the particular mold that is found. For example, there are six or seven species of *Aspergillus* that are potentially very dangerous. There are three or four species of *Penicillium* that are very dangerous and if you do not speciate it, you have no idea what is in the air. So to rely strictly on mold spore counts is a bunch of bull; that's my opinion. Another example is *Stachybotrys*. If you look at *Stachybotrys*, there are two types of *Stachybotrys*. One is called chemo type A, which produces the very toxic trichothecenes such as satratoxin H and G and also stachylysin, which is a hemolytic agent [destroys red blood cells with subsequent release of hemoglobin]. The second is the non-chemo type A that produces a whole variety of other toxic compounds such as spirocyclic drimanes, which are highly toxic both to the nervous system and to the immune system. So just to do a spore count is bull. I look at these reports and shudder at what the so-called current industrial hygienists think and that they think what is going on is appropriate."[41]

We must ask the following logical question: Since the nonviable technique doesn't identify the species within a genus, which tells us if the fungi are mycotoxin-producing or pathogenic, is there value in using the nonviable methodology to indicate the level of risk to human health?

According to Dr. Straus, "Yes, there is a benefit. There are two ways in which fungi can make people sick, aside from infection: inhalation of fungal spores and inhalation of mycotoxins. With the inhalation of fungal spores . . . what happens is you're inhaling high concentrations of living particles that can actually produce biological compounds inside the lungs, some of which can do damage."[42]

Nonviable testing quantifies both living and dead spores. If nonviable test results reveal elevated levels of spores, as Dr. Solomon earlier pointed out, a fair amount of them are going to be living. These living particles can produce harmful biological compounds inside the body, as Dr. Straus explained. The dead particles can also have detrimental effects because they retain their toxic properties. The limitations with nonviable testing are that we do not gain the medically insightful, species-specific information regarding toxic compounds about which Dr. Thrasher has spoken. By analyzing test results from multiple methodologies (such as nonviable testing from the NRDC, viable testing from the CDC/LDHH team, and swab sample testing from Dr. Lipsey), we can gain a more complete assessment of the potential detrimental health effects, in this case, from the post-Katrina mold maze. This published data will undoubtedly add to the historical perspective of the aftermath of Katrina.

The NRDC survey team summarizes some of the negative health effects from fungal exposure. It states, "Filamentous microfungi (mold) can threaten human health through release of spores that become airborne and can be inhaled. Some molds produce metabolites (mycotoxins) that can initiate a toxic response in humans or other vertebrates. Repeated exposure to significant quantities of fungal material can result in respiratory irritation or allergic sensitization in some individuals. Sensitized individuals may subsequently respond to much lower concentrations of airborne fungal materials. Of the thousands of types of fungal spores found in indoor and outdoor environments, adverse health effects in humans have most frequently been associated with *Alternaria*, *Aspergillus*, *Cladosporium*, *Penicillium*, and *Stachybotrys*."[43]

Four of these five fungi were identified post-Katrina in the New Orleans area by the NRDC.[44] The following NRDC findings of averaged mold spore counts by neighborhood are summarized, with portions quoted in part or in full, from the "New Orleans Area Environmental Quality Test Results, Mold Results by Neighborhood Report."[45] Percentages of insignificant fungi were not included. Any rounding up or down of spore count averages were reflected in the original figures.

One outdoor air sample was collected in Bywater November 15, 2005. The previously flooded testing area was designated as "Bartholomew near North Claiborne, Florida Area." The estimated daily spore count average (based on 21 hours of continuous volumetric sampling) was 101,000 spores/m^3, which was classified as *very high*. Of this spore count, 52% was *Cladosporium* and 29% was *Aspergillus/Penicillium*.[46]

One outdoor air sample was collected in Chalmette October 17, 2005. The previously flooded testing area was designated as "Jean Lafitte and Creole." The estimated daily spore count average (based on 6 hours of continuous volumetric sampling) was 77,000 spores/m^3, which was classified as *very high*. Of this spore count, 54% was *Cladosporium* and 16% was *Aspergillus/Penicillium*.[47]

A single outdoor air sample was collected in the French Quarter October 19, 2005. The testing area was designated as "Esplanade and North Rampart." The French Quarter was not flooded during the hurricane. The estimated daily spore count average (based on 6 hours of continuous volumetric sampling) was 26,000 spores/m^3, which was classified as *high*. Of this spore count, 46% was *Cladosporium* and 33% was *Aspergillus/Penicillium*.[48]

Two outdoor samples and one indoor sample were collected in Gentilly. One previously flooded testing area designated as "Elysian Fields and Lombard, Gentilly Terrace" was sampled October 19, 2005. The estimated daily spore count average (based on 6 hours of continuous volumetric sampling) was 30,000 spores/m^3, which was classified as *high*. Of this spore count, 53% was *Cladosporium* and 29% was *Aspergillus/Penicillium*. A second previously flooded testing area designated as "St. Roch near Filmore, Gentilly Terrace" was sampled November 14, 2005. The estimated daily spore count average (based on 20 hours of continuous volumetric sampling) was 63,000 spores/m^3, which was classified as *very high*. Of this spore count, 76% was *Cladosporium* and 16% was *Aspergillus/Penicillium*. The average of the two outdoor samples was "47,000 spores/m^3", which was classified as *high*. A previously flooded, fully remediated indoor testing area designated as "St. Roch near Filmore, Gentilly Terrace" was sampled November 14, 2005. The estimated daily spore count average (based on 24 hours of continuous volumetric sampling) was 45,000 spores/m^3, which was classified as *moldy*. Of this spore count, 38% was *Cladosporium* and 48% was *Aspergillus/Penicillium*.[49]

High spore counts of 45,000/m³ in a *fully remediated* building warrant discussion. "The fully remediated homes had all furniture, carpets, and interior walls removed down to the studs; in some cases the flooring and studs had been sanded and mildewcide may have been applied. Most homes had some windows or doors open, either because they were broken or in an effort to ventilate the interior; the sampling teams caused minimal disturbance of the interior to minimize reaerosolization. When possible, sampling equipment was placed near the center of the first floor of the home, away from open windows or doors."[50]

Dr. Solomon states, "The question is: Were those spores that we were measuring in the fully remediated houses blowing in from neighbors' houses, from outside, or were they indicating that there was still mold growth inside? We couldn't tell."[51]

According to Dr. Straus, "When you remediate a building and pull out all contaminated material, you're throwing spores into the air as well. So, of course, those levels in the newly remediated buildings are going to be high. . . . Just because you have high spore counts in the air does not mean that the building is now in trouble. If the building no longer has any fungal growth on its building surfaces, you can remove fungal spores from the air by using a HEPA filter or a negative air machine. We have machines that essentially filter the air and remove large particles like fungal spores."[52]

In order to remove settled spores, Mr. Pearson explains, "During remediation, lots of airflow is used because we want as much mold as possible airborne so the air scrubbers will remove it from the air. Of course, all surfaces are HEPA vacuumed and moldy non-structural materials are removed too."[53]

"The most important thing people have to worry about in buildings that have fungal problems is fungal growth on building material," states Dr. Straus. "That's the whole cause of all of the problem. If you don't have any more fungal growth in a building and you still have a high spore count, we can clean the air . . ."[54]

It is important to understand that not all fungal growth is readily visible. According to Dr. Puccetti, "A lot of problems in buildings involve hidden mold growth, so you don't really know how big a problem you have until you start investigating it. If you have a flood-like situation from a hurricane or flooding, it's pretty simple to determine how big a problem you have because you're more than likely going to get an extremely visible problem. You also have to understand that the mold is not just in areas where you can see it; it's also in areas that got wet and are hidden from view like wall cavities, ceiling plenums, and other areas of the building that aren't directly observable by doing a walk-through of the occupied space of the building."[55]

In order to determine the origin of the elevated mold spore levels, Dr. Solomon explains, "We did indoor and outdoor samples simultaneously at most of these sites or, at worst, in one quick sequence. One day we went there and sampled indoors, and the next day we sampled outdoors to try to get a point of comparison. In the fully remediated places the levels were generally lower indoors than outdoors, so that was good. It likely meant that the spores we had measured inside had come in from the outdoor air, but we couldn't be sure of that."[56]

The fact that indoor mold spore counts of 45,000 spores/m³ fell within the estimated daily average of outdoor mold spore counts of 30,000 and 63,000 spores/m³ could mean that the high indoor mold spore levels originated from an outside source. However, when we look at the percentage of each genus found outdoors in comparison with the breakdown found indoors, we may question that hypothesis. The first outdoor sample had 53% *Cladosporium* and 29% *Aspergillus/Penicillium*, and the second outdoor sample had 76% *Cladosporium* and 16% *Aspergillus/Penicillium*. If the indoor mold spore counts did indeed come in from the outside, we would expect to see roughly the same percentage breakdown of genuses, but we do not. The indoor air mold spore counts were comprised of 38% *Cladosporium* and 48% *Aspergillus/Penicillium*, which could be cause for concern. If the distribution of indoor spore counts had been 48% *Cladosporium* and 38% *Aspergillus/Penicillium*, it would have been more reflective of the outside air. We must, however, keep in mind that outdoor air samples were not collected directly outside the indoor testing site.

Barely detectable outdoor levels of fungi can explode indoors if given proper conditions. Dr. Puccetti states, ". . . [T]he types of mold spores that germinate on wet building materials and produce mold growth that thrives indoors are spores that are generally present at relatively low concentrations in outdoor air. Some types of spores such as *Stachybotrys*, *Fusarium*, *Trichoderma*, and many others are generally present outdoors at concentrations well below the limit of detection of the common routine methodology used for air sampling. The fact that they often go undetected outdoors does not mean that they are not present outdoors."[57]

To determine if the source of elevated mold spores in fully remediated houses is from the outside, retesting would have to be done after broken windows (that allow airflow in from the outside) were repaired and the inside air was scrubbed to remove existing airborne contaminants. If elevated spores remained or returned, then an indoor source of mold would be suspected.

According to Dr. Puccetti, "Ideally, after remediation the spore levels inside should be much the same as those outside. In fact, in most instances they are lower because the remediation contractors have scrubbed the air with filtered negative-air machines, cleaned everything, and removed all settled dust. I would expect that if a building is properly remediated, it should have airborne spore levels that reflect what is generally found outside at the time of sampling."[58]

"The limiting factor in mold growth is moisture content. If a material has enough moisture content, then mold will grow," states Dr. Puccetti. "The amount of moisture that's required in a given material will vary, depending on the species of mold. Some molds require a relatively small amount of moisture content within a building material, while other molds require a high moisture content. I can give you two specific examples: *Aspergillus versicolor*, a species of *Aspergillus*, can grow on building material with relatively low moisture content; *Stachybotrys chartarum*, on the other hand, requires a very high moisture content on building materials."[59]

Dr. Puccetti reiterates, "It's the moisture content that is the limiting factor. If you remove the moisture, you're not going to have any mold growth. The whole idea behind proper remediation is not to pick and choose what chemicals should be used; it should be to dry things out and physically remove the mold growth from the surface. This is necessary because mold retains its allergenic and toxic properties even if it is dead. Often plumbing leaks or other types of water intrusion problems are repaired without removing any of the mold that grew during the time moisture was present. If you don't eliminate the mold and you've only eliminated the source of water, you still have a mold problem. The mold, if there's enough of it, will still be in the wall cavities and spores from that mold will come out of the wall cavities through the switch plates, power plugs, and other penetrations. These spores settle on surfaces; they collect on carpeting and other surfaces and are re-emitted into the air by occupant activity, resulting in exposure to building occupants."[60]

Further survey results reveal that two outdoor samples and one indoor sample were collected in Lakeview October 18, 2005. One was a previously flooded outdoor testing area designated as "Mouton and Orleans." The estimated daily spore count average (based on 6 hours of continuous volumetric sampling) was 32,000 spores/m³, which was classified as *high*. Of this spore count, 48% was *Cladosporium* and 31% was *Aspergillus/Penicillium*. A second previously flooded outdoor testing area was designated as "Canal and Porteus." The estimated daily average (based on 6 hours of continuous volumetric sampling) was 40,000 spores/m³, which was classified as *high*. Of this spore count, 29% was *Cladosporium* and 57% was *Aspergillus/Penicillium*. The average of the two outdoor spore counts was 36,000 spores/m³, which was classified as *high*. A third partially remediated, previously flooded indoor testing area was designated as "Canal and Porteus." The estimated daily average (based on 6 hours of continuous volumetric sampling) was 638,000 spores/m³, which was classified as *moldy*. Of this spore count, 6% was *Cladosporium* and 83% was *Aspergillus/Penicillium*.[61]

"Partially remediated homes had furniture and carpets removed and some removal of mold on the walls, such as removal of some drywall or visible evidence of scrubbing."[62] The fact that a partially remediated building has 638,000 spores/m³ when the outdoor air in the same general geographic area contained 32,000 spores/m³ and 40,000 spores/m³ undoubtedly indicates that the building has an inside source of fungi. Obviously, the remediation of this building needs to be completed after which retesting must occur.

One outdoor sample and two indoor samples were also collected in the Lower Ninth Ward. One previously flooded outdoor testing area designated as "Tupelo near Bienvenue, Holy Cross" was sampled October 17, 2005. The estimated daily average (based on 6 hours of continuous volumetric sampling) was 67,000 spores/m³, which was classified as *very high*. Of this spore count, 36% was *Cladosporium* and 40% was *Aspergillus/Penicillium*. An indoor, partially remediated, previously flooded testing area designated as "Douglas near Caffin, Holy Cross" was sampled November 15, 2005. The estimated daily average (based on 20 hours of continuous volumetric sampling) was 79,000 spores/m³, which was classified as *moldy*. Of this spore count, 24% was *Cladosporium*, 67% was *Aspergillus/Penicillium*, and less than 1% was *Stachybotrys chartarum*. "Visible *Stachybotrys chartarum* growth [was] detected on surfaces in other rooms but not in the one where air sampling was done."[63] A second indoor, partially remediated, previously flooded testing area designated as "Burgundy near Caffin, Holy Cross" was sampled November 15, 2006. The estimated daily average (based on 22 hours of continuous volumetric sampling) was 414,000 spores/m³, which was classified as *moldy*. Of this spore count, 26% was *Cladosporium*, 65% was *Aspergillus/Penicillium*, and less than 1% was *Stachybotrys chartarum*. "Visible *Stachybotrys chartarum* growth [was] detected on surfaces in other rooms but not in the one where air sampling was done."[64]

By quickly comparing the outdoor levels to the indoor levels, we can see that the building with indoor air samples of 414,000 spores/m³ obviously has an indoor source of mold. The other indoor test site with 79,000 spores/m³ may also have an indoor source of mold since the genus breakdown indoors is not the same as the percentage breakdowns found in the outside ambient air, although total spores/m³ were only 12,000 spores/m³ higher. Retesting at a later date could give final determination.

The fact that *Stachybotrys chartarum* was discovered in the two Lower Ninth Ward indoor air samples further confirms the presence of *Stachybotrys chartarum* in St. Bernard Parish, which was documented in Dr. Lipsey's swab-sampling test results. The growth of *Stachybotrys chartarum* is not surprising as the structures in St. Bernard Parish had been saturated with water for an extended period of time, providing the high moisture content that *Stachybotrys chartarum* needs in order to grow.

So why did Dr. Lipsey's test results find over 8 million cfu/in² of *Stachybotrys* while the NRDC's air sample testing revealed only 1–2 percent? Dr. Lipsey explains, "*Stachybotrys* is a heavy, wet spore that often does not get into the air unless it's disturbed. It's not like some of the other pathogenic molds and bacteria that physically throw their spores into the air to ensure that they get somewhere to find moisture and nutrients so they too can grow. *Stachybotrys* is not like that. *Stachybotrys* will produce a mycotoxin, like trichothecene, to physically kill the other molds around it."[65]

Berlin Nelson, PhD, a professor in the Department of Plant Pathology at North Dakota State University concurs that *Stachybotrys*, when wet, does not readily disseminate into the air as compared to other fungi such as *Aspergillus* and therefore is difficult to detect the presence of by taking air samples. He clarifies, "However, when the fungus and substrate dries and is disturbed by mechanical means or air movement, conidia [asexual spores] can become bioaerosols."[66] Consequently, if the dry dust had been disrupted during the NRDC testing phase, the *Stachybotrys* spore counts would have been much higher.

The fact that *Stachybotrys* was found at all is cause for concern since exposure to even a small amount can lead to undesirable health effects. Numerous news articles have reported ill health effects associated with *Stachybotrys* found in homes and workplaces. One such report states that a 14-month-old girl and her grandmother became sick from what the grandmother described as ". . . a small, round, black circle" that she attributed to a hole in the drywall. As it turned out, this small, round, black circle wasn't a hole at all but rather the mold *Stachybotrys*.[67] Remember, what may appear to be only a small, round, black circle can be the tip of an iceberg-size mold infestation.

A single outdoor sample was collected in Mandeville on the north shore of Lake Pontchartrain November 14, 2005. This outdoor testing area was used as a comparison site as it had not flooded. The location was designated as "Near Intersection of Route 1088 and Route 59." The estimated daily average (based on 24 hours of continuous volumetric sampling) was 21,000 spores/m³, which was classified as *high*. Of this spore count, 76% was *Cladosporium* and 4% was *Aspergillus/Penicillium*.[68]

Two outdoor samples were collected in Metairie. "These samples were collected in areas that were not affected by flooding and were used as comparison samples."[69] One testing area designated as "Elmere near Highway 10" was sampled October 18, 2005. The estimated daily average (based on 6 hours of continuous volumetric sampling) was 25,000 spores/m³, which was classified as *high*. Of this spore count, 41% was *Cladosporium* and 41% was *Aspergillus/Penicillium*. The second testing area designated as "Melody near NRDC collection process Veteran's Blvd." was sampled November 13, 2005. The estimated daily average (based on 24 hours of continuous volumetric sampling) was 21,000 spores/m³, which was classified as *high*. Of this spore count, 69% was *Cladosporium* and 8% was *Aspergillus/Penicillium*. The average spore count of the two outdoor sites was 23,000 spores/m³, which was classified as *high*.[70]

The Mandeville testing area wasn't a flooded area and was even a distance from the flooding, yet mold spore counts were high—21,000 spores/m³. The Metairie comparison sites, also not flooded, reflected about the same spores/m³ as the Mandeville comparison site—25,000, 21,000, and 21,000 spores/m³ respectively— all classified as *high*. The radius of the testing area was not expansive enough to determine how far out from the greater New Orleans area one would have to travel before being able to obtain air samples that reflected moderate to low levels of spore counts.

Two previously flooded outdoor samples and one minimally flooded indoor sample were collected in Mid-City. One outdoor testing area designated as "Telemachus and Canal, Mid-City" was sampled October 19, 2005. The estimated daily average (based on 6 hours of continuous volumetric sampling) was 102,000 spores/m³, which was classified as *very high*. Of this spore count, 22% was *Cladosporium* and 71% was *Aspergillus/Penicillium*. A second outdoor testing area designated as "Hagan near Orleans, Bayou St. John" was sampled November 15–16, 2005. The estimated daily average (based on 21 hours of continuous volumetric sampling) was 83,000 spores/m³, which was classified as *very high*. Of this spore count, 61% was *Cladosporium* and 27% was *Aspergillus/Penicillium*. An indoor, minimally flooded, *inhabited* testing area designated as "Hagan near Orleans, Bayou St. John" was sampled November 15–16, 2005. The estimated daily average (based on 24 hours of continuous volumetric sampling) was 11,000 spores/m³, which was classified as *moldy*. Of this spore count, 39% was *Cladosporium* and 50% was *Aspergillus/ Penicillium*. The average of the two outdoor spore counts was "92,000 spores/m³", which was classified as *very high*.[71]

In regard to the sole inhabited testing area, we must ask the following question: Should people be living in a home with indoor spore counts of 11,000 spores/m³— which is 9,700 spores/m³ over the 1,300 spores/m³ designation of *moldy* for indoor air?

Regarding the areas and houses tested, Dr. Solomon states in an NRDC press release, "The outdoor mold spore concentrations could easily trigger serious allergic or asthmatic reactions in sensitive people . . . The indoor air quality was even worse, rendering the homes we tested dangerously uninhabitable by any definition."[72] As residents returned, how many unwittingly lived in homes with mold levels that would have been classified as uninhabitable had air samples been tested? Worse yet, how many people have been forced to knowingly live in mold-infested homes because insurance coverage was denied, FEMA payouts were limited, and/or other financial resources were exhausted?

Three previously flooded outdoor samples and one previously flooded indoor sample were collected in New Orleans East. One outdoor testing area designated as "Aberdeen and Crowder, Little Woods" was sampled October 16, 2005. The estimated daily average (based on 6 hours of continuous volumetric sampling) was 55,000 spores/m³, which was classified as *very high*. Of this spore count, 43% was *Cladosporium* and 20% was *Aspergillus/Penicillium*. A second outdoor testing area designated as "Eastover Subdivision, Read Blvd. East" was sampled October 16, 2005. The estimated daily average (based on 6 hours of continuous volumetric sampling) was 53,000 spores/m³, which was classified as *very high*. Of this spore count, 51% was *Cladosporium* and 19% was *Aspergillus/Penicillium*. A third outdoor testing area designated as "Aberdeen and Crowder, Little Woods" was sampled November 15, 2005. The estimated daily average (based on 15 hours of continuous volumetric sampling) was 75,000 spores/m³, which was classified as *very high*. Of this spore count, 56% was *Cladosporium* and 23% was *Aspergillus/ Penicillium*. The average for the two outdoor spore counts for the New Orleans East-Little Woods area was 65,000 spores/m³, which was classified as *very high*. An indoor, fully remediated testing area designated as "Aberdeen and Crowder, Little Woods" was sampled November 15, 2005. The estimated daily average (based on four hours of continuous volumetric sampling) was 100,000 spores/m³, which was designated as *moldy*. Of this spore count, 39% was *Cladosporium*, 39% was *Aspergillus/Penicillium*, and less than 1% was *Stachybotrys*.[73]

One of the New Orleans East indoor samples collected from a *fully remediated* building measured 100,000 spores/m³. Since this spore concentration was 25,000– 47,000 spores/m³ higher than the outdoor levels for that same geographic area, a source of indoor mold growth is apparent. Also indicative of an indoor mold source was the detection of *Stachybotrys*—the black mold—which generally produces undetectable to low levels of spores outdoors, according to Dr. Puccetti.[74]

According to Mr. Pearson, "One of the best indicators of a proper job is to have clearance testing performed. The results should show fewer spore counts inside than outside or the same, and you don't want to see anything unusual on the report. For example, if you have no *Stachybotrys* outside but have *Stachybotrys* inside, that's a bad deal, and it tells you there is still a source of mold inside somewhere. If I find the same amount of spores of the same species inside, I don't suspect there is a mold problem in the building."[75] Obviously, the New Orleans East indoor test site has the red flags of indoor mold infestation waving.

Two outdoor samples and two indoor samples were collected in Uptown/Carrollton. One previously flooded outdoor testing area designated as "Dublin near South Claiborne, Leonidas" was sampled October 16, 2005. The estimated daily average (based on 6 hours of continuous volumetric sampling) was 81,000 spores/m³, which was designated as *very high*. Of this spore count, 18% was *Cladosporium* and 61% was *Aspergillus/Penicillium*. A second outdoor testing area, not flooded, designated as "Valence near St. Charles, Uptown" was sampled October 16, 2005. The estimated daily average (based on 6 hours continuous volumetric sampling) was 68,000 spores/m³, which was classified as *very high*. Of this spore count, 46% was *Cladosporium* and 25% was *Aspergillus/Penicillium*. The average of the two outdoor spore counts was "75,000 spores/m³", which was classified as *very high.* An indoor, unremediated, previously flooded testing area designated as "Octavia and Fontainebleau, Broadmoor" was sampled October 16, 2005. The estimated daily average (based on 6 hours of continuous volumetric sampling) was 645,000 spores/m³, which was classified as *moldy.* Of this spore count, 7% was *Cladosporium,* 82% was *Aspergillus/Penicillium,* and 2% was *Stachybotrys.* A second indoor, minimally flooded testing area designated as "Dublin near South Claiborne, Leonidas" was sampled November 14, 2005. The report did not indicate what level of remediation had been completed. The estimated daily average (based on 24 hours of continuous volumetric sampling) was 11,000 spores/m³, which was classified as *moldy.* Of this spore count, 59% was *Cladosporium* and 31% was *Aspergillus/Penicillium.*[76]

The Uptown/Carrollton indoor air sample of an unremediated building was the highest concentration—645,000 spores/m³—of the entire survey. "Unremediated homes contained all contents and were undisturbed since the flooding."[77] An inside source of mold growth is obvious in this particular property as the average of the two outdoor test sites was 75,000 spores/m³. We must ask ourselves, if indoor air spore counts above 1,300 spores/m³ indicate that a building is *moldy,* how do we classify structures with spore counts of 645,000 spores/m³? Dr. Solomon calls these levels "dangerously uninhabitable by any definition."[78]

It is a frightening thought—but plausible—that the air quality of this site very well may be representative of the air quality in other unremediated buildings previously flooded by Katrina. The NRDC qualifies its survey findings, "It is difficult to know how representative these samples are of the conditions inside homes that were not sampled or outdoors at other times or locations and under other meteorological conditions."[79]

The NRDC survey team sums up its results by stating, "The levels of mold spores in the air in New Orleans were extremely high and could pose a serious health threat, especially to anyone who is allergic to molds and to people with asthma and other respiratory disease. We found elevated levels of mold spores both inside homes and outside, especially in flooded areas."[80]

Read the NRDC's following words carefully: ". . . could pose a serious health threat," *especially* to particularly susceptible groups of people. The NRDC isn't saying that the potential for serious health threats exists *only* for people who are allergic to molds or have asthma or other respiratory diseases. The potential for serious health threats can exist for all of us when exposed to elevated mold levels. Don't interpret the NRDC's warning as being applicable to only people allergic to molds and/or with asthma and other respiratory diseases, because it's not. We have to make sure we do not dismiss relevant warnings, because sometimes in the denial of our own predicament—the severity of risk to our own health—we let forewarnings go unheeded or we convince ourselves they don't apply to us. We must not allow our desire for the return of a "normal life" to thwart our vigilance in dealing with excessive mold caused by any source, be it a hurricane, a flood, or a building water leak.

No matter what moisture source causes inside mold growth, it is important to understand that certain factors inherent in air sampling methodologies and some parameters of testing protocols can artificially deflate mold spore counts. A medical report gives an example, "Airborne fungi measurements fail to take into consideration mold contamination in dust or [on] surfaces (often visible to the naked eye) and mycotoxins in air, dust, and on surfaces."[81] This limitation was exemplified in the NRDC's air sampling test results that reflected only 1–2 percent of the spores collected were *Stachybotrys*, yet *Stachybotrys* was readily visible on wall surfaces. A medical study states, "In certain situations, air sampling without . . . [accompanying] surface sampling may not adequately reflect the level of microbial contamination in indoor environments."[82] For these reasons, ". . . testing settled dust for fungi and mycotoxins has been recommended."[83]

Dr. Thrasher reports that even when dust is disturbed and collected using air sampling methodology, mold test results are still gross underestimates of actual air toxicity as they do not measure the presence of mycotoxins, VOCs, or microscopic contaminants such as mold spore fragments, bacteria, endotoxins, bacteria cell wall toxins, and other unknown fine particulate matter and chemicals.[84] Unfortunately, budget restrictions usually limit multifaceted testing because the more inclusive the testing, the more exorbitant the cost, explains Dr. Puccetti.[85]

Whether or not we have testing verification of these microscopic toxins, we need to proceed as if we do because these additional toxins often accompany fungal growth. The obvious concern is that when anyone—a resident, a remediation worker, or whoever—moves through a mold-infested building, dust gets stirred up and inhaled. This toxic dust, as well as the mold itself, must be properly handled and disposed of during the remediation process. "The average person who goes to clean it up, who doesn't do it professionally and doesn't contain the area, will spread the mold and you'll have a higher concentration," cautions George F. Riegel, Jr., MD, who is an indoor air quality medical consultant and president of a Michigan-based environmental inspection company, Healthy Homes.[86]

It is quite possible that much of this remediation work is all for naught. "Scientists worry many poor homeowners will spend tens of thousands of dollars attempting to get rid of mold, only to find out their efforts failed."[87] It very well may be that, in some instances, foregoing any remediation attempt and proceeding directly to the demolition stage could be financially prudent as well as health preserving. Owners need to seek qualified professional assistance to do a cost analysis of both options with best- and worst-case scenarios considered.

Full remediation of a structure does not guarantee that it is, or will remain, mold free. Even in newly remediated buildings, mold spores in the ambient air can re-ignite infestation if sufficient moisture is present to sustain growth. "With enough moisture, mold spores can germinate in just hours and begin eating wood, sheetrock, wallpaper glues, and other organic material[s] that are in the home. Within days, a few spores can produce millions more, which are then carried to other locations by air currents. By the time mold is visible—which can take from a day to several weeks after germination—it often has taken root in walls and may be impossible to get out."[88]

The challenges inherent in mold remediation are evident in a one-year-later revisiting to examine the conditions of remediated properties in the hurricane-affected areas. Wilma Subra, president of Subra Company, a chemical laboratory in New Iberia, Louisiana, performed environmental testing after both hurricanes Katrina and Rita. She expresses concern over the mounting health repercussions residents are experiencing from mold and chemical exposures. She reports that many of the remediated homes that had been stripped to the studs (which were sanded and treated with a fungicide) and reconstructed with new insulation, sheetrock, paint, and flooring had mold growing through the newly replaced sheetrock by the hurricanes' anniversaries.[89]

Another factor that can affect the accuracy of air sample testing is variations in sporulation activity, of which a good example is revealed by the NRDC. According to Dr. Solomon, Mervi Hjelmroos-Koski, PhD, DSc, a mycologist on the NRDC survey team, was able to break collection samples down into hour-long, or even 30-minute, increments. The findings clearly show the value of longer sampling periods.[90]

"In some of these indoor environments, it was incredible what happened. There would be wild fluctuations," states Dr. Solomon. "In one house, the *Stachybotrys* was undetectable for about four out of the six hours of sampling. Then, over a one-hour period, there was a spike in *Stachybotrys* that was incredible. Clearly, it had sporulated during that short period of time and our equipment caught it. When we averaged the results to create the 24-hour average, there was definitely detectable *Stachybotrys*, but the spore counts were not impressive because there was only a brief period of time they were high. But if anybody had been so unfortunate as to have been in there during that period of time (when the *Stachybotrys* was releasing its spores) that would have been terrible. What I think it also shows is that some of these methods that go in for a very short period of time and gather spores over a 15-minute period can really miss serious mold problems."[91]

The NRDC team included Dr. Hjelmroos-Koski's findings in its published report, "Maximal 30-minute spore concentrations ranged from 26 to 251% higher than the mean for the entire sampling period. The highest 30-minute concentrations were 1,002,456 spores/m³ indoors and 259,200 spores/m³ outdoors." The survey results showed, "In one flooded home, the 30-minute maxima revealed a *Stachybotrys* spore concentration of up to 324,648 spores/m³ during daytime hours."[92]

By examining 24-hour sporulation activity in 30-minute and hourly segments, the NRDC was able to clearly show the degree to which fungal sporulations can fluctuate. So how does testing performed by industrial hygienists detect these types of intermittent, elevated sporulations when the most commonly used testing protocols include only a limited number of air samples collected for only short amounts of time?

When asked this question, Dr. Puccetti said, "One of the limitations inherent in current methodology for routine sampling of airborne mold spores is sampling time. The sampling periods can vary from 2 to 15 minutes, depending on the method used. Of course, this limitation can be easily overcome by collecting many samples (for example, ten samples per hour for eight hours). If this is done, the cost to the client skyrockets in proportion to the number of samples and additional field time. Believe me, very few people are willing to pay that kind of money. Since there are no established exposure-response relationships for mold exposure, any additional information obtained from significantly more extensive sampling may not be cost effective."[93]

Therefore, financial constraints of clients are yet another inherent limitation that can affect selection of testing methodology and scope, which can limit the amount of data generated and possibly result in inaccurate conclusions being drawn. In order to maximize knowledge obtained from limited data collection, industrial hygienists compare indoor spore counts to outdoor counts, as did the NRDC. Dr. Puccetti explains, "If the amount of growth is significant, differences in airborne mold spore concentrations between outdoor air and indoor air become detectable."[94]

In 2002, the EPA patented a standardized technology developed to test the moldiness of indoor environments. The Environmental Relative Moldiness Index (ERMI) tests for 36 molds using a DNA-based analysis called Mold Specific Quantitive polymerase chain reaction (PCR).[95] This technology identifies a mold by species not just by genus. Dr. Thrasher gives an example of the value of the ERMI test, "Often, when doing air spore sampling the spores of *Stachybotrys* are missed because they are not readily shed from the colony. However, PCR-DNA analysis of bulk materials of mold growth often reveals the presence of *Stachybotrys*."[96] The ERMI also enables a person to assess the moldiness of their property by comparing their ERMI value to a national database.[97]

By understanding the dynamics of sampling protocols, you will be better equipped to discuss with industrial hygienists the merits of various testing methodologies and select a protocol that will provide the information you are most seeking based on your specific circumstance. The more you understand the process of testing and remediation, the more prudent a consumer you will be. Not all companies that provide mold testing and remediation services are equal. Knowledge will help you spot inexperienced and unskilled remediation "specialists" or, worse yet, con artists who spew empty promises. Remember, fungal testing and remediation, when properly performed, will not be cheap. It is paramount to select an industrial hygienist highly experienced and educated in the design of fungal testing protocols to ensure test results provide enough information to locate the originating source of mold, which then makes possible a successful remediation. Otherwise, your money could be ill spent and your health could suffer.

Most important to understand is that the health effects of mold exposure can be insidious; they can sneak up on you, snatching away your good health before you even realize hidden mold growth exists, for example, in the wall cavity next to your desk at work or in the ceiling plenum under which your couch sits at home. You must remain cognizant of the fact that health can silently decay from exposure to hidden mold and mycotoxins. Even small amounts of moisture accumulation—the precursor to mold—can lead to health-affecting fungal growth, depending on the species. You must pay close attention to any negative change in your health—physically, mentally, and emotionally—and work to identify the cause, if possible.

Above all, no matter how inconvenient or seemingly impossible due to financial limitations, you must immediately remove yourself from health-affecting levels of toxic exposures as soon as you become aware of the exposure. Then you must remain vigilant regarding harmful re-exposures. For some, this will require a temporary move while remediation is taking place. For others, a permanent relocation of home or work will be necessary. Whatever the cost, your health is worth the investment. You must use good ol' common sense: The longer you are exposed to a toxin, the more damage can be done to your health. Prolonged exposure must be avoided. To do otherwise puts your health and the health of your family at stake.

(ENDNOTES)

1   Jessica Bennett, "'It's Incompetence,'" *Newsweek*, August 24, 2006. Available at http://www.msnbc.msn.com/id/14497763/site/newsweek/ (accessed August 2006).

2   Ibid. (Brackets in original article.)

3   Ibid.

4   Bennett, "'It's Incompetence.'"

5   Interview with Dr. Jack Thrasher, April 2006.

6 Lipsey, "The Mold and Bacterial Results from the Sampling Dr. Lipsey did in St. Bernard Parish."

7   Ibid.

8   Interview with Dr. Richard Lipsey, April 2006.

9   Interview with Dr. Gina Solomon, July 2006.

10   National Resources Defense Council (NRDC), "New Orleans Area Environmental Quality Test Results, Mold Results by Neighborhood." Available at http://www.nrdc.org/health/effects/Katrinadata/mold2.asp (accessed March 2006; hereafter cited as NRDC Report).

11   Interview with Dr. Gina Solomon, July 2006.

12   Ibid.

13   Ibid.

14   Ibid.

15 Gina M. Solomon et al., "Airborne Mold and Endotoxin Concentrations in New Orleans, Louisiana after Flooding, October-November 2005," *Environ Health Perspect* 114:1381-1386 (2006). doi:10.1289/ehp.9198. Available at http://www.ehponline.org/members/2006/9198/9198.pdf (accessed June 2006).

16   Ibid.

17   Ibid.

18   Ibid.
19   NRDC Report.
20   Ibid.
21   Interview with Dr. Andrew Puccetti, April 2006.
22   Ibid.
23   Interview with Dr. Gina Solomon, July 2006.
24   NRDC Report.
25   Interview with Dr. Richard Lipsey, April 2006.
26   Interview with Jim Pearson, April 2006.
27   M P Buttner et al., "Monitoring airborne fungal spores in an experimental indoor environment to evaluate sampling methods and the effects of human activity on air sampling." *Appl Environ Microbiol.* 1993 January; 59(1): 219-226. Available at http://www.pubmedcentral.gov/articlerender.fcgi?tool=pmcentrez&artid=202081 (accessed April 2006).
28   EPA, "Indoor Air Quality-Mold." Available at http://www.epa.gov/iaq/molds/moldresources.html (accessed August 2006).
29   NRDC Report citing National Allergy Bureau, "NAB: Reading the Charts." Available at http://www.aaaai.org/nab/index.cfm?p=reading_charts (accessed August 2006).
30   Ibid.
31   NRDC Report.
32   Ibid.
33   Interview with Dr. David Straus, April 2006.
34   Dr. Blanca Cortes, "The Mold After the Storm: A *Stachybotrys* Survey in Central Florida." Available at http://www.ewire.com/display.cfm?Wire_ID=2468 (accessed March 2006).
35   Solomon et al., "Airborne Mold and Endotoxin Concentrations in New Orleans, Louisiana after Flooding, October-November 2005."
36   Interview with Dr. David Straus, April 2006.
37   CDC, "Health Concerns Associated with Mold in Water-Damaged Homes After Hurricanes Katrina and Rita-New Orleans Area, Louisiana, October 2005," p. 42, *Morbidity and Mortality Weekly Report (MMWR),* January 20, 2006, 55(02); 41-44. Available at http://www.cdc.gov/mold/pdfs/mmwr55(2)41-44.pdf (accessed August 2006).
38   Ibid.
39   Solomon et al., "Airborne Mold and Endotoxin Concentrations in New Orleans, Louisiana after Flooding, October-November 2005."
40   Interview with Dr. Jack Thrasher, April 2006.
41   Ibid.
42   Interview with Dr. David Straus, April 2006.
43   Solomon et al., "Airborne Mold and Endotoxin Concentrations in New Orleans, Louisiana after Flooding, October-November 2005."
44   NRDC Report.
45   Ibid. To see a mapping of the NRDC mold data, see http://www.nrdc.org/health/effects/katrinadata/map.asp (accessed August 2006).
46   Ibid.
47   Ibid.
48   Ibid.
49   Ibid.
50   Solomon et al., "Airborne Mold and Endotoxin Concentrations in New Orleans, Louisiana after Flooding, October-November 2005."
51   Interview with Dr. Gina Solomon, July 2006.
52   Interview with Dr. David Straus, April 2006.
53   Interview with Jim Pearson, April 2006.
54   Interview with Dr. David Straus, April 2006.
55   Interview with Dr. Andrew Puccetti, April 2006.
56   Interview with Dr. Gina Solomon, July 2006.
57   E-mail from Dr. Andrew Puccetti, August 2006.
58   Interview with Dr. Andrew Puccetti, April 2006.
59   Ibid.

60   Ibid.
61   NRDC Report.
62   Solomon et al., "Airborne Mold and Endotoxin Concentrations in New Orleans, Louisiana after Flooding, October-November 2005."
63   NRDC Report.
64   Ibid.
65   Interview with Dr. Richard Lipsey, April 2006.
66   Nelson, "*Stachybotrys chartarum*: The Toxic Indoor Mold," p. 9.
67   Joan MacFarlane, "Beware the Mold *Stachybotrys*," CNN, November 5, 1997. Available at http://edition.cnn.com/HEALTH/9711/05/deadly.mold/ (accessed February 2006).
68   NRDC Report.
69   Ibid.
70   Ibid.
71   Ibid.
72   NRDC Press Release, "New Private Testing Shows Dangerously High Mold Counts in New Orleans Air," November 16, 2005. Available at http://www.nrdc.org/media/pressRelease/051116.asp (accessed June 2006).
73   NRDC Report.
74   Interview with Dr. Andrew Puccetti, April 2006.
75   Interview with Jim Pearson, April 2006.
76   NRDC Report.
77   Solomon et al., "Airborne Mold and Endotoxin Concentrations in New Orleans, Louisiana after Flooding, October-November 2005."
78   NRDC, "Mold." Available at http://www.nrdc.org/health/effects/katrinadata/mold.asp (accessed March 2006).
79   Solomon et al., "Airborne Mold and Endotoxin Concentrations in New Orleans, Louisiana after Flooding, October-November 2005."
80   NRDC, "Mold."
81   Curtis et al., "Adverse Health Effects," p. 6.
82   M P Buttner et al., "Monitoring Airborne Fungal Spores in an Experimental Indoor Environment to Evaluate Sampling Methods and the Effects of Human Activity on Air Sampling," *Applied and Environmental Microbiology*, January 1993, 59(1): p. 219. Available at http://www.pubmedcentral.gov/articlerender.fcgi?tool=pmcentrez&artid=202081 (accessed April 2006).
83   Curtis et al., "Adverse Health Effects," p. 6.
84   Interview with Dr. Jack Thrasher, April 2006.
85   Interview with Dr. Andrew Puccetti, April 2006.
86   Quote from MacFarlane, "Beware the Mold *Stachybotrys*." Confirmed with Dr. George Riegel, Jr., May 2007.
87   Daley, "The Next Menace: Mold."
88   Ibid.
89   Interview with Wilma Subra, October 5, 2006. Subra Company, P.O. Box 9813, New Iberia, LA 70562. Phone: (337) 367-2216. E-mail: subracom@aol.com.
90   Interview with Dr. Gina Solomon, July 2006.
91   Ibid.
92   Solomon et al., "Airborne Mold and Endotoxin Concentrations in New Orleans, Louisiana after Flooding, October-November 2005."
93   E-mail from Dr. Andrew Puccetti, August 2006.
94   Ibid.
95   EPA, "EPA Technology for Mold Identification and Enumeration." Available at: http://www.epa.gov/microbes/moldtech.htm (Accessed May 2013).
96   E-mail correspondence with Jack Thrasher, June 2013.
97   EM Lab, "ERMI: Environmental Relative Moldiness Index and ERMI Testing Services." Available at: http://www.emlab.com/s/services/ERMI_testing.html (accessed June 2013).

CHAPTER FOUR:

SEDIMENT AND SEWAGE SLUDGE SPIN

Sediment contaminants can pose potential health risks after displacement of soil and water occurs during and after many types of natural disasters, not just Katrina-size hurricanes. Problems arise when people don't realize the toxicity of what looks like a lot of mud. For this reason, we feel it important to review the Katrina sediment data to provide a strong understanding of the health hazards associated with sediment. Even though the Katrina sediment may consist of a different mix of contaminants than say, for example, sediment from a rural flood, all sediment is vulnerable to the growth of potentially harmful microorganisms. The knowledge gained from this informative review can be applied to sediment anywhere. With adequate understanding of the dangers of sediment toxins, proper steps can be taken to preserve health. We'll review some media coverage of Katrina sediment and then take a look at what the experts have to say about sediment-created toxins. As you will see, many of the forthcoming problems originate from policy issues that affect the entire U.S., not just New Orleans.

The NRDC reports, "The flood waters that inundated New Orleans carried a mixture of soil, sewage, and industrial contaminants. When the flood receded, it left behind a layer of sediment—in some places up to 4 inches thick—that still covers the ground and even coats the interiors of people's homes."[1] Even thicker layers of sediment were reported by Wilma Subra, a chemist known for her environmental activism.[2] Based on her field observations, she states that in some areas this surface layer of sediment was as thick as 6 feet.[3] Testing by various groups including the Subra Company, the NRDC, the EPA, the Louisiana Bucket Brigade,[4] and the Louisiana Environmental Action Network (LEAN)[5] reveals that this sediment is contaminated with heavy metals, petroleum, pesticides, industrial chemicals, and polyaromatic hydrocarbons (PAHs), which are cancer-causing chemicals from soot and petroleum-based products.[6]

Katrina's sea of toxic sediment blanketed everything that was previously flooded—streets, cars, houses, furniture—literally seeping into nooks and crannies. When Fox News' Shepard Smith reported from a hard-hit Katrina area in Mississippi, he said, ". . . [J]ust about everything that had once been submerged in the polluted flood waters was now covered in a thick layer of dusty residue that may contain the same bacteria and toxins that had contaminated the water."[7] As this vile, toxic soup dried, the risk of inhalation rose, increasing health risks.

"Until it is cleaned up, contamination such as this is a serious health concern for residents returning to these neighborhoods. Contaminated sediment can pose a health hazard in both the short-term and the long-term. In the short-term, residents and workers may be exposed to toxic materials by inhaling dust from sediment they are trying to clean up—and that is blowing around the city—or getting it on their hands or in their eyes or mouths. This kind of exposure can cause cough, irritation of the eyes, nose, and throat, and skin rashes. In the long-term, children playing in yards and families living in once flooded neighborhoods may be at significant health risk from exposure to contaminated soil unless it is removed and replaced before people move back to these areas," warns the NRDC.[8]

As the floodwaters receded, health officials were concerned about the "toxic dust"[9] the water left behind. Even ordinary household dust has been found to contain mycotoxins. "Literature reveals that significant amounts of mycotoxins (including ochratoxin, sterigmatocystin, and trichothecenes) are present in indoor dust . . ."[10] If *significant* amounts of mycotoxins are in everyday household dust, what levels could possibly be in the dusty hurricane residue?

Ms. Subra confirms that significant levels of fungi, yeasts, and gram-negative and gram-positive bacteria, among other toxins, are present in the sediment, which is still present in many parts of the hurricane-hit areas even after the one-year anniversary of Katrina and Rita. She reports that residents are still getting infections from coming into contact with the contaminated sediment. Ms. Subra further explains that since the sediment contains both gram-negative and gram-positive bacteria, only broad-spectrum antibiotics designed to kill both will provide effective treatment.[11]

Since the floodwaters were full of sewage, floodwater testing showed a lot of coliform and *E. coli* bacteria, states Dr. Solomon.[12] Thus, these same bacteria ended up in the sediment. Dr. Straus further explains, "When sewage systems flood and get into the general water systems, this will introduce a lot of endotoxins because you normally have gram-negative bacteria, like *Escherichia coli,* in sewage." He continues, "Bacteria also grow in wet buildings, as do fungi; but in most cases, bacteria require even more water than do molds. Most bacteria like to grow in liquid settings, for example, a pool of water. When this occurs, if the bacteria in question are gram-negative, they will produce a lot of endotoxins as only gram-negative bacteria produce endotoxins."[13]

Dr. Solomon stated that the NRDC team knew endotoxins were in the floodwaters because of the positive testing for coliform and *E. coli* bacteria, which are gram-negative bacteria.[14] However, the NRDC findings did not reflect elevated levels. Dr. Solomon and colleagues report, "The lack of association between flooding and endotoxin concentrations as well as between endotoxin and mold concentrations was surprising. Endotoxin does not become airborne as readily as mold spores. Therefore, the lack of elevated air concentrations may reflect the study protocol, which specified minimum disturbance of the area during sampling. A CDC study of 20 New Orleans homes in late October of 2005, found mean endotoxin concentrations of 23.3 EU [endotoxin units]/m³ indoors and 10.5 EU/m³ outdoors. These concentrations were higher than those found in our study [from 0.6 to 8.3 EU/m³] and were consistent with the hypothesis that gram-negative bacterial growth was occurring inside the flooded homes. It remains likely that people doing remediation work in or around flooded homes could cause release of endotoxin into the air, resulting in elevated exposures."[15]

According to Dr. Lipsey, "An endotoxin is similar to a mycotoxin from mold. All gram-negative bacteria produce endotoxins. Some examples of gram-negative bacteria are *E. coli*, *Salmonella*, and *Legionella*, which causes Legionnaires' disease. All of the gram-negative bacteria produce endotoxins and can become airborne. They are chemicals known to be highly toxic, similar to mycotoxins produced by certain pathogenic molds."[16]

Dr. Solomon states, "We knew that the floodwaters had gone everywhere, including into people's homes."[17] This, of course, created an environment conducive to bacterial growth. According to Dr. Lipsey, "Both gram-negative and gram-positive bacteria form in conditions where carpets, couches, wallpaper, etc., have been saturated with moisture for days or longer. Additionally, the rot and decay process from a moldy environment creates an environment conducive to bacteria formation. Some bacteria can double in population every 15 minutes under ideal conditions."[18]

People can develop diseases from just breathing air that contains these types of bacteria, reports Dr. Lipsey. "Legionnaires' disease, for example, comes from breathing bacteria from a heating and cooling system contaminated with *Legionella* bacteria," he explains. "The infectious state depends on which gram-negative bacterium is in the air. *Salmonella*, you know, can be left when you cut up chicken on your drain board. Then the next time you cut anything on your drain board, you can get a serious, life-threatening infection. In fact, you can die from any of those bacteria," cautions Dr. Lipsey.[19]

Dr. Solomon explains, "Endotoxins are a tricky thing. At high concentrations they are known to decrease normal lung function. So if you take healthy, strapping young men and expose them to endotoxins, their lung function will diminish predictably and reliably; they are clearly not good for the lungs. They also create a very powerful inflammatory effect. Basically, our immune system is primed to respond to endotoxins almost more powerfully than to any other substance. Endotoxins will produce major inflammation whenever they are in your body. The interesting and peculiar thing about endotoxins is that some studies indicate that low, low levels of endotoxins can actually sort of help in decreasing risk of asthma, especially in children. Nobody has really resolved that paradox yet. Then there are other studies that indicate endotoxins may increase risk of asthma. Endotoxins are a bit of an enigma at very, very low concentrations. There are definitely people who say they have a role to play in normal immune modulation in early development. That may be true, but high levels of endotoxins are never good. So if there were a lot of gram-negative bacteria in New Orleans, there's no way to frame that as good in any way, shape, or form."[20]

She cautions, "Anybody who has a respiratory problem is going to get sick more easily from either endotoxins or mold. Anybody whose body is less able to bounce back and recuperate will be at higher risk, which can include young children, the elderly, or people with underlying illnesses."[21]

Dr. Solomon continues, "We actually expected to find a lot of endotoxins."[22] However, possibly due to the selected sampling protocol, the NRDC samples did not reflect the anticipated elevated levels of endotoxins. Dr. Solomon states, "We were surprised to find that the levels we were detecting were not higher indoors than out and that endotoxin levels were not high consistently in the flooded areas versus the nonflooded control sites. Interestingly enough, the CDC reported that they had done a little bit of endotoxin testing and they could find high levels of endotoxins, so we are not actually sure whether to trust our results. Designing a method of testing that takes into consideration the different characteristics of organisms is the tough thing about designing this study. One of the things about endotoxins is they don't become airborne as easily as mold. Mold has incentive— the more airborne its spores become, the better their chance of spreading. It's pure, basic biology. The endotoxin is basically just cell wall fragments from bacteria; it's not something that the bacteria are trying to spread. It spreads when it's disturbed."[23]

The NRDC collected samples in undisturbed, uninhabited houses.[24] According to Dr. Solomon, the selected sampling protocol included nondisturbance of the sampling environment—tiptoeing into the houses, not moving anything around, carefully setting up the equipment, and tiptoeing out.[25] Thus, no air movement disturbed and aerosolized settled endotoxins. Likewise, no air movement disturbed and aerosolized settled spores, which would have increased the level of spores in the air to be sampled.

Dr. Solomon surmises that a different sampling protocol would have yielded different results. She reflects, "If instead, we had done a different sampling protocol, and we had put on full protective gear (which we had anyway, in most cases), and if we had been protected from head to foot and had good respirators, and had just gotten in there and started moving furniture, ripping and pulling up the wet carpet, and trying to get it out to the curb—doing all the things that people would do if they were trying to get things cleaned up in their houses (sweep, vacuum, anything like that)—I'll bet we would have found a heck of a lot of endotoxins."[26] This sampling protocol would have also resulted in substantially higher levels of mold spore counts, according to Mr. Pearson.[27]

"We did the same kind of thing outdoors when we sampled people's backyards," continues Dr. Solomon. "We tucked the samplers so that they would be sheltered from the wind, and in a couple cases when it turned out that there was actual disturbance going on nearby, we decided not to sample because we didn't want to mess up our remote samples—if there was tree work or something like that going on. We probably substantially underestimated the real endotoxin risk, but we had only the data that we had to report. We reported what we had and explained the possible reasons why we may have found what we did. Endotoxins are potentially another risk that people may have been facing there. We just don't have the data to show it."[28]

Although the NRDC data does not reflect elevated endotoxin levels, possibly due to its chosen sampling protocol, no question exists that endotoxins were present in elevated levels. According to Dr. Lipsey, "Highly pathogenic endotoxins from gram-negative bacteria were found earlier by NIOSH scientists and personnel from the Louisiana Dept. of Health, and the levels were 20 times above normal on average. The levels were not only high within the flooded homes but also in the ambient air in the neighborhood."[29] Dr. Lipsey states, "The EPA reported that endotoxin levels in the air in St. Bernard Parish were very high."[30]

The CDC reports that air samples collected by the CDC/LDHH assessment team revealed, "In five New Orleans homes, the measured indoor endotoxin levels were comparable to those of certain industrial settings in which declines in pulmonary function have been demonstrated."[31]

Elevated levels of gram-negative bacteria, which produce these endotoxins, were also found. According to Dr. Lipsey, "Most of the homes had extremely high levels of pathogenic mold and bacteria, the highest I have ever seen in my 35 years of testing homes for toxic mold and bacteria. The bacterial counts were as high as 62 million bacteria per square inch of wallpaper and furniture. The most abundant bacteria I found in the homes were *Pseudomonas aeruginosa, P. fluorescens, Enterobacter cloacae, Pantoea, Brevundimonas vesicularis* and *Stenotrophomonas*, all in the tens of millions per square inch of wallpaper and furniture, and gram-negative bacteria of many kinds. These levels of these pathogenic bacteria can cause many diseases and death in humans. The gram-negative bacteria all produce endotoxins, similar to mycotoxins used in germ warfare at very high levels to kill people within minutes, and include *Legionella* bacteria, *Salmonella*, [and] *E. coli* from human sewage, etc."[32]

Dr. Lipsey further states, "Gram-negative bacteria and their spores and the endotoxins they produce got into the ambient air in the neighborhoods, which created sick neighborhoods as well as sick buildings—totally uninhabitable. I would recommend that anyone even walking the neighborhoods in St. Bernard Parish wear HEPA respirators. I developed bronchitis from walking in ten neighborhoods, even though I always put on my HEPA respirator before entering any of the homes."[33]

Dr. Thrasher expands on associated health risks, "The endotoxins and the *Pseudomonas* toxin are the important issues. Endotoxins are very irritating to the mucous membranes of the nose, throat, lungs, etc. [They can] . . . cause sever[e] respiratory distress in asthmatics and may also cause asthma themselves. In addition, it is well accepted that endotoxins open up the blood brain barrier, making it leaky to toxins. Furthermore, animal and test tube studies show that endotoxins act synergistically with mycotoxins, particularly trichothecenes, to enhance the toxicity of both. A quick PubMed search will bring this out to anyone who wishes to look. The other area of concern is that the bacteria also produce solvents."[34]

Dr. Straus confirms that these endotoxins can be inhaled when they become airborne. He reports, "We know that endotoxins get into the air when there are a lot of gram-negatives present, such as *E. coli* contaminating feedlot soil."[35] Dr. Thrasher adds that in addition to many of the gram-negative bacteria that produce endotoxins, "The gram-positives also can produce what we call bacterial cell wall toxins. These also get into the air. The current research, which we have to do in the test tube and with research animals as we can't do it with humans, shows that these compounds are synergistic with the mycotoxins."[36]

Synergistic reactions among contaminants can escalate negative health impacts and, unfortunately, are not limited to endotoxins, bacterial cell wall toxins, and mycotoxins. Anytime exposure to multiple contaminants occurs such as in post-flooding sediment-laden, toxic environments, synergistic reactions are a health threat.

In addition to the sewage, the NRDC team reports, "The floodwaters from hurricanes Katrina and Rita swept a mixture of soil, mud from Lake Pontchartrain, and debris into the greater New Orleans area. When the water receded, it left behind a caked layer of muck on streets, yards, porches, and playgrounds across the region—sediment that was likely contaminated with heavy metals and toxic chemicals swept up from industrial areas. The U.S. Environmental Protection Agency (EPA) collected hundreds of samples of this sediment throughout the New Orleans region from September 10, 2005, to January 15, 2006, and released the data—absent any analysis—to the public on its website (http://oaspub.epa.gov/ storetkp/dw_home) in January 2006."[37]

In an effort to increase public awareness of associated health risks, the "NRDC has now analyzed the EPA's sediment data, and the Greater New Orleans Community Data Center has created maps to demonstrate patterns of contamination in the sediment. Although the EPA tested for a variety of contaminants, this analysis focuses on four toxic contaminants that are most widespread in the sediment samples: arsenic, lead, diesel fuel, and benzo(a)pyrene. All of these contaminants were detected in sediment throughout the greater New Orleans area, often in concentrations in excess of EPA and Louisiana Department of Environmental

Quality (LDEQ) cleanup guidelines for soil in residential areas."[38] [The NRDC's analysis of the EPA sediment data for these four contaminants and the accompanying mapping are available at www.nrdc.org/health/effects/katrinadata/sedimentepa.pdf.]

The NRDC's ". . . analysis of EPA data shows that most districts in New Orleans contain concentrations of arsenic, lead, diesel fuel, or cancer-causing benzo(a)pyrene above levels that would normally trigger investigation and possible soil cleanup in the state of Louisiana. Some hot spots in residential neighborhoods have levels of contamination that are ten times, or even more than a hundred times, normal soil cleanup levels. For example, a location in Gert Town, Mid-City, has arsenic in the soil at a level 6.5 times the Louisiana cleanup level for residential soil and 200 times the federal health-based level of concern for soil in residential neighborhoods. The Lake Terrace neighborhood in Gentilly has lead in the soil at a level three times the Louisiana cleanup level. Locations in Chalmette and in the St. Roch neighborhood have diesel fuel contamination more than 200 times the Louisiana soil screening cleanup level. A hot spot for benzo(a)pyrene contamination is in Bywater, at the Agricultur[e] Street Landfill, where the levels exceed Louisiana soil cleanup levels by more than fiftyfold."[39]

On first hearing of these elevated levels of contamination, people could have reasonably presumed that appropriate agencies would be undertaking cleanup actions. But according to Dr. Solomon, "There's been a lot of buck passing. The EPA says there might be funds to do this cleanup but that the Corps of Engineers would be the agency that would need to do the cleanup. The Corps is saying that they are focusing on the levees, which of course is super important. The Corps also is saying that they need to hear exactly which areas need to be cleaned up, and the EPA says that they're still testing and evaluating the problem."[40]

"Then there's been this issue of sort of moving the benchmarks. This winter the DEQ abruptly came out with a new revised safe level of arsenic, just after arsenic levels over the previous safe level were discovered all over New Orleans," points out Dr. Solomon. "It was just in a one-page memo. They revisited their arsenic standards and declared that a much higher level of arsenic is perfectly safe. That looked suspicious to many people. It seems that the DEQ's obvious solution was to just move the safe level and declare the higher levels that were being discovered in the city safe. There's been a fair amount of funny business. Meanwhile, this winter, the rains did move that sediment around and washed some of it into the storm drains. In many cases, when the EPA went back to do repeat sampling, they said they couldn't find any sediment. The problem with that is that we know that the contaminants in that sediment were probably washing into the storm drains and some were just soaking into the soil. That's the soil that kids may be playing on and that families may be planting vegetables in, so it wasn't very reassuring to know that it was just soaking into the dirt."[41]

The EPA's inability to find elevated levels in repeat samples was undoubtedly affected by its chosen Phase III sampling protocol. The EPA states, "Unlike previous sampling rounds, Phase III samples were collected and mixed together (i.e., composite samples) to characterize the average concentration of chemicals around the original sampling locations. These composite samples were not only of the sediment deposited by floodwaters but also included samples of the underlying [uncontaminated] soil that existed prior to the hurricanes."[42] In other words, the EPA diluted the sediment-contaminated soil samples with dirt—not contaminated by hurricanes Katrina and Rita—before performing Phase III testing.

In an NRDC press release, Dr. Solomon states that if the contamination in the sediment is not cleaned up, ". . . residents could be faced with serious health risks, including cancer, neurological disease, and hormonal and reproductive system problems, over the long term."[43]

Dr. Solomon clarifies her previous statement, "The issues there are about risks to young children who might be crawling around in and in direct contact with the sediment or with sediment-contaminated soil. The major concern here is specifically with the arsenic and petroleum compounds in the sediment. Arsenic is known to cause cancer, neurological problems, and reproductive problems with long-term exposure. This is not something that's likely to happen to adults or after relatively short-term exposure (i.e., over months). The initial signs of concern in a child might be irritant effects from the contaminants in the sediment, such as rashes, skin irritation, or a cough."[44]

So where did all this arsenic come from? And the lead, diesel fuel, and cancer-causing benzo(a)pyrene—where did they come from? According to Ms. Subra, "The contaminated sediment originated in the bottoms of water bodies, lakes, rivers, and estuaries in the paths of the hurricanes [Katrina and Rita]. The sediments were contaminated by many decades of discharges from industry, municipalities, businesses, and agricultural runoff."[45] Essentially, hurricanes Katrina and Rita dredged up and redistributed an age-old pollution problem.

Additional sources of sediment contaminants may have come from the "[t]housands of facilities in the Gulf Coast area—ranging from gas stations to oil refiners to large petrochemical plants . . ."[46] that were in the path of the hurricane. Toxic contaminants from these businesses may have leaked into the floodwaters, which later became part of the sediment. Other likely sources of disrupted toxins may have come from closed landfills and Superfund sites. The EPA confirms, "Flooding can cause the disruption of water purification and sewage disposal systems, overflowing of toxic waste sites, and dislodgement of chemicals previously stored above ground."[47]

Eye-opening is the number of toxic sites in the New Orleans area. For example, the OMB [Office of Management and Budget] Watch, which is a nonprofit research and advocacy organization, compiled information from four EPA databases identifying *57 toxic chemical sites* in New Orleans. Included are major sites that store, use, and/or produce large quantities of toxic chemicals in the Orleans, Plaquemines, and St. Bernard Parishes.[48] The EPA confirms that many of these sites were flooded and disrupted in the wake of Katrina, such as the Agriculture Street Landfill, which is three miles south of Lake Pontchartrain.[49]

The abundance of toxic test results from various sources including the EPA, the NRDC, and independent, nonprofit organizations is proof that the entire Katrina-hit area is a brewing public health disaster. Exposure to these toxins may ignite ill health repercussions felt nationwide since cleanup volunteers and workers have come from all around the country. The longer the abandoned moldy structures and toxic sediments are allowed to go unaddressed, the more likely health repercussions will develop in an increasingly larger number of people. If not properly cleaned up, long-term exposure will continue, putting at risk the future health of residents and workers in the affected areas.

Cleanup efforts have been underfunded, undermanned, and tied up in the bureaucratic arms of the government. Speaking specifically about the hurricane sediment, Ms. Subra reports that the EPA has no plans to remove the contaminated top soil except for a few hot spots it confirmed from her testing.[50] As long as the toxic sediments remain, residents are at risk for future illnesses and diseases that can brew. So why, with unquestionable health risks at stake, was this sediment—loaded with contaminants, many of which are carcinogenic compounds—allowed to simply soak into the soil and wash down the storm drains? This water gets treated and released as municipal water, which is used for both bathing and drinking!

If all these contaminants are being allowed to just wash into storm drains, then perhaps they are not so toxic? Let's see what the EPA has to say about just one contaminant—arsenic. "The EPA has set limits on the amount of arsenic that industrial sources can release to the environment and has restricted or cancelled many of the uses of arsenic in pesticides. [The] EPA has set a limit of 0.01 parts per million (ppm) for arsenic in drinking water."[51] Can current water treatment procedures reduce the increased levels of arsenic in water from hurricane sediment runoff to, or below, the 0.01 ppm limit? And what about the lead, diesel fuel, cancer-causing benzo(a)pyrene, and other toxic contaminants?

The sad reality is that lack of governmental accountability may result in people being exposed to harmful levels of arsenic and other contaminants. Let's take a look at the potential cost to health from exposure to arsenic. Dr. Solomon and colleagues report, "Arsenic is toxic to humans and is known to cause cancer; no amount is considered fully safe. Many scientific studies, including numerous reviews by the National Academy of Sciences [NAC], have determined that arsenic can cause cancer of the bladder, skin, and lungs; likely causes other cancers;

and can cause a variety of other serious health problems including birth defects, cardiovascular disease, skin abnormalities, anemia, and neurological disorders."[52]

The negative health effects of arsenic are not debated. "Several studies have shown that ingestion of inorganic arsenic can increase the risk of skin cancer and cancer in the lungs, bladder, liver, kidney, and prostate. Inhalation of inorganic arsenic can cause increase[d] risk of lung cancer. The Department of Health and Human Services (DHHS) has determined that inorganic arsenic is a known carcinogen. The International Agency for Research on Cancer (IARC) and the EPA have determined that inorganic arsenic is carcinogenic to humans."[53]

Exposure to naturally occurring arsenic can also lead to health concerns. "Organic arsenic compounds are less toxic than inorganic arsenic compounds."[54] However, exposure to ". . . high levels of some organic arsenic compounds may cause similar effects as inorganic arsenic."[55]

Arsenic—organic or inorganic—that doesn't soak into the soil and/or wash down the storm drains has the potential to become airborne as remaining sediment and contaminated soil become disturbed, especially during the long-term rebuilding phase. This continual, long-term disruption of arsenic-contaminated dust can result in increased levels of risk to workers and nearby residents. Arsenic exposure in the workplace is a health risk that has been specifically addressed. "The Occupational Safety and Health Administration (OSHA) has set a permissible exposure limit (PEL) of 10 micrograms of arsenic per cubic meter of workplace air (10 $\mu g/m^3$) for 8-hour shifts and 40-hour workweeks."[56] If the level of health risk from exposure to arsenic is viewed as a risk to limit in the workplace, shouldn't the EPA and other governmental entities be more concerned about exposures to residents, especially children?

The Agency for Toxic Substances and Disease Registry (ATSDR) reports, "Breathing high levels of inorganic arsenic can give you a sore throat or irritated lungs. Ingesting very high levels of arsenic can result in death. Exposure to lower levels can cause nausea and vomiting, decreased production of red and white blood cells, abnormal heart rhythm, damage to blood vessels, and a sensation of 'pins and needles' in hands and feet. Ingesting or breathing low levels of inorganic arsenic for a long time can cause a darkening of the skin and the appearance of small 'corns' or 'warts' on the palms, soles, and torso. Skin contact with inorganic arsenic may cause redness and swelling."[57]

ATSDR further reports, "There is also some evidence that suggests that long-term exposure to arsenic in children may result in lower IQ scores. There is some information suggesting that children may be less efficient at converting inorganic arsenic to the less harmful organic forms. For this reason, children may be more susceptible to health effects from inorganic arsenic than adults."[58]

Pregnant women and fetuses may also be at risk from arsenic exposure. "There is some evidence that inhaled or ingested arsenic can injure pregnant women or their unborn babies, although the studies are not definitive. Studies in animals show that large doses of arsenic that cause illness in pregnant females can also cause low birth weight, fetal malformations, and even fetal death. Arsenic can cross the placenta and has been found in fetal tissues. Arsenic is found at low levels in breast milk," according to ATSDR.[59]

Both children and expectant mothers have been featured by the media in several human interest stories filmed in Katrina's aftermath and rebuilding phase. Not only do these two groups clearly fall into the CDC's designated high-risk groups in regard to mold exposure, but they are also at increased susceptibility if exposed to arsenic and other sediment toxins. According to Dr. Solomon, "The factors influencing susceptibility may be genetic, nutritional, behavioral (referring primarily to children), and also connected to underlying diseases."[60]

Neither the deleterious health effects of arsenic exposure nor the existence of elevated arsenic levels are in dispute; at issue is the question of *unacceptable health risks*. A written DEQ communication, released by the Louisiana DEQ and Governor Kathleen Blanco's office in an attempt to explain the arsenic sampling results and defend their subsequent inaction, states, "One interpretation of the arsenic samples has people saying the levels are 'ten times the EPA standards.' This statement is based on comparison of point-by-point concentrations compared to the EPA screening level of 0.39 parts per million. While it is true that an arsenic reading of 3.9 parts per million is ten times higher than EPA screening levels, that concentration is still well within the acceptable range (as defined by EPA and LDEQ). The range of acceptable arsenic concentrations is actually 0.39 ppm to 39 ppm. A concentration greater than the screening level (the 0.39 ppm at the lower end of the acceptable range) does not mean the levels are going to pose an unacceptable health risk, according to EPA health risk standards."[61]

The release goes on to state, "[The] EPA indicates that exposure to arsenic at 39 ppm or below is not expected to pose unacceptable risk to residents. This health risk level is based on an exposure to 39 ppm (also known as mg/kg) for 350 days for 30 years. In other words, a resident (including children) must be exposed to the concentration of 39 ppm at all times. This includes having it on your skin, inhaling particles, and incidentally ingesting it. And, as we all live our lives, it is impossible to be at that one and only location just about all day, every day for 30 years. The resident would have to be exposed to that level of concentration that often before there would be any increase in potential health risks."[62]

The Louisiana DEQ claims, "Just like the misleading statements regarding 'toxic soup' resulting from the floodwaters, the statements regarding high health risks resulting from elevated arsenic concentrations . . . are a misrepresentation and incorrect interpretation of environmental conditions after Katrina."[63] However, in order for the DEQ's argument to "scientifically" stand, a person would have to be exposed to only *one* contaminant—not the multitude of identified and unidentified contaminants in varying degrees of toxicity, as was the case post-Katrina.

Remember, many of these toxins may synergistically interact, worsening health effects. According to Dr. Solomon, "The problem with environmental health is that people are never exposed to just one thing, and nobody has figured out the health effects of this whole combined mess of toxins to which people are exposed. There are a lot of good reasons to believe that they interact with each other and that the combined exposure may well be far worse than any of those exposures alone. That's never been studied."[64]

Additionally, the LDEQ defends its no cleanup stance by stating, "Arsenic is a naturally occurring element and can be found in varying levels throughout the world. In Louisiana, the average background level for arsenic is 12 ppm. This is based on a comprehensive, statewide sampling effort by Louisiana State University. Based on extensive studies by Dr. [Howard] Mielke ([professor in the College of Pharmacy at] Xavier University in New Orleans), the historic levels of arsenic found in areas of New Orleans (pre-Katrina) are as high as 20 ppm. The average concentration of arsenic found in New Orleans in the sediment sampling effort was 12 ppm, which is well within what would be expected to be found. . . . In Louisiana, the DEQ residential cleanup level for arsenic is 12 ppm."[65] If this defense is valid, then why—minimally—weren't the ". . . serious hot spots of arsenic contamination in several neighborhoods including Lakeview, Gentilly, Mid-City, and New Orleans East"[66] cleaned up?

In a letter dated January 4, 2006, to Mike McDaniel, secretary of the Louisiana DEQ, the NRDC demanded, "Now that thousands of people are returning to the city with their families, it is vital that you provide the public with the specific analysis upon which the LDEQ and EPA relied to conclude that there are 'no unacceptable health risks' to returning residents."[67] Sadly, the unsubstantiated declaration of "no unacceptable health risks" as well as the lack of governmental response appears to be a consistent, reoccurring theme of this natural disaster.

So what exactly does "unacceptable health risks" mean? And how are "acceptable risks" calculated? According to a group of scientists at Cornell University, risk assessments are based on models. "A risk assessment is a model which, like all models, is a simplified simulation of real world conditions that relies . . . [on] many assumptions and subjective judgments. Moreover, a model is only as good as the data from which it draw[s] conclusions. The more complex the system being modeled, the more vulnerable the model and conclusions drawn from it are to errors resulting from the gaps between the model and reality. This is one reason why risk assessments generally fail to effectively evaluate impacts on ecosystems as a whole and do not address synergistic impacts."[68] These Cornell University scientists further point out, "Because of the limitations inherent in a model, results should include an expression of their uncertainty, whether as a range of values or through the application of a safety factor."[69]

Also of concern is that state and federal authorities are not taking into consideration the synergistic impacts of exposure to multiple contaminants present after hurricanes Katrina and Rita. Unfortunately, not addressing the increased health risks of the synergistic effects from multi-toxins is a common practice, even in areas of scientific research. In fact, a published report states, "Most research on the effects of chemicals on biologic systems is conducted on one chemical at a time. However, in the real world people are exposed to mixtures, not single chemicals. Although various substances may have totally independent actions, in many cases two substances may act at the same site in ways that can be either additive or nonadditive. Many even more complex interactions may occur if two chemicals act at different, but related, targets. In the extreme case, there may be synergistic effects in which case the effects of two substances together are greater than the sum of either effect alone. In reality, most persons are exposed to many chemicals, not just one or two, and therefore the effects of a chemical mixture are extremely complex and may differ for each mixture depending on the chemical composition. This complexity is a major reason why mixtures have not been well studied."[70]

Inaction by state and federal entities to address the synergistic impact from multiple toxic exposures does not negate the increased negative health impacts. We have to ask the following question: Why has there been such inaction by the EPA, the Corps of Engineers, and the DEQ to clean up the environmental toxic waste dump left in the wake of Katrina? And why did the DEQ come out with a newly revised *safe* level of arsenic *just after* arsenic levels over the previous safe level were discovered all over New Orleans?

At first glance, it simply appears that New Orleans and surrounding areas are getting a raw deal. However, the truth of the matter is that we are all getting shortchanged. Sewage sludge filled with some of the same contaminants that are in the hurricane sediment, albeit at higher levels, is commonly used as agricultural fertilizer—a practice supported by the EPA.[71]

Let's take a look at the fine print that has made it legally acceptable to "recycle" our sewage sludge and grow our produce in it. "The US EPA adopted regulations in 1993 (40 CFR Part 503, known as Part 503) that establish minimum standards that must be met if sludges are to be land-applied. The regulations include concentration limits for nine metals [arsenic, cadmium, copper, lead, mercury, molybdenum, nickel, selenium, and zinc] and for pathogens, and requirements for vector (flies and rodents) attraction reduction. The regulations establish Class A sludges, which have been treated to essentially eliminate pathogens (disease-causing organisms), and Class B, in which pathogens have been reduced but are still present."[72] In fact, Cornell University scientists report that *significant* levels of pathogens are still present in Class B sludge.[73] "Under the federal 503 rules, certain site restrictions apply to Class B use, but no individual site permits are required for its use."[74] In other words, no record exists of which tracts of land this pathogen-filled "fertilizer" is being applied.

Creative marketing and public relation wizards have coined politically and socially acceptable terms for the toxin-mixed excrement—*biosolids* and *exceptional quality (EQ) biosolids*. However, as we proceed to discuss the quantities and ranges of toxic substances allowed in "exceptional quality" sludge—fertilizer for our food— the thinly veiled marketing ploy becomes apparent.

To best understand why cleanup inaction is occurring in contaminated hurricane-hit areas, let's take a look at the biosolid limits for a contaminant with which we are already familiar—arsenic. Levels of arsenic are allowed in biosolids, ranging from 41 ppm (41 mg/kg) for EQ sludge to 75 ppm (75 mg/kg) as a ceiling limit. The EQ sludge, which contains less than 41 ppm of arsenic, is allowed to be used unrestricted.[75] Sewage sludge that contains up to a ceiling of 75 ppm of arsenic can be applied until a cumulative load is reached.[76]

According to a 2006 guide for integrated field crop management authored by Cornell University scientists, "'Ceiling limits,' significantly less stringent than the EQ limits, have been set by [the] EPA for the same nine metals [as mentioned earlier: arsenic, cadmium, copper, lead, mercury, molybdenum, nickel, selenium, and zinc]. Federal rules allow sludges not meeting the EQ limits but within the ceiling limits to be applied, but the cumulative load of metals applied over time must be calculated and must be less than that allowed under the regulations."[77] This means, using arsenic again as our example, that 75 ppm can be applied to soil until the cumulative load is met or exceeded. Cumulative load is measured in kilograms per hectare (kg/ha).[78] One kilogram equals about 2.2 pounds, and one hectare equals about 2.5 acres.[79] The EPA Part 503 Cumulative Limit for arsenic is 41 kg/ha[80] or, in more familiar terms, 90 pounds of arsenic per 2.5 acres.

According to Ellen Harrison, director of Cornell Waste Management Institute, "The 75 ppm is the maximum concentration that could be allowed in a sludge that . . . [is] applied. The total amount of arsenic that is applied depends not only on the concentration in the sludge but on how much sludge is applied over the years. [The] EPA allows a total amount of arsenic applied to the soil at any site to reach a maximum cumulative limit of 41 kg/ha. That total amount could get there by applying sludge with up to 75 ppm arsenic or by applying a larger quantity of sludge that has a lower concentration. If a total cumulative load of 41 kg/ha is applied, that equates (when it is diluted into the soil) to about 20 ppm concentration of arsenic in the soil resulting from the addition, plus you would add the concentration that was in the soil to begin with."[81]

Sludge contaminants can also be unintentionally applied to soil. According to Ms. Harrison, bagged sludge fertilizer can unwittingly be used in single-household farming and/or lawn care by people not aware of the term *biosolids* or of the practice of sludge fertilizing. One brand in particular about which she expresses concern is Milorganite.[82] The Milorganite label states, "Milorganite products are recycled biosolids. Milorganite products proudly meet U.S. Environmental Protection Agency's 'Exceptional Quality' standards for beneficial use. This designation means that Milorganite products can be used anywhere fertilizer is recommended."[83] Of further concern is that not all commercial farmers and landscapers or purchasing agents at retail health food stores are familiar with the term *biosolids* or of the practice of sludge fertilizing, which can lead to even wider unintentional use of sludge fertilizer.[84]

According to a scientific report, published in 2007, "Milorganite has been reported to be one of the oldest natural *organic* fertilizers. It is derived from sewage sludge that has been dried, heated, and granulated."[85] The Wisconsin State Legislature confirms the production of sludge fertilizer as part of the Milwaukee Metropolitan Sewerage District. It reports, "Since 1925, the District has produced Milorganite, a fertilizer made from the organic sludge that remains after wastewater has been treated."[86] Obviously, not everything that remains in sludge after wastewater treatment is organic. Nevertheless, Milorganite prominently advertises on its product package the words "organic nitrogen fertilizer."[87] Most consumers will not realize that the word "organic" refers only to the origins of the nitrogen, not necessarily all the components in the fertilizer.

The National Biosolids Partnership touts Milorganite as a biosolids success story, "Today, Milorganite is marketed and used for turf-building and landscape maintenance applications, such as retail lawn and garden, professional landscaping, grounds, golf course, and sports field maintenance."[88] The organization states of Milorganite, "Its integration into the consumer marketplace has been so successful that many homeowners are unaware of the product's origins."[89] Sadly, health-focused, yet uninformed, people who fertilize their gardens and yards with "organic" fertilizer sourced from sewage "biosolids" can be unwittingly adding potentially negative-health-impacting contaminants to the soil in which their organic-intentioned produce grow and on which their children may play.

Let's review what we have learned so far: EPA regulations allow arsenic to be *intentionally* added to our agricultural land that when diluted equates to 20 ppm of arsenic, yet the beginning EPA soil screening level for arsenic is 0.39. In other words, the land our food is grown in can be 19.61 ppm higher in arsenic content than the beginning soil screening level for arsenic determined by the EPA. Contaminant screening levels are based on human risk assessment of soil ingestion and groundwater pathways.[90]

According to the EPA, evaluation of an area for related health risks can start when soil levels of arsenic are as low as 0.39, yet it allows levels of arsenic to be added to our farmland, reaching 19.61 ppm higher! When pre-existing background levels are added in, total parts per million of arsenic can reach levels as high as, or higher than, some EPA-designated Superfund sites.[91] Essentially, EPA regulations give permission to create mini "superfund" sites of our country's agricultural land.

So what does the "acceptable" practice of sludge fertilizing have to do with the contaminated sediment in the New Orleans area? Essentially, addressing and cleaning up the hurricane hot spots containing elevated levels of arsenic and other contaminants akin to the levels being dumped on our farmland would be political folly. Let's think about it for a minute. How can the EPA initiate sediment cleanup actions while condoning the practice of sludge fertilizing—a valuable, cost-effective manner of sewage sludge disposal?

In regard to just the one contaminant we've been talking about—arsenic—the New Orleans area clearly has levels far higher than the beginning screening level of 0.39 ppm and even some levels much higher than the EPA's highest acceptable concentration of 39 ppm. (Remember, the range of acceptable arsenic concentrations is actually 0.39 ppm to 39 ppm.) In fact, some EPA Superfund sites contain lower parts per million of arsenic than the highest levels of arsenic found in the New Orleans area post-Katrina.[92]

Averaged, elevated arsenic levels found in the hurricane sediment ranged from 3.1 to 17.2 mg/kg with maximum levels ranging from 11.5 to 78.0 mg/kg. The majority of these levels are not only *lower* than the ceiling limit of arsenic (75 mg/kg) but also *lower* than the allowable limit of arsenic in exceptional quality sludge (41 mg/kg). Therefore, the elevated arsenic levels in hurricane-hit areas look unremarkable to pro-sludge proponents, such as the EPA, the DEQ, and other in-the-know local, state, and federal agencies. On the other hand, most residents of New Orleans and surrounding areas are not aware of the practice of sewage sludge recycling or the resultant "acceptable" levels of contaminants from sewage sludge fertilizer. Thus, government cleanup inaction may appear to citizens to be the inefficient wheels of bureaucracy in motion when, in reality, the government's decision to *not* clean up is likely necessary to keep the lid on Pandora's "sludge" box. Otherwise, the EPA would use the sewage sludge statistics to support its non-cleanup stance of the elevated contaminants in New Orleans and the surrounding areas—an argument that could *only* be used if the practice of spreading "acceptable" levels of contaminants across our farmlands is indeed *not* a health risk. A little bit of common sense tells us otherwise.

Scientists at Cornell University address some additional important points, "Although [the] US EPA asserts that application of sludges is a low risk and thus a low priority for their attention, this seems to ignore the fact that sludges may end up spread over large areas where we grow our food, obtain our water, and where we live and play."[93] The same blind eye is being turned to the contaminants polluting New Orleans and surrounding areas, which is not surprising. To do otherwise, would acknowledge that contamination levels less than, or equal to, the levels being added to our farmlands require cleanup. A governmental cleanup of the elevated levels of contaminants from the hurricane sediment would acknowledge the inherent health risks associated with equal or lower levels of the specified contaminants. Governmental admission of existing health risks could undermine the policy of sludge fertilizing, because anti-sludge proponents would use this admittance to shine the spotlight on the flawed scientific risk assessments that prop up the misleading claim of "no unacceptable health risks" from the use of sludge fertilizer. Instead, the EPA has recommended handling some of these sediment contaminants in much the same manner it promotes with the management of sludge fertilizing—to till the contaminants into existing soil.[94]

According to the EPA, "The New Orleans Health Department and the state of Louisiana have provided general guidance and precautions for returning residents regarding the diesel and oil range organic chemicals detected. These include: Till sediment into existing soil; re-establish and maintain grass and flower beds; remove sediment from driveways and walkways to help minimize wind-blown dust; and/or minimize dirt and dust inside homes. [The] EPA believes the best course of action for diesel and oil range organics is to allow the recommendations from the Health Department and the state of Louisiana to work."[95] In other words, the EPA is passing the buck to the New Orleans Health Department and the state of Louisiana, both of which are doing nothing!

The next line of pro-contaminant propaganda that may be touted by the EPA, New Orleans Health Department, and the state of Louisiana is that the post-hurricane contaminants "stimulate plant growth," which is one of the marketing spiels used to promote sludge fertilizer. "According to the Environmental Protection Agency (EPA), biosolids that meet treatment and pollutant content criteria 'can be safely recycled and applied as fertilizer to sustainably improve and maintain productive soils and stimulate plant growth.' This EPA policy on sewage sludge recycling is highly controversial. Although often thought to consist of only 'human waste,' sewage sludge in fact contains all materials . . ." that city treatment plants remove from wastewater.[96] In other words, sewage sludge contains all the contaminants harmful to health that we do not want in our drinking water—heavy metals, PCBs (polychlorinated biphenyls), PAHs, disease-causing pathogenic organisms (such as bacteria, viruses, and fungi), industrial contaminants, diesel fuels, SVOCs (semivolatile organic compounds), VOCs, pesticides, drugs, pharmaceuticals—the list goes on and on.[97] If we don't want contaminants capable of destructing health in the water we drink, why would we want to dispose of them in the soil in which we grow our food?

The disposal of sewage solids has long been a sizable problem. According to a 1996 report by the National Research Council (NRC), ". . . [S]ewage treatment plants in the United States produce an average of 5.7 million dry tons of sludge per year. The NRC estimated that approximately 36 percent of this sludge—2.1 million dry tons—was applied to land, either as fertilizers for farms, golf courses, cemeteries, or other lands, or was used in reclamation of strip mines."[98] Studies indicate that the 2.1 million dry tons of sludge were filled with high levels of heavy metal contaminants. "In a 1997 evaluation of heavy metals in fertilizer materials, for example, Texas A&M researchers found that sewage sludges had elevated levels of nearly every heavy metal evaluated."[99]

In addition to heavy metal contaminants, detection of carcinogenic toxins has also been documented. In 1998 the Environmental Working Group (EWG) reported, "An EWG analysis of the Toxics Release Inventory (TRI) found that some 2 billion pounds of toxic chemicals—341 different chemicals in all—were transferred to sewage treatment plants between 1990 and 1995. This included 33.6 million pounds of toxic heavy metals like mercury, lead, and cadmium and more than 63 million pounds of carcinogenic substances. Sewage treatment plants report discharges of some pollutants under the Clean Water Act but do not report their discharges to the centrally maintained TRI. Thus, no recent comprehensive national databases are available on the toxic constituents of effluent discharged by sewage treatment plants or on the toxic components of more than 7.5 million tons of sludge generated in the United States annually."[100]

Widespread governmental support of using sludge as an agricultural fertilizer is evident. "In December, 1997, the U.S. Department of Agriculture (USDA) proposed draft[ing] national standards for organic agriculture. As part of this proposal, the department invited the public to comment on the idea of allowing application of municipal sewage sludge on land used to grow organic foods. The Environmental Protection Agency's top sludge regulator urged the department to allow 'high quality biosolids' (i.e., sewage sludge) to be used in organic food production."[101]

Fortunately, this legislation attempt was opposed by the National Organic Standards Board (NOSB).[102] The NOSB is a federally mandated 15-member advisory body established by the Organic Foods Production Act of 1990.[103] The NOSB's main function is to assist the Secretary of Agriculture in ". . . developing standards for substances to be used in organic production."[104] According to standard 205.203, which is the Soil Fertility and Crop Nutrient Management Practice Standard of the National Organic Program, "The producer must not use . . . Sewage sludge (biosolids) as defined in 40 CFR Part 503. . ."[105] The power of the NOSB is limited in its advisory capacity: "Recommendations made by the NOSB are not official policy until they are approved and adopted by USDA."[106]

Preserving optimal health may no longer be synonymous with eating our fruits and vegetables as a human lifetime of ingestion of plants grown in sludge-amended soil is one of the identified pathways of potential exposure to land-applied sludge.[107] In other words, if we eat crops grown in sludge-filled soil, we can be exposed to sludge and all its potentially health-eroding contaminants.

Ms. Harrison and colleagues point out limitations in risk assessment calculations, "The approach taken by US EPA to develop contaminant standards was to identify the various potential routes for exposure to sludge that is land-applied and then to assess the risks posed by each of these exposure pathways. . . . Each pathway was assessed independently and no attempt was made to look at the risk from exposure through several pathways simultaneously or the effects of more than one contaminant at a time."[108]

Governmental policies are based on data created by assessing exposure risks one contaminant at a time via only one exposure route. For example, the "US EPA made a policy decision that a cancer risk of 1-in-10,000 was an acceptable risk resulting from sludge application. For a number of contaminants, cancer risk was determined to be the most significant risk. A cancer risk estimated to lie between 1-in-10,000 and 1-in-1,000,000 is typically used in setting regulations, and in many regulatory contexts (e.g., drinking water regulation), a risk of one excess cancer in one million people exposed is used to establish the standards. Under the Part 503 risk assessment, policy-makers elected to use the less restrictive value."[109] In other words, a 1-in-10,000 cancer risk was used to establish the standards under the Part 503 risk assessment for sludge. Why use a less restrictive value—ten times less restrictive—to set regulation pertaining to food as compared to the value used to set regulation for drinking water?

According to Ms. Harrison, "I believe that the excess cancer risk would be over lifetime exposure."[110] This means, based on the risk assessments, that one additional person out of every 10,000 will get cancer in his or her lifetime from exposure to toxins that originally came from the contaminants in sludge fertilizer. This cancer risk, deemed as "acceptable," may be—unless of course *you* are the statistical *one* out of every 10,000. Based on current U.S. population statistics,[111] this means that more than 30,000 additional people will get cancer in their lifetime because of the use of sludge fertilizer—based on exposure to only *one* contaminant via *one* entry route.

Obviously, single contaminant risk assessment based on one exposure pathway underestimates the number of people who will actually get cancer from exposure to multiple sludge-originated contaminants via multiple routes of exposure. The risk assessment method analyzing a single contaminant via a single exposure does not take into account that sludge does not contain just one contaminant. Sewage sludge can contain *acceptable* levels of *all* nine regulated heavy metals mentioned earlier—arsenic, cadmium, copper, lead, mercury, molybdenum, nickel, selenium, and zinc—plus other unregulated metals such as barium, cobalt, and chromium. This oversimplified risk assessment method does not take into consideration the myriad of other toxins likely present in EQ sludge, such as pharmaceuticals, pesticides, PCBs, diesel fuels, and other undesirable contaminants. Furthermore, the single-contaminant and single-exposure risk assessment does not take into account that exposure is often multifaceted.

Just as sludge contaminants are not limited to one pathway of exposure, neither are the contaminants from the hurricane sediment. Harrison et al. points out, "It is likely that in a number of sludge-use scenarios, a person or animal will be exposed simultaneously through a number of pathways. Thus, the child of a home gardener using sludge will likely eat vegetables from the garden and may ingest soil that has received sludge. They may also drink from a well or eat animals or animal products that have been impacted by sludge use. The US EPA risk assessment calculated 'acceptable risk' according to each of 14 pathways and selected as a standard the lower of those numbers. They did not add exposures from several paths to arrive at a level from the multiple exposures which would result in an 'acceptable' risk. An additive approach is generally used in performing risk assessments. Similarly, the risk assessment did not attempt to address the ways in which the effects of exposure to multiple chemicals simultaneously can affect the toxicity impacts. Although estimating exposure and risk to a single pollutant from a number of pathways simultaneously could be done by summing results from different pathways, our lack of knowledge about how different contaminants interact makes it infeasible to evaluate impacts resulting from exposure to multiple pollutants. There can be synergistic or antagonistic impacts in which exposure to multiple chemicals has a greater or less[er] impact than exposure to each. Our very limited knowledge of how different contaminants may interact is one reason for skepticism regarding risk assessment and for the use of a more conservative approach."[112]

Just as residents in the New Orleans area are exposed through various pathways to sediment contaminants, so are residents or workers near or downwind of sludge contaminants. Victims of sediment exposures may want to read about cases involving ill health effects from extended exposure to contaminated sludge to estimate possible future associated health risks. Although formal epidemiological studies are sparse, Cornell Waste Management Institute is compiling case data regarding "Clustering of Reported Health Incidents by Locality," available at http://cwmi.css.cornell.edu/Sludge/INCIDENTSintropage.htm.[113] The reported cases have not been confirmed by scientific investigation; nevertheless, they give an indication of possible health outcomes associated with these types of toxic exposures.

According to Cornell Waste Management Institute, "To date, there is only one study that has been published that examines any of these cases (Lewis et al., 2002). In that study, interviews were conducted at ten of the sites where neighbors reported symptoms. Land application data and medical records were reviewed. Approximately half of the 48 residents surveyed at those sites reported symptoms consistent with endotoxin exposure and half reported infections. At one site, symptoms decreased linearly with distance from the treated field and increased linearly with duration of exposure to winds blowing from the field. That study also suggests that chemical contaminants in sludges may be responsible for an increased host susceptibility to infection."[114] From these research findings, we can reasonably surmise that exposure to sediment toxins may likewise increase the susceptibility of people exposed to increased levels of infection and disease.

Both the NRDC and Cornell Waste Management Institute advocate for increased research in areas that impact national health. The institute states, "It is important to recognize that the EPA sludge rules (40 CFR Part 503) are not based on an assessment of risks posed by pathogens or biological agents such as endotoxins. Nor are the risks posed by airborne or water runoff transport of sludge constituents or those resulting from any combination of pollutants or of biological and chemical contaminants addressed. Thus, as the National Resource Council (NRC) has stated, there is a need for research to address these and related issues and for a revised risk assessment," according to Cornell Waste Management Institute.[115] Likewise, the declarations of "no unacceptable health risks" by the Louisiana DEQ and the EPA are not based on assessments of risks posed by a multiple pathogens and biological and chemical agents via multiple avenues of exposure.[116]

In summary, Dr. Solomon states in regard to the sediment, "I think that it's clear that there is a health risk. How significant the health risk is, is tough to quantify. All I can say is that if I had a young child, I would think twice before going back and spending time in one of the flooded neighborhoods. If I had an underlying illness—diabetes or a lung disease, for example—I would also think twice about doing it. But there are a lot of things that we all do every day that may or may not shorten our life spans. The thing that bothers me right now is that people don't necessarily understand that there is a risk. If people understand that there is a risk and decide that they want to go back anyway, that's as fine as deciding that you want to go skiing. You might run into a tree, but you wanted to do it and that's fine. People need to know. From my perspective, I'm not going to tell people that they should or shouldn't go back. I just think that it's important to tell them that there is a health risk."[117]

Dr. Solomon adds, "I think people need to be careful, they need to be informed, and the government needs to help clean this problem up. The data do not suggest that New Orleans is 'uninhabitable' or that it will remain 'toxic' over the long term. But the data do show that there is a problem that needs to be addressed."[118]

The question remains: Will these contaminants get properly addressed? EPA Senior Policy Analyst Hugh Kaufman shares insight from a government insider's point of view. He explains, ". . . [I]t's money versus people. The EPA has done limited air, water, and soil sampling. [But] we still don't have an adequate mapping of the contamination on the ground from all the sediments that have settled in the area. So we don't know where all the hot spots are that need removal. We're a year later and that mapping still hasn't been done, because if you do that type of detailed mapping, then you're going to spend more money in more removal actions—and they're already having a difficult time with the rubbish they have to deal with right now. If you do the adequate monitoring to identify all the hot spots for remediation, you're going to have a lot more hazardous material to deal with, and it's going to be a lot more expensive. So again, it's money versus people."[119]

Mr. Kaufman adds, "As a matter of policy, if the federal and state governments put out the true nature of what's going on, it would substantially increase the cost of remediation and increase the time before things are perceived to be back to normal. The state and local governments don't have that kind of money, and the federal money is going overseas. And none of the people who are the decision makers are going to be in office when all these do-gooders start getting cancer . . . The thing is, it's sort of like don't ask, don't tell, because if you do the mapping, that'll put more pressure to do something about it. But none of these contaminants is visible to the human eye."[120] He points out, ". . . [T]he type of work that needs to be done is on a mass scale—requiring the federal government. You're talking about spending a few billion bucks. And that's just not on the agenda."[121]

We must realize that these financially induced cleanup stalemates are not limited to New Orleans or the hurricane-prone Gulf Coast. The money-versus-people dilemma is a challenge that arises any time an area is hit by a sizeable natural disaster. No matter what political party is in control—be it Republican or Democrat—or what persons are in various local, state, and federal offices, all too often cost-exorbitant problems are not made widely known to local residents. By thwarting public awareness of storm contaminants and food-related toxins, Uncle Sam's powers-that-be hope to save government money. Unfortunately, what people don't know can hurt them!

So what's the answer? We must all get involved in our local communities and unite with other residents to reduce health-impacting toxins by leveraging increased citizen involvement into government, corporate, and citizen cleanup efforts. We need to look at the food on our dinner table at night and be brutally honest with ourselves. Do we really know what is in our fruits and vegetables that got absorbed from the soil in which they were grown? We must not only investigate what types of pesticides are being sprayed *on* our produce, but we need to know the quality of soil in which our produce is grown. The best way to find out this information is to locate local and regional farmers. Ask them what fertilizers they use. Ask them to provide documentation to support their no-sludge fertilizer claims. Get to know the produce managers at your local health food stores and conventional grocery stores and ask them to provide the same no-sludge documentation from their produce suppliers. Educate your extended family and friends so they can move to action in their respective areas. As this grass-roots group of health proponents grows, one person at a time, a network of sludge fighters will form and unite across the county sharing vital information.

We must tackle this government-sanctioned land contamination program as if our lives depended on it, because they do. We must realize that if we as individuals do not ban together to demand an end to the practice of applying sludge fertilizer to our agricultural land and pastures, we undoubtedly will face mounting illnesses and diseases in our society. The sludge fertilizer is not limited to affecting just our fruits and vegetables. Our meat, eggs, and milk products can also become contaminated from animals that graze on pastures tilled with sludge fertilizer. Nutrients from uncontaminated food provide the building blocks to restore and retain health and vitality. Food contaminated from sludge fertilizer and other sources can be the seeds of tomorrow's illnesses and diseases. We must each ask ourselves: With what level of "acceptable" health risks are my family and I willing to live the consequences?

(ENDNOTES)

1   NRDC, "Sediment Contamination," Available at http://www.nrdc.org/health/effects/katrinadata/sediment.asp (accessed July 2006).

2   Subra Company, P.O. Box 9813, New Iberia, LA 70562. Phone: (337) 367-2216. E-mail: subracom@aol.com.

3   Interview with Wilma Subra, October 5, 2006

4   For more information on Bucket Brigade see http://www.bucketbrigade.net/ and on Louisiana Bucket Brigade see http://www.labucketbrigade.org/ (accessed August 2006)

5   For more information on Louisiana Environmental Action Network (LEAN) see http://www.leanweb.org/ (accessed August 2006).

6   NRDC, "Sediment Contamination."

7   FOX News, "Public Health Crisis Still Threatens Gulf Coast," September 10, 2005. Available at http://www.foxnews.com/story/0,2933,168974,00.html (accessed November 2005).

8   NRDC, "Sediment Contamination."

9   Ibid.

10   Curtis et al., "Adverse Health Effects," p. 5.

11   Interview with Wilma Subra, October 5, 2006.

12   See Section III, interview with Dr. Gina Solomon, July 2006.

13   See Section III, interview with Dr. David Straus, April 2006.

14   See Section III, interview with Dr. Gina Solomon, July 2006.

15   Solomon et al., "Airborne Mold and Endotoxin Concentrations in New Orleans, Louisiana after Flooding, October-November 2005."

16   See Section III, interview with Dr. Richard Lipsey, April 2006.

17   See Section III, Interview with Dr. Gina Solomon, July 2006.

18   See Section III, interview with Dr. Richard Lipsey, April 2006.

19   Ibid.

20   See Section III, Interview with Dr. Gina Solomon, July 2006.

21   Ibid.

22   Ibid.

23   Ibid.

24   Ibid.

25   Ibid.

26   See Section III, interview with Dr. Gina Solomon, July 2006.

27   See Section III, interview with Jim Pearson, April 2006.

28	See Section III, interview with Dr. Gina Solomon, July 2006.

29	Lipsey, "The Mold and Bacterial Results from the Sampling Dr. Lipsey did in St. Bernard Parish" and e-mail from Dr. Richard Lipsey, October 6, 2006.

30	E-mail from Dr. Richard Lipsey, October 6, 2006.

31	CDC, "Health Concerns Associated with Mold in Water-Damaged Homes After Hurricanes Katrina and Rita-New Orleans Area, Louisiana, October 2005."

32	Lipsey, "The Mold and Bacterial Results from the Sampling Dr. Lipsey did in St. Bernard Parish" and e-mail from Dr. Richard Lipsey, October 6, 2006.

33	Ibid.

34	Ibid.

35	Ibid.

36	See Section III, interview with Dr. Jack Thrasher, April 2006.

37	Gina M. Solomon et al., "Contaminants in New Orleans Sediment," February 2006. Available at http://www.nrdc.org/health/effects/katrinadata/sedimentepa.pdf (accessed June 2006)

38	Ibid.

39	Ibid.

40	See Section III, interview with Dr. Gina Solomon, July 2006.

41	Ibid.

42	EPA, "Summary Results of Sediment Sampling Conducted by the Environmental Protection Agency in Response to Hurricanes Katrina and Rita." Available at http://www.epa.gov/katrina/ testresults/sediments/summary.html (accessed August 2006).

43	NRDC Press Release, December 1, 2005. Available at http://www.nrdc.org/media/ pressReleases/051201.asp (accessed June 2006).

44	See Section III, interview with Dr. Gina Solomon, July 2006.

45	Wilma Subra, "Environmental and Human Health Impacts of the 2005 Katrina and Rita Hurricane Season," March 26, 2006. Available at http://www.leanweb.org/katrina/wilmadata.html (accessed October 2006).

46	OMB, "An OMB Watch Statement in the Aftermath of Hurricane Katrina." Available at http://www.ombwatch.org/article/articleview/3085 (accessed October 2006).

47	EPA, "Superfund-Region 6: South Central." Available at http://www.epa.gov/earth1r6/6sf/ sfsites/default.htm (accessed October 2006).

48	OMB, "Toxic Chemical Sites in New Orleans." Available at http://www.ombwatch.org/ article/articleview/3088 (accessed October 2006).

49	EPA, "Superfund-Region 6: South Central" and EPA, "NPL Site Narrative for Agriculture Street Landfill." Available at http://www.epa.gov/superfund/sites/npl/nar1441.htm (accessed October 2006).

50	Interview with Wilma Subra, October 5, 2006.

51	Agency for Toxic Substances and Disease Registry (ATSDR), "ToxFAQs for Arsenic," September 2005. Available at http://www.atsdr.cdc.gov/tfacts2.html (accessed August 2006).

52	Solomon, "Contaminants in New Orleans Sediment."

53	Agency for Toxic Substances and Disease Registry (ATSDR), "ToxFAQs for Arsenic."

54	Ibid.

55	Ibid.

56	Ibid.

57	Ibid.

58	Ibid.

59	Ibid.

60	Solomon et al., "Contaminants in New Orleans Sediment."

61	LDEQ Communications, "Arsenic Sampling Results Explained," January 10, 2006. Available at http://www.deq.louisiana.gov/portal/Default.aspx?tabid=2245 (accessed August 2006).

62	Ibid.

63	Ibid.

64	See Section III, interview with Dr. Gina Solomon, July 2006.

65	LDEQ Communications, "Arsenic Sampling Results Explained."

66	Solomon, "Contaminants in New Orleans Sediment."

67	From correspondence from Patrice Simms, Senior Project Attorney, NRDC to Mike McDaniel, Secretary of the Louisiana DEQ. Available at http://www.deq.louisiana.gov (accessed August

2006)

68   Ellen Z. Harrison et al., "Land Application of Sewage Sludges: An Appraisal of the US regulations," Int. J. Environmental and Pollution, Vol., 11, No. 1999. Available at http://cwmi.css. cornell.edu/PDFS/LandApp.pdf (accessed August 2006)

69   Ibid.

70   David O. Carpenter et al., "Understanding the Human Health Effects of Chemical Mixtures," Environmental Health Perspectives Supplements, Vol. 110, No. S1, February 2002. Available at http:// www.ehponline.org/docs/2002/suppl-1/25-42carpenter/abstract.html (accessed August 2006).

71   Harrison et al., "Land Application of Sewage Sludges: An Appraisal of the US regulations"; Vonn R. Christenson, "Regulate This: The Politics and Practice of Poo Farming," May 2006. Available at http://leda.law.harvard.edu/leda/data/761/Christenson06.rtf (accessed August 2006); and Environmental Working Group (EWG), "Dumping Sewage Sludge on Organic Farms?" April 30, 1998. Available at http://www.ewg.org/reports/sludgememo/sludge.html (accessed August 2006).

72   Harrison et al., "Land Application of Sewage Sludges: An Appraisal of the US Regulations."

73   Jerome H. Cherney et al., "2006 Cornell Guide for Integrated Field Crop Management." Available at http://cwmi.css.cornell.edu/sludge/Cornellguide2006.pdf (accessed August 2006).

74   Harrison et al., "Land Application of Sewage Sludges: An Appraisal of the US Regulations."

75   Ibid.

76   Ibid.

77   Cherney et al., "2006 Cornell Guide for Integrated Field Crop Management."

78   E-mail from Ellen Harrison, August 26, 2006.

79   Metric Conversions. Available at http://www.extension.iastate.edu/AgDM/wholefarm/html/ c6-80.html (accessed August 2006).

80   Cherney et al., "2006 Cornell Guide for Integrated Field Crop Management."

81   E-mail from Ellen Harrison, August 26, 2006.

82   Interview with Ellen Harrison, August 2006.

83   Milorganite, "Products." Available at http://www.legis.state.wi.us/LaB/reports/97-6summary. htm (accessed February 2007).

84   Interviews with commercial farmers and landscapers and purchasing agents at retail health food stores that requested to remain anonymous, August 2006-June 2009.

85   Robert Green et al., "Development of BMPs for Fertilizing Lawns to Optimize Plant Performance and Nitrogen Uptake While Reducing the Potential for Nitrate Leaching," January 31, 2007. Available at ucrturf.ucr.edu/UCRTRAC/Accumulative%20Report/PDF%20Files%20for%20Down load/CumB05.pdf (accessed February 2007). (Italics added.)

86   Wisconsin Legislative Audit Bureau, "97-6 Milwaukee Metropolitan Sewerage District." Available at http://www.legis.state.wi.us/LaB/reports/97-6summary.htm (accessed August 2006).

87   Milorganite, "Products."

88   Biosolids, "Biosolids Success Stories: Milwaukee Metropolitan Sewerage District Continuing the Tradition of Milorganite." Available at http://www.biosolids.org/docs/source/MilWI.pdf (accessed February 2007).

89   Ibid.

90   Harrison et al., "Land Application of Sewage Sludges: An Appraisal of the US Regulations- Table 6," and DEQ Communications, "Arsenic Sampling Results Explained."

91   CDC, "Public Health Assessment: Agriculture Street Landfill New Orleans, Orleans Parish, Louisiana." Available at http://www.atsdr.cdc.gov/hac/pha/agriculturestreet/asl_p1.html#conta (accessed October 2006).

92   Ibid.

93   Harrison et al., "Land Application of Sewage Sludges: An Appraisal of the US Regulations."

94   EPA, "Summary Results of Sediment Sampling Conducted by the Environmental Protection Agency in Response to Hurricanes Katrina and Rita."

95   Ibid.

96   Christenson, "Regulate This: The Politics and Practice of Poo Farming."

97   Ibid.; Altamont Environmental Inc., "Sediment and Surface Water Sampling and Analyses Eighteen Gulf Coast Locations," November 11, 2005. Available at http://www.sierraclub.com/gulfcoast/ SEDIMENT AND SEWAGE SLUDGE SPIN – 83 testing/ (accessed August 2006); and Solomon, "Contaminants in New Orleans Sediment."

98   EWG, "Dumping Sewage Sludge on Organic Farms?"

99   Ibid.

100   Ibid.

101   Ibid.

102   Ibid.

103   National Organic Standards Board. Available at http://www.ams.usda.gov/nosb/ (accessed May 2007).

104   Ibid.

105   The National Organic Program, "Production and Handling—Regulatory Text." Available at http://www.truthout.org/issues_06/printer_021507HA.shtml (accessed May 2007).

106   National Organic Standards Board.

107   Harrison et al., "Land Application of Sewage Sludges: An Appraisal of the US Regulations."

108   Ibid.

109   Ibid.

110   E-mail from Ellen Harrison, November 1, 2006.

111   The U.S. population was 300,076,365 as of 10:33 Greenwich Mean Time, October 27, 2006. From U.S. Census Bureau, "U.S. and World Population Clocks-POPClocks," which additionally states, "Greenwich Mean Time (GMT) is the Equivalent of Eastern Standard Time (EST) plus 5 hours or Eastern Daylight Saving Time (EDT) plus 4 hours." Available at http://www.census.gov/main/www/popclock/html (accessed October 2006).

112   Harrison et al., "Land Application of Sewage Sludges: An Appraisal of the US Regulations."

113   Cornell Waste Management Institute, "Clustering of Reported Health Incidents by Locality." Available at http://cwmi.css.cornell.edu/Sludge/INCIDENTSintropage.htm (accessed August 2006).

114   Ibid.

115   Ibid.

116   LDEQ, "Sec. McDaniel's response to NRDC letter." Available at http://www.deq.louisiana.gov/portal/portals/0/news/pdf/DEQ-EnvironmentalAssessmentKatrinaRita-NRDCResponse.pdf (accessed June 2006).

117   See Section Three, interview with Dr. Gina Solomon, July 2006.

118   Ibid.

119   Bennett, "'It's Incompetence.'"

120   Ibid.

121   Ibid.

CHAPTER FIVE:

WHAT WE KNOW, SCIENTIFICALLY SPEAKING

In the mind of anyone who has ever become sick from mold exposure, no doubt exists that mold can cause illnesses. To some degree, we are all familiar with the most commonly known health effects of fungal exposure—athlete's foot, toenail and fingernail fungus, jock itch, allergies, and asthma. However, when illnesses develop beyond allergies and asthma, scientific research can become a debated, gray area. In fact, studying what illnesses can and cannot be scientifically correlated to mold exposure has been a focus of scientific research for many years.

Scientifically speaking, several facts have been established in regard to fungal exposure. One undisputed fact is that in order for mold and mycotoxins to be a threat to optimal health, they must have a route of entry into the human host. According to MSUM, "The most obvious pathways are:

- Breathing volatile organic compounds (VOCs), spores and/or fragments of mold is generally considered the primary means of exposure to indoor mold. Inhalation exposes the upper and lower respiratory tracts and allows pathways into the blood system via the lungs.
- Eating food contaminated by mold allows direct exposure of the digestive tract to mold components. Studies of people and livestock that consumed mold-contaminated food has contributed a significant amount of data on infectious and toxic molds.
- Touching mold or items contaminated by mold can provide a pathway [especially through cuts, scrapes, and the eyes]."[1]

Also undisputed by scientific and medical professionals is the fact that health risks from mold and chemical exposures can vary from person to person. For this reason, the CDC has published multiple warnings for various groups identified to be at high risk from mold exposure. According to the CDC, the following groups ". . . may be affected to a greater extent than most healthy adults:

- Infants and children
- Elderly people
- Pregnant women
- People with respiratory conditions (such as asthma) or allergies
- People with weakened immune systems (e.g., chemotherapy patients, organ or bone marrow transplant recipients, people with HIV [human immunodeficiency virus] infections or autoimmune diseases)"[2]

This latter group is often referred to as the immunocompromised, which is a subgroup of the population scientific and medical communities unequivocally agree can become infected with invasive fungi, such as certain species of *Aspergillus*.[3] Medical studies confirm increased susceptibility of high-risk groups. In fact, one study appears to ". . . demonstrate that such fungal exposure can significantly aggravate and complicate pre-existing immunological disease."[4]

The CDC warns, "The level[s] of risk associated with exposure activities and the potential benefit of recommended PPE [personal protective equipment] are unknown for pregnant women, person[s] older than 65 years, and children younger than 12 years. Due caution is recommended. Exposure-reducing behavior and respiratory protection are problems for children younger than 12 years."[5] Young children may not follow exposure-reducing instructions, such as "Don't touch the mold!" or "Leave the mask on your face." Additionally, respiratory masks are not child-size and, therefore, leave gaps, allowing spores and contaminant-filled air to enter. A respiratory mask on a child creates the illusion of respiratory protection— no more than a façade.

In another warning the CDC states, "The following people should *avoid mold-contaminated environments entirely:*

- Transplant recipients, including those who received organ or hematopoietic stem cell transplants within the last six months or who are undergoing periods of substantial immunosuppression
- People with neutropenia (neutrophil count < 500/µL) due to any cause, including neoplasm, cancer chemotherapy, or other immunosuppressive therapy
- People with CD4+ lymphocyte counts < 200/µL due to any cause, including HIV infection
- Other individuals considered by their physicians to have profoundly impaired antifungal host defenses due to congenital or acquired immunodeficiency
- *Infants and children*"[6]

How many parents returned post-hurricane to mold-contaminated areas completely unaware of the health risks to their children? The CDC's warning should have been covered on every media outlet—television, radio, and newspapers—immediately following Hurricane Katrina. There is no acceptable excuse for not widely publicizing this statement other than the priority of economics over health.

The CDC issues additional warnings, "The following people may be able to tolerate limited exposure, but they should consult with their physicians and should consider avoiding areas where moldy materials are disturbed:

- People receiving chemotherapy for cancer, corticosteroid therapy, or other

immunosuppressive drug therapy, as long as neutropenia or CD4+
lymphopenia are not present
- People with immunosuppressive diseases, such as leukemia or lymphoma, as long as there is not marked impairment in immune function
- Pregnant women"[7]

We must read between the lines of all these warnings. To "avoid mold-contaminated environments entirely" or "to limit exposure" while in the midst of a mold haven, these groups of people must evacuate the contaminated areas entirely until testing confirms air quality has returned to a safe level. Also, remember to exercise good judgment when interpreting test results as tests are contaminant specific, not all inclusive. For example, viable and nonviable testing methodologies will not reflect volatile air contaminants, such as mycotoxins, VOCs, and smoke, or air contaminants other than spores, such as fungal fragments, fine particulates, dust, bacteria, endotoxins, bacterial cell wall toxins, and other airborne toxic debris.

Evacuation procedures for those at high risk from fungal exposure are included in a publication by The New York City Department of Health and Mental Hygiene, which was the original industry guideline for mold, according to Mr. Pearson.[8] The report states, "Infants (less than 12 months old), persons recovering from recent surgery, or people with immune suppression, asthma, hypersensitivity pneumonitis, severe allergies, sinusitis, or other chronic inflammatory lung diseases may be at greater risk for developing health problems associated with certain fungi. Such persons should be removed from the affected area during remediation . . ." Furthermore, it states, "Persons diagnosed with fungal-related diseases should not be returned to the affected areas until remediation and air testing are completed."[9] This information is clearly stated under the section titled "Medical Relocation."

People, including medical doctors, remain uninformed about the dangers of mold exposures. They continue to believe the fallacy "Mold cannot hurt you" in part because of improper media coverage. For example, during Katrina the term *medical relocation* and the phrase *avoid mold-contaminated environments entirely* were not heralded to the forefront of public awareness by media networks, FEMA, the CDC, the EPA, other governmental agencies (local, state, and federal), or even the Red Cross. Furthermore, no public service announcements were made to warn high-risk groups about the health risks of returning to a mold-infested environment. Instead, frequent news footage showcased children returning to water-damaged schools laden with mold, which served to further uphold the pretense of a safe environment and perpetuate the belief "Mold cannot hurt you." The lack of public dissemination of post-hurricane warnings by FEMA, the CDC, and other governmental agencies put many people needlessly at risk. Leaving these warnings lying silently, buried in the fine print of governmental publications and websites, is not going to control or prevent any disease!

The fact fungal exposure can cause illness and disease is not disputed, but whether mycotoxin inhalation can cause human disease is controversial. Dr. Straus, who coauthored the first findings revealing the ability to measure trichothecenes in the air of contaminated buildings and the resultant trichothecene levels in human blood sera, sums up his research, "In a nutshell, we believe, at least from the fungal point of view, that we understand what causes sick building syndrome. It's really pretty simple. We believe that the inhalation of high concentrations of fungal spores causes respiratory diseases in human beings, and most people who work in this field agree with that statement. There's not much controversy about that. We also believe that when people inhale the trichothecenes, it also causes human disease, and that's very controversial, as I'm sure you probably have read. We've come a long way in the last 12 years toward proving all of this. What we know at this point is that when *Stachybotrys* grows in buildings, you oftentimes find people who are sick. We know that when *Stachybotrys* grows in buildings, it produces these trichothecene mycotoxins. There's no question about that, so there's no question that you'll find sick people in buildings where there's *Stachybotrys* growing. (Sometimes you don't find sick people, but oftentimes you do.) There's no question that *Stachybotrys* produces these trichothecene mycotoxins on building materials in these buildings."[10]

"The next thing we know is that when *Stachybotrys* grows inside buildings on building surfaces, the trichothecene mycotoxins get into the air on fragments smaller than spores," continues Dr. Straus. "I'm talking about all stuff that is true and that we've published. We also know that these trichothecene mycotoxins, because they're floating around in the air and are highly respirable, get into the human body—no question about that."[11]

Dr. Straus sums up the highly controversial issue of mycotoxin-induced diseases while specifically addressing trichothecenes, the mycotoxin produced by *Stachybotrys*. He states, "Now, here's really the last question to be answered: Do the trichothecenes, which float around in the air in these buildings and actually get into human beings, get into them in concentrations sufficient to cause the human diseases that we see? We don't know the answer to that, and that really is the last question yet to be addressed."[12]

As you will recall from an earlier chapter, Dr. Straus explains the ways in which exposure to fungi can make people sick: infection, inhalation of fungal spores, and inhalation of mycotoxins.[13] Understanding the multiple health ramifications of fungal and mycotoxin exposures is paramount, especially for people in high-risk groups. The potential cost to their health can be much greater. For example, the NRDC points out that the mold concentrations detected in its study raise ". . . potentially significant concerns for long-term health effects in the pediatric population that may be returning to the flooded areas."[14] Children do not have fully developed immune systems and therefore are more susceptible to the development of ill health effects from fungal exposure. This increased susceptibility is why the CDC designates children as a high-risk group.[15]

The NRDC reports, "In outdoor air, elevated concentrations of fungal spores are associated with allergic and asthmatic responses in humans. A large Canadian time-series study reported that daily fluctuations in ambient mold spores are directly associated with childhood asthma attacks . . . [that require visits] to an emergency department. Researchers in Southern California reported an association between ambient mold spore concentrations and childhood asthma attacks even in areas where the airborne spore concentrations are relatively low (12-hr daytime mean spore concentration of approximately 4,000 spores/m³)."[16] If children are known to have increased asthma attacks when the ambient air contains low concentrations of mold spores—for example, 4,000 spores/m³—then what is the health impact on children when outdoor concentrations of mold spores range from 21,000 to 102,000 spores/m³, as was the situation in the New Orleans area post-Katrina?

In addition to the increased risk of asthma, inhalation of elevated levels of fungi can lead to colonization and infection. Although most people understand that airborne particles, such as pollen, ragweed, and fungi, can trigger asthma attacks, the concepts of colonization and infection are less clear. According to Dr. Straus, "A colony of something doesn't necessarily mean that it's causing pathology or causing infection. It can, but it doesn't necessarily. Colonization just means that it's there. Infection, or pathology, means that it's beginning to produce compounds, maybe even infect tissue and cause disease."[17]

Dr. Gray points out, "If you're colonized, the funguses are growing, they're reproducing, and there is a battle going on between them and your immune system." He further clarifies, "It's an infection from the moment it starts growing. Your defenses against it begin to mobilize and you're fighting that infection from the moment your cells discover that the infection is there. By infection we mean that some foreign organism (fungus, parasite, virus, or bacteria) is growing on or inside of you."[18]

According to Dr. Straus, when fungal spores are inhaled, the spores do not grow inside the lungs, unless the organism inhaled is certain species of *Aspergillus* or other invasive fungi that can colonize and/or cause infection.[19]

It is important to note not all species of the *Aspergillus* genus cause infection. Some are more harmful than others, as Dr. Puccetti points out, ". . . [T]here are only a few species of *Aspergillus* that can grow in human tissue at the temperature of the human body. When they're growing, they're infecting the tissue. They're basically living off the tissue. They're not throwing spores anywhere; they're just growing. The mold is producing spores which germinate because they are using the human tissue for metabolism." Dr. Puccetti adds, "The parameters under which the mold will grow [in the human body] depend on the species and on the host individual."[20]

Dr. Straus concurs, stating that *Aspergillus* is the one organism prevalent in damp buildings that can colonize and cause infection within the human body. Other fungi can colonize but won't necessarily cause pathology. In other words, colonization of a fungus can cause infection but doesn't necessarily, clarifies Dr. Straus.[21]

Although we cannot determine from the NRDC test results if an invasive species of *Aspergillus* was present in the testing areas (because the spores were not cultured), the presence of *Aspergillus flavus* and *Aspergillus fumigatus* was confirmed by other testing sources. Ginger Chew, ScD, assistant professor at Columbia University Mailman School of Public Health, and colleagues collected air samples from houses that had between 0.3 and 1.8 meters of flooding. Samples were cultured and test results revealed the presence of these two toxigenic, mycotoxin-producing species.[22]

The potential widespread presence of *Aspergillus fumigatus* is noteworthy as exposure to *A. fumigatus* increases health risks, especially for anyone with an impaired immune system. A published, peer-reviewed study states, "*A. fumigatus* is associated with several human health issues. It is the most common airborne fungal pathogen of humans. It can cause invasive aspergillosis in immunocompromised individuals, and the resulting mortality rate is >50%. In *immunocompetent* individuals, this fungus can colonize pre-existing cavities in the lungs or sinuses without penetrating into tissues, a condition known as aspergilloma or fungus ball."[23] In other words, healthy (immunocompetent) people can become colonized with *Aspergillus fumigatus*.

Furthermore, scientific studies prove that high levels of certain ergot alkaloids, which are mycotoxins produced by *Aspergillus fumigatus,* are present in or on the conidia (asexual spores) of *A. fumigatus*. Researchers report, ". . . [T]he conidia of *A. fumigatus* are smaller, lighter, and less dense than those of closely related species. These physical properties may promote the aerosolization and buoyancy of the conidia, which serve as vehicles for the alkaloids."[24] Spores of *Aspergillus fumigatus* are also more easily aerosolized than the larger, heavier, and wetter toxigenic mold spores of other genuses, such as *Stachybotrys*. Once airborne, these respirable-sized, mycotoxin-containing spores can become inhaled, after which the mycotoxins become soluble. Once the mycotoxins dissolve, they release into the body. The 2005 study addresses the potential health effects, "Ergot alkaloids are mycotoxins that interact with several monoamine receptors, negatively affecting cardiovascular, nervous, reproductive, and immune systems of exposed humans and animals."[25] The potential for these serious health effects from exposure to *Aspergillus fumigatus* remains in hurricane-hit areas as long as the toxigenic fungi exists in indoor environments.

Another genus prevalent in wet-building materials, *Stachybotrys,* is believed not to grow inside humans or cause human infections, although it can exist inside the human body, according to Dr. Straus.[26] "Let's say, for example, that you walk into a room and you inhale *Stachybotrys* spores. They have the potential to make you sick because when you inhale the spores, the mycotoxins are on the spores. The mycotoxins then become solubilized, get into your blood stream and begin to kill cells. But *Stachybotrys* does not grow, does not multiply, inside the lungs, so you will never see a *Stachybotrys* lung infection," states Dr. Straus.[27]

Dr. Straus points out an important fact, "You don't have to be immunocompromised to be poisoned by mycotoxins or to have a respiratory disease due to inhalation of high concentrations of fungal spores." Dr. Straus adds, however, "You do have to be immunocompromised, in most cases, to have a fungal lung infection."[28]

It is a widely held belief in scientific and medical fields that *only* the immunocompromised are at risk for invasive fungal infections. However, the CDC report published in the aftermaths of hurricanes Katrina and Rita states, "Invasive fungal infections can occur in individuals with normal host defenses and in certain situations can even be life threatening."[29] In other words, the CDC is saying that healthy people can become infected with invasive fungi and can even succumb to it in certain situations.

Dr. Straus explains the CDC's statement, "There are some fungi that can cause primary human disease in healthy people. An example would be a *Coccidioides immitis*, which is a fungus that can cause a human pneumonia in a healthy person. That's a fungus, for example, you or I could become infected with just by inhaling the arthroconidia [asexual spores], even if we were healthy—but that organism does not grow in water-damaged buildings. That organism can be found in the soil."[30] Thus, risks to our health from exposure to fungi are not limited to incidences of damp buildings from hurricanes, floods, leaks, and other sources of indoor moisture.

Dr. Ammann further explains, "There are other fungi that cause systemic infections, such as *Coccidioides*, *Histoplasma*, and *Blastomyces*. These fungi grow in soil or may be carried by bats and birds but do not generally grow in indoor environments. Their occurrence is linked to exposure to wind-borne or animal-borne contamination."[31]

Several molds grow quite well in the human body (for example, *Aspergillus fumigatus* causing aspergillosis, *Coccidioides immitis* causing coccidiomycosis, and *Zygomycetes* causing zygomycosis), according to Dr. Chew.[32] Further documentation of systemic fungal infections comes from the CDC. It states that almost ". . . 200 fungal species have been described as human pathogens that can cause infections."[33]

We are all at risk from exposure to mold in our environments, whether they be the molds that flourish in damp buildings—as has occurred in New Orleans and other parts of the country from hurricanes, floods, and/or leaks—or the more invasive, dirt-inhabiting fungi that can be transmitted by birds and bats. Both can become airborne—the former from indoor air movement and the latter from winds.

Remember, health risks are not limited to infection. Illness can develop from inhalation of fungal spores and mycotoxins as well, as earlier discussed.[34] According to Dr. Straus, the more widespread negative health implications from inhalation of fungal spores occur because the spores are living particles that can produce biological compounds inside the lungs, some of which can do damage.[35] Dr. Straus discusses an example in regard to *Stachybotrys*, "We also know that some of the compounds produced by *Stachybotrys* cause what we call immunosuppression. These compounds—for example, cyclosporin A—cause inhibition of the production of interleukin 2 by activated T cells. Interleukin 2 is necessary for the development of T cell immunologic memory. This stops expansion of the number and function of antigen-selected T cell clones, resulting in immunosuppression."[36] Antigens are foreign substances that have invaded the body.[37]

Dr. Thrasher reports clinical findings of decreased natural killer cell activity in people who have been mold exposed. He explains that some people are detoxification compromised when it comes to fungal and mycotoxin exposure, based on their individual genetic predisposition. Dr. Thrasher clarifies, "They would not be immunocompromised. They are detoxification compromised, which would lead to immunocompromised, because if they can't detoxify and get rid of these free radicals, these free radicals are going to bother and depress or affect the immune system."[38] In other words, if a person's body cannot sufficiently detoxify fungal and mycotoxin contaminants, they can become immunocompromised. It is firmly established in scientific and medical literature that immunocompromised people are at much greater risk for developing invasive fungal infections.

Dr. Thrasher clarifies that people genetically less able to detoxify fungi and fungal metabolites would not be considered immunosuppressed. He explains, "What we can say is that they are immune disregulated. Sometimes they can show suppression, depending on what arm of the immune system you're looking at, while other portions of the immune system are enhanced. I'll give you an example. We see that some 40 percent of the people exposed to molds and toxic chemicals, in general, but also molds or fungi, have decreased natural killer cell activity, which means that they are immunocompromised. On the other hand, when we look at their T and B cells, they are going to have elevated T and B cells, which means that portion of the immune system is enhanced. Thus, their immune systems are disregulated."[39] For a full, yet simplified, explanation of the immune system, please refer to an educational health book, such as *Alternative Medicine: The Definitive Guide*.[40]

In review, Dr. Straus has explained that exposure to some mycotoxins from *Stachybotrys* causes immunosuppression. Clearly, a suppressed immune system leaves a person more vulnerable to illness, infection, and disease. Therefore, an important question to ask is the following: Do all mycotoxins suppress the immune system?

According to Charles Bacon, PhD, research leader and supervisory microbiologist for the USDA Toxicology and Mycotoxin Research Unit, "They all probably do for the sake that the immune system can manifest in several forms as humoral immunity . . ."[41] Humoral immunity relates to, or is part of, immunity or the immune response that involves antibodies secreted by B cells, which circulate in bodily fluids such as blood.[42]

Jia-Sheng Wang, MD, PhD, a professor in and head of the Department of Environmental Sciences, College of Public Health at the University of Georgia, specifies, "Aflatoxins suppress the immune system. The molds that produce trichothecenes are actually a problem in the U.S., in the Western society. Trichothecenes suppress the immune system, in particular, T-2 toxin and deoxynivalenol. For a long time we have known these trichothecene toxins are all strong immunosuppressors. They can really suppress protein synthesis, which will also suppress the immune system."[43] [See interview with Dr. Wang in Section Three.]

Dr. Ammann concurs, "It is important to note that almost all mycotoxins have an immunosuppressive effect, although the exact target within the immune system may differ. Many are also cytotoxic [toxic to cells] so that they have route-of-entry effects that may be damaging to the gut, the skin, or the lung. Such cytotoxicity may affect the physical defense mechanisms of the respiratory tract, decreasing the ability of the airways to clear particulate contaminants (including bacteria or viruses), or damage alveolar macrophages, thus preventing clearance of contaminants from the deeper lung. [Macrophages contain granules or packets of chemicals and enzymes, which serve the purpose of ingesting and destroying microbes, antigens, and other foreign substances]. The combined result of these activities is to increase the susceptibility of the exposed person to infectious disease and to reduce his defense against other contaminants. They may also increase susceptibility to cancer."[44] Most people are completely unaware of the serious health damage that can occur from exposure to toxigenic fungi and their mycotoxins.

Wet building materials inherently create an indoor environment in which multiple airborne contaminants can be inhaled. However, without extensive testing, it is not possible to determine the actual toxicity level of air and the degree to which health will be affected from exposure to such air. Dr. Ammann explains, "Because indoor samples are usually comprised of a mixture of molds and their spores, it has been suggested that a general test for cytotoxicity be applied to a total indoor sample to assess the potential for hazard as a rough assessment."[45] The level of cell destruction caused by the cytotoxic substances present in the air sample would then be known.

Dr. Straus gives an example of cytotoxicity, ". . . [W]e could take heart cells and grow them in a petri dish. Then we could expose them to the air, say from New

Orleans, and see if that air does any damage to the cells or kills them. That's cytotoxicity testing." Dr. Straus further explains, "Cytotoxicity tells you how toxic the air is against a particular cell line; it doesn't tell what the air sample does to a human being because cells in a culture are very different from cells in human beings. Also, it doesn't tell you what you're measuring. It doesn't tell you if you're measuring mycotoxins, fungal toxins, bacterial toxins, or chemical toxins. It just tells you that there's something in the air that's killing these cells, but that's all it tells you. . . . Cytotoxicity doesn't tell you what the toxin is or even if the stuff would be dangerous to humans."[46]

According to Dr. Straus, if a toxic substance is found to damage or kill tissue-cultured cells, we can't draw the correlation that the toxic component would also do the same damage to tissue cells inside the human body. Dr. Straus points out, ". . . [B]ut you can make the assumption that if there's something toxic in the air to tissue-cultured cells, then it's not going to be good for humans, but you can't make the correlation." He explains, "There's a difference between an assumption and a correlation. Anybody can make an assumption. A correlation is something we have to prove scientifically."[47]

Calculating the toxicity of air is complex because multiple contaminants can exist, especially in a wet-building environment. According to Dr. Thrasher, it is exposure to the entire mix of toxins, not just one or two components, that cause the health effects often experienced by occupants of water-damaged buildings. Dr. Thrasher explains, "I want to make sure that people understand that it's not only mold; it's all the other stuff that's going on with it that we talked about. The indoor air of water-damaged buildings contains a variety of biological matter, such as spores, hyphae, fine particulates, endotoxins, 1-3 beta glucans, and bacterial fragments, among others. In addition, both bacteria and molds produce a variety of VOCs. All of these impinge upon the occupants, creating an environment that is probably additive as well as synergistic in toxic effects. Remember, we are dealing with a variety of toxins that affect all organs of the body, particularly the nervous system, the immune system, and the upper and lower respiratory tract."[48]

Dr. Thrasher points out these fungal fragments and other fungal-related microscopic air contaminants are not measurable with traditional air sampling and testing methodologies. However, to estimate the additional level of toxicity in air, for example, in the post-hurricane New Orleans area, he says to multiply the spore counts found in the NRDC air samples by 300, "[b]ecause that is the effective particulate load of the fine particulate plus the spores."[49]

In regard to this estimated 300 times increased level of toxicity, Dr. Straus agrees that Dr. Thrasher is ". . . probably right. The only problem is we will never know because there's no way to measure things that you don't know are there. For example, we know we can measure bacteria; we can measure bacterial products; we can measure fungi; and we can measure fungal products. I'm sure there are toxins in the air in New Orleans that we don't even know are there, say, from the oil and chemical industries. If you don't know what's there, you don't know how to measure for it."[50]

So what level of spores/m³ can people be exposed without their health being negatively impacted? Dr. Thrasher states, "There is absolutely no data to tell us what concentration is safe or not safe. The thing that we have to consider is colonies of fungi. (I prefer to call them fungi rather than molds.) Colonies of fungi shed different types of particulate matter into the air. Let me explain what I mean by that: They shed spores, and the spores range from 2 microns in diameter up to 7–9 microns in diameter. They also shed fragments of their cell bodies that we call hyphae, so we've got hyphae fragments in the air as well."[51]

Dr. Thrasher continues, ". . . [C]olonies also shed what we call very fine particulate matter. . . . This particulate matter is smaller than the size of the spores; it goes from about less than 2 microns in diameter down to 0.2 microns in diameter. *That fraction is some 300–400 times more concentrated than the spore count.* That fraction also contains all of the toxic chemicals that are produced by the molds and the spores. We are looking at an environment that is not only mold spores but also mold spores plus the fine particulate matter. Furthermore, the fine particulate matter is shed by frequencies of 1– 20 hertz, which are the frequencies associated with normal human activity, e.g., television, radio, conversation, walking, dancing, etc. Therefore, the shedding of this material is independent of mold spore release."[52]

"Thus, the indoor air is a soup of antigenic and chemical materials. When inhaled, the fine particulates get deep into the alveoli and release their contents (chemicals, mycotoxins, etc.) by simple diffusion into the blood stream," states Dr. Thrasher.[53]

In regard to the health ramifications of inhaling this airborne antigenic and chemical mix, Dr. Thrasher explains, "Let's take the fine particulate matter—we know that the spores can get into the nasal cavity and some down into the lungs. There's no question about that because they're small enough that they can get down into the lungs. They are from 2–7 microns in diameter. What do you think is happening with something much smaller than that, say down to 0.1–0.2 microns in diameter? It is getting deep into the airways and, therefore, is being taken up by the macrophages and absorbed into the blood and everything."[54]

According to Dr. Thrasher, Rafal Górny, MD, PhD, from the Department of Biohazards, Institute of Occupational Medicine and Environmental Health in Sosnowiec, Poland, was the first to study and characterize how fungal colonies shed fine particulate matter.[55] Dr. Górny reports, "In the human respiratory tract, the penetration depth and behavior of bioaerosol particles depends on their size, shape, density, chemical composition, and reactivity."[56]

Dr. Górny expounds, "The presence of fragments in the air is well documented with pollen exposure, when seasonal asthma attacks begin several weeks before the exact period of pollen grain dissemination is detected in the air. In contrast, the role of fragments in fungal exposures has not been sufficiently recognized. The reason for this may be that fine and ultrafine fragment propagules cannot be detected with traditional bioaerosol sampling."[57]

Another study conducted by Madelin and Madelin showed, ". . . [T]he mycelium fragments are often aerosolized from microbiologically contaminated surfaces, and some of these pieces preserve their viability and are capable of starting a new seat of growth. It is also possible that the fragments are pieces of spores and fruiting bodies or are formed through nucleation from secondary metabolites of fungi, such as volatile organic compounds (VOCs)."[58]

Dr. Górny concluded from his studies, "Fungal fragments, so far not measured in indoor air environments, are aerosolized in high numbers and may thus contribute to adverse health effects. They can, at least partially, be responsible for the symptoms observed among inhabitants with a mould problem and/or water damage in buildings."[59] He recommends future exposure assessment studies contain, as their immanent part, the measurements of the load of fungal and actinomycetal [gram-positive bacteria of the genus *Actinomyces*] fragments in the indoor air.[60]

Dr. Straus and colleagues, who have studied and expanded Dr. Górny's research, further report, "Recent research has shown that, along with airborne conidia, highly respirable fungal fragments can lead to human exposure because the fragments can be aerosolized simultaneously with conidia. The amounts of these fragments could be as large as 320 times the amounts of conidia."[61] In other words, the air can contain as many as 320 times more fungal fragments than spores, which substantially adds to the potential health risks. Remember, as Dr. Thrasher pointed out earlier, these airborne fungal fragments contain all the same toxic chemicals as do the molds and spores.[62]

The research conducted by Dr. Straus's team at Texas Tech University Health Sciences Center found data that show, ". . . *S. chartarum* trichothecene mycotoxins can become airborne in association with intact conidia or smaller particles." Dr. Straus and colleagues point out, "To date, however, there are no data describing what airborne concentrations of these toxins are necessary to adversely affect human health. Most indoor air quality investigations focus on surface growth and airborne conidium concentrations. As Górny et al. concluded, spore counts do not adequately represent the amount of fungal fragments present in the air at any given time. In fact, fragments and particles that are the same size greatly outnumber intact fungal spores. Based on our results, we feel that air sampling in *S. chartarum*-contaminated indoor environments should include a means of collecting particulates smaller than conidia, followed by a specific and sensitive test for mycotoxins. . . . Finally, it should be noted that background normal levels of such toxins . . . have yet to be determined."[63]

So not only are spores and fungal fragments in the air in mold-infested environments, we have airborne mycotoxins. The mycotoxins become airborne when the spores and fungal fragments on which they exist become airborne. Thus, not only are spores and fungal fragments inhaled but also the mycotoxins.

To demonstrate direct human exposure to these mycotoxins in the air, Dr. Straus and group investigated the presence of trichothecene mycotoxins in sera from individuals exposed to *Stachybotrys chartarum*. In these research findings published in June of 2004, Dr. Straus and colleagues report, "In this study, we were successful in demonstrating the presence of trichothecene mycotoxins in serum samples from individuals exposed to mold (primarily *Stachybotrys*) in water-damaged indoor environments. Our findings indicate that these highly toxic compounds can actually be found in people exposed to these environments and, therefore, have the potential to negatively affect the health of such individuals. This relationship is further strengthened by our previous investigation, which demonstrated that *Stachybotrys* trichothecene mycotoxins can become airborne in a controlled situation and have the potential to do so inside a building."[64]

A later study published in August of 2006 by Dorr Dearborn, MD, PhD, and colleagues at Case Western Reserve University investigated the formation of adducts of satratoxin G (SG) with serum albumin in vitro (in an artificial environment outside the living organism, for example, in a test tube or culture plate) and searched for similar adducts formed in vivo (in a living body) using human and animal serum.[65] Satratoxin is one of the macrocyclic trichothecenes produced by *Stachybotrys chartarum*. Macrocyclic trichothecenes are the most potent members of a large family of trichothecenes.[66] (Macrocyclic means the compound contains or is a chemical ring that consists usually of 15 or more atoms.) Some of the nonmacrocyclic trichothecenes are T-2 toxin, verrucarol, nivalenol, and deoxynivalenol, which many studies have shown are metabolized rapidly in animal models.[67]

Data from the Dearborn group study confirmed that toxic compounds from *Stachybotrys chartarum* can be found in humans and animals exposed to *Stachybotrys*. The study concluded, "These data document the occurrence of SG-albumin adducts in both in vitro experiments and in vivo human and animal exposures to *S. chartarum*."[68]

The Straus and Dearborn studies prove not only can people inhale *Stachybotrys* spores and fragments—both on which mycotoxins can be found—but also that the toxic compounds from such organisms and particulate get into human and animal body fluids. Based on these scientific findings, we can reasonably assume that spores, fragments, and mycotoxins from other genuses also become airborne, get inhaled by humans and animals, and get into body fluids. (Remember, *Stachybotrys* spores are larger, heavier, wetter, and are not as easily aerosolized as smaller, lighter, and less dense spores from other genuses, such as *Aspergillus*.)

Straus and colleagues expanded their research to examine whether mycotoxins are present in the body tissue of people exposed to mycotoxin-producing fungi in their surroundings. Their findings were published in 2009 and stated, "Mycotoxins, specifically trichothecenes, aflatoxins, and ochratoxins, can be detected in human tissue and body fluids in patients who have been exposed to toxin-producing molds in their environment."[69] Not only were the three mycotoxins found in the body fluids of people exposed to indoor toxigenic molds, but they were also found in the peoples' body tissues. Human tissue specimens tested included the brain, liver, lung, bladder, ovary, muscle, and skin.[70]

In a later study Dr. Struas and colleagues demonstrate the detrimental effects of mycotoxins. The study reports, "Damage to the neurological system can result from exposure to trichothecene mycotoxins in the indoor environment. This study demonstrates that neurological system cell damage can occur from satratoxin H exposure to neurological cells at exposure levels that can be found in water-damaged buildings contaminated with fungal growth."[71]

The findings of another study in which both Dr. Straus and Dr. Thrasher participated suggest a link between mycotoxin exposure and Chronic Fatigue Syndrome (CFS), "Mycotoxins can be detected in the urine in a very high percentage of patients with CFS. This is in contrast to a prior study of a healthy, non-water damaged building (WDB) exposed control population in which no mycotoxins were found at the levels of detection. The majority of the CFS patients had prior exopsure to (WDB). Environmental testing in a subset of these patients confirmed mold and mycotoxins exposure."[72]

Dr. Straus poses the last question to be answered in this highly complex scientific puzzle: "Do the mycotoxins that get into human beings get there in concentrations sufficient to cause human disease?"[73]

Deductive reasoning will lead us to an affirmative answer. Consider the following findings of current scientific research: 1) Mycotoxin exposure negatively affects multiple systems within the bodies of humans and animals. Case in point, the ergot alkaloids negatively affect the cardiovascular, nervous, reproductive, and immune systems. 2) Mycotoxin exposure causes immunosuppression. Case in point, the mycotoxins cyclosporin A, aflatoxin, and trichothecenes are all proven immune suppressors. 3) Reduced function in multiple body systems, including the immune system, can lead to the formation of diseases.

The CDC confirms mycotoxins can be contributing factors to human disease, "Fungi can cause a variety of infectious and noninfectious conditions. Several basic mechanisms can underlie these conditions, including immunologic (e.g., IgE-mediated allergic), infectious, and *toxic*. Several of these mechanisms contribute to pathogenesis of a fungal-induced disease."[74] The "toxic" component of mold exposure to which the CDC refers is exposure to fungal mycotoxins.

(ENDNOTES)

1   Department of Environmental Health and Safety, MSUM, "About Mold."
2   The CDC Mold Work Group, "Mold Prevention Strategies," p. 17.
3   See Section Three, interviews with drs. Thrasher, Straus, and Solomon, April 2006, April 2006, and July 2006 respectively.
4   Eckardt Johanning et al., "Clinical Experience and Results of a Sentinel Health Investigation Related to Indoor Fungal Exposure." *Environmental Health Perspectives Supplements*, Volume 107, Number S3, June 1999. Available at http://www.ehponline.org/docs/1999/suppl-3/489- 494johanning/abstract.html (accessed February 2006).
5   The CDC Mold Work Group, "Mold Prevention Strategies," p. 18.
6   Ibid., p. 36-37. (Italics added.)
7   Ibid., p. 37.
8   See Section Three, interview with Jim Pearson, April 2006.
9   The New York City Department of Health and Mental Hygiene, "Guidelines on Assessment and Remediation of Fungi in Indoor Environments."
10   See Section Three, interview with Dr. David Straus, April 2006.
11   Ibid.
12   Ibid.
13   See Section Three, interview with Dr. David Straus, April 2006.
14   Solomon et al., "Airborne Mold and Endotoxin Concentrations in New Orleans, Louisiana after Flooding, October-November 2005."
15   The CDC Mold Work Group, "Mold Prevention Strategies," p. 17 and 37.
16   Solomon et al., "Airborne Mold and Endotoxin Concentrations in New Orleans, Louisiana after Flooding, October-November 2005."
17   Ibid.
18   Interview with Dr. Michael Gray, March 2007.
19   See Section Three, interview with Dr. David Straus, April 2006.
20   See Section Three, interview with Dr. Andrew Puccetti, April 2006.
21   Ibid.
22   Ginger L. Chew et al., "Mold and Endotoxin Levels in the Aftermath of Hurricane Katrina: A Pilot Project of Homes in New Orleans Undergoing Renovation," *Environmental Health Perspectives*. Available at http://www.ehponline.org/members/2006/9258/9258.pdf (accessed August 2006).
23   Daniel G. Panaccione et al., "Abundant Respirable Ergot Alkaloids from the Common Airborne Fungus *Aspergillus fumigatus*." *Appl Environ Microbiol*. 2005 June; 71(6): 3106-3111. Available at http://www.pubmedcentral.nih.gov/articlerender/fcgi?artid=1151833 (accessed August 2006). (Italics added on the word *immunocompetent*.)
24   Ibid.
25   Ibid.
26   See Section Three, interview with Dr. David Straus, April 2006.
27   Ibid.
28   Ibid.
29   The CDC Mold Work Group, "Mold Prevention Strategies," p. 25.
30   See Section Three, interview with Dr. David Straus, April 2006.
31   Ammann, "Is Indoor Mold Contamination a Threat to Health?"
32   E-mail from Dr. Ginger Chew, July 24, 2006.
33   The CDC Mold Work Group, "Mold Prevention Strategies," p. 1.
34   See Section Three, interview with Dr. David Straus, April 2006.
35   Ibid.
36   Ibid.
37   Burton Goldberg, *Alternative Medicine*, p. 21, (California, Celestial Arts, 2002).
38   See Section Three, interview with Dr. Jack Thrasher, April 2006.
39   Ibid.
40   Burton Goldberg, *Alternative Medicine*, (California, Celestial Arts, 2002).
41   Interview with Dr. Charles Bacon, August 2006.
42   Merck, "B Cells and Humoral Immunity," The Merck Manual (copyright 1995-2006 Merck & Co., Inc., Whitehouse Station, NJ, USA). Available at http://www.merck.com/mrkshared/mmanual/

section12/chapter146/146c.jsp (accessed August 2006).
43   See Section Three, interview with Dr. Jia-Sheng Wang, July 2006.
44   Ammann, "Is Indoor Mold Contamination a Threat to Health?"
45   Ibid.
46   See Section Three, interview with Dr. David Straus, April 2006.
47   Ibid.
48   See Section Three, interview with Dr. Jack Thrasher, April 2006.
49   Ibid.
50   See Section Three, interview with Dr. David Straus, April 2006.
51   See Section Three, interview with Dr. Jack Thrasher, April 2006.
52   Ibid. (Italics added.)
53   E-mail from Dr. Jack Thrasher, September 2006.
54   Ibid.
55   See Section Three, interview with Dr. Jack Thrasher, April 2006.
56   Górny, "Filamentous Microorganisms and Their Fragments in Indoor Air-A Review."
57   Ibid.
58   Ibid.
59   Ibid. ("Mould" is the British spelling of the word *mold*.)
60   Ibid.
61   T.L. Brasel et al., "Detection of Airborne *Stachybotrys chartarum* Macrocyclic Trichothecene Mycotoxins on Particulates Smaller than Conidia," *Applied and Environmental Microbiology*, January 2005, 114-122. Available at http://aem.asm.org/cgi/content/abstract/71/1/114 (accessed March 2006).
62   See Section Three, interview with Dr. Jack Thrasher, April 2006.
63   Brasel et al., "Detection of Airborne *Stachybotrys chartarum* Macrocyclic Trichothecene Mycotoxins on Particulates Smaller than Conidia."
64   Brasel TL et al., "Detection of Trichothecene Mycotoxins in Sera From Individuals Exposed to *Stachybotrys chartarum* in Indoor Environments," *Archives of Environmental Health*, June 2004, vol. 59, no. 6, 317-23. Available at http://www.medscape.com/medline/abstract/16238166?prt=true (accessed March 2006).
65   Iwona Yike et al., "Mycotoxin Adducts on Human Serum Albumin: Biomarkers of Exposure to Stachybotrys chartarum," *Environmental Health Perspectives*, vol. 114, no 8, August 2006. Available at http://www.ehponline.org/members/2006/9064/9064.pdf (accessed August 2006).
66   Ibid.
67   Brasel TL et al., "Detection of Trichothecene Mycotoxins in Sera From Individuals Exposed to *Stachybotrys chartarum* in Indoor Environments."
68   Yike et al., "Mycotoxin Adducts on Human Serum Albumin: Biomarkers of Exposure to *Stachybotrys chartarum*."
69   Dennis G. Hooper et al., "Mycotoxin Detection in Human Samples from Patients Exposed to Environmental Molds," *Int J Mol Sci.* 2009 April; 10(4): 1465–1475. Available at http://www.pubmedcentral.nih.gov/articlerender.fcgi?artid=2680627.
70   Ibid.
71   Enusha Karunasena et al., "Building-Associated Neurological Damage Modeled in Human Cells: A Mechanism of Neurotoxic Effects by Exposure to Mycotoxins in the Indoor Environment," Mycopathologia (2010) 170:377–390 Available at: http://www.ncbi.nlm.nih.gov/pubmed/20549560 (accessed May 2013).
72   Joseph H. Brewer et al., "Detection of Mycotoxins in Patients with Chronic Fatigue Syndrome" Toxins 2013, 5, 605-617. Avalable at: http://www.mdpi.com/2072-6651/5/4/605 (accessed May 2013).
73   See Section Three, interview with Dr. David Straus, April 2006.
74   The CDC Mold Work Group, "Mold Prevention Strategies."

CHAPTER SIX:

MOLD *CAN* HURT YOU

We live in a day and age where health consequences are exploding from allergic and environmental causes. In order to regain and/or maintain our health, we nearly have to become medical cause-and-effect detectives to provide our doctors with information that may uncover the underlying cause(s) of our symptoms. For example, what environmental factors, foods, and/or products cause a reactive response?

In addition, we have to search for people battling similar symptoms and illnesses from allergens and environmental contaminants and look for parallel exposures. This type of comparative research may yield information that could prove valuable in discussions with our medical professional(s). Remember, we shouldn't hesitate to share information with our doctors(s) as we never know which "clue" might uncover the culprit(s) eroding our health. Even specialists in allergy and environmental medicine, such as Doris Rapp, MD, acknowledge the field is not a finite science and is an ever-changing area of study. In order for medical professionals to keep abreast of new understandings uncovered by scientists in the field of biology, doctors must persistently and consistently pursue applicable continuing education and medical paradigms of the medical community must change and keep pace with new scientific discoveries.[1] In other words, doctors must keep an open mind and be willing to alter existing beliefs as progress in science outdates priorly accepted medical paradigms and the research from which they were formed. If a doctor does not actively practice this philosophy, we need to look for another who does—our health is worth it.

At this point, everyone should realize that, yes, mold *can* make you sick. However, the question remains: In what manner does mold and mycotoxin exposure manifest into illness and disease? Some mold-related illnesses and diseases are widely accepted and understood by professionals in the fields of science and medicine. Others are hotly debated, refuted, and dismissed. We share much of the following information so you, the reader, can understand the "climate" in which medical doctors are trained in regard to the health effects of mold and mycotoxin exposure. Some doctors are more open-minded and educated about this subject; others are not, stuck in a time warp of fallacies void of up-to-date scientific proof. In fact, it was the words of one doctor who adamantly stated, "Mold cannot hurt you" that drove us to research in the first place. Now, should you be told "Mold cannot hurt you," you will know that the person making this erroneous claim is uneducated and ignorant of the facts.

Mold and chemical exposures are not just a hurricane-size problem, although hurricanes, floods, and other natural disasters certainly do create environments conducive to mold growth and chemical disruption. Everyday exposures of mold and chemicals can also occur from exposure to sick buildings, industrial pollutants, and contaminated food. Ill health effects can occur no matter the circumstance in which we are exposed. We should each have at least a basic understanding of some of the signs and symptoms of toxic mold and chemical exposures, as mold and toxic exposures can silently invade anyone's life, anytime. We want to avoid the inception of what we call *The War Within*: The human host battling against a living, biological enemy—internal fungi. For this reason, let's review some facts.

A recent study at Michigan State University acknowledges the existence of fungal-related illnesses from hurricanes Katrina and Rita. It states, "Numerous adverse human health effects have been attributed to damp indoor air environments generated by aberrant water exposure due to excessive condensation and failure of water-use devices, as well as building envelope breach during heavy rains or flooding, as occurred during hurricanes Katrina and Rita on the Gulf Coast of the United States."[2]

The fact people can get sick from exposure to damp/wet environments is not debated. However, the extent to which mold can make people sick is, as mentioned earlier, a point of controversy. In fact, according to the CDC, "In recent years, the health effects of exposure to mold in built environments have been a subject of intense public concern. . . . [T]he issue of how damp indoor spaces and mold contamination affect human health has been highly controversial. In response, CDC commissioned the IOM (Institute of Medicine) to perform a comprehensive review of the scientific literature in this area. The resulting report, 'Damp Indoor Spaces and Health,' was published in 2004."[3] A later scientific study notes that the IOM ". . . expert panel concluded that an association exists between damp buildings and upper respiratory tract symptoms, wheeze, cough, and exacerbation of chronic lung diseases such as asthma . . ."[4] Not surprising, these symptoms are some of the same experienced by people who had the Katrina Cough.

The CDC documents the IOM findings, "There is sufficient evidence linking upper respiratory tract symptoms (such as nasal congestion, sneezing, runny or itchy nose, and throat irritation) to damp indoor environments and mold (with exposure to mold often determined by self-report). Similarly, there is sufficient evidence for a link with the lower respiratory tract symptoms of cough and wheeze. Sufficient evidence was also found for a link between damp indoor environments, mold, and asthma symptoms in sensitized people with asthma. There is also sufficient evidence for an association between mold exposure and hypersensitivity pneumonitis in a small proportion of susceptible people, invasive respiratory and other fungal infections in severely immunocompromised people, and fungal colonization of the respiratory tract or infection in individuals with chronic pulmonary disorders."[5]

Dr. Ammann states, "The most common response to mold exposure may be allergy. People who are atopic, that is, who are genetically capable of producing an allergic response, may develop symptoms of allergy when their respiratory system or skin is exposed to mold or mold products to which they have become sensitized. Sensitization can occur in atopic individuals with sufficient exposure."[6] According to the EPA, allergic reactions can range from ". . . rhinitis, nasal congestion, conjunctival inflammation [inflammation of the mucous membranes that line the inner surface of the eyelids and continues over the forepart of the eyeball], and urticaria [hives] to asthma."[7]

The *IICRC S520 Standard and Reference Guide for Professional Mold Remediation,* a reference guide published for people in the field of industrial hygiene that is coauthored by Mr. Pearson and referenced by OSHA, states, "Allergic respiratory disease can occur when indoor microbial contaminants, primarily fungal and bacterial spores and growth fragments, are deposited in the nasal or sinus cavities and upper and/or lower airways. Symptoms may develop rapidly or may be delayed, depending upon the extent of the exposure and an individual's sensitivity to the agent(s). Allergic reactions occur only in selected genetically susceptible individuals (as opposed to the entire population) and require prior exposure for sensitization. Once sensitized, occupants' symptoms may be initiated by very low exposures."[8] Our family has also experienced this type of heightened fungal sensitivity and resultant allergic reactions since our initial fungal exposure post-hurricane.

According to Dr. Ammann, "Allergic reactions can range from mild, transitory responses, to severe, chronic illnesses. The IOM estimates that one in five Americans suffers from allergic rhinitis, the single most common chronic disease experienced by humans. Additionally, about 14 percent of the population suffers from allergy-related sinusitis, while 10 to 12 percent of Americans have allergically related asthma. About 9 percent experience allergic dermatitis. A very much smaller number, less than one percent, suffer serious chronic allergic diseases such as allergic bronchopulmonary aspergillosis (ABPA) and hypersensitivity pneumonitis. Allergic fungal sinusitis is . . . [not an] uncommon illness among atopic individuals residing or working in moldy environments. There is some question whether this illness is sole[ly] allergic or has an infectious component. Molds are just one of several sources of allergens, including house dust mites, cockroaches, effluvia from domestic pets (birds, rodents, dogs, cats), and microorganisms (including molds)."[9] Medical studies confirm, "At least 70 allergens have been well characterized from spores, vegetative parts, and small particles from fungi (0.3 mm and smaller)."[10]

According to the CDC, "Fungi can cause a number of infectious and noninfectious conditions."[11] The ill health effects from mold exposure are generally addressed in three categories: allergy and irritation, toxicity, and infection. The CDC reports, "It is common for several of these mechanisms to contribute to pathogenesis of fungal-induced disease. The types and severity of symptoms and diseases related to mold exposure depend in part on the extent of the mold present, the extent of the individual's exposure, and the susceptibility of the individual (e.g., people who have allergic conditions or who are immunosuppressed are more susceptible than those without such conditions)."[12] In other words, allergic coughs, wheezes, and respiratory difficulties—a.k.a. the Katrina Cough—can be forerunner symptoms of more serious fungal-induced diseases.

Fungal colonization can occur as allergic fungal sinusitis and allergic bronchopulmonary aspergillosis. The latter ". . . occurs when the airways of individuals with obstructive pulmonary diseases such as asthma, cystic fibrosis, or chronic obstructive pulmonary disease (COPD) become colonized with *Aspergillus fumigatus* or other *Aspergillus* species. . . . Airways colonization with other fungal species can result in a similar clinical picture," reports the CDC.[13]

According to the EPA, "Another class of hypersensitivity disease is hypersensitivity pneumonitis . . ."[14] It is a severe debilitating lung disease caused from inhalation of airborne antigens.[15] The CDC states, "A wide range of materials, including fungi, can be inhaled and thus sensitize susceptible people by inducing both antibody and cell-mediated immune responses."[16] These processes refers to the two major branches of the adaptive immune responses: humoral immunity and cell-mediated immunity.[17]

According to the Department of Biology at the University of New Mexico (UNM), "Humoral immunity involves the production of [an] antibody in response to an antigen and is mediated by B lymphocytes."[18] Health literature states, "Antibodies are protein molecules set in motion by the immune system against a specific antigen. Also referred to as immunoglobulins, antibodies occur in the blood, lymph, colostrum, saliva, and gastrointestinal and urinary tracts, usually within three days of the first encounter with an antigen. The antibody binds tightly with the antigen as a preliminary for removing it from the body or destroying it."[19]

On the other hand, cell-mediated immunity refers to any immune response against organisms or tumors in which antibodies play a subordinate role, according to the Department of Biology at UNM.[20] Other health literature states cell-mediated immunity involves T cells, macrophages, and other leukocytes.[21]

What is most alarming about hypersensitivity pneumonitis is that published reports state it can be caused from just ordinary residential exposures.[22] In fact, "Hypersensitivity pneumonitis (extrinsic allergic alveolitis) has now been associated with over 50 inhaled environment[al] substances. Most are biological materials such as fungi, bacteria, and animal proteins, while a few industrial chemicals have been found to cause this immunologic lung disease."[23]

Dr. Gray reports, "Hypersensitivity pneumonitis is an inflammatory condition which involves inflammation in the smallest of the airways in the lungs, triggered by exposure to commonly encountered insoluble volatile organic compounds of a chemical nature, as well as several types of biological dusts, pollens, mold spores, and mycotoxins 'packaged' within and on the surface of spores and cell wall fragments. The ensuing inflammation results in small airways spasms and obstruction occurring in regular and repetitive episodes. This condition causes shortness of breath and often severe debilitating chest pain and is only preventable by avoidance of exposure to the triggering agents, such as mold spores and other antigenically active debris."[24] According to the EPA, "Outbreaks of hypersensitivity pneumonitis in office buildings have been traced to air conditioning and humidification systems contaminated with bacteria and molds."[25] Diagnoses can be difficult as medical research indicates, "Hypersensitivity pneumonitis may initially be clinically mistaken for acute pneumonia, asthma . . . or other forms of interstitial lung disease."[26]

Two types of inhalation fever (humidifier fever and organic dust toxic syndrome) can occur from inhalation of airborne materials, ". . . including fungi, bacteria, and microbial constituents such as endotoxins," according to the CDC.[27] The CDC reports it can cause fever and flu-like symptoms.[28]

One doctor cites clinical findings, "Home dampness with resulting mold growth may be associated with several medical conditions (one or sometimes all) including immediate hypersensitivity reaction, hypersensitivity pneumonia, or what has been described as 'humidifier fever.' Clinically, I see these patients with recent onset asthma, recent onset sinusitis, and/or recent onset skin rashes. Several studies have shown a clear correlation and association between the occurrence of molds in the inside air environment, dampness in the indoor environment, and the symptomology of the skin and respiratory tract, especially in children."[29]

Scientific, medical, and governmental literature reflects agreement among professionals that these aforementioned negative health conditions can result from fungal exposure. Also agreed on is the risk of infection to people who are immunocompromised. However, regarding the more serious health ramifications possible from mold exposure, a variety of positions are documented in scientific, medical, and governmental reports. As we review some highlights of the governmental reports, you will see that some of the data are already outdated by nongovernmental scientific research.

A 1994 EPA report addresses the possibility of disease formation stemming from mycotoxin exposure, "Another class of agents that may cause disease related to indoor airborne exposure is the mycotoxins. These agents are fungal metabolites that have toxic effects ranging from short-term irritation to immunosuppression and cancer. Virtually all the information related to diseases caused by mycotoxins concerns ingestion of contaminated food."[30] The latter statement does not reflect the latest studies by Dr. Straus and group. Their findings confirm mycotoxins present in the air on respirable-sized particles and mycotoxins present in human blood serum.

Therefore, the ever-changing innate nature of science requires us to pay close attention to not only the publication dates of reported findings but also to the dates during which the studies actually took place.

The CDC quotes the outdated findings of the IOM's 2004 report, "Almost all of the known effects of mycotoxin exposures are attributable to ingestion of contaminated food. Health effects from inhalational exposures to toxins are not well documented. The IOM found inadequate or insufficient evidence for a link between exposure to damp indoor environments and molds with a variety of conditions that have been attributed to toxicity [from mycotoxins]."[31]

The CDC's report regarding mold prevention strategies published shortly after Katrina and Rita listed the following as conditions that the IOM documents as having inadequate and/or insufficient evidence of being caused by fungal exposure: mucous membrane irritation syndrome, airflow obstruction in otherwise healthy persons, chronic obstructive pulmonary disease, respiratory illness in otherwise healthy adults, shortness of breath (dyspnea), asthma, and acute idiopathic pulmonary hemorrhage in infants. Conditions for which limited or suggestive evidence was determined were asthma development, shortness of breath, and respiratory illness in otherwise healthy children. Interestingly, some conditions were listed in both categories.[32]

The 2005 CDC report points out that the IOM's qualification of determination was based on a review of available scientific literature at the time of the IOM panel's review. The CDC states, "It should be noted that 'inadequate or insufficient evidence to determine whether an association exists' does not rule out the possibility of an association. Rather, it means that no studies examined the relationship or that published study results were of insufficient quality, consistency, or statistical power to permit a conclusion about an association."[33] In other words, the CDC acknowledges that existing published studies or future studies may disprove some of the conclusions drawn by the IOM panel. Thus, the possibility remains that some of these health effects deemed to have "inadequate or insufficient evidence" can indeed be caused by exposure to fungi and fungal metabolites—mycotoxins.

Further scientific studies soon challenged the IOM's findings in regard to mycotoxin inhalation. A 2006 study notes that its findings refute the following IOM conclusion, ". . . [S]upportive data for other reported outcomes such as neurocognitive dysfunction, mucous membrane irritation, fatigue, fever, and immune disorders are lacking." The published findings shed new scientific light on the health effects of mycotoxin inhalation. The scientists state, "These findings suggest that neurotoxicity and inflammation within the nose and brain are potential adverse health effects of exposure to satratoxins and *Stachybotrys* in the indoor air of water-damaged buildings."[34]

It should also be noted the CDC's report inaccurately states, "Currently no commercial clinical diagnostic tools are available to determine whether an individual's health effect is related to exposure to mycotoxins."[35] This report, just as the IOM report, does not take into consideration the technology used by Dr. Straus and colleagues. They used new testing methods for mycotoxin detection in human blood sera. The findings were published in June 2004.[36] Furthermore, commercial clinical diagnostic tools are available at limited laboratories with patented processes that can detect the presence of mycotoxins in blood and/or urine.

The CDC heavily cites the IOM data and bases many of its conclusions on the outdated IOM report. This overreliance on the IOM's data could explain the CDC's oversight. As for the IOM report, Dr. Thrasher points out possible reasons why it is inaccurate, "If you look at that report very closely, the IOM stopped reviewing the literature as of December 2003, so its report doesn't include the last two to three years of scientific and medical literature that's been coming out. By the time it was published, the information was already outdated."[37]

The CDC does acknowledge in its post-hurricane report the medical risks of mold exposure to both the immunocompromised as well as the immunocompetent. It states, "Invasive fungal infections can . . . occur in individuals with *normal host defenses* and, in certain situations, can even be *life threatening*." The CDC states that individuals with impaired host defenses are at greatest risk for developing invasive fungal infections from heavy fungal contamination after hurricanes Katrina and Rita. It reports, "*Severely immunosuppressed individuals*, such as solid-organ or stem-cell transplant recipients, or those receiving cancer chemotherapy agents, *corticosteroids*, or other agents inhibiting immune function are at much higher risk for these infections: locally invasive infections of the lungs, sinuses, or skin; and systemic infections. . . . These serious infections are *often fatal*, even with aggressive antifungal therapy."[38]

We, as authors, hope you, the reader, process the full impact of the CDC's last few statements. We, ourselves, had to read these statements several times. To read that a person taking corticosteroids is included on the same "immunosuppressed" list as persons having received an organ transplant jolted us into a new level of reality. Doctors further reveal that an immunosuppressed state can render someone immunocompromised.[39]

We finally had the explanation as to why Kurt had gotten near fatally ill. For years he had been prescribed steroids for allergy management, which is a common practice because the major function of steroids or cortisone drugs is to ". . . diminish inflammation by decreasing swelling and redness . . ."[40] Sadly, we had always been assured by doctors that use of a nasal steroid inhaler and routine, intermittent use of oral steroids for allergy/asthma management was safe. However, this assertion is far from the truth. In fact, if doctors told patients that steroids would *inhibit their immune function* to a level that gets them included in the same high-risk category as someone who has undergone an organ or cell transplant or someone who is receiving chemotherapy for cancer, who would take the steroids?

Kurt certainly wouldn't have taken years of steroids had he had full disclosure of the associated health risks. We also want to point out that these repeated assurances and lack of disclosures regarding the safety of steroid use came from many, many doctors—not just one or two—over a period of several years. How many other doctors fail to disclose to their patients this potentially life-threatening "side effect" of steroids? How many people lose their lives each year simply because their doctor(s) neglected to fully inform them of this potentially fatal side effect? Furthermore, how many people in hurricane-hit areas, complete with high mold spore counts, were prescribed steroids for allergy management? When the potential price to pay for taking a prescription exceeds the potential benefit (which we believe is the case with steroids), we must look for less life-threatening alternative solutions. Just think about it for a moment. Steroid use can reduce your immune function to the point that your life can be at risk from fungal exposure, which is a frightful thought since mold is everywhere!

The risks associated with allergy management are especially important to understand in the aftermath of any natural disaster when storm-related contaminants can increase incidences of allergies, asthma, and other respiratory conditions. In fact, post-Katrina moldy structures are essentially ticking time bombs with the potential to ignite a range of health problems in anyone who enters without proper protective equipment. Because of health concerns over potential exposure levels and safe remediation practices, the Louisiana Department of Health and Hospitals (LDHH) enlisted the CDC to assist in assessing the existing health risks. The joint survey efforts on October 6, 2005, revealed, ". . . 46% of inspected homes had visible mold growth and . . . residents and remediation workers did not consistently use appropriate respiratory protection. . . . Despite their awareness of health effects associated with mold, one third of a convenience sample of residents could not identify an appropriate respirator, and the majority of those participating in mold-remediation activities reported doing so without consistently using respiratory protection."[41]

In fact, the LDHH and the CDC state that even when remediation workers, homeowners, and volunteers are aware of the potential health risks, they don't always make the right choice to wear personal protective equipment. The following are the results from the CDC and the LDHH post-Katrina survey: "Of 159 residents interviewed, 82 (51.6%) were male; the overall mean age was 51 years (range: 18–81 years). Nearly all (96.2%) residents responded affirmatively to the question, 'Do you think mold can make people sick?' One hundred eight (67%) correctly identified particulate-filter respirators as appropriate respiratory protection for cleaning of mold. Sixty-seven (42.1%) had cleaned up mold; of these, 46 (68.7%) did not always use appropriate respirators. Reasons for not using respirators included discomfort (10 [21.7%] respondents) and lack of availability (10 [21.7%]

respondents). For public communications about potential risks from exposure to mold and the use of personal protective equipment, 139 (87.4%) respondents recommended the use of television or radio."[42]

In addition to the health risks to residents and volunteers, the LDHH and the CDC acknowledge that remediation workers are a group with high potential for exposure. Additional survey findings reflect the following: "Seventy-six persons who self-identified as remediation workers were interviewed. Of these, 14 (18.4%) were self-employed, and 62 (81.6%) worked for a company doing remediation. Of the 76 workers, 70 (92.1%) were male; the mean age of respondents was 33 years (range: 18–57 years); 40 (52.6%) spoke only Spanish. Seventy-two (94.7%) thought mold causes illness. Sixty-five (85.5%) correctly identified particulate-filter respirators as appropriate protection for cleaning of mold. Sixty-nine (90.7%) had already participated in mold remediation activities at the time of the interview. Of these, 34 (49.3%) had not been fit-tested for respirator use and 24 (34.8%) did not always use appropriate respirators; 13 (54.2%) cited discomfort as the reason for not using respirators. For worker communications about potential risks from exposure to mold and the use of personal protective equipment, 36 (47.4%) recommended use of television or radio and 17 (22.4%) recommended communication through employers."[43]

The fact that 96.2 percent of the residents and 94.7 percent of the self-identified remediation workers interviewed responded that mold *can* make people sick, yet 21 percent of the residents and 54.2 percent of the remediation workers still cited *discomfort* as the reason for not wearing a respirator, clearly indicates the need for a public awareness campaign. By educating the public about the health risks related to mold exposure and emphasizing that these health risks can be lowered by using adequate personal protective equipment, mold-related illnesses would be reduced and lives could be saved. People need to understand that the discomfort experienced from wearing personal protective equipment pales in comparison with the level experienced from mold-related illnesses.

The CDC and the LDHH acknowledge the need for increased awareness regarding the use of personal protective equipment. The joint project reports, ". . . [M]easures to increase awareness of appropriate respiratory protection among the public are warranted."[44] Furthermore, public awareness programs should include warning notices regarding the limited effectiveness of respirators.

Recent published research indicates that even if people wear N95 respirators or, better yet, the elastomeric respirators that provide an increased level of protection over the N95, they may still be subject to the ill health effects of fungal inhalation. A pilot project evaluating the renovation of homes in New Orleans and the sufficient effectiveness of two types of respirators determined that the protection offered by the N95 filtering facepiece—or even the elastomeric respirator—is questionable.[45] These findings were based on counts of microscopic intact spores on the outside of the respirators in comparison with the counts of microscopic intact spores that penetrated through to the inside of the respirators.[46]

Health risks may be even greater than what is indicated by the level of intact spore counts that penetrated the respirators. The report states, "Recent studies have shown that exposure to fungi occurs also through submicrometer fungal fragments. These particles may penetrate at even higher rates as intact spores because the filter materials commonly used in N95 respirators have maximum particle penetration around 0.03–0.07 µm [which allows penetration of the smaller fine particulates]."[47] In other words, in addition to the intact spores that penetrate the respirators, submicrometer fungal fragments also penetrate the respirators—at a much higher rate. Thus, people wearing the N95 filtering facepiece and the elastomeric respirator are subject not only to the ill health effects from inhalation of intact spores but also from inhalation of fungal fragments. This study did not address the possible added health risks from mycotoxin exposure, which occurs anytime toxigenic mold spores and fragments are inhaled.

Mr. Pearson, a 20-plus-year veteran in the field of industrial hygiene, explains an additional reason why the N95 respirators do not provide sufficient protection. He has observed that people rarely pinch the nosepiece, which is required to obtain a proper fit and maximize protection afforded by the N95. Mr. Pearson notes that the elastomeric respirator (half or full face) is the form of personal protective equipment worn by those in the industrial hygiene profession.[48] The findings of the aforementioned study state, ". . . [T]he results suggest that the elastomeric half-facepiece respirators offer at least 10 times the protection against fungal spores."[49] However, as just mentioned, even the higher grade of respirators may not provide sufficient protection. The report also addresses the regular occurrence of increased aerosolization of fungal spores, fragments, and other contaminants during remediation. It states, "Because the fungal spore concentrations were extremely high during the renovation, we question whether the protection offered by the N95 filtering facepiece or the elastomeric respirators is sufficient."[50]

The possibility of limited respirator efficiency and the increased health risks from inhalation of fungal fragments should be of heightened concern to people in the remediation field, many of whom work five or more days a week. It is impossible for remediation workers to strip a building to the two-by-four studs without aerosolizing debris, fungal spores, fungal fragments, fine particulates, bacterial organisms and compounds, and other contaminants into the very air they are breathing.

No one disputes the fact that the remediation process increases the level of airborne spores, fungal fragments, and fine particulates. In fact, increased levels of airborne contaminants were evidenced in the same pilot project of homes undergoing renovation in New Orleans post-Katrina.[51] Obviously, the potential risk to health increases as the level of aerosolized spores, fragments, fine particulates, and mycotoxins (which are in/on the spores and fragments) increase as they are disrupted during the remediation process. Furthermore, lengthy remediations can also increase the risk of possible health effects. It can often take homeowners weeks or even months to complete a remediation, especially if the "official" cause of loss was not covered by their homeowners insurance.

With such a potential risk to health from fungal exposure, we must all be proactive in protecting our health when in environments with suspected elevated fungal activity. Unfortunately, people's lack of diligence in wearing personal protective equipment could be caused and/or reinforced by the following statement published by the CDC: "People not involved in activities that disturb mold-contaminated materials have a lower risk of inhalation exposure relative to people performing those types of activities. People collecting belongings, visually inspecting homes or buildings, or doing basic cleanup for short periods in a previously flooded home or building will not usually need to use a respirator."[52] Essentially, the CDC is telling people they don't need to wear personal protective equipment if they are cleaning for only "short periods." But what if someone is cleaning for only "short periods" in a building with average airborne spore counts of 645,000 spores/m^3 or during a half-hour time period of high sporulation, such as 324,648 spores/m^3?

Dr. Thrasher sets the record straight. He adamantly states, "The CDC is quite mistaken. The CDC has not considered the fine particulate matter . . ." He explains that not only does this fine particulate have the same toxic properties as intact spores, but it is also released at an increased rate as compared to intact spores.[53] Thus, the concentration levels of fine particulate matter released is greater than the concentration levels of intact spores released.

According to Dr. Thrasher, the increased concentration of fine particulate was first studied by Dr. Górny and group.[54] As reported in their published report, Dr. Górny and colleagues ". . . discovered that fungal fragments are aerosolized simultaneously with spores from contaminated agar and ceiling tile surfaces. Concentration measurements with an optical particle counter showed that the fragments are released in higher numbers (up to 320 times) than the spores. . . . The most interesting finding of this study was that a significant amount of immunologically reactive particles having sizes considerably smaller than those of the spores was released from surfaces contaminated with fungi."[55] In other words, negative health effects can occur from exposure to fungal fragments and fine particulates, as well as from exposure to spores.

An important point to clarify is the fact that mycotoxins are not just randomly released into the air by fungal spores and/or colonies. Dr. Straus explains, "The mycotoxins are in the spores and are released from the spores only when the mycotoxins become solubilized in water. Spores don't release mycotoxins into the air. The mycotoxins get into the air when they are in the dust and picked up by air currents."[56] Mycotoxins also get into the air when the spores, fungal fragments, and fine particulate on which they exist become airborne. Once these fungal components are inhaled, the mycotoxins solubilize inside the human body, get into the blood, and begin to kill cells.[57]

Dr. Gray describes how the inhalation of toxigenic mold spores and the associated mycotoxins can lead to systemic conditions, "While the weight and physical properties associated with spherical particles dictates that only particles between 0.005 and 5 micron are capable of penetrating to the alveoli (the lung's air sacs), the size range of mold *spores* uniquely capable of penetrating to the alveoli is from 0.5 to 7 micron. Once having arrived in the alveoli, they stimulate a dramatic immune response. This is also the site in which under the influence of pulmonary surfactant they are able to release their mycotoxins, allowing them to be absorbed into the blood flowing through the prolific capillary beds found adjacent to the air sacs. The mycotoxins then circulate throughout the body."[58]

A question that warrants asking: Once toxigenic molds are inhaled, do they produce their poisonous mycotoxins while inside the human and/or animal host? Although this question is debated in scientific and medical communities, many doctors believe they do. According to Dr. Thrasher, mycotoxins can be found in human tissue, organs, and blood serum, not only because mycotoxins are inhaled but also because mycotoxins are produced by toxigenic spores inside the human body after the spores are inhaled. He states, "I'll give an example of people with aspergillosis caused by, I think, *Aspergillus fumigatus*. *Fumigatus* causes aspergillosis of the lungs. *Fumigatus* also produces a mycotoxin called gliotoxin. When researchers look at the blood of people with aspergillosis caused by this particular species of *Aspergillus*, they find gliotoxin in the blood. That tells me that in the infectious stage, that particular organism is producing a mycotoxin. There are several studies that point to this; unfortunately, we have not been looking at all the other mycotoxins as to what's going on."[59]

Dr. Gray concurs that mycotoxin-producing fungi continue to produce mycotoxins once inside a human host. He states, "Certainly, anything that is colonizing you and growing inside of you that has the potential of producing mycotoxins will do it. . . . It is a reasonable medical and scientific certainty that these organisms will produce these mycotoxins wherever they grow. They don't have a way of knowing that they're growing in your body versus somewhere else and they're simply going to produce the poisons."[60]

However, other scientists are not as certain. As mentioned earlier, Dr. Straus and colleagues at Texas Tech University Health Sciences Center documented trichothecene mycotoxins in human blood sera. When asked whether the trichothecenes were found in the blood serum of study participants because the subjects had inhaled the trichothecenes, which then absorbed into their blood, or because the trichothecenes were produced inside the people's bodies by inhaled spores, Dr. Straus said, "You can't tell."[61] One thing we do know for certain is whether mycotoxins are inhaled or produced internally by inhaled fungi—or both—they can pose serious health risks.

In summary up to this point: We have learned that fungal colonies release not only spores but also fragments and fine particulates. Although smaller than spores, the fragments can be immunologically reactive like intact spores because "[t] hat fraction also contains all of the toxic chemicals that are produced by the molds and the spores," according to Dr. Thrasher.[62] In order to more accurately measure the toxicity of air, we must consider not only the intact spores but also the fungal fragments and fine particulates, including dust. A broader consideration is especially necessary because mycotoxins are present on these fragments, fine particulates, and dust, as well as on the spores, all which can become airborne and be inhaled. Dr. Straus and colleagues have proven the presence of trichothecenes, which is the mycotoxin produced by the fungi *Stachybotrys*, in the sera of people who occupied buildings contaminated with *Stachybotrys*.[63] Dr. Gray states, "Once inhaled, mycotoxins solubilize inside the human body, are absorbed into the blood, and circulate throughout the body exerting their cytotoxic effects on cells."[64] According to Dr. Thrasher, the effective particulate load of the fine particulate plus the spores is 300 times that of the spores alone.[65] These fungal fragments, fine particulates, mycotoxins, and VOCs are not measurable in traditional air sampling and testing methodologies.[66]

So how do these mycotoxins that are on respirable particles affect our health? According to Dr. Ammann, "Possible results of mycotoxin exposure to multiple molds indoors are:

- Vascular system (increased vascular fragility, hemorrhage into body tissues, or from lung, e.g., [from] aflatoxin, satratoxin, roridins [types of mycotoxins])
- Digestive system (diarrhea, vomiting, intestinal hemorrhage, liver effects, i.e., necrosis [decaying of tissue], fibrosis [fibrous growth]: [from] aflatoxin; caustic effects on mucous membranes: [from] T-2 toxin [a mycotoxin]; anorexia: [from] vomitoxin [a mycotoxin])
- Respiratory system: respiratory distress, bleeding from lungs, e.g., [from] trichothecenes [a mycotoxin]
- Nervous system, tremors, incoordination, depression, headache, e.g., [from] tremorgens, trichothecenes [types of mycotoxins]
- Cutaneous system: rash, burning sensation, sloughing of skin, photosensitization, e.g.,[from] trichothecenes [a mycotoxin]
- Urinary system: nephrotoxicity [kidney toxicity], e.g., [from] ochratoxin, citrinin [types of mycotoxins]
- Reproductive system: infertility, changes in reproductive cycles, e.g., [from] T-2 toxin, zearalenone [types of mycotoxins]
- Immune system: changes or suppression: [from] many mycotoxins"[67]

As you can see, mycotoxins can have detrimental effects on multiple systems in the human body. Multiple symptoms can manifest. These health effects from mycotoxin exposure can be further compounded by the negative effects from exposure to the mold itself—the spores, fungal fragments, and fine particulates—and other possible existing contaminants, such as bacterial microorganisms and VOC off-gassing from organisms and building materials. According to doctors Thrasher and Lipsey, gram-negative and gram-positive bacteria, along with the endotoxins and bacterial cell wall toxins they respectively produce, synergistically interact with the mycotoxins.[68]

It is no wonder with such a toxic mix of biologically active contaminants that multiple symptoms often present in people exposed, making diagnoses difficult and/or resulting in misdiagnoses. Identifying the cause of ill health from mold and mycotoxin exposure is often difficult, as is pin pointing the source of many environmental illnesses. For example, as Dr. Rapp points out, "The symptoms caused by foods, molds, pollen, dust, yeast, pets, chemicals, toxic metals, nutritional deficiencies, hormone imbalances, and stress can all be similar."[69]

Medically speaking, reports generally address two categories of diagnoses when discussing ill health effects from mold and mycotoxin exposure—mycosis and mycotoxicosis. In other words, the scientific and medical communities acknowledge that people can get sick not only fungal exposure but also mycotoxin exposure.

Mycosis is a fungal infection in or on a part of the body and/or a disease caused by fungi. In a published report, Dr. Thrasher states, "Human mycotic infections (mycoses) are grouped into superficial, subcutaneous, and systemic (deep) mycoses. Superficial mycoses (skin, hair, nails), may be chronic, resistant to treatment, seldom debilitating, and can be transferred from human to human or from pets to humans. The deeper lying mycoses (subcutaneous and systemic) are usually acquired infections. The mold spores enter the body through injured skin, invading subcutaneous tissues, spreading systemically through lymphatic vessels. Infection from fungi occurs from inhalation, penetration of the skin, and ingestion."[70]

Mycotoxicosis, on the other hand, is a disease or poisoning caused by mycotoxins. Published literature states, "The majority of mycotoxicoses . . . result[s] from eating contaminated foods. [However,] skin contact with mold-infested substrates and inhalation of spore-borne toxins are also important sources of exposure." The same report further states, "Mycotoxicoses, like all toxicological syndromes, can be categorized as acute or chronic. Acute toxicity generally has a rapid onset and an obvious toxic response, while chronic toxicity is characterized by low-dose exposure over a long time period, resulting in cancers and other generally irreversible effects. . . . Almost certainly, the main human and veterinary health burden of mycotoxin exposure is related to chronic exposure (e.g., cancer induction, kidney toxicity, immune suppression). However, the best-known mycotoxin episodes are manifestations of acute effects . . ."[71] A prime example is the turkey X syndrome from which aflatoxins were discovered.

In summary, multiple species of fungi can grow on the same substrate. Thus, the presence of indoor fungal growth can result in exposure (be it via inhalation, touch, or ingestion) to one or more species of fungi. Fungal exposure may or may not include mycotoxin exposure, as not all fungi are mycotoxin-producing. However, mycotoxin exposure (be it via inhalation, touch, or ingestion) always occurs in combination with some level of fungal exposure. Depending on the species, exposure to one or more type of mycotoxin can occur as some species are capable of producing more than one type of mycotoxin.

Exposure to multiple molds and mycotoxins is common when mold overgrowth occurs indoors from adequate moisture buildup or water intrusion (be it from a hurricane, a flood, a leak, or a construction defect). The NRDC report illustrates this point; *Aspergillus, Penicillium, Cladosporium,* and *Stachybotrys* are all identified as cohabitating. Even under noncatastrophic conditions, multiple species of fungi often grow on the same substrate in indoor environments when enough moisture is present. This multi-species mold infestation occurs because spores from a variety of species float in from the outside and germinate when conditions are conducive. This, of course, can result in the occupants being exposed to spores from multiple species, the accompanying fungal fragments and fine particulates, a variety of mycotoxins (if toxigenic molds are present), VOCs, bacterial microorganisms, and related compounds. The inhalation, touch, and ingestion of these airborne contaminants can affect multiple systems within the bodies of the occupants.

Clinical observations of mold-exposed patients have led to documented symptoms of mold and mycotoxin exposures and the various systems that are affected in the body. These clinical profiles are valuable sources for people seeking medical information regarding mold and mycotoxin exposures. In the future we can likely expect an increase in this type of data because the mold explosion caused from hurricanes Katrina and Rita has skyrocketed the number of people suffering ill health effects from mold exposure. In the meantime, let's look at some of the data available now.

A comprehensive review of current published literature and conference reports was conducted by several medical doctors. It states, "Indoor airborne mold exposure has been associated with adverse human health effects in multiple organs and body systems, including respiratory, nervous, immune, hematological, and dermatological systems. Indoor mold exposure can also lead to life-threatening systemic infections in immunocompromised patients."[72]

The report continues, "Exposure to high levels of indoor mold can cause injury to and dysfunction of multiple organs and systems, including respiratory, hematological, immunological, and neurological systems, in *immunocompetent* humans." This statement, based on a comprehensive review of literature and conference reports, clearly dispels the myth that only people who are immunocompromised are severely affected by exposure to mold. Whether you are immune compromised or not, mold exposure can clearly adversely affect your health.[73]

The report includes clinical profiles of patients. It states, "Patients have been reporting multiple ill health effects linked to exposures to mold. Studies of more than 1,600 patients suffering ill effects associated with fungal exposure were presented at one meeting in Dallas in 2003 (21st Annual Symposium of Man and His Environment, Dallas, Texas, 19–22 June 2003). To cite a few studies: [Dr. Allan] Lieberman examined 48 heavily mold-exposed patients who had the following health problems: muscle and/or joint pain (71%), fatigue/weakness (70%), neurocognitive dysfunction (67%), sinusitis (65%), headache (65%), gastrointestinal problems (58%), shortness of breath (54%), anxiety/depression/ irritability (54%), vision problems (42%), chest tightness (42%), insomnia (40%), dizziness (38%), numbness/tingling (35%), laryngitis (35%), nausea (33%), skin rashes (27%), tremors (25%), and heart palpitations (21%). [Dr. William] Rea et al.'s study of 150 heavily indoor mold-exposed patients found the following health problems: fatigue (100%), rhinitis (65%), memory loss and other neuropsychiatry problems (46%), respiratory problems (40%), fibromyalgia (29%), irritable bowel syndrome (25%), vasculitis (4.7%), and angioedema (4.0%). These clinical reports suggest that there can be multisystem adverse effects of airborne mold. All reported cases had environmental mold exposure consistent with toxic mold exposure."[74]

The report further states, "In recent years, the incidence of life-threatening infections in immunocompromised patients from *Aspergillus* and other common fungi has been growing rapidly. . . . Even with strong antifungal drugs and intense hospital treatment, mortality rates from invasive aspergillosis range from 50% to 99% in the immunocompromised."[75] Again, we see the bleak outcome projected for anyone immunocompromised. These incredibly high mortality rates—ranging from 50% to 99%—are odds none of us want to face.

Although many health effects are attributed to mold exposure, a fact clearly documented in the earlier referenced IOM report, several remain fiercely disputed by factions within scientific and medical communities. A quick browse through newspaper or TV archives will reveal that some of these battles of scientific interpretation have been publicly played out in the press. Let's take a look at one of the most hotly debated cases, which centered on the cause of bleeding lungs in several infants in Cleveland, Ohio. The case involved an initial cluster of eight incidences of pulmonary hemorrhage/hemosiderosis identified in November 1993. The number of cases rose to ten by December 1994. The initial cases were reported to the CDC by Dr. Dearborn of the Department of Pediatrics at the Rainbow Babies & Children's Hospital and Case Western Reserve School of Medicine.[76] The CDC reports, "The children resided in seven contiguous postal tracts and had had one or more hemorrhagic episodes, resulting in one death during January 1993–December 1994."[77] An investigation was launched with the original hypothesis that exposure to *Stachybotrys chartarum* and other hydrophilic molds might have caused the severe lung disease in the infants.[78]

According to the CDC, "Preliminary results of a CDC case-control study indicated that hemorrhage was associated with 1) major household water damage during the six months before illness and 2) increased levels of measurable household fungi, including the toxin-producing mold *S. chartarum*."[79] These findings were published in articles in the CDC's *Morbidity and Mortality Weekly Report* (*MMWR*) in December 1994 and January 1997.[80] The CDC stated, "These findings and the observation that trichothecene mycotoxins were produced in the laboratory by some *S. chartarum* isolates recovered from the homes of study subjects have been published and referenced in peer-reviewed scientific literature. The hypothesis from the findings of the investigation was that infant pulmonary hemorrhage may be caused by exposure to potent mycotoxins produced by *S. chartarum* or other fungi growing in moist household environments."[81]

Dr. Straus confirms that compounds produced by *Stachybotrys* are what caused the infants' lungs to bleed. He explains, "*Stachybotrys* produces a compound that's been named stachylysin, which is a hemolysin. Hemolysin lyses [disintegrates] red blood cells. The spores of *Stachybotrys* either have on their surface or produce these hemolysins, so when these infants inhaled the *Stachybotrys* spores, the hemolysin then dissolved or diffused into the lungs of these infants and began to lyse red blood cells. What that means is red blood cells began to pop, so then the physicians began to see the disease called pulmonary hemosiderosis—the presence of heme inside of macrophages, which are the cells in the lungs that phagocytize [engulf and absorb] things that shouldn't be there. Now you have red blood cells popping; you have a lot of heme, which is the red portion and one of the most important components of red blood cells, getting into the macrophages—that's what pulmonary hemosiderosis is. So you now have a lot of pulmonary bleeding due to the fact that these red blood cells were lysing, and there was heme just going everywhere due to the compounds that these organisms were producing."[82]

Dr. Thrasher concurs that stachylysin produced by *Stachybotrys* is a hemolytic agent, which destroys red blood cells and causes the subsequent release of hemoglobin.[83]

Furthermore, a NIOSH scientist cites the 1997 published findings of animal studies that demonstrate the compounds from *Stachybotrys chartarum* cause incidences of hemorrhage in mice. The animal studies indicate that these same agents could have caused the incidences of infant pulmonary hemorrhage in Cleveland, Ohio. The NIOSH scientist states, "Further support for the proposed mycotoxin etiology [cause of disease] of infant pulmonary hemosiderosis comes from animal studies in which mice were treated intranasally with both nontoxic and toxic spores of *S. chartarum*. Severe inflammatory changes, including hemorrhage, were observed in mice receiving toxic spores of this species."[84]

In addition, a published 2002 report documents stachylysin exposure as a possible cause of hemorrhaging in humans. The authors, one of whom works for the EPA, state, "We believe that stachylysin exposure may be proposed here as a reasonable mechanism to explain the bloody noses in adults and hemorrhaging in infants exposed to *S. chartarum*. This exposure model is now being tested by measuring the antibody response to stachylysin in exposed individuals. It is also possible that other indoor fungi also produce hemolysins with potentially similar effects."[85] The possibility that other indoor fungi produce compounds capable of causing hemorrhaging in humans is significant to note, as multiple species of fungi often grow in indoor environments given adequate moisture buildup.

Also of concern is the fact that pulmonary bleeding in infants may not always be properly diagnosed. In published study findings regarding the Cleveland, Ohio, incidents, Dr. Dearborn and colleagues report there may be infants who have pulmonary hemorrhage that escape clinical detection.[86] They state, "This [conclusion] is supported by the retrospective observation of six sudden infant death syndrome victims from this same geographic area and time period who had extensive pulmonary hemosiderosis on reexamination of their lung sections with iron staining."[87] Iron staining reveals the presence of hemosiderin.

Scientific proof appears to sufficiently correlate *Stachybotrys* exposure to incidents of hemorrhage. However, further investigation into the Cleveland incidents and review of the original investigation and resultant published findings led to questions regarding the scientific soundness of the initial investigation. Ultimately, the reinvestigation undermined the conclusions drawn by the original investigation team. Members of the original investigation team also reexamined the causation. In a 1997 published report, the authors (which included four of the seven original investigators) stated, "The cluster of IPH [idiopathic pulmonary hemosiderosis] cases that we examined, although not directly attributable to any one specific environmental factor, strongly points toward an agent-host-environment interaction that is influenced by several risk factors. The strength of the association between home water damage and IPH suggests that the primary environmental risk factor for IPH may be associated with water damage."[88]

In March 2000 the CDC published concerns over its original, published findings. It stated, "The findings . . . were cited in environmental health guidelines, congressional testimony, and the popular media, and have been debated among industrial hygienists and other occupational and environmental health scientists. Despite caution that 'further research is needed to determine . . . causal[ity],' the findings have influenced closure of public buildings, cleanup and remediation, and litigation."[89]

In the same March 2000 report, the CDC retracted its prior published, peer-reviewed findings. The CDC claimed, "A review within CDC and by outside experts of an investigation of acute pulmonary hemorrhage/hemosiderosis in infants has identified shortcomings in the implementation and reporting of the investigation described in *MMWR* and detailed in other scientific publications authored, in part, by CDC personnel. The review led CDC to conclude that a possible association between acute pulmonary hemorrhage/hemosiderosis in infants and exposure to molds, specifically *Stachybotrys chartarum* (commonly referred to by its synonym *Stachybotrys atra*) was not proven."[90]

By 2002 additional studies were published authored again by some members of the original investigation team. The report states that although *Stachybotrys* toxins can be suggested as a causative factor, ". . . additional evidence to support causation is needed. The possibility remains that the presence in the infants' homes of *Stachybotrys*, a high water-requiring fungus, may simply be an indicator of water intrusion, and other unknown, related factors (in addition to ETS [environmental tobacco smoke]) may play primary or secondary causative roles."[91]

As of October 2005 the CDC continues to hold the position that the association between *Stachybotrys* and acute pulmonary hemorrhage/hemosiderosis was not proven. The CDC reiterated this stance in the earlier mentioned report published in the aftermath of Katrina and Rita. It states, "Mycotoxins were prematurely proposed as the cause of a disease outbreak of eight cases of acute pulmonary hemorrhage/hemosiderosis in infants in Cleveland, Ohio, in 1993 and 1994. The cluster was attributed to exposure to mycotoxins produced by *Stachybotrys chartarum*. Subsequent reviews of the evidence concluded that insufficient information existed and no such association was proven."[92]

The overturning of the original findings was based on the review of two teams that re-analyzed the original case information. One was an internal CDC review team comprised of CDC personnel; the other was an external team of consultants, not part of the CDC.

The original investigation team, which included Dr. Dearborn, responded to the report published in the *MMWR*. They pointed out, ". . . neither review group included any member of the initial study team and neither review group visited Cleveland to see the buildings or talk to the families, community groups, Cuyahoga County Health Officials and building inspectors, nor the physicians at the Rainbow Babies & Children's Hospital."[93]

In their response, the original investigation team further pointed out, "The *MMWR* states that the infant hemorrhage was possibly caused by unmeasured confounders [unknown or unaccounted for factors], a potential problem in any epidemiologic study."[94]

Dr. Thrasher expands on this theory. He states, "It's not only the black mold and all the other molds; it's also the entire mixture to which people are being exposed." Dr. Thrasher has concerns about not only fungal inhalation but also other air contaminants commonly found in environments with damp building materials, such as mycotoxins produced by toxigenic molds, gram-negative and gram-positive bacteria, the endotoxins and bacterial cell wall toxins they respectively produce, and the synergistic effects of these compounds with the mycotoxins.[95] Dr. Thrasher points out, "They [Dr. Dearborn and the original CDC investigation team] did not look at all the other things that were in the indoor air at that time . . ."[96] The significant point to note is all these air contaminants can stem from the accumulation of indoor sources of moisture.

The CDC did not undertake future studies or monitoring efforts to identify the prevalence of infant pulmonary hemorrhage or further explore the role of confounding factors. According to the original study team, "Plans to continue the Cleveland case-control study prospectively and expand it nationally were submitted [to] and reviewed by the CDC. The CDC elected not to pursue this work." The team additionally states, "There is no national surveillance for AIPH [acute idiopathic pulmonary hemorrhage], and physicians who have tried to report cases to [the] CDC in 1998 and 1999 were told that CDC was no longer collecting this information."[97]

The Ohio cases did, however, lead to ". . . the examination of all infant coroner cases over a three year period, 1993–1995. This revealed seven 'SIDS' (sudden infant death syndrome) cases with evidence of pre-existing major pulmonary bleeding. All but one of these infants had lived in the ten zip code cluster area," reports Case Western Reserve University.[98] The CDC review teams later acknowledged the existence of pulmonary bleeding in SIDS cases and published the following recommendation: "Because of the overlap between AIPH in infants and sudden infant death syndrome (SIDS), investigators also were advised to consider risk factors for SIDS."[99]

The fact that infant pulmonary bleeding was found in several cases of SIDS, yet no formal, nationwide case-control study was initiated to further investigate the possible causes of infant pulmonary bleeding, is disconcerting. Furthermore, evidence of additional cases of pulmonary bleeding in infants supports the need for future research. According to reports by professionals at Case Western Reserve University, "An informal national survey of all pediatric pulmonary centers and continued reporting has identified over 100 similar cases of pulmonary hemorrhage in infants across the country over the last seven years."[100]

Although the CDC did not commence a nationwide study, by 1998 pediatricians were being instructed to report cases of idiopathic pulmonary hemorrhage and hemosiderosis to state health departments, not to the CDC, and to order autopsies of any infant who died suddenly without a known cause. Pediatricians were also instructed to include as part of the autopsy a Prussian blue stain of lung tissue, which would reveal the presence of hemosiderin if present.[101] Additionally, the American Academy of Pediatrics (AAP) addressed the issue of mold exposure in the following published policy statement: "Until more is known about the etiology of idiopathic pulmonary hemorrhage, prudence dictates that pediatricians try to ensure that infants under one year of age are not exposed to chronically moldy, water-damaged environments."[102]

Obviously, the AAP has not discounted the scientific evidence that links exposure to mold and other contaminants inherent in water-damaged environments to infant pulmonary bleeding, as apparently did the CDC. Unfortunately, the AAP's warnings are ineffectively tucked away in a carefully worded policy statement and are not being aggressively disseminated to parents and other people who care for infants. Thus, most care providers of infants are not aware that exposure to water-damaged environments, such as mold-laden, post-hurricane areas, can cause such serious health repercussions for infants.

The debate over whether *Stachybotrys* was the sole causative factor of the infant pulmonary bleeding in Cleveland, Ohio, or was just a contributing factor should not be our focus. The critical piece of the puzzle to understand is that incidents of pulmonary bleeding in infants and SIDS have occurred in environments that have suffered a form of water intrusion. We must understand that wet building materials inherently create an environment that fosters the growth of the same types of microorganisms, such as toxigenic molds like *Stachybotrys* and gram-negative and gram-positive bacteria. Also present with these organisms are chemical compounds, such as mycotoxins, bacterial toxins, and VOCs off-gassed from both the organisms and building materials.

Most of the homes in which the infants in Cleveland, Ohio, resided had reports of recent water damage; most of the buildings in the Katrina-hit areas suffered some level of water damage. We have a parallel of environmental conditions; thus we must recognize the possible parallel risk of infant pulmonary hemorrhage/ hemosiderosis and SIDS. In fact, given that many homes in hurricane-hit areas likely sustained more structural water damage than the level that existed in the homes in Cleveland, Ohio, why weren't warnings of the possible increased health risks to infants publicly disseminated pre- and post-Katrina via multiple avenues of media by the CDC, AAP, and/or other governmental or medical entities aware of the potential dangers?

If we were told "Mold cannot hurt you" and the Habitat for Humanity workers were told "Mold cannot hurt you," parents of infants and small children also may be being told "Mold cannot hurt you. Mold cannot hurt your baby." At best, these assertions are ignorant perpetuations of misinformation. At worst, they are intentional, blatant lies supporting some political agenda and/or protecting various industry pocketbooks. If we became so sick from only two weeks of mold exposure, how can the health of infants and young children—who are also considered at high risk from mold exposure—not be negatively impacted?

One scientist points out, ". . . [A]lthough there are many unanswered questions about the effects of *S. chartarum* on human health, the accumulation of data (from observations and research) over the past 65 years tells us that one should not handle materials contaminated with *S. chartarum* (without proper safety procedures) and strongly indicates that indoor environments contaminated with *S. chartarum* are not healthy, especially for children, and may result in serious illness."[103] Thus, mold— especially *Stachybotrys*—can hurt you.

Although the initial assessment of *Stachybotrys* as the causative factor in the Ohio infant incidents may have awakened some in the medical community to the dangers of exposure to *Stachybotrys* and possibly other molds prevalent in water-damaged buildings, it did little to increase awareness in the general public. Hydrophilic molds are health concerns anytime building materials get wet, yet for the most part, parents, grandparents, uncles, aunts, and other caregivers responsible for the health of infants and small children are unaware of the potential dangers of mold exposure, even though various newspapers and television media across the U.S. have reported stories of mold plight.

For example, "The *New York Times Magazine*, August 12, 2001, ran a front-page story on toxic mold. Newspaper articles such as 'Fungus in "Sick" Building' (*New York Times*, May 5, 1996) or 'Mold in schools forces removal of Forks kids' (*Fargo Forum*, June 1997) are eye-catching news items. The nationally syndicated comic strip *Rex Morgan* ran a series on *Stachybotrys*, and television news shows have run entire programs on *Stachybotrys* contamination of homes."[104] The media has focused on the human interest element—the destruction of both physical property and health from mold growth and exposure—and high profile, multi-defendant lawsuits. One scientist notes of the fungus *Stachybotrys*, "The fungus has resulted in multimillion dollar litigations and caused serious problems for homeowners and building managers who must deal with the human issues and remediation."[105]

Since *Stachybotrys* flourishes in water saturated building materials, it is one of the contaminants that has been found ". . . inhabiting buildings with major problems in mechanical system design, construction, and operational strategies, leading to excess indoor moisture."[106] Additionally, over the years *Stachybotrys* has been repeatedly linked to sick building syndrome. One report confirms, "Indeed, over the past 15 years in North America evidence has accumulated implicating this fungus as a serious problem in homes and buildings and one of the causes of the 'sick building syndrome.'"[107]

Research into incidents of illnesses related to sick buildings has also revealed that the following other molds have been found structurally in damp buildings: *Aspergillus, Penicillium, Alternaria,* and *Cladosporium.*[108] According to the CDC, in the U.S. alone, "More than 1,000 kinds of indoor molds have been found in . . . homes."[109] Because indoor mold growth results in exposure to occupants, incidents of negative health ensue. Unfortunately, structural mold is reportedly on the rise. One medical report states, "Indoor air contamination with toxic opportunistic molds is an emerging health risk worldwide."[110] In fact, "Various surveys of homes in North America and Europe have reported that visible mold and/or water damage are common, found in 23–98% of all homes examined."[111]

It is important to understand that mold growth is not limited to areas with humid climates. Nor is it limited to areas ravaged by floods or hurricanes. In fact, even in dry, northern states, buildings are being designated as "sick" from the health havoc wreaked by fungal growth. For example, in Kalispell, Montana, indoor mold growth has been a prevalent problem. According to Gene Thompson, western vice president of the Montana Landlords Association, "epidemic levels" of mold are being experienced in newer, multilevel apartment buildings in Kalispell due to tight building construction methods in combination with high water levels that lie beneath the foundations of these buildings.[112]

According to Jim Pearson, newer airtight building methods prohibit airflow and don't allow structures to "breathe," as do older methods of construction.[113] Obviously, the more air that can flow through a structure, the less likely moisture accumulation will occur from condensation and the more likely small amounts of moisture will evaporate. However, it is important to understand both newer and older construction can fall prey to mold infestation. Dr. Straus points out, ". . . [N]ew buildings will have just the same problems that old buildings have if they get wet. Water is the common denominator. The age of the building does not matter at all. It's the amount of water that gets into places that it shouldn't be that causes the problem."[114]

An additional problem can occur in cold climates when people turn on their heat and condensation builds up from inadequate air flow. The risk of condensation buildup increases when occupants seal their windows with plastic in an attempt to retain heat in the structure and reduce heating costs. When moisture builds up inside the walls, mold and other biologicals can form. Chemical reactions can also ensue, including off-gassing of formaldehyde and other VOCs from building materials. Research has shown, ". . . sick building syndrome may result from several chemical and physical factors as well as biological factors, including moulds [molds]."[115] Dr. Thrasher supports this theory, emphasizing that not just one component creates the adverse health effects experienced from fungal exposure.[116] For example, in regard to the highly studied mold *Stachybotrys*, medical literature suggests, ". . . *Stachybotrys* may play a role in development of sick building syndrome but most probably together with other factors."[117]

Critically important to understand is the fact that building materials can become contaminated with mold growth at any point during manufacturing, storage, transport, and/or construction. The widely publicized *Gorman* case that went to trial in 2005 is a prime example. The case settled mid-trial for $22.6 million—$13 million of which was paid by the lumber company that reportedly improperly stored the wood.[118] We all must remember that mold spores can lie dormant in and on building materials until enough moisture accumulates, at which point they will begin to germinate. Oftentimes indoor mold growth is hidden, and symptoms of ill health are the first indication of a hidden mold infestation. Unfortunately, if no visible mold growth is apparent, homeowners often fail to connect their health problems to mold exposure.

Additional challenges arise for tenants who become sick from mold exposure while living in mold-infested buildings, as no governmental agencies assist renters (who are often unable to move due to limited financial resources). Remember, government agencies and private industries have not set air quality guidelines in regard to mold exposure limits. Therefore, no government agency provides oversight and/or enforcement in this area. Renters who call local, state, and/or federal agencies are repeatedly told that issues of mold infestation and the ensuing ill health effects are civil matters between the renter and the landlord.

In the workplace, ill health attributed to indoor air quality issues is also making its way into the legal system. "Physicians and industrial hygienists may be asked to contribute reports to assist the courts in settling suits. In 2002, an estimated 10,000 mold-related cases were pending in U.S. courts. Also in 2002, the insurance industry paid out $2 billion in mold-related claims in Texas alone."[119]

Even with mounting lawsuits and media attention spawned by issues of black mold and sick building syndrome, mold awareness is still limited. One mycologist comments on the limited awareness of *Stachybotrys*, regardless of its prevalence, "I have been impressed with the common occurrence and extensive growth of *S. chartarum* in homes and buildings damaged by flood waters or other types of water incursions and the lack of knowledge by the general public and public and private institutions about this fungus."[120] This same lack of knowledge also exists in regard to other toxigenic fungi, the dynamics of moisture buildup and water intrusion, and the need for consistent use of adequate personal protective equipment.

Another area in which most people are lacking knowledge is in the selection process of professional assistance when facing a mold infestation. Many people incorrectly believe a one-stop-shopping concept can be applied to the detection and remediation of mold. Dr. Puccetti points out, "One thing that needs to be understood is that mold remediation is an interdisciplinary activity. Not only do you need someone such as me [a CIH] and a mold remediation contractor, but also, in many instances, you need a building expert—a contractor or an architect—who can understand or help to understand the dynamics of the water intrusion or moisture problem that caused the mold growth. You have to address the cause of the water damage before you can even hope to permanently get rid of the mold. If you haven't removed the problem that allowed the mold to grow in the first place, then once the mold is removed, it'll come back again. It takes only about 24 to 72 hours for mold growth to occur if building materials get wet enough."[121] Most people are not aware of how quickly mold can take hold. Prior to our post-Katrina mold experience, we had no idea how quickly mold could infest a house when building materials became damp/wet, or how quickly health could deteriorate from exposure to the resultant airborne fungal contaminants, other accompanying biologicals, and VOC off-gassing.

In summary, "Many studies link exposure to damp or moldy indoor conditions to increased incidence and/or severity of respiratory problems, such as asthma, wheezing, and rhinosinusitis . . ."[122] Specifically, exposure to indoor fungal growth can lead to serious, negative health repercussions. Scientists point out in a published report, "Indoor mold exposure can alter immunological factors and produce allergic reactions. Several studies have indicated that indoor mold exposure can alter brain blood flow, autonomic nerve function, [and] brain waves and worsen concentration, attention, balance, and memory."[123]

Dr. Thrasher explains how the inhalation of airborne contaminants can affect the brain, ". . . [W]hen you inhale through the nose, there is a strong likelihood that the inhaled toxins will go directly into the brain. I don't know if you're aware of this, but the olfactory neurons are in direct communication with the brain, and there is no blood-brain barrier. Anything you have inhaled and smelled, whether you detect an odor, or even if it doesn't have an odor, goes directly into the brain through the olfactory neurons."[124] Inhaled toxins affecting the brain is a disturbing fact given all the airborne contaminants that have been, and continue to be, inhaled in post-Katrina and other heavily air-polluted areas. Dr. Thrasher points out an even more unsettling thought in regard to the lack of a blood-brain barrier. He asserts, "The government doesn't want people to know these things."[125]

The lack of a blood-brain barrier is confirmed in recent scientific studies, according to Dr. Thrasher. He states that it was demonstrated in a mice study at Michigan State University led by James Pestka, PhD, that documented the health effects of satratoxin G (a trichothecene mycotoxin) produced by *Stachybotrys*. Dr. Thrasher explains, "They instilled the mycotoxins into the olfactory neurons in the nasal cavities and demonstrated cell death and inflammatory conditions in the olfactory neurons, the olfactory tract, and the olfactory lobe of treated mice."[126] The report was published in *Environmental Health Perspectives* in July 2006 and, as earlier mentioned, states, "These findings suggest that neurotoxicity and inflammation within the nose and brain are potential adverse health effects of exposure to satratoxins and *Stachybotrys* in the indoor air of water-damaged buildings."[127] This study demonstrates the level of seriousness with which indoor water intrusion and/or accumulation should be addressed—lest we end up with toxic effects in our brains.

Remember, molds and mycotoxins are consistently non-discriminative. Even if you don't *think* mold can hurt you, it still can. Even if you don't *think* mold can hurt your child, it still can. Even if you don't *think* mold can hurt your baby, it still can. Remember, *thinking* can be dangerous to your health if it is not based on fact. Researching what is scientifically proven and what can be surmised from clinical data is necessary to formulate wise, health-enhancing decisions. Do not let some uninformed medical doctor's rhetoric lull you to death! Take the initiative to protect and preserve your health and the health of your family, because no one else will—not the Centers for Disease Control and Prevention, the American Academy of Pediatrics, or any other governmental or medical entity. Self-education can help you locate a licensed doctor(s) educated in the latest research and treatment options and can lead to self-preservation.

(ENDNOTES)
1 Rapp, Is This Your Child? p. 19-20 (William Morrow, New York, 1991).
2   Zahidul Islam et al., "Satratoxin G from the Black Mold Stachybotrys chartarum Evokes Olfactory Sensory Neuron Loss and Inflammation in the Murine Nose and Brain," Environ Health Perspect. 2006 July; 114(7): 1099-1107. Available at http://www.pubmedcentral.nih.gov/articlerender.fcgi?tool=pubmedid&pubmedid=16835065 (accessed August 2006).
3   The CDC Mold Work Group, "Mold Prevention Strategies," p. 21.
4   Islam et al., "Satratoxin G from the Black Mold Stachybotrys chartarum Evokes Olfactory Sensory Neuron Loss and Inflammation in the Murine Nose and Brain."
5   The CDC Mold Work Group, "Mold Prevention Strategies," p. 21.
6   Ammann, "Is Indoor Mold Contamination a Threat to Health?"
7   EPA, "Indoor Air Pollution: An Introduction for Health Professionals," U.S. Government Printing Office Publication No. 1994-523-217/81322, 1994, EPA 402-R-94-007, 1994. Also available at http://www.epa.gov/cgi-bin/epaprintonly.cgi (accessed March 2006).
8   Institute of Inspection, Cleaning and Restoration Certification (IICRC), Standard and Reference Guide for Professional Mold Remediation, S520, p. 43 (Vancouver, Washington, IICRC, December 2003) Hereafter referred to as IICRC S520.
9   Ammann, "Is Indoor Mold Contamination a Threat to Health?"
10   Curtis et al., "Adverse Health Effects," p. 3.
11   The CDC Mold Work Group, "Mold Prevention Strategies," p. 20.
12   Ibid.
13   Ibid., p. 22.
14   EPA, "Indoor Air Pollution," p. 9.
15   Gray, "Molds, Mycotoxins, and Human Health."
16   The CDC Mold Work Group, "Mold Prevention Strategies," p. 23.
17   Department of Biology, UNM, "Cell-Mediated Effector Responses." Available at http://biology.unm.edu/cadavid/Immunology/Notes/Lecture15.pdf (accessed August, 2006).
18   Ibid.

19   Goldberg, Alternative Medicine, p. 21.
20   Department of Biology, UNM, "Cell-Mediated Effector Responses."
21   Goldberg, Alternative Medicine, p. 21.
22   M J Apostolakos et al., "Hypersensitivity Pneumonitis From Ordinary Residential Exposures," Environmental Health Perspectives, September 2001, 109(9), p. 981. Also available at http://www.pubmedcentral.gov/articlerender.fcgi?tool=pmcentrez&rendertype=abstract&artid=1240451 (accessed April 2006).
23   M J Apostolakos et al., "Hypersensitivity Pneumonitis From Ordinary Residential Exposures."
24   Interview with Dr. Michael Gray, March 2007, and Gray, "Molds, Mycotoxins, and Human Health."
25   EPA, "Indoor Air Pollution," p. 9.
26   Apostolakos et al., "Hypersensitivity Pneumonitis From Ordinary Residential Exposures," p. 980.
27   The CDC Mold Work Group, "Mold Prevention Strategies," p. 24.
28   Ibid.
29   Peraica et al., "Toxic Effects of Mycotoxins in Humans."
30   EPA, "Indoor Air Pollution," p. 10.
31   The CDC Mold Work Group, "Mold Prevention Strategies," p. 24.
32   Ibid., p. 31.
33   The CDC Mold Work Group, "Mold Prevention Strategies," p. 21.
34   Islam et al., "Satratoxin G from the Black Mold Stachybotrys chartarum Evokes Olfactory Sensory Neuron Loss and Inflammation in the Murine Nose and Brain."
35   The CDC Mold Work Group, "Mold Prevention Strategies," p. 25.
36   Brasel TL, et al., "Detection of Trichothecenes Mycotoxins in Sera From Individuals Exposed to Stachybotrys chartarum in Indoor Environments."
37   See Section Three, interview with Dr. Jack Thrasher, April 2006.
38   The CDC Mold Work Group, "Mold Prevention Strategies," p. 25. (Italics added.)
39   PBS, "Is It Safe to Return?" September 19, 2005, PBS. Available at http://www.pbs.org/newshour/bb/weather/july-dec05/return_9-19.html (accessed August 2006); and See Section Three, interview with Dr. Jack Thrasher, April 2006.
40   Rapp, Is This Your Child? p. 523-524.
41   CDC, "Health Concerns Associated with Mold in Water-Damaged Homes After Hurricanes Katrina and Rita-New Orleans Area, Louisiana, October 2005," p. 43.
42   Ibid., p. 42-43.
43   Ibid.
44   Ibid., p. 44.
45   Chew, "Mold and Endotoxin Levels in the Aftermath of Hurricane Katrina: A Pilot Project of Homes in New Orleans Undergoing Renovation."
46   Ibid.
47   Ibid.
48   From written communication from Jim Pearson, September 2006, and IICRC S520, appendix
49   Chew, "Mold and Endotoxin Levels in the Aftermath of Hurricane Katrina: A Pilot Project of Homes in New Orleans Undergoing Renovation."
50   Ibid.
51   Ibid.
52   The CDC Mold Work Group, "Mold Prevention Strategies," p. 17.
53   See Section Three, interview with Dr. Jack Thrasher, April 2006.
54   Ibid.
55   Rafal L. Górny et al., "Fungal Fragments as Indoor Air Biocontaminants," Applied and Environmental Microbiology, July 2002, p. 3522-3531, Vol. 68, No. 7. Available at http://aem.asm.org/cgi/content/full/68/7/3522?view=long&pmid=12089037#R2 (accessed August 2006).
56   E-mail from Dr. David Straus, September 2006.
57   See Section Three, interview with Dr. David Straus, April 20; and Anyanwu et al., "The neurological significance of abnormal natural killer cell activity in chronic toxigenic mold exposure."

58   Interview with Dr. Michael Gray, March 2007; and Gray, "Molds, Mycotoxins, and Human Health."
59   See Section Three, interview with Dr. Jack Thrasher, April 2006.
60   Interview with Dr. Michael Gray, March 2007.
61   See Section Three, interview with Dr. David Straus, April 20.
62   See Section Three, interview with Dr. Jack Thrasher, April 2006.
63   See Section Three, interview with Dr. David Straus, April 2006.
64   Interview with Dr. Michael Gray, March 2007.
65   See Section Three, interview with Dr. Jack Thrasher, April 2006.
66   Interview with Jim Pearson, April 2007.
67   Ammann, "Is Indoor Mold Contamination a Threat to Health?"
68   See Section Three, interviews with drs. Jack Thrasher and Richard Lipsey, April 2006.
69   Rapp, Is This Your Child? p. 273.
70   Jack D. Thrasher, "Molds and Human Diseases." Available at http://www.drthrasher.org/ Molds_and_Human_Disease.html (accessed April 2006).
71   Bennett et al., "Mycotoxins," p. 2 and 4.
72   Curtis et al., "Adverse Health Effects," p. 2 and 6.
73   Ibid., p. 2-4.
74   Ibid., p. 2 and 6
75   Ibid.
76   Dorr G. Dearborn et al., "Pulmonary Hemorrhage and Hemosiderosis in Infants." Available at http://gcrc.cwru.edu/stachy/default.htm#THE%20CLEVELAND%20OUTBREAK (accessed August 2006); Ruth A. Etzel et al., "Investigator Team's Response to MMWR Report." Available at http://gcrc. cwru.edu/stachy/InvestTeamResponse.html (accessed August 2006); and CDC, "Acute Pulmonary Hemorrhage/Hemosiderosis among infants-Cleveland, January 1993-November 1994," MMWR December 9, 1994; 43: 881-3. Available at http://www.cdc.gov/mmwr/preview/mmwrhtml/00033843. htm (accessed March 2006).
77   CDC, "Update: Pulmonary Hemorrhage/Hemosiderosis Among Infants-Cleveland, Ohio, 1993-1999." Morbidity and Mortality Weekly Report, 2000, 49: 180-184. Available at http://www.cdc. gov/mmwr/preview/mmwrhtml/mm4909a3.htm (accessed August 2006).
78   Ruth A. Etzel et al., "Investigator Team's Response to MMWR Report."
79   CDC, "Update: Pulmonary Hemorrhage/Hemosiderosis Among Infants-Cleveland, Ohio, 1993-1999."
80   CDC, "Acute Pulmonary Hemorrhage/Hemosiderosis among infants-Cleveland, January 1993-November 1994," and CDC, "Update: Pulmonary Hemorrhage/Hemosiderosis among infants-Cleveland, Ohio, 1993-1996," MMWR 1997; 46:33-5. Available at http://www.findarticles.com/p/ articles/mi_m0906/is_9_49/ai_60895180/pg_2 (accessed March 2006).
81   CDC, "Update: Pulmonary Hemorrhage/Hemosiderosis Among Infants-Cleveland, Ohio, 1993-1999."
82   See Section Three, interview with Dr. David Straus, April 2006.
83   See Section Three, interview with Dr. Jack Thrasher, April 2006.
84   Sorenson, "Fungal Spores: Hazardous to Health?"
85   Vesper et al., "Stachylysin May Be a Cause of Hemorrhaging in Humans Exposed to Stachybotrys chartarum." Infect Immun. 2002 April; 70(4): 2065-2069. Available at http://www. pubmedcentral.nih.gov/botrender.fcgi?blobtype=html&artid=127818 (accessed August 2006).
86   Dorr Dearborn et al., "Clinical Profile of 30 Infants with Acute Pulmonary Hemorrhage in Cleveland." Pediatrics, vol. 110 No. 3 September 2002, pp. 627-637. Available at http://pediatrics. aappublications.org/cgi/content/full/110/3/627 (accessed August 2006).
87   Ibid.
88   Eduardo Montaña et al., "Environmental Risk Factors Associated with Pediatric Idiopathic Pulmonary Hemorrhage and Hemosiderosis in a Cleveland Community." Pediatrics, Vol. 99 No. 1 January 1997, pp e5. Available at http://pediatrics.aappublications.org/cgi/content/full/99/1/e5 (accessed August 2006).
89   CDC, "Update: Pulmonary Hemorrhage/Hemosiderosis Among Infants-Cleveland, Ohio, 1993-1999."
90   Ibid.

91   Dearborn et al., "Clinical Profile of 30 Infants with Acute Pulmonary Hemorrhage in Cleveland."

92   The CDC Mold Work Group, "Mold Prevention Strategies," p. 31.

93   Etzel et al., "Investigator Team's Response to MMWR Report."

94   Ibid.

95   See Section Three, interviews with drs. Thrasher, Straus, and Solomon, April 2006, April 2006, and July 2006 respectively.

96   See Section Three, interview with Dr. Jack Thrasher, April 2006.

97   Etzel et al., "Investigator Team's Response to MMWR Report."

98   "The Cleveland Outbreak." Available at http://gcrc.cwru.edu/stachy/default.htm (accessed August 2006).

99   CDC, "Investigation of Acute Idiopathic Pulmonary Hemorrhage Among Infants-Massachusetts, December 2002-June 2003," MMWR, September 10, 2004/53(35); 817-82 Available at http://www.cdc.gov/MMWR/preview/mmwrhtml/mm5335a4.htm (accessed August 2006).

100   Dearborn et al., "Pulmonary Hemorrhage and Hemosiderosis in Infants."

101   Committee on Environmental Health, American Academy of Pediatrics (AAP), "Toxic Effects of Indoor Molds."

102   Ibid.

103   Nelson, "Stachybotrys chartarum: The Toxic Indoor Mold."

104   Ibid., p. 6.

105   Ibid., p. 6.

106   Dr. Fungus, "Stachybotrys sp."

107   Nelson, "Stachybotrys chartarum: The Toxic Indoor Mold," p. 1.

108   Dr. Fungus, "Stachybotrys sp."

109   The CDC Mold Work Group, "Mold Prevention Strategies," p. 1.

110   Anyanwu E et al., "Brainstem Auditory Evoked Response in Adolescents with Acoustic Mycotic Neuroma Due to Environmental Exposure to Toxic Molds," Center for Immune, Environmental and Toxic Disorders. Available at www.pubmed.com (accessed February 2006).

111   Curtis et al., "Adverse Health Effects," p. 3.

112   Interview with Gene Thompson, Western Vice President of the Montana Landlords Association, Kalispell, Montana, October 2005.

113   Interview with Jim Pearson, November 2005.

114   See Section Three, interview with Dr. David Straus, April 2006.

115   Dr. Fungus, "Stachybotrys sp."

116   See Section Three, interview with Dr. Jack Thrasher, April 2006.

117   Dr. Fungus, "Stachybotrys sp."

118   ABCnews, "Family Settles for 22 Million Over Moldy House," November 8, 2005. Available at http://abcnews.go.com/GMA/OnCall/story?id=1291345 (accessed August 2006).

119   Curtis et al., "Adverse Health Effects," p. 2.

120   Nelson, "Stachybotrys chartarum: The Toxic Indoor Mold," p. 2.

121   See Section Three, interview with Dr. Andrew Puccetti, April 2006.

122   Curtis et al., "Adverse Health Effects," p. 1.

123   Ibid.

124   See Section Three, interview with Dr. Jack Thrasher, April 2006.

125   Ibid.

126   Ibid.

127   Islam, et al., "Satratoxin G from the Black Mold Stachybotrys chartarum Evokes Olfactory Sensory Neuron Loss and Inflammation in the Murine Nose and Brain."

CHAPTER SEVEN:

SCIENTIFIC CORRELATIONS OR ASSUMPTIONS?

In the months following Hurricane Katrina, many of us watched national news coverage of residents and volunteer cleanup crews remediating structures wearing no personal protective equipment—no respirators, no eye protection, and no gloves. We saw footage of entire high school football teams performing remediation without any type of personal protective equipment whatsoever. These well-intentioned boys ripped out mold-laden sheetrock, carpets, and cabinets disrupting mold and toxic dust. They unknowingly aerosolized mold spores, fungal fragments, fine particulates, and biological contaminants and compounds into the very air they were breathing. We watched heartwarming stories of volunteers who had come from northern churches to help their southern neighbors rebuild— without personal protective equipment. We saw news reporters and accompanying camera people, also without personal protective equipment, interview and capture on film these acts of benevolent compassion. While we admire these acts of selfless humanitarianism, without the proper use of adequate personal protective equipment, these heartfelt actions could very well cost many of these volunteers a portion of, or all of, their good health.

Some will get sick, some will not—why? This disparity is simple to understand when someone who becomes sick falls into one of the high-risk categories. But what about when two healthy, nonhigh-risk individuals work side by side inhaling all the same toxins, and one gets sick and one does not. Why? In addition to the increased susceptibility of those in high-risk groups, individual genetics also play a role in who falls ill and who remains strong and healthy. In other words, exposure to the same level of mold spores, chemicals, and other toxins will affect people differently based on genetic differences. For example, exposure to the same contaminants may affect some people very little, yet induce physical and psychological symptoms, illness, and even disease in others.

These inherent biological differences in people compound the difficulty in drawing scientific dose-response correlations between exposure to molds and the associated health outcomes. Dose-response correlations determine that exposure to "x" dose (amount) of mold spores will elicit "y" response (symptoms, illness, and/or disease). Some dose-response data have been correlated by scientists in regard to exposure to mold spore counts and related negative health effects. For example, "Some studies have shown that outdoor levels of mold spores are directly associated with childhood asthma attacks requiring a visit to an emergency room. Studies that have reported links between outdoor mold spore levels and childhood asthma attacks have found these respiratory effects even in areas where the daily airborne spore counts were relatively low (around 2,000 spores per cubic meter)."[1] However, in regard to human mycotoxin exposure and the induction of multiple symptoms, illnesses, and diseases, no scientific dose-response correlations have been determined. One scientist explains the inherent challenges, "In order to demonstrate that a disease is a mycotoxicosis [a disease or poisoning caused by mycotoxin exposure], it is necessary to show a dose-response relationship between the mycotoxins and the disease. For human populations, this correlation requires epidemiological studies. Supportive evidence is provided when the characteristic symptoms of a suspected human mycotoxicosis are evoked reproducibly in animal models by exposure to the mycotoxins in question. Human exposure to mycotoxins is further determined by environmental or biological monitoring. In environmental monitoring, mycotoxins are measured in food, air, or other samples; in biological monitoring, the presence of residues, adducts, and metabolites is assayed directly in tissues, fluids, and excreta."[2]

Another factor that hampers the scientific and medical communities' ability to draw direct scientific correlations between mold and mycotoxin exposure and the associated symptoms is the phenomenon of mixed mold mycotoxicosis. Dr. Ammann explains, ". . . [R]ecently concern has arisen over exposure to multiple mycotoxins from a mixture of mold spores growing in wet indoor environments. Health effects from exposures to such mixtures can differ from those related to single mycotoxins in controlled laboratory exposures. Indoor exposures to toxigenic molds resemble field exposures of animals more closely than they do controlled experimental laboratory exposures. Animals in controlled laboratory exposures are healthy, of the same age, raised under optimum conditions, and have only the challenge of known doses of a single toxic agent via a single exposure route. In contrast, animals in field exposures are of mixed ages and states of health, may be living in less than optimum environmental and nutritional conditions, and are exposed to a mixture of toxic agents by multiple exposure routes. Exposures to individual toxins may be much lower than those required to elicit an adverse reaction in a small controlled exposure group of ten animals per dose group. The effects from exposure may, therefore, not fit neatly into the description given for any single toxin or the effect from a particular species of mold."[3]

James Craner, MD, MPH, describes clinical symptoms presented by patients exposed to mixed molds and their mycotoxins. He reports, "Occupants developed their illnesses shortly after their homes had been water damaged. A few occupants had a specific building-related illness, such as hypersensitivity pneumonitis or asthma exacerbation, but most had a 'sick building syndrome' symptom complex involving irritation/inflammation of the mucous membranes, respiratory tract, and skin; fatigue; and/or neurocognitive dysfunction. All cases required months or years to correctly diagnose. Air, surface and/or bulk microbiological sampling in most of the homes yielded high concentrations of toxigenic fungi, including *Stachybotrys chartarum*, *Penicillium*, and *Aspergillus* species, emanating from water-damaged building materials."[4]

Dr. Thrasher, Dr. Gray, and colleagues explain in a published report on the symptoms of mycosis and mycotoxicosis, "Exposure to mixed molds and their associated mycotoxins in water-damaged buildings leads to multiple health problems involving the CNS [central nervous system] and the immune system, in addition to pulmonary effects and allergies. Mold exposure also initiates inflammatory processes."[5] Because of this multifaceted factor involving mold and mycotoxin exposure, Dr. Thrasher, Dr. Gray, and colleagues have proposed the term *mixed mold mycotoxicosis* to identify the multisystem symptoms and illnesses that can develop from exposure to multiple molds and mycotoxins.[6]

Multiple symptoms have been experienced by people in the mold-laden, Katrina-hit areas to such a degree that both the media and medical professionals have dubbed the multiplicity of symptoms the Katrina Cough. "The *Los Angeles Times* reported [that] a large number of people have developed an illness that many are calling the 'Katrina Cough.' The ailment could be the result of mold, dust, and debris spawned by the storm. Health officials said they are trying to determine how widespread the problem is, though the *Los Angeles Times* quoted physician Dennis Casey, who said the condition is 'very prevalent.'"[7] The article goes on to describe the symptoms of the Katrina Cough as ". . . scratchy eyes, sore throat, and excessive coughing."[8] The Katrina Cough was not limited to these seemingly harmless symptoms. More serious upper and lower respiratory-related illnesses and diseases developed. For example, Dr. Lipsey reports both he and a colleague developed bronchitis from just breathing the outside air in St. Bernard Parish six months after the hurricane (February 2006).[9]

According to Dr. Thrasher, the Katrina Cough symptoms are to be expected in this type of wet, moldy environment, as coughing is the lungs way of trying to expel foreign particles that have been inhaled. The foreign particles are the aerosolized debris, fungi, fungal fragments, fine particulates, mycotoxins, and all the other toxins that accompany wet-building environments, such as gram-negative and gram-positive bacteria, the endotoxins and bacterial cell wall toxins they respectively produce, and off-gassing VOCs. "People say they have this 'Katrina Cough' and try to laugh it off, but it's got to be a lot more serious than that," adamantly states Dr. Thrasher. He explains, the Katrina Cough is "[i]nflammatory conditions of the lungs. No question about that."[10]

As we know from earlier discussions, both the mycotoxin exposures and the initial allergic reactions can weaken the immune system, making the body more susceptible to invasive fungal infections and other potentially life-threatening diseases. Furthermore, common sense tells us the longer the duration of exposure, the more likely these allergic-type reactions will develop into sinus and respiratory infections and even more serious mold-related illnesses. In fact, a FEMA public health news bulletin cautions, "Exposure to the mold flourishing in flooded buildings can pose health problems, especially for those who have weakened immune systems or mold allergies, said Dr. Stephen Redd of the Centers for

Disease Control and Prevention's National Center for Environmental Health. Symptoms include coughing, wheezing, stuffy nose, and skin rashes."[11] Again, these are symptoms of the Katrina Cough.

When we add in the human genetic factor with multiple genetic components that affect a person's immune system, a dose-response relationship becomes even more convoluted and difficult to assess. For example, why do some people experience allergies from mold spores in the ambient air, while others do not? We are all unavoidably exposed on a daily basis to airborne fungal spores, fragments, and fine particulate matter because they are an innate, integral part of the environment. In addition, many of these species produce mycotoxins that are in/on the toxigenic spores, fungal fragments, and fine particulates, all of which float around in the ambient air and get inhaled. Once inhaled, these mycotoxins dissolve from internal sources of moisture, get absorbed into the blood, and circulate throughout the body, which can cause systemic effects.[12] So why, when we all are exposed regularly to also some level of mycotoxins, aren't negative health effects experienced by all? Again, a major determining component is genetic differences. Genetic variations are one reason some people develop fungal infections and/or fungal- and mycotoxin-related illnesses and diseases when exposed to indoor elevated levels of fungi and mycotoxins, while others do not.

According to medical professionals, several contributing factors affect a person's physiological response to mycotoxin exposures. One doctor states the severity of toxic effects from mycotoxin exposure ". . . depends on the toxicity of the mycotoxins, the extent of exposure, age and nutritional status of the individual, and possible synergistic effects of other chemicals to which the individual is exposed."[13] Another doctor concurs, expanding on a few additional contributing factors, "The symptoms of a mycotoxicosis depend on the type of mycotoxins; the amount and duration of the exposure; the age, health, and sex of the exposed individual; and many poorly understood synergistic effects involving genetics, dietary status, and interactions with other toxic insults. Thus, the severity of mycotoxin poisoning can be compounded by factors such as vitamin deficiency, caloric deprivation, alcohol abuse, and infectious disease status. In turn, mycotoxicosis can heighten vulnerability to microbial diseases, worsen the effects of malnutrition, and interact synergistically with other toxins."[14]

A 2003 published medical report points out, "It is difficult to prove that a disease is a mycotoxicosis. Molds may be present without producing any toxin. Thus, the demonstration of mold contamination is not the same thing as the demonstration of mycotoxin contamination. Moreover, even when mycotoxins are detected, it is not easy to show that they are the etiological [causal] agents in a given veterinary or human health problem. Nevertheless, there is sufficient evidence from animal models and human epidemiological data to conclude that mycotoxins pose an important danger to human and animal health, albeit one that is hard to pin down. The incidence of mycotoxicoses may be more common than suspected. It is easy to attribute the symptoms of acute mycotoxin poisoning to other causes; the opposite is true of etiology. It is not easy to prove that cancer and other chronic conditions are caused by mycotoxin exposure."[15]

The difficulty in correlating negative health outcomes to mycotoxin exposure has resulted in less than conclusive data regarding the health effects of mycotoxin exposure. More clearly understood are the health risks associated with exposure to fungal spores. This more understood risk spurred the NRDC, CDC, LDHH, and other entities to perform sampling surveys of ambient spore levels in the post-hurricane air, both indoors and outdoors. These test results, which confirm the presence of specific levels of spore counts and sometimes species, give a general overall picture of the amount of mold present post-hurricane. This type of documentation may, over time, prove to add insight into the health effects of mold exposure. For example, future review of the illnesses and diseases documented in the medical records of people present during and after Katrina may reflect patterns of health conditions similar to those known to be caused from exposure to elevated levels of spores and/or specific species of fungi. Thus, some of these post-hurricane developed health conditions may be attributed to exposure to the elevated mold spore counts and/or specific species documented post-Katrina. However, whether large scale patterns of illness and disease will indicate mold as a contributing factor is questionable because of the presence of so many other post-Katrina factors such as exposure to other contaminants and increased stress.

It is unknown at this time if the medical records of people whose health was negatively impacted by toxic exposures during and after the hurricane are being stored in some huge government database for future study. We do know, however, in order to receive medical reimbursement from FEMA, people are required to submit pertinent medical records, data which is then input into computerized FEMA files. It could be that tens of thousands of people are essentially involuntarily partaking in a live human experiment in which their medical records are documented and stored for future review to provide insight into patterns of post-hurricane illness and disease caused by mold and chemical exposures.

Dr. Solomon concurs, "New Orleans was a vast experiment, I suppose, except that all the people who were involved in that experiment were rather unwilling and unwitting. It is probably the first time that there's ever been such a huge number of people exposed to this particular mix of contaminants at these kinds of high doses."[16] If the compiled FEMA data were reviewed, large-scale patterns could point to mold exposure as a possible cause of long-term degenerative diseases and cancers. However, even if the data did indicate mold as a degenerative disease source, the data would not equate to or be considered scientifically correlated data because it was not drawn from controlled studies. Positive indicators could, however, lead to confirmation studies using animal models.

Regardless of the results from any forthcoming study of government-collected data, we all must use our own common sense when it comes to mold exposure, as no government agency is going to tell us *how much is too much*. To date, there are no set mold exposure limits. The CDC clearly states, "No health-based standards or recommended exposure limits (e.g., OSHA, EPA, NIOSH) for indoor biologic agents (airborne concentrations of mold or mold spores) exist."[17] Although no threshold limit values exist in the U.S. in regard to mold exposure, Dr. Chew and colleagues report, "Excessive mold can be cited as a health concern by the U.S. OSHA under their general duty clause. Still, no defined level of mold is listed that warrants a specific level of respiratory protection."[18] In other words, no government agency is going to tell us *what* level of mold exposure warrants *what* level of respirator protection.

For this reason, it is imperative people educate themselves regarding the health risks of mold exposure and make prudent decisions regarding exposure limits and the level of personal protective equipment that will provide adequate protection. Even if there were study-based recommended exposure limits, government agencies are reticent and slow to act, at best, in alerting the public to potential health risks associated with at-risk exposure levels. A prime example is the local, state, and federal government's handling of the post-hurricane elevated arsenic levels, as was addressed earlier. Unfortunately, this blind eye being turned by government agencies, in due time, will undoubtedly result in health repercussions for people exposed.

The lack of air quality regulations regarding mold is why the DEQ was able to issue on November 1, 2005—the same time frame in which the NRDC collected air samples for testing—the following statement: "Air samples show southeast Louisiana air meets federal standards. The Department of Environmental Quality has analyzed more than 30 air samples taken in the area impacted by Hurricane Katrina. The results show that ambient air has returned to pre-hurricane quality in most areas. DEQ scientists and toxicologists have also studied data from the U.S. Environmental Protection Agency's air-canister samples and mobile air-sample lab, known as TAGA [Trace Atmospheric Gas Analyzer]. Both agencies' data show air quality results meet all federal standards, according to DEQ."[19]

We cannot take these types of "rose-colored" governmental statements at face value. We must research matters thoroughly to uncover what is not being said that perhaps should be being said! For example, the DEQ website states, "'Our ambient monitoring to date indicates the air quality is fine,' said DEQ Administrator Chris Roberie. 'There are still large amounts of debris inside and outside of people's homes, as well as other safety issues not related to air quality that people should be aware of when returning.'"[20] What is important to understand is that the DEQ can make the statement, "Air quality is fine," because the assessment of mold spore counts aren't included in federal air quality standards—no federal standards exist. We must read between the lines and ask ourselves the following question: Is the DEQ's warning of "other safety issues not related to air quality that people should be aware of when returning" referring to the large amounts of mold growing on the "still large amounts of debris inside and outside of people's homes"[21]

Just because mold spore counts aren't regulated by federal guidelines doesn't mean the government isn't interested in tracking the health effects of such exposure. The CDC, in its post-hurricane report, "MOLD: Prevention Strategies and Possible Health Effects in the Aftermath of Hurricanes Katrina and Rita," states, "It is important that clinicians report cases of mold-induced illness to local health authorities to assist in health surveillance efforts."[22]

The CDC further states, "Because of the large number of flooded and mold-contaminated buildings in New Orleans and the repopulation of those once-flooded areas, a large number of people are likely to be exposed to mold and other microbial agents. Efforts to determine the health effects of these exposures and the effectiveness of recommendations to prevent health effects from mold exposure require a public health surveillance strategy. Developing such a strategy requires that federal and local health agencies work together to monitor trends in the incidence or prevalence of mold-related conditions throughout the recovery period. Tracking different health outcomes that may be caused by mold exposure requires different surveillance methods. In some cases, follow-up research will be needed to verify that surveillance findings and health outcomes are the result of mold exposure."[23]

Tracking different health outcomes of those exposed to mold post-Katrina is a sizable task to undertake. Returning residents, as well as exposed evacuees who permanently relocated elsewhere, are unwitting candidates for this live human experimentation to monitor the health implications from mold exposure. Non-returning residents may be sick because the immediate post-hurricane air was filled with concentrated levels of multiple species of mold spores, as billions of dormant mold spores were hydrated and aerosolized during the hurricane.[24] Thus, even if a person evacuated shortly after the hurricane, he/she could develop negative health repercussions from having had inhaled these elevated levels of spores and other contaminants. Remember, Dr. Lipsey said he and a colleague developed bronchitis from inhaling just the air while walking through the neighborhoods five months post-hurricane![25]

Also remember, the post-hurricane air was heavily contaminated with chemicals and aerosolized debris that took weeks to settle. Inhalation of the contaminated air, as well as exposure to contaminated water, could have planted the spores, so to speak, for long-term health implications even in evacuees who did not return to the affected areas. Even limited exposure to these contaminants could have been health devastating, especially for anyone in the high-risk groups.

These high-risk individuals were of special interest to the CDC. It specifies, "Providers caring for patients at high risk for poor health outcomes related to mold exposure could be targeted."[26] The CDC further instructs, "The surveillance data should be used to identify increases in disease that are significant enough to trigger public health interventions or follow-up investigations to learn the reason for the increase and establish targeted prevention strategies."[27]

This type of data collection may aid in formulating future government responses to hurricane-size mold levels and may affect future treatment plans for mold-related illnesses based on treatments that proved effective in combating illnesses induced by Katrina exposure. Unfortunately, it will do little to assist people already exposed to the multi-mix of molds, mycotoxins, and other contaminants that have individually and synergistically adversely affected human health post-Katrina. Without a doubt, the post-hurricane levels of elevated spore counts created a human laboratory full of potential test subjects for doctors, scientists, and researchers. The Katrina natural disaster has, undoubtedly, given local, state, and federal entities a never-seen-before opportunity to evaluate and track the health of mold victims—to document the resultant health conditions, the responses to medical treatments, and the casualty rate.

The unsettling question we have to ask is: Why didn't government entities warn people about the dangers of mold exposure BEFORE the hurricane hit—before many areas had no TVs or radios because of loss of electricity? For days prior to the hurricane's landfall, most Gulf Coast residents were glued to their TVs watching hurricane coverage waiting to see if their respective communities were slated to receive a direct hit. Disseminating warnings during this pre-landfall period would have reached a captive audience, educating hundreds of thousands of people with health-preserving and lifesaving information. There is no acceptable "excuse" for this lack of action. Mold has one very consistent, innate property—it will *always* grow in wet-building materials after 24 to 72 hours. With the vast size of Katrina filling the entire radar screen, water damage and mold disruption was inevitable. Even more inhumane was the lack of health warnings disseminated after the hurricane—after the extent of the mold infestation and chemical toxicities were readily apparent. How many people's health could have otherwise been preserved?

We cannot turn back the clock of time and undo exposure that, in hindsight, could possibly have been avoided, but we can certainly become more proactive in guarding against future exposures. We, as individuals and families, must safeguard our health by becoming aware of current scientific data in regard to the health effects of mold and mycotoxin exposure; we must use basic common sense to logically reason from "assumptions" the health risks that have yet to be scientifically proven. For example, scientifically, we know fungal exposure can negatively affect our health, yet mycotoxin inhalation has yet to be *scientifically* correlated to human illness and disease. So, without scientifically proven correlations of deleterious health effects from inhalation of mycotoxin, do we *just not worry* about them? Hardly. We must use wise judgment and protect ourselves today from exposures to toxins that will be scientifically correlated tomorrow, next year, or ten years from now. We must use our God-given ability of reason to preserve our health and the health of our families. We must not be beguiled by carefully worded statements issued by governmental agencies.

For example, CDC literature clearly states, "There is currently no conclusive evidence of a link between indoor exposure to airborne mycotoxin and human illness."[28] This statement is based on the fact that scientifically controlled clinical studies have yet to produce *conclusive* evidence. This lack of conclusive evidence does not mean research studies do not "suggest" or "indicate" that human illness and disease *may* be caused by indoor exposure to airborne mycotoxins. It just means the integrity of the findings have been questioned or discounted because of confounding factors, as was the case in the infant pulmonary hemorrhage incidents in Cleveland, Ohio.

From a scientific point of view, Dr. Straus concurs that more research is needed to *scientifically correlate* mycotoxin inhalation to human illness and disease.[29] But we as laypeople, just interested in preserving our health, must be very careful not to let scientific technicalities create a false sense of complacency, lest it quicken our demise. For example, the CDC, Dr. Straus, and others in the scientific world have concluded that the scientific threshold of "conclusive evidence" has not yet been met in regard to mycotoxin inhalation causing human illness and/or disease. But let's apply a little common sense and think about the conclusive evidence we do have. It is a scientifically proven fact that indoor environments containing wet-building materials will contain a variety of airborne contaminants: fungi, fungal fragments, fine particulates, mycotoxins, gram-negative and gram-positive bacteria, endotoxins, bacterial cell wall toxins, and off-gassing of VOCs from both organisms and building material. It is a scientifically proven fact that these air contaminants get inhaled. So whether negative health effects, such as infant pulmonary hemorrhage/hemosiderosis, are caused by one specific contaminant, by a mixture of contaminants, or by a synergistic interaction between certain contaminants, is irrelevant—nonscientifically speaking.

The important concept to understand and remember is that indoor environments containing wet-building materials, wet furniture, indoor plants in pots filled with mold-contaminated dirt and/or microorganism-filled sitting water, waterlogged concrete foundations, and water pooled beneath the foundations, just to name a few, are all detrimental to health. Let's not get hung up on the scientific technicalities of which contaminant is responsible for health deterioration when all these components, and probably more, are innately inherent in wet-building environments. Let's not fool ourselves—the biological organisms and chemical compounds inherent in indoor environments with wet-building materials *will* adversely affect human health. Both short-term and long-term exposure can lead to illness, disease, and possibly even death for those at high risk, such as infants. As Dr. Thrasher points out over and over again—it is the entire mix of contaminants with which we must be concerned.[30]

This determination of lack of "conclusive evidence" enables the CDC to publish statements that, for all intents and purposes, go against basic common sense. For example, the CDC states, "Although the potential for health problems is an important reason to prevent or minimize indoor mold growth and to remediate any indoor mold contamination, currently there is inadequate evidence to support recommendations for greater urgency of remediation in cases where mycotoxin-producing fungi have been isolated."[31] Given the previously reviewed health risks possible from mycotoxin exposure, this statement is based not in logic but has drowned in scientific technicalities.

Furthermore, we must keep in mind the snail's pace of governmental bureaucracy. Dr. Lipsey points out, "By the time an agency comes forth with information for the public on the health dangers of a new chemical or mycotoxin, etc., they may have known about the dangers for many years but could not prove it with epidemiological data."[32]

The CDC acknowledges this lack of epidemiological data in regard to the health effects of mold exposure. It states, "For most adverse health outcomes related to mold exposure, higher levels of exposure to live molds or higher concentrations of allergens on spores and mycelia are thought to result in a greater likelihood of illness. However, there is currently no standardized method to measure the magnitude of exposure to molds. For this reason, it is not possible to sample an environment, measure the mold level in that sample, and make a determination as to whether the level is low enough to be safe or high enough to be associated with adverse health effects."[33]

Some medical and science professionals believe mold exposure limits will never be established because health effects vary so widely depending on individual genetics. They further cite the difficulties in creating scientific dose-response data in a controlled environment. Dr. Straus explains some inherent limitations regarding human studies. He states, "The truth of the matter is—and this is why there are no state or federal standards saying what levels of fungal spores in the air are safe—that nobody knows what those levels are. No one will ever know what those levels are because no one will ever be able to do experiments on human beings—put them in a room, make them inhale high concentrations of fungal spores, and see when they begin to get sick."[34]

The logical question to ask is: Why can't risk assessment models be used to determine exposure limits for fungi as they are used to determine exposure limits for arsenic and other heavy metals? Dr. Straus explains, "Arsenic and other heavy metals are just one entity. Fungi are alive and no one is exposed to just one fungus." Even if someone was exposed to just one fungus, Dr. Straus explains that fungi can be very complex organisms. "Then you add mycotoxins to the mix and it is now complicated beyond belief," says Dr. Straus.[35] The infeasibility of using risk assessment models to assess exposure limits to complex mixes of organisms and compounds is addressed by scientists at Cornell University. As earlier noted, they state, "The more complex the system being modeled, the more vulnerable the model and conclusions drawn from it are to errors resulting from the gaps between the model and reality." They noted these difficulties as reasons why risk assessments ". . . do not address synergistic impacts."[36]

Preserve your health. Do not allow the lack of scientifically set fungal exposure limits to lull you into complacency, believing there are no health risks from mold and mycotoxin exposures. Exposure limits simply have not been scientifically correlated yet; thus common sense must prevail. For example, Dr. Straus gives a rough guideline when dealing with mold exposure. He estimates, "If the levels are higher than we normally see (10 or 20 times higher), then obviously that is a cause for concern."[37] He concludes, ". . . [P]eople should not be exposed to growing fungi in buildings because we cannot predict what will happen to them."[38]

Since no set exposure limits based on scientifically proven risk assessments are available, we must all take personal responsibility for our own health. We must avoid exposure to elevated levels of fungi and the mycotoxins some species produce, especially when confined in indoor spaces, as the health dangers of indoor mold exposure have long been recognized. Dr. Thrasher and colleagues report, "The potential harmful effects of exposure to molds in inhabited buildings were recognized in early biblical times."[39] Dr. Górny adds, "Probably the first reference about the destructive influence of fungal flora on human dwellings and clothes is found in the 3rd Book of the Bible, Leviticus [chapter 14, verses 34–47, King James version]."[40]

By now, we shouldn't need to have headlines screaming *Biological Warfare Agents* for mold to get our attention. Common sense should tell us the potential exists for long-term health consequences anytime someone is exposed to elevated levels of molds and mycotoxins—especially for anyone at high risk. Pediatrician Scott Needle, MD, of Waveland, Mississippi, who has treated many post-hurricane pediatric residents, confirms there are possibilities of long-term health repercussions.[41]

We must be aware that seemingly innocent allergies and sensitivities can be indicative of a future filled with more serious mold-induced illnesses and diseases. Then the big question of *Cancer* looms. Will exposure to post-hurricane levels of mold, mycotoxins, fungal fragments, fine particulates, chemicals, and other toxic and biological contaminants lead to increased levels of cancers?

Dr. Thrasher offers personal insight into this question. He states, "It's too early to say. When you get into cancer, cancer is the end stage of a disease process. My wife, for example, just died last July of kidney cancer. That's an end stage. The question is: What led up to the end stage? She was an ex-smoker. Was it smoking? Was it the mycotoxins in the tobacco or what? We don't know, but her end stage was cancer. Up until she got to her end stage and death, she went through all the symptoms and health problems that you [the Billings family] have. She went into olfactory sensitivity; she went into chronic fatigue; she went into fibromyalgia. She went into all these health problems before she succumbed to the cancer, so it's a continuous process. The answer to your question, I'll say, within a reasonable scientific probability, would be yes. We will see an increase, but what types of cancers, I don't know."[42]

To realize that many of the symptoms our family was experiencing could be considered precancerous was very pro-activating. In fact, the continual stream of knowledge and inspiration we gained from the professionals we interviewed while researching this book served as a constant source of strong motivation that encouraged us to consistently make life-enhancing choices, as much as possible. We could not change the fact that we had been exposed to toxic contaminants, but we could move forward in as positive a health-supporting lifestyle as possible, minimizing future harmful exposures via inhalation, touch, and ingestion.

Dr. Santella supports this type of philosophy. She warns, "Stress has been shown to have an impact on cancer and disease risk in general. I think people have to be realistic about what their total exposures are and obviously minimize exposures, but not go into a panic mode especially about bad things that it is too late to do anything about, and from then forward, just do the best they can in terms of a healthy lifestyle."[43]

A positive, proactive approach is also encouraged by Udo Erasmus, PhD, an expert in the field of nutrition, specifically fats and oils. He states, ". . . [T]he stress, the lack of help from the emergency agencies, the depression, confusion, anger, and fear also knocked down the immune system. We are not talking just about physical causes of illness. There are mental and emotional causes of illness that also play a very powerful role, can lead to the same diseases, and will certainly enhance the damage done by toxic microorganisms, toxic molecules, and other physically toxic agents. In addition, poor eating habits, poor lifestyle practices, and general neglect of our health and fitness also play a role in the extent to which we become vulnerable and then become victims in a crisis, whatever the nature of that crisis."[44]

No matter the circumstances in which you encounter mold—crisis or not—the choice is quite simple: Remove yourself from the mold exposure and the accompanying toxins and preserve your health, or don't remove yourself and risk launching a downward spiral of your health. Remember, the healthy choice may not be the convenient choice, but it is the only wise choice. It is also the only financially prudent decision. Loss of health can lead to mounting medical bills and possibly inhibit your ability to work.

The loss of health and property resulting from mold growth and exposure has long been a focus of the insurance industry. In fact, health liability issues relating to mold exposure have been being played out in the legal battle field, putting scientific and medical conclusions through the process of legal scrutiny. The OSHA-referenced *IICRC S520 Standard and Reference Guide for Professional Mold Remediation* published to educate people in the field of industrial hygiene notes, "Current controversy about the validity and severity of mold health effects (largely debated in the context of construction defects litigation and insurance-related flooding events in the U.S.) stems from several sources: a lack of scientific information about the pathophysiology and mechanism(s) of illness and limitations in quantifying exposure and correlating that with specific health effects. Further, there is the question of whether health effects are transient or permanent and whether certain populations are predisposed or more susceptible to certain health effects."[45]

Insurance companies are fiercely fighting the validity of claims regarding loss of structure, personal property, and physical and mental health resulting from structural mold growth and the resultant exposure to occupants, the insureds. Attorney Cynthia Mulvihill states, "I see the bigger carriers drafting mold exclusions as tight as they possibly can . . . With the mold exclusions, what the insurers are trying to do, of course, is protect themselves from losses. . . . So, if someone makes a claim today for a mold problem, he or she is likely to have either very limited coverage or no coverage at all."[46] [See interview with Ms. Mulvihill in Section Three.]

Issues of liability in regard to losses incurred from mold growth and exposure are currently being shaped in legal battles. These financial war games involve not only insurance companies but also powerful industry groups. According to Melinda Ballard, who founded Policyholders of America (POA) to provide information to policyholders to assist in facilitating claim payments, "Builders, insurance companies, and large landlords do not want this issue front and center in medical research or have the CDC say it's toxic. This is an issue fraught with political interests, not public interests."[47] [See interview with Ms. Ballard in Section Three.]

Ms. Ballard may very well be right. For example, the accuracy of some medical position papers that appear to minimize the health effects of mold exposure have been scrutinized and challenged by professionals in the fields of medicine and science. The reports were authored by highly credentialed scientific and medical professionals and published in peer-reviewed journals. In question are alleged large donations made by insurance companies to various teaching universities where some of the authors work. Also in question are alleged monetary payments made by insurance companies directly to some of the authoring doctors. Most of these alleged financial ties are not disclosed in the respective position papers, although one position paper does openly state it was commissioned by an insurance company.

This alleged quid pro quo leads to the following question: Are the powers that be within the insurance industry using potentially biased, commissioned, and allegedly paid-for reports (directly or indirectly) to legally counter sound scientific and medical studies that indicate negative health effects from mold and mycotoxin exposures? Essentially, is the insurance industry, in an attempt to reduce potential liabilities from unfavorable rulings in the courtroom, creating defendant-friendly reports? [A defendant in a mold case could be, for example, the owner (the insured) of an apartment building. Depending on policy coverage, the insurance company insuring the building could be required to pay for the defense. A plaintiff would be the party who originated the lawsuit, for example, the tenant.] Answers to these questions may come in shades of gray, rather than definite black and white. Thus, we must keep in mind that big bucks are at stake. Therefore, big-business-type tactics could be in play. We must weigh not only article content but also possible author bias and conflict of interest—even when position papers are published and peer-reviewed. We must confirm the source of research funding.

The power of these published, peer-reviewed position papers should not be underestimated. In order to be admissible in a court of law, scientific and medical evidence must meet the requirements set forth by case law. According to Ms. Mulvihill, one such case law is *Daubert*, which is a U.S. Supreme Court decision that involves state court laws and provides a checklist for the admissibility of evidence. The *Daubert* checklist is as follows: "Whether the theories and techniques employed by the scientific expert have been tested; whether they have been subjected to *peer review and publication*; whether the techniques employed by the experts have a known error rate; whether they are subject to standards governing their application; and whether the series and techniques employed by the experts enjoy widespread acceptance."[48]

According to Ms. Mulvihill, another test for admissibility of expert witness testimony is the *Kelly-Frye*[49] test, which is still used in many jurisdictions. She reports, "That test is 'The proponent of evidence derived from a new scientific methodology must satisfy three prongs, by showing, first, that the reliability of the new technique has gained general acceptance in the relevant scientific community, second, that the expert testifying to that effect is qualified to do so, and third, that correct scientific procedures were used in the particular case.'"[50]

Ms. Mulvihill explains the significance of the term *general acceptance*. She states, "On appeal, the 'general acceptance' finding under prong one of *Kelly* is a mixed question of law and fact subject to limited *de novo* review [*de novo* is a Latin term for *anew*, which means 'starting over']. The appellate court reviews the trial court's determination with deference to any and all supportable findings of 'historical' fact or credibility and then decides as a matter of law, based on those assumptions, whether there has been general acceptance."[51] In other words, if a methodology that has been used to derive evidence presented in court has been published and peer reviewed, it will more likely be viewed as "generally accepted."

Ms. Mulvihill further explains, "The first thing I look for is that not only have the theories been tested in accordance with scientific procedures but also whether they've been peer reviewed and published. If something is peer reviewed and published, the testing procedures have usually been verified, and that's why I try to base what I can on peer-reviewed publications."[52]

Thus, the published, peer-reviewed position papers that allegedly downplay the negative health effects of mold exposure would most likely be ruled as admissible in a court of law. Obviously, defense attorneys will use these defendant-friendly position papers to undermine plaintiffs' cases. These pro-defense reports can potentially influence jurors to find a defendant-favorable verdict and/or reduce the amount of a defendant's judgment. Depending on policy coverage, a defense may be paid for by a defendant's insurance company. When a defendant is found liable, the awarded judgment may be paid in whole, or in part, by the defendant's insurance company. Therein lies the insurance industry's potential motivation.

Issues of admissibility can also influence lawsuit outcomes. According to Katrine Stevens, formerly of Gloucester, Massachusetts, issues of admissibility of evidence often reduce the number of complaints that are able to be filed in a lawsuit. Ms. Stevens faced the same difficulty in her case, which she filed after realizing a waterfront condominium she purchased in 1995 was mold infested.[53]

"'It was a dream home in an unbeatable location,' said [Ms.] Stevens. 'But within days, I was itchy, my face turned red, it felt like needles were stuck into my skin, and I had difficulty breathing.'"[54] After just six weeks, she was forced to evacuate her new dream home after experiencing flu-like and asthmatic symptoms from toxic mold. Thus began an eight-year, precedent-setting lawsuit against the condominium trust. In November 2003, the *Boston Globe* reported, ". . . [A]n Essex County Superior Court jury in Salem awarded [Ms.] Stevens $285,000. With interest, the amount totals $549,326."[55]

According to the *Boston Globe*, "Legal specialists say the Gloucester case is the first jury award for a toxic mold case in Massachusetts and one of an increasing number of lawsuits nationwide. While the exact number of such cases cannot be determined, the Insurance Information Institute, a New York-based industry group, estimates legal claims involving mold have tripled nationwide in the past three years with $3 billion paid out in homeowners' policies last year for mold-related cases, up from $1.4 billion in 2001. In 1999, such claims were virtually nonexistent, the trade group said."[56]

"Ultimately, the plaintiff's [Stevens'] attorney said he believes that through the evidence presented and the testimony of his client and other witnesses, he was able to make the connection that not only was there mold in his client's condo unit, but it was the mold that caused the loss of her property and the deterioration of her health," reported the *Massachusetts Lawyers Weekly*.[57]

Although Ms. Stevens received a sizable verdict in her lawsuit, the bottom line is no amount of money can sufficiently compensate her, or others, when loss of health occurs from mold and mycotoxin exposure; no amount of money can restore the health—physical, mental, and emotional—that was damaged. Oftentimes, medical injuries do not even end up in front of a jury because medical evidence is ruled inadmissible due to lack of supporting epidemiological studies. Ms. Ballard faced admissibility difficulties in her case when she legally battled an insurance giant over a mold-related loss. After the toxic exposure, her husband, Ronnie, was diagnosed with toxic encephalopathy—brain damage—said to have been caused by the mold exposure, according to Ms. Ballard.[58] However, the merits of the personal injury portion of the case were never heard by the jury.

The Texas Court of Appeals ruled in regard to the personal injury portion of the *Ballard* case, "The district court found that although the experts' foundational data to prove general causation met the requirements of *Daubert* and *Robinson*, the data was unreliable according to the factors discussed in *Havner*. . . . The *Havner* court determined that for an epidemiological study to be a reliable foundation, it must be unbiased in its design, otherwise properly designed, properly executed, and show that exposure to the substance more than doubles the risk of injury. Another factor for determining if an epidemiological study is reliable is that it must be capable of repetition with the same results ninety-five percent of the time, known as a confidence interval of ninety-five percent."[59]

In other words, in order for a personal injury case to meet the requirements set forth in *Havner*, epidemiological studies would have to show a doubling of symptoms with a margin of error less than 5 percent.[60] Ms. Ballard gives an example in regard to epidemiological studies relating to encephalopathy. She explains, "If it were legal to intentionally expose 1000 people to the exact same species of mold and quantity of mold for the same time period and expose another 1000 people to nothing, and if the exposed group was 100 percent more likely (with a confidence ratio of 5 percent) to develop brain damage, then the jury might be able to hear the merits of the personal injury case. Such research cannot legally and ethically be conducted; therefore, it does not exist."[61]

Obviously, prevention and avoidance of fungal contamination is, undoubtedly, first and foremost. Unfortunately, as is often the case, by the time mold exposure is identified as the health-robbing baron, it is usually too late, as health damage has already occurred. Many mold victims continue to wrestle with mold-related health problems long after the initial exposure. Thus, if we find ourselves working or living in a mold-infested environment, the only health-preserving decision is to immediately remove ourselves from the exposure, no matter how inconvenient or costly. For example, if we are property owners, we should stay with friends or relatives or in a hotel until remediation is complete. If we are renters, we should do the same until our rental unit has been remediated or a rental unit free of mold infestation has been located. Even after remediation is complete, we must remain vigilant about possible fungal regrowth. The price of not doing so could destroy our most valuable and irreplaceable asset—our health.

Education and vigilance regarding mold is a necessity. Even a few weeks of exposure—as occurred in our case and Ms. Stevens'—is enough to pummel previously sturdy good health. As just mentioned, some people recover, while others keep fighting *The War Within*, battling a daily host of mold-related health challenges. Although some mold victims receive compensation for their losses and receive large legal settlements and/or judgments, most do not. Furthermore, victims who do receive financial settlements are often restricted from discussing their cases by the terms of their legal settlements. Therefore, we must look at the health and property losses incurred by people who are able and willing to come forward with their stories and use the wisdom we can glean from their unfortunate experiences. Let not the price they, and others, have paid go unheeded.

We must realize exposure to mold truly equates to playing genetic Russian roulette with our health. In order to win at this game, we must take individual responsibility to educate ourselves so we can make decisions that will preserve and enhance the longevity of our lives—not get shot down in the crossfire from one of Mother Nature's own. The legal, medical, and scientific issues relating to mold are complex, and the degree to which mold can hurt us will be played out in courts of law and in the scientific and medical communities for many, many years to come. However, we must not rely on what is, or isn't, proven in a court of law or in the inherently slow-moving fields of science and medicine to protect us.

Additionally, we must look to the past to benefit from a historical perspective of similar events, as history has a tendency of repeating itself, and hindsight is often 20/20. Let's take a look at 9/11, for example. Five and a half months after 9/11, Hugh Kaufman, then chief investigator for the EPA's Ombudsman Office ". . . accused the EPA and other government agencies of deliberately not testing the air quality in the World Trade Center area properly and possibly covering up the reasons why," reported CNN New York.[62]

Mr. Kaufman told CNN New York, "'I believe EPA did not do that because they knew it would come up not safe and so they are involved in providing knowingly false information to the public about safety . . . Not just EPA, [but] the state and the city, too,' he said. 'We also had testimonies that all the agencies—local, state, and federal—have been consorting together every week to discuss these issues.'"[63]

According to CNN New York, Mr. Kaufman said that he believed the air quality at Ground Zero was worse than the EPA would admit and that the agency had mislead the public about the inherent risks for residents and workers in the area.[64] We must ask, using a term Dr. Solomon voiced earlier: Did this same type of governmental "funny business" take place in the New Orleans area and the other areas hit by Katrina? Is history repeating itself?

Unfortunately, we can already see some disturbing parallels. For example, the fact that the EPA maintained its testing showed safe air quality levels at and around Ground Zero[65] lured trusting people into an area where exposures to toxic contaminants would ultimately prove harmful to many of the people's health. This same ploy was used post-Katrina to reassure residents, workers, and tourists and entice them back to the New Orleans area—regardless of the cost to their personal health. The EPA declared only two and a half months after Katrina that it was safe to return.[66] But was it?

During this same time period, the NRDC collected air samples that reflected high and very high levels of mold spore counts. Nevertheless, other governmental agencies backed the EPA's claims, not revealing the existence of other hazardous contaminants present during this same time period. For example, Tulane University reported, "The Louisiana Department of Environmental Quality (DEQ) has analyzed air samples taken by DEQ and the U.S. Environmental Protection Agency (EPA) in the area impacted by Hurricane Katrina. The results show that ambient air has returned to pre-hurricane quality and meets all federal standards."[67] Remember, as mentioned earlier, federal air quality standards have no regulations in regard to fungal spore counts. In addition to the existence of mold contaminants, the following were also present in elevated levels in many hurricane-hit areas during this same time period: gram-negative and gram-positive bacteria, endotoxins, bacteria cell wall toxins, arsenic and other heavy metals, PCBs, diesel fuel, and other industrial chemicals. This list does not include the many contaminants that weren't measured, yet were still present.

Nevertheless, further statements from Tulane University included, "The air quality in New Orleans is very safe according to comprehensive testing by the EPA, DEQ, CDC, and the New Orleans Department of Health. Dr. Kevin Stephens, Director of the New Orleans Department of Health, stated that the Department has tested, retested, and continues to test land, sea, and air, and these tests repeatedly confirm that the city is safe for residents and visitors."[68]

In addition, the LDEQ claims, "Based on our initial assessment and the environmental data we have gathered and reviewed since, LDEQ and its partner environmental and public health agencies continue to support the statement that there are generally no *unacceptable* long-term health risks directly attributable to environmental contamination resulting from the two hurricanes [Katrina and Rita]."[69]

If there are no "unacceptable long-term health risks," we must ask ourselves, what are the "acceptable" long-term health risks? If *you* ultimately develop a disease induced from the environmental contamination resulting from Katrina and/or Rita that eventually proves fatal, and thus *you* become a historical acceptable health statistic, will you feel losing your life was an *acceptable* long-term health risk?

We must not assume we will not become one of the acceptable health statistics. Instead, we should *all* assume we *will* become one of the acceptable health statistics because then our desire for health preservation might spur us to action, implementing a health-restoring diet and lifestyle. Don't get caught up in the game of statistical calculations. They are based on inherently flawed models that can be agenda driven by industry and/or governmental groups. Additionally, they may dangerously underestimate the added health risks from synergistic interactions among contaminants.

The Katrina debacle is a clear demonstration that government agendas—for example, to rebuild the economic infrastructure of New Orleans—include acceptable health risks to the workers and other people whose presence is needed in order to accomplish the designated goals. History shows that larger societal goals are often at odds with individual health preservation. It is the age-old issue of economics versus health. To illustrate this disparity, we only have to ask: If the government had been more forthright in disseminating information of the associated health risks that could be experienced by returning residents, paid and volunteer remediation workers, and people necessary to facilitate the rebuilding phase, what would have been the economic impact to the New Orleans area? Now we have to ask: What will be the long-term health consequences for the residents who returned and the other people who helped revitalize the area?

Environmental expert Hugh Kaufman feared that Gulf Coast residents and volunteers exposed to deadly toxins could suffer health effects similar to those of 9/11 workers, albeit to a lesser degree.[70] After 9/11, Mr. Kaufman also predicted negative health repercussions for the workers at Ground Zero.[71] He stated, ". . . I predicted it then, and it turned out even worse."[72] Only time will tell the long-terms health effects caused by exposures from Hurricane Katrina. We must not be Pollyannas. We must, in order to preserve our health, understand that politics and big business are intricately intertwined and money is the bottom line—not health.

Wherever we live and whatever our individual circumstances, we need to be proactive and build personal confidence in our own abilities to assess through self-education what is best for us and our families. Although documentation published by both industry and governmental groups is often agenda-based, it is balanced by literature authored by free-thinking scientists and medical doctors. We must learn to read between the lines and weigh information from many different sources in order to best protect ourselves and our families from the perverse amount of toxic contaminants that pervade our world. Sadly, our world is over inundated with toxins, be them toxic concoctions stirred in the wake of devastating hurricanes or floods, indoor concentrations of biological microorganisms and toxins from structural moisture, or decades of industrial-generated pollutants. Most of all, remember, it is not the government's job to educate us. As Dr. Lipsey points out, "Regulatory agencies are not educational institutions; people need to take the initiative to educate themselves."[73]

(Endnotes)

1 NRDC, "Mold."

2   Bennett et al., "Mycotoxins," p. 5.

3   Ammann, "Is Indoor Mold Contamination a Threat to Health?"

4   James Craner, M.D., M.P.H., "Building-Related Illness in Occupants of Mold-Contaminated Houses: A Case Series." In: Johanning E, ed. *Bioaerosols, Fungi and Mycotoxins: Health Effects, Assessment, Prevention and Control*. Albany, NY: Eastern New York Occupational & Environmental Health Center; 1999; 146-57. Available at http://drcraner.com/articles_presentations.htm (accessed March 2006).

5   Gray et al., "Mixed Mold Mycotoxicosis: Immunological Changes in Humans Following Exposure in Water-Damaged Buildings," *Arch Environ Health*. 2003 Jul; 58(7):410-20. Available at http://www.ncbi.nlm.nih.gov/entrez/query.fcgi?db=pubmed&cmd=Retrieve&dopt=AbstractPlus&list_uids=15143854&query_hl=4&itool=pubmed_docsum (accessed February 2006).

6   Ibid.

7   WTOP, "'Katrina Cough' Plagues Gulf Residents," November 4, 2005. Available at http://www.wtopnews.com/?sid=612794&nid=104 (accessed March 2006).

8   Ibid.

9   Lipsey, "The Mold and Bacterial Results from the Sampling Dr. Lipsey did in St. Bernard Parish," and e-mail from Dr. Richard Lipsey, October 6, 2006.

10   See Section Three, interview with Dr. Jack Thrasher, April 2006.

11   Stephanie Nano, "Headlines: Public Health News," Metropolitan Medical Response System, November 12, 2005. Available at http://www.mmrs.fema.gov/news/publichealth/2005/nov/nph2005-11-12.aspx (accessed February 2006).

12   Interview with Dr. Michael Gray, March 2007; and Gray, "Molds, Mycotoxins, and Human Health."

13   Peraica et al., "Toxic Effects of Mycotoxins in Humans."

14   Bennett et al., "Mycotoxins," p. 1-2.

15   Ibid., p. 24.

16   See Section Three, interview with Dr. Gina Solomon, July 2006.

17   The CDC Mold Work Group, "Mold Prevention Strategies," p. 7.

18   Chew, "Mold and Endotoxin Levels in the Aftermath of Hurricane Katrina: A Pilot Project of Homes in New Orleans Undergoing Renovation."

19   DEQ Louisiana, "Air Samples Show Southeast Louisiana Air Meets Federal Standards," November 1, 2005. Available at http://www.deq.louisiana.gov/portal/portals/0/news/pdf/airsamples111DEQ.pdf#search=%22Air%20Samples%20Show%20Southeast%20Louisiana%20Air%20Meets%20Federal%20Standards%22 (accessed June 2006).

20   Ibid.

21   Ibid.

22   The CDC Mold Work Group, "Mold Prevention Strategies," p. vi.

23   Ibid.

24   Daley, "The Next Menace."

25   Lipsey, "The Mold and Bacterial Results from the Sampling Dr. Lipsey did in St. Bernard Parish," and e-mail from Dr. Richard Lipsey, October 6, 2006.

26   The CDC Mold Work Group, "Mold Prevention Strategies," p. 41.

27   Ibid.

28   Ibid., p. 8.

29   See Section Three, interview with Dr. David Straus, April 2006.

30   See Section Three, interview with Dr. Jack Thrasher, April 2006.

31   The CDC Mold Work Group, "Mold Prevention Strategies," p. 8.

32   See Section Three, interview with Dr. Richard Lipsey, April 2006.

33   The CDC Mold Work Group, "Mold Prevention Strategies," p. 3.

34   See Section Three, interview with Dr. David Straus, April 2006.

35   E-mail from Dr. David Straus, February 2006.

36   Ellen Z. Harrison et al., "Land Application of Sewage Sludges: An Appraisal of the US regulations."

37   See Section Three, interview with Dr. David Straus, April 2006.

38   E-mail from Dr. David Straus, February 2006.
39   Campbell AW et al., "Mold and Mycotoxins: Effects on Neurological and Immune Systems in Humans." In: *Sick Building Syndrome, Advances in Applied Microbiology*, volume 55, Straus, DC, editor (Elsevier Academic Press, San Diego, CA, 2004). Article available at http://www.drthrasher.org/Molds_Mycotoxins.html (accessed August 2006).
40   Górny, "Filamentous Microorganisms and Their Fragments in Indoor Air-A Review."
41   Interview with Dr. Scott Needles, October 11, 2006.
42   See Section Three, interview with Dr. Jack Thrasher, April 2006.
43   See Section Three, interview with Dr. Regina Santella, July 2006.
44   Interview with Dr. Udo Erasmus, May 2006.
45   *IICRC S520*, p. 43.
46   See Section Three, interview with Cynthia Mulvihill, April 2006.
47   See Section Three, interview with Melinda Ballard, April 2006. For more information on POA, see www.policyholdersofamerica.org.
48   See Section Three, interview with Cynthia Mulvihill, April 2006. (Italics added.)
49   People v. Kelly (1972) 17 Cal.3d 24; Frye v. United States (D.C.Cir.1923) 293 F.
50   See Section Three, interview with Cynthia Mulvihill, April 2006.
51   Ibid.
52   Ibid.
53   Interview with Katrine Stevens, April 2006.
54   Thomas Grillo, "After 8 Years, a Milestone in Battle Over Mold," *Boston Globe*, November 25, 2003.
55   Ibid.
56   Ibid.
57   Jason M. Scally, "Scientific Evidence Key to Mold Verdict, Says Lawyer," *Massachusetts Lawyers Weekly*, December 22, 2003.
58   See Section Three, interview with Melinda Ballard, April 2006.
59   Ballard Appellate Decision, Texas Court of Appeals, Third District, at Austin. Available at http://www.heindllaw.com/Ballard_v._Farmers_Insurance_appellate_decision_toxic_mold.htm (accessed August 2006).
60   See Section Three, interview with Melinda Ballard, April 2006.
61   Ibid.
62   CNN, "Investigator: EPA not properly testing WTC air. One of the agency's own says EPA providing false info," February 24, 2002. Available at http://archives.cnn.com/2002/TECH/science/02/24/rec.wtc.air.quality/ (accessed August 2006).
63   Ibid.
64   Ibid.
65   Ibid.
66   Tulane University, "The New Wave." Available at http://www.tulane.edu/newwave/environmental_update.html (accessed December 2005).
67   Ibid.
68   Tulane University, "New Orleans Metropolitan Convention & Visitors Bureau." Available at http://www.cme.tulane.edu/nofaq-mcvb.pdf#search=%22EPA%20air%20quality%20returns%20New%20Orleans%22 (accessed August 2006).
69   LDEQ, "Sec. McDaniel's response to NRDC letter." (Italics added.)
70   Bennett, "'It's Incompetence.'"
71   Ibid.
72   Ibid.
73   See Section Three, interview with Dr. Richard Lipsey, April 2006.

SECTION TWO

HEALTH MATTERS

CHAPTER ONE:

THE MEDICAL MINE FIELD OF MOLD

Before our family experienced ill health from mold exposure, we didn't give *mold* much thought. We kept our shower clean of it and threw out any moldy food. However, after months of battling to regain our health that was derailed by mold and chemical exposures, we have had to become vigilant regarding exposure to mold, as any sizable amount of re-exposure can trigger an array of allergic symptoms. Having been sensitized to mold in the aftermath of Katrina, we now must avoid it—in the food we eat and in the air we breathe. For this reason, we continue to monitor our food selections and avoid any building where the smell of mold previously greeted us at the door. However, we no longer have to be as hypervigilant as we did during our initial healing period when even reports of high spore counts could alter outdoor plans such as playing with our children in the yard, going to the grocery store, or running a short errand. Altering our activities was inconvenient, but necessary, at the time. Now our recovery has progressed to the point that even re-exposure reactions have begun to subside.

How much re-exposure triggers a mold reaction is not a question with a calculable answer. However, the CDC states, "Re-exposure of sensitized individuals leads to lung inflammation and disease."[1] In our personal experiences we have found that re-exposure can trigger the same physical reactions, such as hives, chest pain, respiratory problems, headaches, and other symptoms from which we initially suffered. Although these incidences of re-exposure are difficult to avoid entirely, when they do occur, we can negate the re-exposure reactions, usually within hours, by drawing from the arsenal of vitamins, herbs, and nutritional resources that initially restored our family's health. Getting to this point took time, patience, and research, but the work and effort was worth the end result. We now have a simple yet effective treatment plan that not only restored our family's health from the initial mold and chemical exposures, but it continues to keep us healthy in spite of accidental and unavoidable re-exposures.

In the beginning, we didn't realize the medical field was not knowledgeable of proven, effective, and nonlife-threatening treatment options for mold and chemical exposures. We did, however, quickly come to the realization that no quick fixes existed, and the pseudo quick fixes that were offered came with a price—called side effects—with no guarantees. As we gathered information on medical treatment options from various doctors, we soon found that opinions varied among health care providers as to whether mold exposure can even make people sick, the degree to which mold exposure can make people sick, effective treatment options, and the degree to which the side effects of treatment methods can be health threatening.[2]

Because many of the treatment options for mold exposure include the use of pharmaceuticals—many of which come with warnings of liver toxicity—treatment decisions must not be made lightly, no matter how well-intentioned the prescribing doctor. Patients must carefully research options offered and weigh the associated risks, because the patient pays the price—not the doctor—when the best-case scenario does not happen.

Hindsight is often 20/20, which was the case in our journey of health recovery. Had we known in the beginning what we now know after months of research, our recovery time would have been much shorter. As it was, our journey to recovery was filled with false starts and the time-consuming process of trial and error. We scoured scientific and medical literature to understand why initial treatment methods were health degrading or just not effective and, ultimately, why others restored our health. We interviewed many scientific and medical professionals along the way, clarifying information in published reports, seeking answers to questions we could not locate in published literature, and tailoring questions specific to address the health needs of mold survivors. We share the information we gathered during this discovery process, as well as the health decisions we made along the way—both the ineffective and the health-restoring—to enlighten others.

The treatment methods that ultimately proved effective for our family may or may not provide the same results for others sick from mold and chemical exposures. Many factors—types of contaminant exposure(s), duration of exposure(s), individual health histories, genetics, and a host of other dynamics—play a role in who becomes ill from mold and chemical exposures, the extent of illness, the ease of recovery, and what is—or isn't—effective in treatment. Our medical and nutritional decisions were made in consultation with our primary health care provider, based on each family member's individual health history. Likewise, consideration to use any treatment option we share while telling our story of recovery should be evaluated by each individual's licensed physician. We provide this information solely for educational purposes.

Certainly, by no means do we discuss all possible treatment options for mold and chemical exposures. That is neither the scope nor the purpose of this book. As well, the information we share is not medical advice and should not be construed as medical advice. Rather, we simply allow you, the reader, to gain knowledge from our experiences—the appointments with multiple doctors, the emergency room visits, our quest for answers, and ultimately, our journey of healing and health restoration.

We had no idea in the beginning that we would literally have to maneuver through a medical mine field while seeking medical treatment for exposure to mold and chemicals. Others may or may not experience similar challenges while seeking treatment. We are not implying that our experience is the norm, although it very well could be. As you will see, we consulted many doctors, including specialists, and had many false starts before finally commencing on a path that ultimately led to the restoration of our health. Likewise, each person should seek a licensed physician(s) who will work with him/her to develop an effective treatment plan tailored to his/her individual health history and current health situation.

Even before our family was evacuated from Louisiana to Montana by the Red Cross two weeks after Hurricane Katrina, we had experienced mold- and chemical-related symptoms, such as acute bronchitis, sinusitis, wheezing, coughing, dizziness, chest pain and congestion, heart palpitations, thyroid irregularities, hives, and rashes, to name just a few.[3] We were most concerned with the thyroid irregularities since thyroid function helps regulate the heart. We reasoned the thyroid irregularities were the cause of some of the other symptoms, such as the heart irregularities. According to the Mayo Clinic, thyroid hormones ". . . maintain the rate at which your body uses fats and carbohydrates, help control your body temperature, influence your heart rate and help regulate the production of protein."[4] The thyroid gland also produces a hormone that regulates the amount of calcium in your blood.[5] As you can see, a healthy thyroid function is very important.

The CDC states in regard to mold-related illnesses, ". . . [O]ptimal treatment is elimination of causative exposures."[6] In other words, do whatever is necessary to immediately stop further exposure from occurring. Although eliminating exposure is a critical, necessary first step that must be taken as even short durations of exposure can be detrimental to health, elimination of exposure does not necessarily guarantee the immediate return of pre-exposure health. For example, we incorrectly thought that once we were removed from the mold, chemicals, and other toxins stirred in the wake of Hurricane Katrina, our symptoms would clear up and our pre-hurricane health would return. We were sadly mistaken. Our symptoms did not improve but instead worsened, which often can be the case with illnesses stemming from fungal and chemical exposures.

According to Aerias AQS IAQ Resource Center,[7] the health-related content of which is reviewed by Joseph Jarvis, MD, MSPH, "Health problems caused by mold can be acute, which occur immediately or within a few days of exposure. Health problems may also be chronic, which are long-term health effects that might not occur immediately. . . . These diseases may permanently worsen the health of persons affected, even after they have been removed from exposure. Mycotoxin-induced diseases have been postulated, but scientific consensus about the nature of these diseases has not been reached."[8]

As we now know, exposure to mold affects each person differently depending on age; genetics; strength of immune system; history of allergies, asthma and/or respiratory-related illnesses; current medications; amount of exposure; species of mold; and other factors.[9] These factors influenced the health of our family members as well. Kurt, who was borderline asthmatic and was already on steroids for allergy management at the time of the hurricane, became ill first and was by far the sickest member of our family. We did not know at the time that steroids weaken the immune system and put a person at high risk when exposed to mold.[10] In fact, according to Fred Lopez, MD, an infectious disease specialist at Louisiana State University's School of Medicine in New Orleans, steroids are an immunosuppressive medication, which can render a person taking them immunocompromised, leaving him/her more likely to suffer complications from being exposed to mold.[11]

Prior to researching why Kurt had become so ill, we had no idea of the degree to which steroids weaken the immune system. Had the prescribing doctors fully informed us of all the risks associated with steroid treatment of allergies, we most likely would have searched for alternative treatment methods. But as it was, we had no idea that taking steroids would leave Kurt's health in an immunosuppressed/immunocompromised state, making him more susceptible to disease, illness, and fungal exposure with its myriad of health effects.

By the time we reached our evacuation destination (Billings, Montana), we had one very sick family member with the rest of us trailing close behind. Shortly thereafter, we started our journey from doctor to doctor, specialist to specialist, naïvely trusting and believing that these initially friendly doctors would be able to scientifically and medically back up their statements. We soon found our initial expectations were far from what would transpire.

The first doctor we saw was a doctor of internal medicine. We made the first appointment for just Kurt, whose health was on the quickest decline. Our thinking was, once a successful treatment plan was identified, we would make appointments for the rest of us. Since we all had become sick from exposure to the same thing—post-hurricane chemicals and mold, the latter which took root in the air system and on the damp walls of our home—we reasoned that whatever treatment was effective for Kurt would work for the rest of us as well. At this point, we really did not realize getting healthy was going to be such a challenge. We soon found otherwise.

Blood tests were run on Kurt, and the doctor diagnosed him with bronchitis, acute sinusitis, and extreme hypothyroidism (low thyroid), none of which he had before the hurricane. (In fact, he had a slightly high thyroid before the hurricane.) The physical exam revealed that Kurt had only 25 percent of his lung capacity. The doctor prescribed an antibiotic he jovially described as the strongest antibiotic he could get his hands on. In addition, he prescribed steroids to "restart" Kurt's thyroid and reduce the allergic reactions he was experiencing from exposure to the molds and other toxins inhaled post-hurricane.

We questioned not only the diagnoses but also the prescribed treatments because Kurt's blood work did not reveal an elevated white count, which always had been the case in the past when he had been diagnosed with a respiratory infection.

A simple explanation regarding white blood cells comes from the Florida State University Research Foundation, "White blood cells circulate throughout your body like an army. They protect you from an attack by germs. White blood cells are larger than red blood cells. They have funny shapes [when viewed under a microscope] and can attach to the germs they find in your blood. [Y]our body needs to have enough white blood cells so you can fight off infections and stay healthy. If you don't have enough white blood cells, it is easier for you to become sick. When you get sick, for example like when you have a sore throat, your number of white blood cells goes up higher than normal. This is because your body makes more white blood cells to help you get well. A high amount of white blood cells is a sign that there is an infection in the body. . . . The normal adult range is 4,500 to 11,000 cells per cubic millimeter of blood . . . [A] white blood cell count of 13,000 white blood cells per cubic millimeter . . . indicates that there is an infection in the body. Something very serious like cancer or other less serious infections such as *bronchitis* could cause this increase in white blood cells."[12]

When Kurt questioned the doctor regarding the normal white count, the doctor stated he was treating Kurt as he would treat someone exposed to biological warfare because he equated the post-hurricane toxins to which we had been exposed to biological warfare agents. The antibiotic he prescribed was Avelox, 400 mg. The pharmacy information printout given to patients lists the side effects as ". . . nausea, stomach pain, diarrhea, mild dizziness, or headache."[13] Unlikely but possible side effects are ". . . unusual vaginal itching or discharge [for female patients], yellowing eyes or skin, dark urine, white patches in the mouth, seizures, mental/mood changes, [and] chest pain."[14] Highly unlikely side effects are ". . . fainting, [and] fast/slow/irregular heartbeat."[15] Symptoms of an allergic reaction include ". . . rash, itching, swelling, dizziness, and trouble breathing."[16]

Kurt was already experiencing over half of these symptoms and he hadn't even taken the antibiotic yet! We were concerned Kurt would have an allergic reaction to this antibiotic, which would then exacerbate the symptoms with which he was already struggling. Furthermore, how would we know whether he was reacting to the antibiotic or if his existing symptoms were just getting worse? These questions were posed to the prescribing doctor, who dismissed our concerns.

Additionally, we were wary of Kurt's taking such a strong antibiotic with so many possible side effects without proof of a bacterial infection indicated by an elevated white count.[17] We were already aware at the time of what Dr. Gray later confirmed, "White counts elevate with bacterial infections."[18] However, we did not know at the time what we later learned from Dr. Gray, "White counts do not generally elevate with fungal infections."[19] Kurt did not have an elevated white count because his lung infection was fungal, not bacterial.

The fact that a person's white count does not generally elevate with fungal infections can be a bit confusing without clarification. Dr. Gray explains that although fungi don't necessarily raise the white count in the blood, they will sometimes cause an increase in eosinophils, which are a type of white blood cell associated with allergies and parasites. He points out, however, that eosinophils will often increase in the area where there is local infection, such as in nasal mucosa.[20] Localized elevation of eosinophils was proven in an earlier discussed study by the Mayo Clinic.

Since the treating doctor had sidestepped our questions and concerns in regard to the antibiotic, we put the decision regarding the antibiotic treatment on the back burner and focused on the steroids. We thought, at the time, we understood the concept of treating allergies with steroids. This doctor's recommendations regarding treatment of Kurt's allergic reactions to the mold and chemical exposures were in line with prior treatments prescribed by previous doctors for allergy management. For that reason, we decided the wisest course of action would be for Kurt to take first the steroid treatment pack the doctor had prescribed, after which we would evaluate whether he should take the antibiotic. We did not want Kurt to take the two drugs together because of the possibility of an allergic drug interaction—even though the doctor stated it was not a concern to him. The doctor gave no special instructions or warnings and said to come back in two weeks.

The doctor's disregard of possible increased allergic reactions made no sense to us because if Kurt took both pharmaceuticals at the same time and did have an allergic reaction, how would we know to which one he had reacted? We couldn't understand why the doctor was simultaneously prescribing two drugs to a person with a history of allergies and drug reactions, especially when the person was already in a state of medical distress. We did not feel comfortable with this Russian roulette philosophy of medicine.

In addition to the steroid pack, the doctor instructed Kurt to continue taking the prescription Nasonex steroid inhaler he had been using pre- and post-hurricane for allergy management. We did not know at the time what Dr. Gray later explained, "The treatment that is traditionally used, which is the use of inhaled nasal steroids, actually suppresses the immune response and while it does reduce inflammation it also encourages the continued growth of the fungus. So the treatment of the inflammation symptomatically ends up perpetuating the problem, which is the underlying fungal infection."[21] In hindsight, Kurt should never have taken the steroid pack or continued using the nasal inhaler. However, as mentioned earlier, we did not know that steroids are proven agents that weaken the immune system, which in Kurt's already delicate and precarious condition proved almost fatal.

After Kurt began taking the steroid pack, he started coughing up blood and other disgusting, never-seen-before "stuff" and broke out in a severe case of hives. Even more perplexing was that his pulse was racing and his heart was pounding, both of which are symptoms of hyperthyroidism (elevated thyroid), not hypothyroidism (low thyroid) with which he had just been diagnosed. As is often the case with illnesses, this turn for the worse happened at night when the doctor's office was closed, so we called the doctor's on-call line. We spoke with one of the prescribing doctor's partners who stated that Kurt needed to be on a higher dosage level of steroids. We disagreed. Kurt told the on-call doctor that he was experiencing the same types of symptoms he had felt in the past when his thyroid was elevated. Knowing that elevated thyroid levels can lead to very serious health repercussions, Kurt told the doctor that he needed to be seen so his thyroid levels could be checked. The doctor adamantly stated that Kurt's thyroid could not possibly have flipped to an elevated state and he refused to see him. The doctor ended the conversation by rudely laughing and restating that Kurt simply needed more steroids.

Shocked by the unprofessional bedside manner of the on-call doctor, but not knowing what else to do (not understanding the detrimental effects of steroids on the immune system), Kurt continued taking the steroid pack and using the Nasonex inhaler. We didn't realize that the steroids were the cause of the increased health problems. Furthermore, by morning the unexplained, reactive episode had subsided. We did, however, call the prescribing doctor to discuss the incident and re-inquire whether Kurt should take a scheduled trip to collect some furniture, a trip for which he had already been medically released. The doctor offered no explanation as to what had occurred the night before but reconfirmed that Kurt was medically released to take the trip. Trusting the doctor's judgment, we set off on the trip. Two days later, while on the road, Kurt experienced a reoccurrence of the bizarre symptoms, and we had to rush him to the nearest emergency room. The emergency triage nurse immediately admitted him as a cardiac emergency.

Four hours later, test results revealed his heart was fine, but not his thyroid. It had indeed flipped from an extreme hypothyroid level (low thyroid) to an extreme hyperthyroid (high thyroid) state. Kurt had been right two nights earlier; his thyroid was elevated.

The ER doctor said he had never before seen a thyroid roller coaster like this—four days earlier tests had revealed Kurt was extremely hypothyroid. Now he was in an extreme hyperthyroid state. The ER doctor stated that he believed that the toxins we had been exposed to post-hurricane were synergistically interacting. The escalation of the thyroid was not attributed to the steroids, and Kurt was instructed to continue taking the steroid pack as prescribed. The ER doctors administered some thyroid suppressing medicine, which stabilized Kurt's thyroid within a few hours, and he was discharged. Unknown to us at the time, the steroids had caused Kurt's thyroid to not just restart, but skyrocket.

After this close encounter from steroid use, we began to earnestly research the associated health risks of prescription steroids, ignoring all the assurances we had received over the years from previous doctors and pharmacists. An example of a typical reassurance is one that we recently received from a pharmacist after asking about the possibility of lowered immunity from steroid use. The pharmacist's response was as follows: "Yes, anytime you take a corticosteroid, or any steroid for that matter, you suppress your immune system, but not to the level you think. It's not like you are going to get cancer or something like that. It simply means that if you are around someone with chicken pox (and have not had them before), you would have a higher likelihood of coming down with chicken pox. Or if you were near someone with a cold, the flu, or strep throat, it means you may get sick more easily." Then came the qualifying reassurance: "My own daughter is on a steroid inhaler because of allergies, and I wouldn't put her on them unless I thought they were safe."[22]

Patients, more often than not, receive verbal rhetoric when they question doctors regarding the safety and side effects of pharmaceuticals. What is more accurate is the really tiny, fine print on the package inserts that most of us don't read—because we allow our doctors to placate us with their reassurances. What we read on the package insert was disconcerting, given that for years we had repeatedly heard similar dismissive reassurances. The package insert of the steroid pack that Kurt was prescribed states, ". . . [T]hey modify the body's immune responses to diverse stimuli. . . . Children who are on immunosuppressant drugs are more susceptible to infections than healthy children. Chicken pox and measles, for example, can have a more serious or even fatal course in children on immunosuppressant corticosteroids. In such children, or in adults who have not had these diseases, particular care should be taken to avoid exposure."[23]

Two things to take away from this small portion of the product literature are 1) the manufacturer refers to the steroids as immunosuppressant agents, and 2) the manufacturer points out that someone taking these steroids can die from chicken pox, which is nowadays a relatively nonlife-threatening childhood disease. The same warning was given regarding exposure to measles for both adults and children. Thus, these steroids can lower the immune system to such a degree that normal childhood diseases can be fatal.

Manufacturers of pharmaceuticals document fine-print warnings to limit their liability. Documenting any contraindication, which is a symptom or condition that makes treatment with a particular substance inadvisable, is routine. What is most revealing is the following contraindication for oral steroids documented by this particular manufacturer: "Systemic fungal infections and known hypersensitivity to components."[24] In other words, the manufacturer states that someone with a systemic fungal infection should not be prescribed these steroids!

The Nasonex steroid inhaler literature states much the same information as the literature in the steroid pack regarding a lowered immune system and exposure to chicken pox or measles—both could be fatal. In addition, the literature states, "Intranasal corticosteroids may cause a reduction in growth velocity when administered to pediatric patients."[25] This risk is also detailed in the pharmacy patient information sheet that states, "Corticosteroids may affect growth rate in children and adolescents in some cases."[26] In other words, by using a steroid inhaler, a child's growth could become "stunted," which is the term a national-chain pharmacist used to describe the resultant state. The patient information sheet also confirms that a child taking steroids could die from a normal childhood disease, such as chicken pox and/or measles.

Information not included on the pharmacy patient information sheet, but hidden in a sea of minuscule print on the package insert, states, "Nasal corticosteroids should be used with caution, if at all, in patients with active or quiescent tuberculous infection of the respiratory tract, or in *untreated fungal*, bacterial, [or] systemic viral infections, or ocular herpes simplex."[27] The pharmaceutical company itself makes clear that someone with an untreated fungal infection should not be prescribed nasal corticosteroids!

Obviously, with post-hurricane fungal exposure, a physical exam revealing only 25 percent functioning lung capacity, and a diagnoses of sinusitis and bronchitis—with no elevated white count indicating a bacterial infection—the prescribing doctor should have at least suspected a fungal lung infection. Due consideration of a fungal-induced illness did not occur because this doctor held (and still holds) the preconceived idea that mold exposure cannot make you sick—an ignorant, inaccurate belief clearly dispelled by just reading the information in the steroid package inserts.

Other possible dangers of corticosteroids are discussed by Dr. Rapp, a board certified environmental medical specialist and pediatric allergist who started studying allergies in 1958. She states, "Cortisone [a steroid] or other strong drugs are often prescribed when a child is very ill. Sometimes they are helpful, *but only on a temporary basis*."[28] Additionally, she states, "Cortisone tends to allow infections that normally would be confined to a certain body area to spread to other parts of the body. While this happens, the patient usually has an associated feeling of well-being. This combination can be very dangerous. It means that infection can spread, and the parents have little warning because their child feels fine. For this reason, it must be stressed that children on cortisone should be closely and regularly supervised by their physician *regardless* of how well they feel."[29] Additional medical literature states that steroid use is listed as one of the causes of environmental sensitivities.[30]

Looking back over the years, we have to wonder why the many doctors who prescribed steroids for management of Kurt's allergies and the pharmacists who filled the prescriptions never once explained the medical risks associated with steroid use. We were never informed of the possible increased risk of infection and disease as a result of a lowered immune system. How many people taking prescription steroids for allergy and asthma management are likewise not informed? Furthermore, how many returning residents to the Katrina-hit areas were prescribed steroids for similar post-hurricane allergies—while living in an environment laden with elevated mold spore counts, highly toxic substances, and bacteria?

Dr. Rapp, founder of the Founder of Health Research Foundation and past president of the American Academy of Environmental Medicine, raises an alarming point, "Recent studies . . . indicate that the death toll (now 1.9 deaths/100,000 [as of the 1991 publish date]) from asthma has tripled since 1976. The death rate in the United States for children aged five to nine years increased fivefold from 0.1 to 0.5 per 100,000 from 1979 to 1987, while in those aged fifteen to nineteen years, the rate doubled during that same time period. At about that time, the use of theophylline and cortisone became preferred methods of treating asthma in the United States. Could there be a relationship? The cause of this mortality increase, however, is usually attributed to a delay in therapy or inappropriate treatment."[31]

These words were written over 15 years ago by Dr. Rapp. We must ask, why has the practice of steroid treatment for allergy management and asthma control continued when "inappropriate treatment" has been designated as a cause of mortality when steroids were the preferred method of treatment?

Dr. Rapp theorizes, "Maybe the use of powerful steroids or other drugs might be diminished if more emphasis was placed on *why* children are wheezing. Maybe changes in diet, home factors, and limiting exposures to the ubiquitous array of chemicals in today's world would be helpful."[32] In other words, let's identify and remove the allergens, as steroids can only relieve—not cure—allergies, as Dr. Rapp points out.[33]

The elimination of physical symptoms has long been the primary focus of care followed by practitioners of conventional medicine.[34] The philosophy of the first six doctors with whom we consulted was no different. Unfortunately, this one-dimensional approach does not take into consideration what could be going on in the entire body or what factors could be causing the symptoms in the first place. In our case, for example, the initial doctors did not consider the nature of mold itself: Depending on the species, fungi can produce toxins and have the capability of colonizing inside and/or infecting a person's body. Other toxins—such as arsenic, lead, diesel fuel, and asbestos—cannot colonize and/or infect. The doctors also did not consider that chemical exposure can set the stage for fungal growth by altering the body's natural biological terrain. When in correct balance, intestinal flora supports growth of good bacteria and yeasts, which help keep fungal overgrowth in check. Furthermore, the doctors did not consider that antibiotics can create a fungal-friendly intestinal environment as well. The simple truth is, without repopulation of good bacteria, the body cannot effectively combat fungi.

By providing superficial treatment of symptoms without addressing the underlying causes, doctors are doing nothing more than prescribing a Band-Aid—a stopgap solution—guaranteed to bring them repeat business down the road. "Most over-the-counter and almost all prescribed drug treatments merely mask symptoms or control health problems or in some way alter the way organs or systems work. Drugs almost never deal with the reasons why these problems exist, while they frequently create new health problems as side effects of their activities," states John R. Lee, MD.[35]

This Band-Aid approach almost cost Kurt his life. The steroids created a life-threatening situation with Kurt's thyroid and further undermined his immune system, leaving him even more susceptible to disease, illness, and increased fungal growth. By the time Kurt's two-week doctor appointment arrived, he was far worse than he had been before starting the steroids. And to think that the on-call doctor wanted to prescribe an increased dose!

Before we move on, let's take a look at the other pharmaceutical—Avelox—that was prescribed for Kurt, but which we eventually elected for him not to take. Not only would have use of the Avelox indirectly promoted fungal growth by destroying "friendly" bacteria and yeasts, it could have proven fatal. On the website of the law firm Sheller, Ludwig & Badey, Avelox is listed as a defective drug with severe adverse reactions such as ". . . liver failure including death."[36] The law firm additionally states, "Avelox is an antibiotic in the class of drugs known as fluoroquinolones . . . [and] that there are numerous drugs which should not be taken in combination with fluoroquinolones. There are increased risks of injury when fluoroquinolones are taken in combination with *corticosteroids*."[37] Sheller, Ludwig & Badey handle cases involving loss incurred from treatment with Avelox and other fluoroquinolones.[38]

This information—that taking Avelox, alone or together with a prescribed steroid, could potentially cause liver failure or even death—was easily accessed via a quick internet search, yet the prescribing doctor either was unaware of or disregarded these warnings. In addition, he ignored our concerns when we questioned him regarding possible negative drug interactions and/or allergic drug reactions. Not only that, but he prescribed Avelox and a steroid pack to a patient already using a nasal steroid. By carelessly combining two different forms of steroids with the Avelox, this doctor even further increased the risk to Kurt's health.

Furthermore, the information that antibiotics are not recommended for treatment of fungal infections because they target bacteria, not fungi, which we learned from the conclusions of the 1999 Mayo Clinic study[39], was also easily accessed via a quick internet search. The fact that fungal infections are often mistreated with antibiotics is confirmed by Timothy Callaghan, MD, DC, who has treated mold-exposed patients at the Center for Occupational & Environmental Medicine (COEM) in North Charleston, South Carolina. He reports that by the time patients came to COEM, they often had been prescribed antibiotics multiple times for sinus infections yet had had no significant improvement. He attributes their lack of healing progress to the fact that antibiotics won't kill mold in sinus cavities.[40] This cycle of misdiagnosing fungal infections as bacterial infections and mistreating them with multiple rounds of antibiotics could be taking place with patients who do not even have elevated white counts, as was the situation in Kurt's case.

The common mistreatment of sinus fungal infections with antibiotics can create systemic havoc within a patient's body. If truth be told, treating fungal infections with antibiotics is like throwing gasoline on a fire. Antibiotics actually enable the fungal infection to spread because antibiotics kill the body's natural level of good bacteria and yeasts that help combat opportunistic fungi.[41] According to Dr. Rapp, "In our intestines we naturally have microorganisms or bacteria called lactobacilli, as well as yeast called monilia. They coexist in harmony unless we upset the natural balance by eating too much of certain food items or taking drugs such as antibiotics. Any type of significant imbalance of the normal flora or germs in any body area can lead to illness."[42]

According to published health literature, "Inside each of us live vast numbers of bacteria without which we could not remain in good health. There are several thousand billion in each person (more than all the cells in the body) divided into over 400 species, most of them living in the digestive tract. Certain of these bacteria help to maintain good health, while others have a definitive value in helping us regain health once it has been upset."[43]

We must protect these microorganisms that help protect, promote, and restore our health by eating probiotic-rich foods such as yogurts and fresh vegetables and by choosing alternative treatments to antibiotics when possible. Although there are certainly times when antibiotics are necessary and even lifesaving, the Mayo Clinic study states that they are often prescribed for conditions they will not effectively treat, for example, fungal conditions.[44]

Had Kurt taken the Avelox, he would have had a double hit to his intestinal flora because, unknown to us at the time, "*[s]teroids* (cortisone, ACTH [adrenocorticotropic hormone], prednisone, and birth control pills) also cause great damage to the bowel flora."[45] When the balance of intestinal flora is upset, for example, from the use of antibiotics and/or steroids, the immune system can become compromised.[46] According to medical sources, "Lowered host resistance due to such factors as underlying debilitating disease, neutropenia, chemotherapy, *disruption of normal flora*, and an inflammatory response due to the use of *antimicrobial agents and steroids* can predispose the patient to colonization, invasive disease, or both."[47] Given that Kurt had been treated pre- and post-hurricane with two different forms of steroids and was later treated with an antimicrobial agent, we were thankful (in hindsight) that he did not take the Avelox. Its use likely would have further compromised Kurt's health, potentially with deadly consequences.

Our prescription fiasco is a prime example of the level of caution that must be taken when evaluating the use of prescribed drugs. Through research, consultations with your primary doctor, and securing multiple opinions, you have to decide whether components of a treatment plan will effectively restore your health or be akin to supplying arms to the enemy—who in this case is fungi! For example, given the fact that the use of antibiotics, antimicrobial agents, and steroids all lower host resistance, they can predispose a person to colonization, invasive disease, or both. We have to ask: On which side of the fungal war are pharmaceutical companies?

At the second appointment, Kurt again questioned the doctor regarding the mold exposure in our Louisiana rental home. We were trying to identify *what* we had been exposed to in hope that identifying the cause would lead to some answers, explain our mysterious symptoms and illnesses, and enable the doctor to identify an effective treatment plan. Understandably—in hindsight—Kurt's health was still on the decline. Unfortunately, the children and I were not regaining our health either. Most concerning was that I was experiencing heart palpitations and irregularities, although they were not the life-threatening heart fluctuations that had landed Kurt in the emergency room. We were further concerned that the continual decline of Kurt's health was a future indicator of what lie ahead for the children's and my health. One thing we knew for sure, we had all been exposed to the same toxins and were experiencing most of the same symptoms, albeit in varying degrees.

When questioned at this appointment, the doctor adamantly stated, "Mold cannot make you sick." He said it is impossible for mold to get into the body, the sinuses, and the lungs. This steroid-prescribing doctor then informed Kurt that he was setting a surgery date to remove Kurt's thyroid. He said he had made the decision in joint consultation with one of the other endocrinologists (in the same practice) who Kurt had seen earlier. The doctor then stated he was prescribing antidepressants for post-traumatic stress disorder (PTSD) caused by the trauma of the hurricane.

It was at this time that the Katrina Syndrome was being televised nationwide, featuring antidepressants as the answer to post-hurricane depression and trauma. Kurt politely told the doctor that he was not depressed, but sick. Kurt again explained that he believed he was suffering from a fungal infection that was somehow affecting his thyroid. He proposed that once the fungal infection was effectively treated, his thyroid function would return to its pre-hurricane state. Kurt asserted he didn't need surgery and certainly didn't need antidepressants. The doctor didn't take kindly to this patient "mutiny." He followed Kurt out to the front desk where he began screaming at Kurt in front of the entire waiting room, ranting that Kurt needed antidepressants and was crazy for believing mold could get into his body and make him sick. Kurt walked away, telling the doctor that he was the one who was crazy and needed antidepressants. Needless to say, Kurt did not return to this doctor, his partners, or the affiliated hospital.

At the time, we didn't realize that two basic factions exist within the medical community: one that understands that fungal spores can enter the body and cause ill health effects and another one that refutes the possibility—even though many of the health effects of fungal inhalation are well documented in published, peer-reviewed scientific and medical journals. This division within the medical community was later explained to us by Curt Kurtz, MD, an environmental specialist. According to Dr. Kurtz, he repeatedly encounters the common misperception that *Mold cannot make you sick*, a belief ardently held by many of his peers regardless of the significant amounts of scientific and medical literature to the contrary.[48]

These misperceptions can lead to misdiagnoses. According to Dr. Callaghan, "Mold and the mycotoxins that they produce can cause multiple symptoms in the body and often leave the patient confused and the examining physician making an incorrect diagnosis—quite often a psychiatric one—in order to have something to treat."[49] We must ask: Are these misdiagnoses being made by treating physicians because of ignorance; lack of continuing education; blatant rejection of published, peer-reviewed findings; or overt, greed-based allegiance to the pharmaceutical industry instead of loyalty to the patient?

What is most disconcerting to us is the closed-mindedness of the many doctors who treated us. For example, even though a medical study from a major teaching hospital documents that antibiotics are not effective in treating fungal infections, some doctors, such as Kurt's, adamantly deny that fungal infections can even exist. How can doctors deny the existence of a condition proven to exist by studies conducted at a major teaching university hospital? Furthermore, not only are these doctors denying the existence of fungal infections, but they are prescribing antibiotics for illnesses induced by fungal exposure, which is the very treatment the Mayo Clinic advises against. Our experience clearly illustrates we must all be stewards of our own health and not assume our treating physician(s) is up-to-date on current, or even semi-recent, medical studies. Nearly six years to the date after the Mayo Clinic published its findings, a doctor prescribed antibiotics for Kurt to treat fungal-related sinus and lung infections.

Kurt's first treating physician prescribed antibiotics and steroids—two drugs that predispose a person's body to "colonization, invasive disease, or both"—when he was already suffering from fungal-related illnesses. This medical mistreatment is a prime example that illustrates the necessity of thoroughly checking and understanding the side effects of medications hastily prescribed by doctors.

The last drug this doctor attempted to prescribe—an antidepressant—is in a category of drugs that the U.S. Food and Drug Administration has recognized as causing potential mental health risks. On March 22, 2004, the FDA released a public health advisory stating, "Today the Food and Drug Administration (FDA) asked manufacturers of the following antidepressant drugs to include in their labeling a Warning statement that recommends close observation of adult and pediatric patients treated with these agents for *worsening depression or the emergence of suicidality.* The drugs that are the focus of this new Warning are: Prozac (fluoxetine); Zoloft (sertraline); Paxil (paroxetine); Luvox (fluvoxamine); Celexa (citalopram); Lexapro (escitalopram); Wellbutrin (bupropion); Effexor (venlafaxine); Serzone (nefazodone); and Remeron (mirtazapine)."[50]

This, again, is a prime example that confirms patients must check the side effects of prescribed drugs because, although these side effects may be documented in the fine print, prescribing doctors often do not verbally inform their patients. Just think about it for a moment: Why would a person who is truly suffering from PTSD take a prescription drug that can increase depression and/or suicidality?

The topic of flailing mental health—mental instability, depression, PTSD, and suicidality, regardless of the cause—is a serious, legitimate social issue that is often not a well-received topic of conversation. However, the truth of the matter is that the human mind can become overloaded for a variety of reasons to the point it cannot function and cope as intended. According to Harvard Medical School's Consumer Health Information, "In post-traumatic stress disorder (PTSD), a group of distressing symptoms occur after a frightening incident. A person must have directly experienced or witnessed the event, which must have involved serious physical injury or the threat of injury or death. By definition, the trauma must cause a strong experience of intense fear, horror, or helplessness. Some psychological and physiological arousal seems to be a key to developing this disorder."[51]

Harvard Medical School states, "Some common PTSD stressors include:

- Serious motor vehicle accidents, plane crashes, and boating accidents
- Industrial accidents
- Natural disasters (tornadoes, *hurricanes*, volcanic eruptions)
- Robberies, muggings, and shootings
- Military combat (PTSD was first diagnosed in soldiers and was known as shell shock or war neurosis)
- Rape, incest, and child abuse
- Hostage situations and kidnappings
- Political torture
- Imprisonment in a concentration camp
- Refugee status"[52]

Although Harvard Medical School identified hurricanes as a PTSD stressor and stated that people affected by Hurricane Katrina in Louisiana and Mississippi *may* develop PTSD, it predicted that most survivors would not.[53] This prediction may—or may not—have been accurate. Let's take a look at multiple media accounts of the psychological aftermath of Katrina and make up our own minds.

The California Nurses Association has addressed the issue of PTSD post-Katrina. It reports, "William Gasparrini, a Biloxi clinical psychologist, calls it 'Post-Katrina

Stress Disorder,' in which residents suffer bouts of grief, shock, rapid mood shifts, confusion, anger, marital discord, guilt, escape fantasies, and substance abuse. 'The effects are lasting longer than I suspected,' Gasparrini said. 'I thought everything would be back to normal in three to four weeks. Now, three months later, it looks like it'll be one to two years—if we are lucky.'"[54]

Harvard Medical School explains that PTSD can develop shortly after the trauma or even many months later. It states, "Acute PTSD is the term used when symptoms develop within the first one to three months after a traumatic event. The term PTSD with delayed onset is used when symptoms surface six months or more after the traumatic event."[55]

Usually, "After other disasters, between 7 percent and 12 percent of the people directly affected eventually suffered PTSD symptoms," stated Irwin Redlener, MD, director of the National Center for Disaster Preparedness at the Columbia University Mailman School of Public Health.[56] "Because Katrina victims number in the hundreds of thousands—all the people who lost homes, lost relatives, or were forced into temporary shelters—the mental toll could be huge," he said.[57] However, "'Because of the prolonged nature of this disaster, it's impossible to guess what rate of PTSD (post-traumatic stress disorder) we will see. It may be much higher than we would normally expect,'" stated Dr. Redlener, who has spent time in New Orleans and Mississippi since the hurricane.[58]

Evidence of post-Katrina PTSD-related behavior is abundant. A December 2005 article in the *Washington Post* discusses the turbulent psychological aftermath of Katrina. It reports that with nearly 2 million people displaced by Hurricane Katrina, entire neighborhoods eradicated, families separated, and racial issues heightened, the emotional scars are far beyond what mental health experts in this country have ever confronted. The *Washington Post* further reports, "In the extreme cases—and there have been many—they have hanged themselves, overdosed, and put guns to their heads. The number of suicides in neighboring Jefferson Parish is more than double what it was in the fall of 2004. In the first days of the crisis, coroner Robert Treuting saw five suicides in three days. In the two months since, there have been 11, compared with five a year ago. Two New Orleans police officers have taken their lives, and at least one more has attempted suicide."[59]

The article continues, "Orleans Parish coroner Frank Minyard said he does not have statistics for the city because many deaths—including nine by gunshot—remain a mystery. He knows of at least one woman who killed herself recently. New Orleans emergency personnel have responded to at least six suicides and nearly two dozen suicide attempts since Katrina. The tightly knit community of Academy of the Sacred Heart, the Rosary, is coping with two suicides, headmaster Timothy M. Burns said. Shortly before Thanksgiving, a woman with young children took her life. Last week, the father of a Sacred Heart student was buried."[60]

The *Washington Post* further reports, "Calls to a national suicide-prevention hotline skyrocketed from the typical 100 to 150 a day to more than 900 in the immediate aftermath of Katrina before leveling off to about 210 a day now, said Charles G. Curie, administrator of the federal Substance Abuse and Mental Health Services Administration."[61] To put these figures into proper perspective, we need to remember that post-hurricane populations are a fraction of what they were before the hurricane.

The *Washington Post* article also cites evidence of psychological trauma documented by the CDC. "In a clinical survey of Orleans and Jefferson parishes, the Centers for Disease Control and Prevention found that 45 percent of the residents were experiencing 'significant distress or dysfunction' and 25 percent had an even 'higher degree of dysfunction,' said Dori Reissman of the CDC."[62]

In August 2006, Time/CNN reported, "While the physical devastation of New Orleans from Hurricane Katrina has been well documented, the psychic toll is just becoming clear. The suicide rate has nearly tripled, depression is common, domestic abuse is on the rise, and self-medicating with booze is a favored method of forgetting."[63] Because of the post-hurricane lack of mental and medical health care facilities, doctors, and necessary medications, people have not had the resources normally available to address mental and/or physical health needs.

Time/CNN reports, "As bad as it is right now, the real crisis will come if the city can't resolve the post-Katrina lack of primary care and rising depression. 'In five years, we'll be the stroke capital of the world, the heart disease capital of the world,' warns [Peter] DeBlieux [MD]. 'We're going to see long-range complications from diabetes and heart disease and stress because people are neglecting primary care now.'"[64]

People who are able to find mental health care often seek treatment for depression. Monir Shalaby, MD, medical director of one of the rare community health centers to reopen post-Katrina (EXCELth), estimates ". . . 40% of the adults they see are taking medication for depression."[65]

Time/CNN further reports, "Firm statistics on the mental health situation are admittedly hard to come by. Demographers can't even agree on how many people live in New Orleans now, but best estimates put [the population] at less than 200,000 vs. 450,000 people pre-Katrina. The coroner's office recently told the *Times-Picayune* that suicides had gone up from 8 to 26 per 100,000 people. 'On a per capita basis, we've seen an increase in suicides, depression, substance abuse, and domestic violence. If you've driven the city, you see why. We've not made a lot of progress,' says cardiologist Pat Breaux, past head of the Orleans Parish Medical Society."[66] Included in this lack of progress are the remains of moldy structures, exposure to which can directly, and indirectly, affect people's psychological health.

The fact that exposure to fungi can psychologically affect people, to date, has not been a major focus of the media. The issue, however, has been examined by medical and scientific professionals. For example, a 2003 medical study assessed the ". . . psychological, neuropsychological, and electrocortical effects of human exposures to mixed colonies of toxigenic molds." The study revealed, "Most of the patients were found to suffer acute stress, adjustment disorder, or post-traumatic stress."[67] The findings of this study bring up a critical question that must be asked: How many of the post-Katrina PTSD cases are actually misdiagnosed symptoms of mold exposure?

Dr. Callaghan further details the mental and psychological impact of mold exposure. He explains, "Mold is a neurotoxic. At our detoxification facility we see scores of mold patients who are brain fogged, confused, unable to perform cognitive tasks that they easily could before their exposure, unable to get their words out or comprehend properly (aphasia), and have short-term memory loss, irritability, fatigue, and depression."[68] In other words, post-Katrina residents are trying to cope under extraordinarily stressful conditions, while possibly suffering from neurotoxic disorders from mold exposure.

In order to evaluate a true diagnosis of PTSD, doctors should also consider exposure to any contaminant that can psychologically influence a person, such as a biotoxin and/or a synthetic chemical. Medical literature states, "Bio-accumulation, a buildup in the body of foreign substances, seriously compromises physiological and psychological health. Over the last ten years, hundreds of studies have demonstrated the dangers to health from toxic bio-accumulation. . . . Certain pesticides like malathion, diazinon, and Dursban accumulate in the nervous system and can cause brain disease, motor dysfunction, and psychological disturbances. Studies have shown that environmental pollutants can cause cancer as well as neurotoxic diseases including depression, apathy, and a diminished capacity to think."[69]

Dr. Erasmus explains, "If the physical toxins poison the brain, then they will also affect our psychology."[70]

Dr. Rapp concurs, "Because chemicals affect the brain and nervous system, the symptoms they cause may appear to be emotional or psychiatric in nature. Some affected individuals are therefore treated solely for an emotional disturbance, when in reality they have an unrecognized—and possibly treatable—medical problem."[71] Dr. Rapp makes an important, valid point, which leads us to ask: How many returning residents to Katrina-hit areas have not only been misdiagnosed with PTSD but also have been unnecessarily treated with antidepressants? Not only will their mental states likely further deteriorate because antidepressant drugs have been known to worsen depression and the emergence of suicidality, but the underlying cause(s) of their physiological and psychological conditions will remain untreated.

Prescribing antidepressants for conditions caused by fungal and chemical exposures is medical injustice.

According to Dr. Callaghan, "Doctors who are mold literate test for mold sensitivity, antibody levels, mycotoxin levels, and autoimmune disease like thyroiditis and look for autoimmune markers in the brain and peripheral nervous system. They use multiple strategies to treat their patients, including four-week in-house bio-detoxification, mold/fungus medication, immune-enhancing nutrients and nutritional IVs, and mold allergy neutralization."[72] (As we mentioned earlier, patients must thoroughly research the many treatment options available to determine which treatment therapies and components will likely be the most effective based on their individual health histories and circumstances.)

Obviously, multiple factors contribute to the human psychological decay in biologically and chemically complex environments such as Katrina-hit areas. What must be determined is whether an individual's PTSD, mental instability, depression, and/or suicidality is due to the actual trauma of the natural disaster or if the exhibiting symptoms are by-products of exposure to the massive amounts of mold, other biological contaminants, chemicals, and synergistic effects of the toxins. Obviously, many people likely suffer PTSD symptoms caused by a combination of both trauma and toxic exposures. However, without accurate diagnosis, proper treatment will not be forthcoming.

The causative factors of psychological conditions must be addressed on an individual basis in order to identify effective treatment, because if a person's PTSD symptoms are caused from fungal and chemical exposures, prescribing an antidepressant will not detoxify, neutralize, or alleviate the actual toxic exposures causing the psychological symptoms. On the other hand, if PTSD symptoms do stem from the trauma itself, people must realize that antidepressants are not the quick fix they may appear to be. Worse yet, antidepressants can promote life-threatening behavior.

Alternative treatment options—such as psychological therapy, ministerial counseling, naturopathic remedies, and nutritional therapies along with removal from the affected area and/or exposure contaminants—inherently have less risk and, in our opinion, have the capacity to be more effective. One thing is for certain, if not addressed, these underlying causes—be them purely mental stressors and/or toxin-induced psychological behaviors—will ultimately take a toll on a person's physical health.

Dr. Erasmus points out, "We are not talking just about physical causes of illness. There are mental and emotional causes of illness that also play a very powerful role, can lead to the same diseases, and will certainly enhance the damage done by toxic microorganisms, toxic molecules, and other physically toxic agents."[73]

Remember also, how we react to external stress can affect our health. As Dr. Santella stated earlier, "Stress has been shown to have an impact on cancer and disease risk in general."[74] Therefore, we must strive to not only achieve a healthy living environment—free of mold and chemicals—but also create and maintain a positive mindset, in spite of stress from possible unhealthy living conditions, faltered insurance coverage, the lack of governmental response, grief, loss, and complete disruption of societal norms and individual lifestyles.

Dr. Erasmus empathizes, "If I thought I lost my house to Katrina, and I'm thinking, 'Thank God, it's still standing,' but then I have to take it down because the mold is making me sick, that would be pretty hard on my stress level, unless I knew how to stay inspired." He cautions, "We have to ask ourselves, 'What are we doing in a house full of mold?' We have to look at that because that's an environmental influence in which we are living. The first step is to get rid of the mold. This can be tough."[75]

Removing yourself from mold and chemical exposures has to be the first priority, which can be a monumental task after a natural disaster given limited resources in an area of widespread destruction. Most people cannot begin to understand this type of challenge. Coping under this level of stress can test people with even the strongest fortitudes. Special focus needs to be paid to the health repercussions of decisions because long-term health outcomes can be affected—positively or negatively—by choices made in times of dismay. No matter where we live, we face the sad reality that in today's world no place is completely safe, free from risks of natural disasters, industrial toxins, environmental catastrophes, or terrorism.

There are, undoubtedly, legitimate cases of PTSD post-Katrina because the events that occurred during and after the natural disaster are historically known to cause PTSD. It behooves us all to realize the limitations of the human mind, because if we really open our eyes to the anguish, personal despair and loss, and mental plight of others—our hearts will open to the compassion and empathy within and melt the hardness many of us have created to shield ourselves from the grief, loss, and unfortunate circumstances that others endure and from which they suffer. We should not be afraid to reach out to individuals and families affected by tragedies lest next time we see our own reflection in the looking glass of despair and misfortune.

We must not turn a blind eye to the domino effect of mental anguish caused by hurricanes Katrina and Rita—or any other natural disaster. We must not embrace the philosophy *Ignorance is Bliss*, as it does not relinquish us from our social responsibility. We should reach out to people who are suffering—physically, mentally, and emotionally—as we would want others to reach out to us if we were victims of similar circumstances. Their plight could be ours *next time*. Remember, even the strongest iron will can begin to show fissures when under incredible stress.

In times of extraordinary circumstances, we must draw strength from places within that we might not otherwise. Dr. Erasmus offers insight, "There is something in every human being that is not affected by crisis. Whether we call that God, peace, contentment, feeling taken care of by life, or whatever we call it, it is a feeling. It is something inside every human being that cannot be affected by stress, trauma, drama, success, or failure. It is independent of gender, age, color, religion, culture, education level, status, or bank balance. If we cultivate that feeling in ourselves, we have fewer problems in stressful situations. If we cultivate that feeling, then when things get intense, we are much less likely to shoot each other, steal from each other, beat each other up, or destroy more than has already been destroyed by the natural disaster."[76]

Dr. Erasmus reflects, "The really cool thing about it is that everyone has that inside; we just don't go there much. Honestly, in the end, that's the best place to be. One day our bodies will fail, our medicines will fail, our food will fail, our minds won't work, our knees will buckle; but the feeling of being taken care of that is present in the core of our being will not be affected by any of that and will always be a source of inspiration in the midst of everything. I don't know how else to say it. I think it is the most overlooked, the most important thing we have to do for building ourselves, our families, our communities, our nation, and our world. We can cultivate inspiration in peacetime so that we can access it in times of crisis, and for those who accept the challenge of taking responsibility for their healing and health, this is just the beginning of a learning journey into better quality of life."[77]

(Endnotes)
1 The CDC Mold Work Group, "Mold, Prevention Strategies," p. 23.
2   Our findings were confirmed by Dr. Curt Kurtz in an interview, April 2006.
3   These mold-related symptoms are confirmed in scientific and medical literature, for example: The CDC Mold Work Group, "Mold, Prevention Strategies," p. 29-32; and Ammann, "Indoor Mold Contamination."
4   Mayo Clinic, "Thyroid nodules."
5   Ibid.
6   The CDC Mold Work Group, "Mold, Prevention Strategies," p. 23.
7   Aerias AQS [Air Quality Sciences] IAQ [Indoor Air Quality] Resource Center is available at http://www.aerias.org (accessed August 2006).
8   Aerias AQS IAQ Resource Center, "Health Problems Associated with Mold." Available at http://www.aerias.org/DesktopModules/ArticleDetail.aspx?articleId=52 (accessed August 2006).
9   The CDC Mold Work Group, "Mold, Prevention Strategies," p. 27; and Peraica et al., "Toxic Effects of Mycotoxins in Humans."
10   The CDC Mold Work Group, "Mold, Prevention Strategies," p. 37.
11   PBS, "Is It Safe to Return?"

12   Florida State University Research Foundation, Inc., "White Blood Cells." Available at http://scienceu.fsu.edu/content/virtuallab/hematology/docs/whitebloodcells.html (accessed April 2006). (Italics added.)

13   Avelox, 10/07/05, patient information sheet, Pharmacy 1, no expiration date, copyright by First Data Bank-The Hearst Corporation.

14   Ibid.

15   Ibid.

16   Ibid.

17   Florida State University Research Foundation, Inc., "White Blood Cells."

18   Interview with Dr. Michael Gray, March 2007.

19   Ibid.

20   Ibid.

21   Ibid.

22   From a discussion with a pharmacist at a national chain pharmacy.

23   From a Methylprednisolone Tablets package insert.

24   Ibid.

25   From a Nasonex steroid inhaler package insert.

26   Nasonex, 10/27/06, patient information printout, Wal-Mart Pharmacy, information expires 11/16/06.

27   From a Nasonex steroid inhaler package insert. (Italics added.)

28   Rapp, *Is This Your Child?* p. 528.

29   Ibid., p. 526.

30   Goldberg, *Alternative Medicine*, p. 219-220.

31   Rapp, Is *This Your Child?* p. 528.

32   Ibid.

33   Ibid., p. 524.

34   Goldberg, *Alternative Medicine*, p. 8.

35   Ibid., p. 9.

36   For more information on Sheller, Ludwig & Badey, see http://www.sheller.com/Practice.asp?PracticeID=21 (accessed April 2006).

37   Ibid. (Italics added.)

38   Interview with Stephen Sheller, March 2007.

39   Science Daily, "Mayo Clinic Study Implicates Fungus as Cause of Chronic Sinusitis," September 10, 1999. Available at http://www.sciencedaily.com/releases/1999/09/990910080344.htm; and Mayo Clinic Press Release, "Mayo Clinic Study Implicates Fungus as Cause of Chronic Sinusitis."

40   Interview with and e-mails from Dr. Timothy Callaghan, January 2007. For more information on the Center for Occupational & Environmental Medicine, see www.coem.com/ or call 843-572-1600.

41   Goldberg, *Alternative Medicine*, p. 36 and 620.

42   Rapp, *Is This Your Child?* p. 439.

43   Goldberg, *Alternative Medicine*, p. 35.

44   Science Daily, "Mayo Clinic Study Implicates Fungus as Cause of Chronic Sinusitis," and Mayo Clinic Press Release, "Mayo Clinic Study Implicates Fungus as Cause of Chronic Sinusitis."

45   Goldberg, *Alternative Medicine*, p. 36. (Italics added.)

46   Ibid., p. 22 and 620.

47   Dr. Fungus, "Aspergillosis." Available at http://www.doctorfungus.org (accessed February 2006). (Italics added.)

48   Interview with Dr. Curt Kurtz, April 2006.

49   Interview with and e-mails from Dr. Timothy Callaghan, January 2007.

50   U.S. FDA, "FDA Public Health Advisory: Worsening Depression and Suicidality in Patients Being Treated With Antidepressant," March 22, 2004. Available at http://www.fda.gov/cder/drug/antidepressants/AntidepressanstPHA.htm (accessed April 2006). (Italics added.)

51   Harvard Medical School, "Post-Traumatic Stress Disorder." Available at http://www.intelihealth.com/IH/ihtIH/WSIHW000/8271/21849/187897.html?d=dmtHealthAZ (accessed August 2006).

52   Ibid. (Italics added.)

53   Ibid.
54   California Nurses Association, "CNA Press Releases." Available at http://www.calnurses.
org/media-center/in-the-news/news2005/december/page.jsp?itemID=27505685 (accessed April 2006).
55   Harvard Medical School, "Post-Traumatic Stress Disorder."
56   California Nurses Association, "CNA Press Releases."
57   Seth Borenstein and Chris Adams, "Health problems abound months after Katrina roared
ashore," November 29, 2005. Available at http://www.realcities.com/mld/krwashington/13285938.htm
(accessed April 2006).
58   California Nurses Association, "CNA Press Releases."
59   Cecil Connolly, "Katrina's Emotional Damage Lingers," *Washington Post*. Available at
http://www.washingtonpost.com/wp-dyn/content/article/2005/12/06/AR2005120601594.html (accessed
June 2006).
60   Ibid.
61   Ibid.
62   Ibid.
63   Cathy Booth Thomas, "The Storm Lingers On: Katrina's Psychological Toll," Time/CNN,
August 28, 2006. Available at http://www.time.com/time/nation/article/0,8599,1417780,00.html
(accessed August 2006).
64   Ibid.
65   Ibid.
66   Ibid.
67   Crago, B. et al., "Psychological, Neuropsychological, and Electrocortical Effects of Mixed
Mold Exposure," *Archives of Environmental Health*; 58(8); 452-63, August 2003. Available at http://
www.ncbi.nlm.nih.gov/entrez/query/fcgi?cmd=Retrieve&db=PubMed&list_uids=15259424&dopt=Abst
ract (accessed April 2006); Ammann, "Indoor Mold Contamination."
68   Interview with and e-mails from Dr. Timothy Callaghan, January 2007.
69   Goldberg, *Alternative Medicine*, p. 168 and 226.
70   Interview with Dr. Udo Erasmus, May 2006.
71   Rapp, *Is This Your Child?* p. 310.
72   Interview with and e-mails from Dr. Timothy Callaghan, January 2007.
73   Interview with Dr. Udo Erasmus, May 2006.
74   See Section Three, interview with Dr. Regina Santella, July 2006.
75   Interview with Dr. Udo Erasmus, May 2006.
76   Ibid.
77   Ibid.

CHAPTER TWO:

THE PROCESS OF TRIAL AND ERROR

At about the same time Kurt had his confrontation with the mold-cannot-make-you-sick doctor, we realized we had entered a new world of multiple chemical sensitivities, which can often result from mold and/or multi-toxin exposures.[1] Not only were we reacting to elevated levels of ambient mold spores and getting sick from even minute traces of mold in our food, but we had also become allergic to many personal hygiene products, such as lotions, shampoos, and soaps.

The same products we had used for years were now causing ugly red chemical burns that would remain for days. We wondered if the chemical ingredients in these products were interacting with chemicals our bodies had absorbed post-hurricane or if our bodies were so toxically overburdened that they couldn't detoxify the same chemicals they did before the hurricane.[2] We didn't know that the exposure to the hurricane toxins would begin a domino effect of multiple sensitivities.

In the beginning, we were looking for a common denominator in the products to which we were reacting. We didn't realize that multiple chemical sensitivities aren't necessarily triggered by any similar component. In fact, according to published health literature, "A distinguishing feature of multiple chemical sensitivity (MCS) is its 'spreading' effect. After the sensitizing event or chronic low-level exposure to one or more chemicals, an individual becomes sensitive to an increasing number of substances. This may include chemicals that had never triggered reactions before the sensitization, such as glue, perfumes, air fresheners, or gasoline. It may also spread to include food, drugs, alcohol, caffeine, or even airborne allergens. 'Switching' is another phenomenon associated with MCS in which one symptom is replaced by another. For instance, whereas the smell of paint used to make you dizzy, it now gives you headaches."[3]

We began to eliminate use of the offending products, trying to eliminate the health challenges created by exposure to post-Katrina mold and chemicals. At this time, we had not yet confirmed our suspicions that the exposure to the mold inside our rental home had been the main culprit that had derailed our health. All we knew is that we were getting increasingly sick each day. Then we met Jim Pearson, who explained that certain mold species can cause infection. He cited several documented sources of information regarding the health effects of mold exposure, such as allergy, asthma, and other respiratory conditions. Mr. Pearson also recommended we seek treatment at a naturopathic clinic.

The realization that a certified mechanical hygienist, albeit one who coauthored a highly recommended reference guide on mold, knew more about the documented health effects of mold exposure than did the multiple doctors we had already seen was eye-opening and shocking. As adamant as the mold-cannot-make-you-sick doctors had been, we did not expect to find an abundance of scientifically and medically documented literature that clearly refuted their statements.

We took Mr. Pearson's suggestion and visited a naturopathic clinic. We were hoping there would be a quick solution via a naturopathic and/or homeopathic remedy, but we soon found that the initial prescribed treatment plan had to be replaced by a time-consuming system of trial and error. This process was frustrating and challenging because we not only had to deal with our initial symptoms but also new discomforts brought on by reactions to unsuccessful treatments. At times, we felt as though there were no parts of our bodies left unaffected—skin, hair, eyes, ears, intestines, lungs, muscles, bones…the list goes on. Also, since we were experiencing multisystem symptoms, widespread relief was elusive. Only after months of trial and error and intensive research was an effective treatment solution identified and our health restored. Only in hindsight are answers quick. For this reason, we share the results of our research and our journey of healing.

In the following chapters we share our medical and nutritional discoveries—what worked, what didn't, and why—based on relevant scientific and medical documentation. Remember, just as mold can affect people differently because of genetics and many other influencing factors, such as duration of exposure, species, synergistic effects of other contaminants, age, nutrition, medical history, current prescriptions, and health status,[4] the effectiveness of treatment plans can also vary because of these same components. Thus, people may not necessarily respond in the same way when given the very same treatment. However, becoming aware of treatment options that were effective for others may help people gather information regarding alternative options, which they can then evaluate with their treating physician(s). If certain vitamins, herbs, nutritional supplements, and food selections helped us overcome mold- and chemical-related illnesses and re-establish strong immune systems, possibly these same treatment options will help others as well. Certainly, no guarantees exist that the combined components of our treatment plan will be effective for anyone else.

We share this information in an educational format. It is not provided or intended to be used as medical advice, nor should it be misconstrued as medical advice. People need to consult with their personal health care provider(s) and individually address any use of the products we mention in this book. Note that some products may require restricted usage by certain people, depending on existing health conditions. For example, people with low blood pressure should consult with their doctor(s) before increasing consumption of garlic, which is known to lower blood pressure.[5] Also, interactions can occur between pharmaceuticals and natural products. Thus, all components of a treatment plan must be discussed with a person's treating physician so he/she can evaluate any possible interactions. The potential benefits and risks associated with the use of any product, pharmaceutical or natural, needs to be evaluated by each individual and his/her doctor. Likewise, we consulted our primary care physician regarding the use of each product. Each product we used was prescribed by our treating physician, who monitored our progress and made dosage adjustments as necessary.

For alternative products that were ineffective or to which we had a negative reaction, we have not included the specific brand names. Other people may have successes with these products even though we did not. For the products that were effective in reducing symptoms, we also have not provided the specific brand names. Although we found these products, along with diet changes, to be responsible—collectively—for our recovery, we are not, per se, recommending these products. Over time, the manufacturers of these products may change the ingredients or the supply source of the ingredients, which could affect product quality. We are simply sharing our experiences of the positive impact that some products had on the health of our family and documenting why they may have been effective from a scientific medical standpoint. We are not paid to endorse any of these products. We do not sell any of these products, nor do we plan to sell any of these products in the future. Again, all our medical decisions regarding treatment options were made under the care and prescription of our primary care doctor, our naturopath (ND).

Medically treating people exposed to and sick from the Katrina-aerosolized debris, chemicals, and fungi was a formidable challenge with no blueprint to follow. Because of Katrina's force and magnitude, unprecedented levels and combinations of contaminants, mold spores, fungal fragments, and particulates were disrupted and aerosolized. We, as others in the disaster area, were exposed to not only multiple species of mold spores, mold fragments, and mycotoxins, but also to everything else that was aerosolized during and after Hurricane Katrina—bacteria, chemicals, smoke, fine particulate from pulverized debris...the list goes on. Even though many contaminants undoubtedly settled and/or dissipated by the time they reached our town, the air was still thick with visible particulates, as well as particulates not visible to the naked eye.

The spores in the ambient air germinated in the heating and cooling system of our rental house, as well as on the damp walls, wet from water leakage during the hurricane. This structural mold created additional toxic exposures for our family. As you will recall from Section I of this book, if indoor fungal growth is toxigenic, concentrated levels of mycotoxins can occur. Other toxic contaminants that can occur from wet building materials include gram-negative and gram-positive bacteria, the endotoxins and bacterial cell wall toxins they respectively produce, and VOC off-gassing from both organisms and building materials. These additional toxins combined with the fungal contaminants—all confined within the walls of our house—undoubtedly compounded the negative health effects for our family.

By the time we were looking into naturopathic care, the children and I were starting to experience increasing ill health effects, although Kurt's health was still deteriorating at a faster rate. We set appointments for each member of our family. We were all experiencing multisystem symptoms, yet we didn't necessarily exhibit the same ones. Several of us experienced the same symptoms in varying degrees of intensity, while just one or two of us exhibited other symptoms. We did not know at this point that exposure to mixed molds and mycotoxins can produce such a myriad of symptoms. When we did come to this realization from our research, it was actually reassuring. The "known" is much less frightening than the "unknown."

If you have never established medical care with an ND or another type of alternative health care provider, the first main difference you will notice is the length of the appointment! Our initial visit lasted approximately 40 minutes for each person. The ND reviewed medical history and current symptoms. Most of the symptoms we experienced included the following: skin rashes, hives, wheezing, intestinal/digestive problems (including diarrhea interspersed with constipation), loss of hair, irregularity of menstrual cycle, vision problems (including inability to focus at times and discharge coming from behind our eyes), ringing in the ears, ear pain, fungal infection of the sinuses, sinus bleeding, extreme headaches, extreme pain in the sinuses, visible fungal infections (on feet, toenails, fingernails, tongue, mouth, and private areas), food intolerances/allergies, chemical intolerances/allergies, vomiting, shakiness, mood changes, numbness in limbs, twitching of nerves, thyroid irregularities, heart palpitations, heart arrhythmias, increased pulse, joint pain and inflammation, lack of concentration, memory loss, sleep disturbances, and lack of energy. Needless to say, we were not having a good time!

These symptoms are the same types of symptoms that have been reported by other people affected by mold exposure from Hurricane Katrina.[6] As well, these symptoms are the same types exhibited by people with mold-related illnesses from fungal exposure in water-damaged buildings.[7]

Dr. Thrasher, Dr. Gray, and colleagues confirm, "Exposure to mixed molds and their associated mycotoxins in water-damaged buildings leads to multiple health problems involving the CNS [central nervous system] and the immune system, in addition to pulmonary effects and allergies. Mold exposure also initiates inflammatory processes." They propose a new term *mixed mold mycotoxicosis* to describe these multisystem illnesses.[8]

According to clinical studies, multiple symptoms are often exhibited with fungal-related illnesses. One treating physician reports, "During the last five years, I have treated patients with various mold-related illnesses contracted at either industrial buildings such as old buildings, schools, and governmental offices, as well as residences, all of which have suffered either faulty ventilation, water damage, or both. The most common presenting symptoms are those of 1) cough, 2) asthma, atypical asthma, 3) nasal congestion, 4) sinusitis/rhinitis, 5) skin rashes, and

6) generalized fatigue. On many occasions the patients presented with neurological symptoms such as headaches, reduced concentration ability, and memory loss. The patients may present with only one symptom (such as sinusitis) or a combination of symptoms."[9]

It is easy to start questioning yourself when so many parts of your body are simultaneously going awry. This self-doubt can be compounded when doctors ignorant of the facts—that fungal exposure can cause multiple symptoms and systemic illnesses—are consulted. When we finally read reports of doctors' clinical findings of mold-exposed patients, we found a level of reassurance in just knowing that other people had experienced this same phenomenon of multisystem symptoms and illnesses from mold exposure. Others suffering the systemic effects of mold exposure may also find this type of information reassuring. For this reason, we detail our symptoms and those documented in reports to provide a "sanity" check for others experiencing multisystem symptoms from exposure to fungi and the inherent accompanying contaminants.

The treatment plan our ND initially prescribed included diet changes and multiple homeopathic products. According to published health literature, "Homeopathic remedies are generally dilutions of natural substances from plants, minerals, and animals. Based on the principle of 'like cures like,' these remedies specifically match different symptom patterns or 'profiles' of illness and act to stimulate the body's natural healing response."[10] In other words, homeopathic medicine relies on the ability of a person's immune system to react to "like" stimuli, a theory based on the Law of Similars.

As explained in the book *Alternative Medicine*, ". . . [T]he Law of Similars is a principle first recognized in the fourth century B.C.E. by Hippocrates, who was studying the effects of herbs upon disease. The Law of Similars was also the theoretical basis for the vaccines of physicians Edward Jenner, Jonas Salk, and Louis Pasteur. They would 'immunize' the body with trace amounts of a disease component, often a virus, to strengthen its immune response to the actual disease. Allergies are treated in a similar fashion by introducing minute quantities of the suspected allergen into the body to bolster natural tolerance levels."[11]

We were sent home with an array of homeopathic products, detoxification products, and a restricted diet of no sugar, no simple carbohydrates (such as refined white bread, refined pasta, and crackers), limited fruits, and limited dairy products (such as milk, cheese, and ice cream). Our diet was to be limited to whole foods, heavy on the vegetables.

These diet recommendations are similar to the diet guidelines outlined in *The Yeast Connection Handbook*[12] and *The Yeast Connection Cookbook*.[13] These books focus on the treatment of Candidiasis, which is an infection or disease caused by the fungus of the genus *Candida*. *Candida* are yeasts, which is why Candidiasis is sometimes referred to as a yeast infection. Although yeasts are not molds and vice versa (molds are not yeasts), both are fungi. According to Dr. Erasmus, "Sugars feed *Candida* (yeasts), fungi, other pathological (toxin-producing) organisms, and cancer cells."[14] The similarities in the anti-*Candida* diet and the one prescribed by our naturopath made sense to us since yeasts and molds are both fungi, and both were flourishing in our bodies.

It sounded simple in the beginning—withhold their food supply (sugar) and starve the fungi out! We quickly found it wasn't quite that simple. Sugar is in everything, or so it seemed. Let's take a look.

Simple carbohydrates, which are sometimes called simple sugars, are processed as a sugar by the body. Simple sugars include ". . . fructose (fruit sugar), sucrose (table sugar), and lactose (milk sugar), as well as several other sugars. Fruits are one of the richest natural sources of simple carbohydrates."[15] We had to limit our intake of fruit and milk products, foods normally considered healthy, because the sugars in these foods would feed the fungi. Another important reason to exclude simple sugars from the diet is that they lower the immune system. According to published health literature, "Thirty minutes after consuming sugar, the immune system is measurably suppressed, losing as much as 50% of its capacity. Sugar consumption is linked to a variety of health problems, such as mood swings, heart disease, gallbladder disease, ulcers, colitis, overeating, and addictive behaviors such as alcoholism."[16]

Dr. Erasmus reports, "Sugars inhibit the functions of our immune system and increase diseases caused by poor immune function, such as colds and flu."[17]

Germs are everywhere, but for the most part, healthy immune systems ward them off. We are all familiar with the phrase *I caught a bug*. This expression refers to the vast array of germs that exist in our environment that sometimes take hold in our bodies, causing colds, flus, and/or other illnesses. But unless we are immune compromised or become seriously ill, most of us don't consciously think about the microscopic world of organisms that surround us, both externally and internally. We certainly did not before becoming ill.

Dr. Erasmus explains, "Our world is teeming with one-celled natural creatures: yeasts, fungi, bacteria, and protozoans. They are present in our soils, on our foods, in our water, and in the air we breathe. We find them on our skin and inside our intestinal tract. They are everywhere. Whether they harm us, help us, or simply leave us alone depends to a large extent on factors that are within our control."[18]

He further explains, "In our intestines, yeasts, fungi, and bacteria fulfill normal, natural, even beneficial roles, and most of the time cause us no harm. Natural intestinal mechanisms control their growth and numbers. But bad things can happen when these natural creatures get out of control."[19] For example, yeasts produce toxins, also called mycotoxins, much like molds produce mycotoxins.[20] (Remember, yeasts and molds are both fungi.) So when intestinal yeasts grow out of control, toxin levels increase, adversely affecting a person's level of health.

In an effort to enhance the viability and population and maintain the proper balance of these good microorganisms, many manufacturers have developed products called probiotics, which are designed to supply millions or billions of beneficial microorganisms to the intestinal tract. A healthy flora is important because "[a] healthy intestinal wall, one coated primarily with 'friendly' bacterial microorganisms, provides the protective lining that is necessary to keep damaging substances out of the body's circulation while letting helpful ones in."[21]

In an increasingly toxic world, we all need to focus on how to strengthen our bodies' defenses against toxins and microorganisms, because the future of our health depends on the ability of our bodies to detoxify and properly dispose of foreign toxins and cells. If our bodies can detoxify contaminants and keep unhealthy levels of foreign microorganisms in check, we stay healthy. If our bodies cannot adequately detoxify toxins, they accumulate—in our blood, in our fat and muscle tissues, and possibly even in our bones and organs. This toxic accumulation can promote overpopulation of foreign microorganisms. Eventually, various forms of chronic, degenerative diseases can manifest. Therefore, detoxification and establishment of a healthy flora must be a focus in order to maintain future health.

Published health literature states, "In healthy individuals, the body's detoxification system is able to neutralize and eliminate toxins, thereby minimizing tissue damage and preventing illness. But the detoxification system, including the liver, the intestines, and the lymphatic system, can become overwhelmed by toxins. Toxic overload causes congestion in the lymphatic system, in which thickened lymph accumulates in the nodes without being emptied into the blood for removal from the body, and may also involve chronic intestinal constipation and liver dysfunction. The body's inability to remove toxins is a major cause of accelerated aging and a primary contributor to chronic, degenerative disease process."[22]

The body is naturally designed to protect itself against toxins. Let's first take a look at the gastrointestinal tract. It usually ". . . serves as the first line of defense against toxins entering the body. When it becomes compromised, it also affords disease-causing agents a place to fester, sometimes to the point where they eventually break through the intestinal membrane and enter the bloodstream. Once the bowel is toxic, the entire body soon follows. When undigested food particles, bacteria, and other substances normally confined to the intestines escape into the bloodstream, they trigger the immune system and inflammation ensues. If the intestines continue letting toxins through, then the liver, lymph, kidneys, skin, and other organs involved in detoxification become overwhelmed."[23]

The catalyst of our illnesses was essentially disruption of healthy intestinal flora.

The fungal and chemical exposures from Hurricane Katrina disrupted the natural flora of our intestinal tracts, thus allowing the fungal and chemical toxins to pass through the normal intestinal barriers and into our bloodstreams. Once in the bloodstream, these toxins can travel to and affect any part of the body, causing multisystem symptoms.

"At this point, the front line of digestive defense has fallen, and abnormal proteins and toxic particles begin passing through the intestinal membrane into the bloodstream, causing what is called 'leaky gut syndrome.'"[24] A leaky gut is a condition that people are often unaware of and one that conventional medical doctors often don't address. In addition, antibiotics and steroids are often prescribed, as occurred in our case. They can accelerate a leaky gut condition, as both can severely upset the delicate balance of the gastrointestinal tract and cause great damage to the bowel flora.[25]

An imbalance of flora within the gastrointestinal tract allows overgrowth of yeasts, fungi, and bacteria that, when existing at normal levels, are necessary for a healthy, optimally functioning gastrointestinal tract. However, when overgrowth occurs, ill effects can ensue from these unchecked microorganisms, causing a variety of symptoms and worsening existing health conditions.

The body is designed to dismantle the millions of toxic compounds that enter it each day.[26] It has many detoxification pathways with the main detoxifying organ being the liver. According to health sources, "The liver bears most of the burden for eliminating toxins once they have entered the bloodstream. All foreign substances are carried to the liver to be filtered and expelled from the body. Using enzymes and antioxidants, the liver chemically transforms toxins into harmless substances that can be excreted via the urine or stool. Other toxins are eliminated through the lymphatic system, the kidneys, the skin (through perspiration), and the respiratory system."[27]

"When the liver's ability to detoxify becomes impaired due to toxic overload, it becomes more difficult for toxins to be eliminated. This causes them to circulate in the blood and accumulate in fat and muscle tissue."[28] Furthermore, "When imbalances occur in the detoxification system, the result can be poor digestion, poor assimilation of nutrients, constipation, bloating and gas, immune dysfunction, reduced liver function, and a host of degenerative diseases."[29] Remember, when our first line of defense—our gastrointestinal tract—becomes compromised, the ensuing leaky gut condition will increase the body's toxicity, as toxins that would normally be eliminated travel through the intestinal wall and absorb into the blood. The toxin-filled blood then circulates, re-exposing the body to the same toxins it was trying to eliminate in the first place.[30]

To help people understand what actually happens in the body from fungal exposure, Vincent Marinkovich, MD, a leading specialist in allergy and immunology, has given a succinct description—based on clinical findings—of the functional changes that occur in the body from fungal exposure: "Exposure to high levels of fungi can be a health threat from inhalation, ingestion, or skin contact involving tiny mold spores (invisible to the eye), mold toxins (mycotoxins), or mold bodies themselves. Initial symptoms seem to be the result of inhalation, such as sore throats, hoarseness, cough, and nasal congestion. With time, symptoms can progress to include headaches, fatigue, rashes, dizziness, shortness of breath, sinus infections, ear pain, muscle and joint pain, and fever. These symptoms are the result of direct mycotoxin exposure and the effects of an overactive immune system trying desperately to overcome what it perceives to be an overwhelming infection. The immune system generates antibodies to the absorbed mold materials (antigens). These antibodies react with the antigens to form immune complexes, which is all part of the body's normal immune elimination function. When the immune clearance machinery is on overload, the complexes remain in the blood stream causing myriad symptoms, known to clinical immunologists as serum sickness, and appearing to the patient as a severe, unrelenting flu syndrome. Exposure to certain mycotoxins can result in brain damage seen as short-term memory loss, cognitive dysfunction, inability to concentrate, and 'fuzzy thinking.' *These changes seem to be reversible*, at least in part, but they can take years to resolve."[31]

To read that some of the symptoms we were experiencing—"short-term memory loss, cognitive dysfunction, inability to concentrate, and 'fuzzy thinking'"—were signs of brain damage got our attention, to say the least. Of course, the significant, encouraging point to remember is that Dr. Marinkovich also states, "These changes seem to be reversible . . ."[32] This ray of hope was highly motivating to us. We logically reasoned that since our fuzzy thinking, lack of concentration, and other signs of cognitive disfunction were relatively new conditions (it had been only several months since the hurricane), we had a better chance of full recovery than if our conditions had been ongoing for a longer duration. We do not know if this logic is correct, but it was our thinking at the time. Our goal was to get rid of the *cause* of the symptoms as fast as possible to reduce any chance of long-term brain damage.

The fact that Dr. Marinkovich points out that it can take years to resolve, or reverse, these cognitive symptoms may indicate that it is possible for people who have experienced these symptoms for even years to also see improvement in brain function. So don't get stuck in the mental quicksand pit of believing you can't get any better because you've had your symptoms for years—and think it's "too late" for you. It is true that we did not experience these symptoms for years, but by the time we read Dr. Marinkovich's published clinical findings, we had been "brain wrestling" with these symptoms for more than a few months. Furthermore, it took months of effective treatment before we recovered our original clarity and sharpness of mind. It was a process during which we saw gradual, consistent progress.

The other issue that Dr. Marinkovich addresses is re-exposure. "Once the patient has become hypersensitive to the mold in their environment, they have also become overly reactive to all molds in their life including those they breathe elsewhere, those they eat, and those that may be colonizing their tissues. Relief of symptoms can only come with a significant reduction in exposure including a mold-free diet, avoidance of mold-ridden environments, and treatment of mold colonization."[33]

The ramifications of re-exposure made complete sense to us, as we were having severe reactions to mold in the outside ambient air, inside stores, and in our food, and were also experiencing layers of visible mold working its way out of our tissues. We quickly realized we had to eliminate mold, as much as possible, from our lives. One, we needed to breathe in as little fungi as possible. Two, we needed to eat as little fungi as possible. And three, we needed to rid our bodies of all fungi colonizing our tissues. These goals were at the forefront of our minds, constantly, as we made daily choices. Our focus was, as Dr. Erasmus states, to ". . . remove toxins from our bod[ies] at a faster rate than we take them in."[34]

Another influencing factor of which we became keenly aware was that our pre-hurricane state of health affected the ability of our bodies to battle the fungi, which turned out to be a formidable foe. Although our family had a clean bill of health pre-hurricane, as shared earlier, we each fell into one of the high-risk groups in regard to mold exposure.

Kurt was at high risk because he had pre-existing allergies and borderline asthma, which the CDC states as conditions of high risk.[35] This increased risk was additionally compounded by the fact that Kurt had been using, both pre- and post-hurricane, prescription steroids for allergy management, which weaken the immune system.[36] Steroid package inserts clearly state that steroids are immunosuppressants. In fact, according to the CDC, medical doctors, and scientists, people using steroids are considered immunosuppressed, a condition which can further render them immunocompromised.[37]

Our children were at high risk because of their ages.[38] The CDC states that children ". . . may be affected to a greater extent than most healthy adults."[39] In fact, the CDC states that infants and children ". . . should *avoid mold-contaminated environments entirely*."[40] And I was at high risk because I had recently undergone major surgery. The New York City Department of Health and Mental Hygiene, which published the first industry guidelines regarding mold (*Guidelines on Assessment and Remediation of Fungi in Indoor Environments*), states, ". . . [P]ersons recovering from recent surgery . . . may be at greater risk for developing health problems associated with certain fungi."[41]

So just how much fungal exposure can people in high-risk categories safely tolerate? According to Dr. Thrasher, "People in the high-risk groups should stay away from *any* exposure."[42] The New York City guidelines concur, stating that high-risk individuals should be medically evacuated and not return until remediation is complete and air testing reveals safe levels.[43]

Exposure to contaminants created in the wake of natural disasters can significantly add to a person's pre-existing level of toxic body burden. A strong detoxification system is imperative, because exposure to abnormally high levels of multiple toxins can further overburden the body's detoxification pathways. For example, if a person's body is already not properly detoxifying existing toxic exposures and is therefore accumulating toxins, his/her body will most likely be unable to detoxify any additional toxins. More toxins are then stored in muscle and fat tissues,[44] creating an internal toxic-waste dump. If left untreated, this toxic condition will ultimately lead to accelerated aging and contribute to chronic, degenerative disease processes.[45]

We have to remember, too, that toxins can come in many forms. Although sometimes necessary and lifesaving, pharmaceuticals are toxic to our bodies. Dr. Erasmus states, "No drug is a good option because they are all synthetic. They have never been in nature, which means the genetic program doesn't know what to do with them. That's why there are side effects. The side effects are disease caused by molecules that don't belong in the human body."[46] Dr. Erasmus adds, "Drug-oriented, high-tech medicine seriously undermines the foundations of human health, adds to our toxic load, and produces unwanted side effects that are themselves full-blown diseases."[47]

"For this reason, alternative physicians often employ detoxification therapies to reduce or eliminate the body's 'toxic load,' restore the proper function of the immune and other body systems, and help alleviate age-related illnesses."[48] Following this philosophy, our ND outlined a treatment plan and overall health regime that focused on detoxification and restoration of a proper immune function.

As we embarked on the aforementioned, prescribed homeopathic treatment, we attempted to keep each day as fungal free as possible in regard to both food and environmental exposures. Having not used homeopathic medicine before, we had no idea what to expect. It was certainly simple enough. The homeopathic remedies came in the form of drops that we were to add to water, juice, or protein shakes. We were encouraged and highly optimistic. However, our high hopes were quickly deflated when Kurt immediately and violently reacted to the first sip of the homeopathic solution. To confirm it was the homeopathic drops to which he was reacting, we added the drops to only distilled water the next time. He still had a severe reaction, which told us that, yes, it was the homeopathic drops to which he was having a reaction. After two to three swallows of the drop-filled water, he was off to the bathroom and out of commission for days. Our ND immediately took Kurt off the homeopathic components of the treatment plan.

The rest of us continued to use the homeopathic solutions, and although we were able to keep them down (unlike Kurt), they did not alleviate any of our symptoms. In fact, over time, the homeopathic drops made us worse, forcing us to abandon this particular treatment method after several weeks. We were discouraged to have invested so much time in a treatment plan that, in theory, should have worked, but didn't. In fact, homeopathy treatment not only didn't work, but made us worse! Why?

To answer that question, we have to remember that homeopathic remedies are designed to stimulate the immune system with "like" stimuli. In hindsight, we have two theories. One, if a person does not have a strong enough immune system intact, homeopathic remedies may be much like the straw that breaks the camel's back and serve to only worsen the person's symptoms and conditions. Remember, if fungal exposure got you sick, a little more of the same (i.e., like cures like) could very well make you violently ill, as was Kurt's reaction, or just slowly make you worse, as was my and our children's reactions.

The other plausible theory as to why homeopathy agitated our conditions has to do with the fact that many people sick from fungal exposure are immune disregulated. According to Dr. Thrasher, a disregulated immune system means the immune system is both suppressed and enhanced, depending at which arm of the immune system you're looking. For example, one arm of the immune system can show suppression while other portions are enhanced. Dr. Thrasher gives an example, "We see that some 40 percent of the people exposed to molds and toxic chemicals, in general, but also molds or fungi, have decreased natural killer cell activity, which means that they are immunocompromised. On the other hand, when we look at their T and B cells, they are going to have elevated T and B cells, which means that portion of the immune system is enhanced. Thus, their immune systems are disregulated."[49] Thus, stimulating an already disregulated immune system with homeopathic medicines may not be effective.

Homeopathic treatment was the first of many false starts. Some of the other treatment components did nothing, but many worsened our conditions, which, in part, motivated us to write this book to possibly shorten the learning curve for others suffering from fungal exposures.

In sharing our journey of healing, we note products that, logically, should have helped, but didn't. Although ineffective, we list these products because they are likely first-choice treatment options that would be prescribed by other naturopaths. We also specify which products we felt were helpful support products and which ones we felt were directly instrumental in battling the fungi and restoring our immune systems. We do not list the individual brand names of alternative products that were ineffective, but rather list the type of product.

First, let's talk about the non-homeopathic products initially prescribed that were ineffective and/or proved to agitate our conditions.

To aid in detoxification, a bentonite supplement was prescribed. Bentonite is advertised to aid in detoxification by absorbing toxins, which then can be eliminated through the alimentary canal. It was a logical first-choice option to try as literature states, ". . . [B]entonite has been increasingly prescribed by practitioners of alternative medicine as a simple but effective internal cleanser to assist in reversing numerous health problems. . . . Bentonite is not a mineral but a commercial name for montmorillonite, the active mineral in many medicinal clays, which comes from weathered volcanic ash."[50]

The theory as to the effectiveness of this type of product is that "[l]iquid clay contains minerals that, once inside the gastrointestinal tract, are able to absorb toxins and deliver mineral nutrients. Liquid clay is inert, which means it passes through the body undigested. Technically, the clay first adsorbs toxins (heavy metals, free radicals, pesticides), attracting them to its extensive surface area; then it absorbs the toxins, taking them in the way a sponge mops up water. The clay's minerals are negatively charged while toxins tend to be positively charged; hence the clay's attraction works like a magnet. According to the *Canadian Journal of Microbiology*, bentonite can absorb pathogenic viruses, aflatoxin (the toxic by-product of a mold), and pesticides and herbicides. The clay is eventually eliminated from the body with the toxins bound to its surfaces."[51] None of these claims came to fruition for our family when we were treated for mold- and chemical-related conditions.

While taking bentonite, I experienced increased fatigue, lethargy, and other symptoms. It felt almost as though I wasn't getting the nutrients from the foods I was eating. Kurt could not tolerate this product at all, and the children refused to drink it. In hindsight, we have wondered if the bentonite absorbs valuable nutrients in addition to toxins. Also, with the presence of a leaky gut (created by the disruption of intestinal flora from fungal and chemical exposures), could the bentonite clay bound with toxins enter the bloodstream via compromised intestinal walls, causing increased allergic reactions and further depleting the immune system?[52]

Psyllium husks powder, a natural source of dietary fiber, was another product prescribed. Again, this supplement was a logical choice to add to a treatment plan for mold and chemical exposures, as literature states, "Fiber helps flush wastes from the body. . ."[53] However, this benefit did not prove to be the case for our family. I experienced exacerbation of symptoms while taking this product; Kurt could not tolerate it at all, and the children refused to drink it. In hindsight, as we became more experienced in recognizing symptoms of fungal re-exposure, we began to suspect that fungal antigens were present in this particular brand of psyllium husks. According to scientific literature, psyllium is derived from the husks of the seeds of the plant *Plantago ovata* or *Plantago psyllium*. As we already know, fungi can grow on plants.

An adult rice protein supplement was also prescribed. Product claims included nutritional support for various conditions (including allergies and inflammation) from which we were suffering, which made this supplement a logical addition to our detoxification plan. However, these product claims were not experienced by our family. We tried mixing it with a variety of liquids: water, milk, and soy milk (with and without a few strawberries for added flavor) as recommended by our ND.

For awhile I was able to tolerate this supplement, but over time, it made me sick and worsened my symptoms. Kurt did not respond positively to this product, even with just the first sip. Noticing that we felt best when we ate vegetables, which are alkaline, we contacted the manufacturer of this rice protein supplement and inquired about the alkalinity/acidity content. The manufacturer confirmed that the rice protein powder itself, without influence of any added liquid, was definitely acidic. We believe that the acidity level of the product explains my increasingly negative reaction over time: By consuming the acidic rice protein supplement two to three times a day as prescribed, the biological terrain of my body became progressively more acidic. This experience was further proof that our bodies were responding better to alkaline foods, but why?

We researched the alkalinity/acidity issue for many months, looking for a scientific-based reason that explained why we felt better when consuming an alkaline-based diet. We were also searching for why any sizable consumption of acidic foods created a draining, lethargic feeling in our bodies and sometimes even induced stomach and intestinal wrenching, headaches, and other symptoms. Let's take a look at what we learned.

According to health literature, "The term pH, which means 'potential hydrogen,' represents a scale for the relative acidity or alkalinity of a solution. Acidity is measured as a pH of 0.1 to 6.9, alkalinity is 7.1 to 14, and neutral pH is 7.0."[54] According to Dr. Chew, "Molds usually require an acidic pH . . ."[55] The logical question is: Does consumption of acidic foods in sufficient amounts create an internal acidic pH breeding ground for fungi? Robert Young, PhD, confirms the fact that an acidic biological terrain can increase the growth of fungi in the human body.[56] Therefore, a diet high in acidic foods can foster the growth of fungi and result in a noticeable increase in ill health effects for people, such as our family, who already battle fungal-related illnesses.

In addition, health literature indicates that a more acidic biological terrain can increase absorption of toxins through the intestinal tract because acid foods more slowly process. "The relative acidity or alkalinity of food is also important to diet and health. Foods that are too acidic—such as meats, dairy, and sugar—can interfere with the function of the colon, while alkaline foods (like vegetables and fruits) have a natural laxative effect upon the digestive tract. 'An acid environment in the body increases transit time while an alkaline one decreases it,' says Rima Laibow, M.D., founding Medical Director of the Alexandria Institute of Natural and

Integrative Medicine, in Croton-on-Hudson, New York. 'Excessively rapid transit time means poor nutrient absorption, while a transit time that is too slow leads to autotoxicity and increases colon cancer risk.'"[57] Thus, an acid-based diet can lead to autotoxicity, an increase in colon cancer risk, and the development of other degenerative diseases.[58]

A children's rice protein supplement also proved ineffective. Our son refused to drink it, and after several weeks of use, our daughter became increasingly nauseous, lethargic, and intestinally distressed. Essentially, she and I had the same response to the acid-building protein drinks.

Consumption of flax seeds was also prescribed. Flax seeds, which are rich in omega-3 essential fatty acids, magnesium, and potassium, are a natural source of fiber and, logically, should have helped. However, raw product quality, manufacturing methods, quality control measures, and packaging materials can all affect the quality of the end product. Over time, we found that if we ate flax seeds packaged in plastic, we would have a negative reaction, but if we ate flax seeds purchased in bulk or packaged in paper, we would not experience an adverse reaction.

According to Dr. Erasmus, "Research has shown that plastic leaches into foods that come in contact with it and the more oil/fat that is in the food, the faster is the leaching of plastic into the food." He states that oil makes up one third of flax seeds.[59] Although we did not experience adverse reactions when we ate flax seeds not packaged in plastic, we also did not notice any positive effect either.

Following are products prescribed that alleviated some symptoms but did not address the underlying fungal issues causing our symptoms.

To help with dehydration, Kurt was prescribed a rehydration supplement that was a balanced electrolyte concentrate to add to water. A rehydration product is a logical component of a treatment plan for anyone losing fluids due to stress, poor diet, exercise, illness, or episodes of diarrhea and vomiting, as was the case with Kurt.

Medical sources state, "Electrolytes are minerals in your blood and other body fluids that carry an electric charge. It is important for the balance of electrolytes in your body to be maintained because they affect the amount of water in your body, blood pH, muscle action, and other important processes. You lose electrolytes when you sweat, and these must be replenished by drinking lots of fluids."[60] The body loses electrolytes anytime it becomes dehydrated. Since Kurt had not been able to keep down sufficient fluids for some time, this electrolyte supplement was an effective way to restore his mineral balance. He immediately responded in a positive manner as this product rehydrated his body and replenished electrolytes lost from prolonged, repeated episodes vomiting and diarrhea.

The children and I were initially prescribed a homeopathic mineral supplement. When it didn't provide noticeable results (negative or positive), our doctor switched us to the rehydration concentrate that had been successful for Kurt. Right away we felt a noticeable improvement as our bodies began to rehydrate. Our family has continued to use this product as needed.

A dietary supplement that contained a combination of calcium and bovine liver fat extract was prescribed to help with the severe rashes and hives from which our entire family suffered for months. Not only were these rashes and hives extremely itchy, but they were also painful and created bleeding sores on our skin. Although this supplement only slightly minimized the extent of the rashes and hives, it did reduce the itching to a more tolerable level. After the internal fungi were effectively treated, we no longer needed to use this product because the rashes and hives were gone. It took months, however, for the scarring from the open sores to fully heal.

In addition to not responding favorably to most of the components of the initial treatment plan prescribed by our naturopath, we were having severe fungal re-exposure reactions to many foods and locations. We would become violently ill from these re-exposure incidences. The children and I would feel as though we had severe cases of the flu—body aches, headaches, intestinal pain, nausea, and sometimes vomiting and diarrhea. Kurt, on the other hand, could be bed ridden for two to three days, unable to eat, battling severe episodes of vomiting and diarrhea.

We soon came to realize that the days of eating without careful planning were gone. One wrong food choice could put us out of commission for days. We were also reacting to the outside ambient air on days when mold spore counts were high[61] and to the ambient air inside several stores that had poor air quality due to structural mold. To make matters worse, we also had heightened sensitivities to other air contaminants as well. For example, a quick errand to the local shopping mall that had roped-off areas under construction would leave us suffering for days. Or a trip to the library, filled with musty old books, would send us running for the door. Because of the possibility of violent illness and a two- to three-day convalescence anytime we were re-exposed, we began to limit our outings to only those absolutely necessary, such as trips to the grocery store. Even then, usually I would be the one who went into the store since Kurt paid the highest price for these re-exposures. To put it lightly, we were not having fun.

Food was the big issue. We could limit going out, but we couldn't eliminate food. We had realized for some time that many foods were making us sick, sometimes even violently ill, but we didn't understand why. We were eating *healthy* whole foods—the reactions to which were making us regret we ever ate! We reacted to refried beans, almonds, canned tomato products, wheat products, corn and corn products, milk (even in small quantities), eggs, vinegar, mushrooms, jellies (even in very small quantities), pickled jalapeños, soy sauce, whole grain breads, oats, and peanut butter—the list goes on.

We shared these challenges with our ND, who in turn told us that some of these foods very well may be contaminated with mold from fungal growth that occurred in the field and/or during storage. In addition, our naturopath said that some of these foods contained mold for flavoring! We were stunned. The fact that small amounts of fungal contamination exist in many foods had never occurred to us. Although this knowledge put our problem into perspective, we still did not have an answer as to how to cope with this challenge. There were *so many* foods to which we were reacting. How were we to locate foods that would be free of fungal contaminants?

We posed this question to our naturopath. The answer was not one that we really wanted to hear. We had already been told to remove from our diet simple carbohydrates and all foods containing sugar. Now, we were being told to cut out (what felt like, at the time) most of the remaining food options because they had fungal contamination! After assimilating this information (and recovering from the shock), we cut our diet to the few foods that we had found to be "safe". Then we began researching fungal contamination in foods, searching for foods as fungal free as possible so we could eat—without the food making us sick! As we researched, we came to realize that foods could contain not only various levels of fungal contamination, but also various levels of mycotoxins if the fungal contamination was from a toxigenic source. Not only must we be aware of possible fungal contamination, but also mycotoxin contamination. Again, we found there were no quick or easy answers.

Most of us, when we think about moldy food, think of the block of cheese that got pushed to the back of the refrigerator and became moldy. We think of the leftovers that got moldy, sitting waiting to be eaten. We think of blue cheese, Brie cheese, and overripe moldy fruit. Most of us don't think of whole wheat bread, cereals, flour, vinegar, beans, canned tomato products, and corn products. However, these foods and many, many more can be contaminated with fungi, mycotoxins, and/or fungal antigens. An antigen is a foreign protein or substance that elicits an antibody response. According to Dr. Erasmus, antibodies are proteins produced by the body to neutralize the antigens that your body perceives as foreign proteins.[62] Thus, antigenic foods had to be identified and avoided.

Dr. Erasmus further explains what takes place in the body when an allergic reaction occurs, "Let me start with allergies. Allergic reactions in nature generally come from poor digestion, especially of proteins, because what makes creatures different and what tells the immune system that the body is being attacked by something is the introduction into the body of proteins that are not our own proteins. That reaction is the reason why when you get a transplant, there is the concern about rejection. Your system will say, 'Wait a minute. That's not our protein. We are under attack. We must destroy this.' That leads to rejection of a transplanted organ because the proteins of another person are different from yours. We are that unique. It's rare, even if you get a transplant from a father or a mother, that it will not be rejected (although sometimes it happens that an organ is not rejected)."[63] Dr. Erasmus explains that in the rare occasion when a transplanted organ is not rejected, it is because the donor's and recipient's proteins are so similar that the immune system of the recipient accepts the proteins in the transplanted organ as his/her body's own proteins.[64]

He continues, "The reaction of the immune system to foreign proteins is our body's defense against attack by foreign organisms—and that could even be protein in your digestive tract that was not properly digested, which is usually because you didn't have enough enzymes to properly digest it. Or it could be because of stress. Stress inhibits your digestive performance. It could be because the foods were cooked, and the enzymes that are present in raw foods were destroyed. Therefore, you had more digestive work than your system could handle. Or it could be because you've gotten older, and your digestive tract has deteriorated, as it does naturally as you get older. Or you had injury in the digestive tract, and so you got protein in touch with the tissues. There is supposed to be a barrier on the inside of the digestive tract, but when that barrier is gone and protein can touch the tissues, you get an immune reaction."[65]

Dr. Erasmus further explains, "The immune system makes a protein, which is called an antibody, to neutralize the foreign protein, which is called an antigen, that got into the tissues. That's the way your system fights foreign invasion and puts the antigens out of commission. The antigens are the undigested proteins that got in touch with the tissues. Antigens and antibodies react together, which puts the antigens out of commission, but since both are proteins, you now have even more proteins to digest. If you couldn't digest the food proteins present in the first place, you are likely not to be able to deal with these additional antibody proteins either. Then the immune system or your circulation trucks these proteins off to somewhere else in the body, and you get inflammation in that place. That is why an allergy caused by undigested or poorly digested protein in your digestive tract can cause inflammation and pain in nearly any part of the body."[66]

He continues, "The other possible problem comes from the fact that an antibody, although the way that it is made against an antigen is quite efficient, is made against only a short stretch of the antigen protein. Proteins are made up of amino acids. If you have a body-own protein with that same short stretch of amino acids—about 10–15 amino acids in a protein that may have 250, 300, or 500 amino acids in its total length—then the antibody made against the short stretch of amino acids in the foreign protein can also attack the body's own proteins that have that same sequence of amino acids. That's how an autoimmune disease happens. So it is very important to make sure that you digest (completely break down) the food proteins. If you break them down completely into their amino acids, which are the building blocks of the proteins made by your body, then you cannot be allergic to that protein anymore. This is because you cannot be allergic to the amino acids in the proteins. You can only be allergic to the way that they are strung together."[67]

In the beginning, we may not have understood the complexities of this antigen/antibody cycle, but we did know that it was paramount to avoid the foods making us so sick. We proceeded, methodically, with a system of trial and error, testing each food for a fungal (re-exposure) reaction. We quickly found lack of consistent product quality in regard to fungal contamination, even when purchasing the same brand of the same product. For example, we might not have experienced a reaction the first ten times we ate a particular brand of bread, yet the eleventh time would cause a severe reaction.

These inconsistencies in product quality, of course, have to do with fluctuations in the quality of the raw product (e.g., wheat), storage, and processing. We discovered varying levels of fungal contamination in many different raw food products as well as foods made from the raw products: wheat, wheat products, corn, corn products, peanuts, peanut products, milk, milk products, and more. Even when the raw product is homogenously blended during the manufacturing process, such as when peanuts are blended into peanut butter, it is difficult to evenly distribute the fungal and mycotoxin contamination that is inherent in the raw product.

Laboratories that test product samples for mycotoxin content also report challenges in obtaining a uniform blend when working with even small sample sizes. According to Christina Brewe, an analytical chemist at Romer Labs, samples to be tested are homogenously blended to ensure that test results will reflect the level of mycotoxins present in the entire product sample. She states that even small samples of product can be difficult to homogenously blend.[68]

When we had a fungal reaction to a food, we researched each ingredient to try to identify the offending culprit. We found two main reasons why food can have fungal and mycotoxin contamination. One, the crop from which the food originates is inherently prone to fungal growth in the field and/or in storage. Two, fungal-derived additives are added during the manufacturing process. Many foods contain fungal contaminants because they are produced from crops susceptible to fungal growth, such as wheat, corn, and peanuts. Other foods contain potentially reactive additives derived from fungi. For example, authentic soy sauce heavily used in Asian cuisine contains fungal-derived ingredients. According to Tom Volk, PhD, professor of mycology at the University of Wisconsin-La Crosse, "Authentic soy sauce is fermented with the fungus *Aspergillus oryzae* . . . The fungus gives soy sauce its distinctive flavor. Unfermented soy sauce is just not as tasty."[69] Another source concurs, "Soy sauce is made from a mixture of soy beans and rice fermented by a variety of bacteria and fungi. These include *Lactobacillus delbrueckii, Aspergillus oryzae, Aspergillus soyae*, and *Saccharomyces rouxii*."[70] These fungal components of food manufacturing may or may not be listed on food labels. When they are, they appear quite innocuous.

At first, learning all this information was overwhelming because there were so many ingredients to research. In fact, it took months to identify a substantial number of safe foods because we had to try each food, one at a time, and if we had a fungal reaction, we had to wait several days before reintroducing another food. Otherwise, we wouldn't be able to tell if the next food we were testing was giving us a fungal reaction or if we were still suffering effects from the previously tested food.

We had to become serious food detectives—a term we describe as a step above a label reader. You can read all the labels you want, but if you haven't initially invested the time to learn which food groups are more susceptible to fungal growth, which additives come from fungal sources, and which manufacturing processes increase or decrease the level of fungal contaminants, you will not have the knowledge to decipher what you are reading on the labels. For this reason, we eventually limited our food selections to unprocessed, fresh or fresh frozen whole foods, which enabled us to avoid the complex issue of food additives. We identified whole foods as free as possible from fungal and mycotoxin contamination, foods to which we didn't react, and used them as the basis to form our daily meal plans. We viewed these health-enhancing food selections not as a diet or a deprivation, but rather as health-preserving choices to embrace.

As we continued to research fungal and mycotoxin contamination in foods, we learned a disconcerting, yet motivating fact: Not only are certain crops (such as wheat, corn, and peanuts) prone to fungal growth, but the ingestion of various mycotoxins produced by certain species of fungi that grow on crops have been linked to multiple forms of cancer. Yet another reason (in addition to allergic reactions) to sustain our discipline in avoiding fungal- and mycotoxin-contaminated foods!

One mycotoxin in particular—aflatoxin B1—has been extensively studied. One medical report states in regard to aflatoxin B1, ". . . [T]here is no other natural product for which the data on human carcinogenicity are so compelling."[71] In fact, aflatoxin B1, which has been identified as the predominant mycotoxin in food, is ". . . the most potent natural carcinogen known . . ."[72] For this reason, naturally occurring aflatoxins have been classified by the International Agency for Research on Cancer (IARC) as a Group I carcinogen.[73] The IARC is part of the World Health Organization (WHO). Its mission is to ". . . coordinate and conduct research on the cause of human cancer, the mechanisms of carcinogenesis, and to develop scientific strategies for cancer control."[74] As a part of its research, the IARC evaluates the carcinogenicity of mycotoxins.

Grasping the concept that fungal contaminants, including mycotoxins, exist in/on many foods and that there is a possible associated risk of cancer from ingesting these contaminated foods was eye-opening. Also disturbing was to realize that most of the long-term scientific and medical studies regarding aflatoxin B1 have been conducted in developing countries, such as those in Africa and Asia—not in developed countries, such as the United States.

Studies reveal that many of these Third World countries have increased rates of liver cancer from ingestion of aflatoxin B1 in combination with high hepatitis B and C viral infection rates.[75] According to Dr. Santella, "The biggest risk is with the combination of hepatitis virus and the aflatoxins."[76] An article in Oregon State University's research magazine *Oregon's Agricultural Progress* concurs, stating that studies of human populations around the world reveal, "Continued exposure to aflatoxin in a person's diet increased the likelihood of liver cancer by about four times, but aflatoxins in combination with hepatitis B or C increased the risk of liver cancer as much as 60 times."[77] Thus, these geographic areas that have both high aflatoxin B1 consumption and high incidences of hepatitis B or C are prime locations for aflatoxin studies.

Although these Third World aflatoxin studies produce valuable data, they (of course) do not address the level of risk existing for people in the United States from chronic ingestion of aflatoxins in the U.S. food supply. This issue, essentially, has not been touched by the scientific community. Dr. Wang states that not much data of this type exists, if any.[78] The question to ask is: Are these studies not being conducted in the United States because there truly is minimal aflatoxin exposure in the U.S., as George Bailey, PhD, a distinguished professor emeritus of the Department of Environmental and Molecular Toxicology at Oregon State University, claimed in 2002[79]—or are governmental and industry groups keeping the lid on Pandora's "mycotoxin" box?

Also concerning is that limited studies have been conducted in regard to other mycotoxins that are produced by fungi that also inherently grow on certain food crops. Scientific studies have proven that these other mycotoxins become ingested, which can also lead to associated health risks.[80] Studies regarding these other mycotoxins are in their infancy compared with the expansive research that has been conducted on aflatoxin B1. However, even with limited scientific research, some of these other mycotoxins have been classified as possible human carcinogens by the IARC. For example, The IARC has designated the mycotoxin ochratoxin A as a Group 2B carcinogen. It states, "Ochratoxin A is possibly carcinogenic to humans (Group 2B)."[81] The IARC has also designated as Group 2B carcinogens the mycotoxins derived from *Fusarium moniliforme*: fumonisin B1, fumonisin B2, and fusarin C. It states, "Toxins derived from *Fusarium moniliforme* are possibly carcinogenic to humans (Group 2B)."[82]

Dr. Wang states in regard to fumonisin B1, "Animal studies prove it is a carcinogen to animals, but at this point, human epidemiology studies have not been done, and there are no convincing human data to prove that it's a human carcinogen."[83] Unfortunately, with no results from human epidemiological studies, there are no convincing human data either to prove that fumonisin B1 is *not* a human carcinogen.

As science progresses, we will learn more regarding the extent to which exposure to fumonisin B1 and other mycotoxins can affect our health. Meanwhile, remember, just because scientific research has not progressed to the point of scientifically correlating the associated health risks in humans from exposure (ingestion, inhalation, and touch) to these other mycotoxins, does not mean that negative health outcomes are not occurring—specifically in regard to chronic ingestion of low levels of these mycotoxins. Remember, as Dr. Lipsey cautioned earlier, "By the time an agency comes forth with information for the public on the health dangers of a new chemical or mycotoxin, etc., they may have known about the dangers for many years, but they could not prove it with epidemiological data."[84]

We must each realize we can exercise a certain amount of control over preserving, protecting, and restoring our health. By altering our food selections and avoiding ingestion of known toxic carcinogens and other potentially health-destroying mycotoxins *today*, we are proactively fighting the diseases of *tomorrow*. What level of health do you want to be enjoying ten years from now? How about twenty years from now? Thirty years from now? Undeniably, today's food choices either invest in—or bankrupt—tomorrow's health. The choice is yours.

ENDNOTES
1 This concept is discussed in Goldberg, *Alternative Medicine*, p. 226-227.
2   Ibid., p. 23.
3   Goldberg, *Alternative Medicine*, p. 226-227.
4   Peraica et al., "Toxic Effects of Mycotoxins in Humans," and Bennett et al., "Mycotoxins," p. 1-2.
5   Goldberg, *Alternative Medicine*, p. 777.
6   Stephanie Nano, "Headlines: Public Health News," November 12, 2005, AP. Available at https://www.mmrs.fema.gov/news/publichealth/2005/nov/nph2005-11-12.aspx (accessed February 2006).
7   Nachman Brautbar, MD, "Toxic Molds—The Killer Within Us: Indoor Toxic Molds and Their Symptoms." Available at http://www.experts.com/sourceshowArticle.asp?id=55 (accessed February 2006).
8   Gray, MR et al., "Mixed Mold Mycotoxicosis: Immunological Changes in Humans Following Exposure in Water-damaged Buildings," *Arch Environ Health*, 2003, 58(7): 410-20. Available at http://www.medscape.com/medline/abstract/15143854?prt=true (accessed April 2006).
9   Brautbar, "Toxic Molds—The Killer Within Us."

10 Goldberg, *Alternative Medicine*, p. 270.
11 Ibid., p. 271.
12 William G. Crook, MD, *The Yeast Connection Handbook*, (Jackson, Tennessee, Professional Books, 2002).
13 William G. Crook, MD, and Marjorie Hurt Jones, RN, *The Yeast Connection Cookbook*, (Jackson, Tennessee, Professional Books, 2005).
14 Udo Erasmus, PhD, *Fats That Heal Fats that Kill*, p. 37 (BC, Canada, Alive Books, 1993).
15 James F. Balch, MD, and Phyllis A. Balch, CNC, *Prescription for Nutritional Healing*, Third edition, p. 3 (New York, Avery Publishing Group, 2000). According to Avery Publishing, the first half of the Third Edition was printed with author names Balch and Balch; the second printing of the Third Edition and the printing of the Fourth Edition were printed with Phyllis A. Balch listed as the sole author.
16 Goldberg, *Alternative Medicine*, p. 182.
17 Erasmus, *Fats That Heal Fats that Kill*, p. 358. Ibid.
18 Ibid.
19 Ibid.
20 Rapp, *Is This Your Child?* p. 439.
21 Goldberg, *Alternative Medicine*, p. 24.
22 Ibid., p. 23.
23 Ibid.
24 Ibid., p. 25.
25 Ibid., p. 36.
26 Ibid., p. 24
27 Ibid., p. 23.
28 Ibid., p. 24
29 Ibid.
30 Ibid.
31 Vincent Marinkovich, MD, "Fungal Hypersensitivity Pathophysiology." Available at http://cnri.edu/drwilson/toxins-mold-brain-effects.htm (accessed March 2006). (Italics added.)
32 Marinkovich, "Fungal Hypersensitivity Pathophysiology."
33 Ibid.
34 Erasmus, *Fats That Heal Fats that Kill*, p. 410.
35 The CDC Mold Work Group, "Mold Prevention Strategies," p. 17.
36 PBS, "Is It Safe to Return?"
37 The CDC Mold Work Group, "Mold Prevention Strategies," p. 25 and 37; PBS, "Is It Safe to Return?"; Goldberg, *Alternative Medicine*, p. 22 and 620; See Section Three, interview with Dr. Jack Thrasher, April 2006; and Dr. Fungus, "Aspergillosis."
38 The CDC Mold Work Group, "Mold Prevention Strategies," p. 17 and 37.
39 Ibid., p. 17.
40 Ibid., p. 37. (Italics added.)
41 The New York City Department of Health and Mental Hygiene, "Guidelines on Assessment and Remediation of Fungi in Indoor Environments."
42 See Section Three, interview with Dr. Jack Thrasher, April 2006. (Italics added.)
43 The New York City Department of Health and Mental Hygiene, "Guidelines on Assessment and Remediation of Fungi in Indoor Environments."
44 Goldberg, *Alternative Medicine*, p. 24.
45 Ibid., p. 23.
46 Interview with Dr. Udo Erasmus, May 2006.
47 Erasmus, *Fats That Heal Fats that Kill*, p. 358.
48 Goldberg, *Alternative Medicine*, p. 24.
49 See Section Three, interview with Dr. Jack Thrasher, April 2006.
50 Goldberg, *Alternative Medicine*, p. 690.
51 Ibid.
52 Ibid., p. 685.
53 Goldberg, *Alternative Medicine*, p. 829.
54 Goldberg, *Alternative Medicine*, p. 584.

55   E-mail from Dr. Ginger Chew, July 2006.
56   Robert O. Young and Shelley Redford Young, *The pH Miracle*, p. 18 (New York, NY, Warner Books, Inc., 2002).
57   Goldberg, *Alternative Medicine*, p. 684.
58   Young et al., *The pH Miracle*, p. 17 and 18.
59   Interview with Dr. Udo Erasmus, May 2006.
60   Medline Plus, "Electrolytes." Available at http://www.nlm.nih.gov/medlineplus/ency/article/002350.htm (accessed June 2006).
61   For mold spore counts in your area, see the Mold and Pollen Report under Allergies at www.weather.com.
62   Interview with Dr. Udo Erasmus, May 2006.
63   Ibid.
64   Ibid.
65   Ibid.
66   Ibid.
67   Ibid.
68   Interview with Christina Brewe, February 2006. For more information on Romer Labs, see www.romerlabs.com/romer.htm (accessed February 2006).
69   Tom Volk, "Tom Volk's Fungus of the Month."
70   "Food Production." According to Dr. Fungus, *Saccharomyces rouxii* is the obsolete synonym for *Zygosaccharomyces rouxii*. From Dr. Fungus, "Synonym and Classification Data for *Zygosaccharomyces* spp." Available at http://www.doctorfungus.org/imageban/synonyms/zygosaccharomyces.htm (accessed February 2006).
71   Bennett et al., "Mycotoxins," p. 9.
72   Ibid., p. 6 and 9.
73   WHO/IARC, "Volume 56 Some Naturally Occurring Substances: Food Items and Constituents, Heterocyclic Aromatic Amines and Mycotoxins." Last updated August 21, 1997. Available at http://monographs.iarc.fr/ENG/Monographs/vol56/volume56.pdf (accessed August 2006).
74   From the International Agency for Research on Cancer website. Available at http://www.iarc.fr/index.html (accessed August 2006).
75   See Section Three, interview with Dr. Regina Santella, July 2006.
76   Ibid.
77   Peg Herring, "Cancer and Chlorophyllin," Oregon's Agricultural Progress, winter/spring 2002. Available at http://extension.oregonstate.edu/oap/story.php?S_No=8&storyType=oap&cmd=pf (accessed July 2006).
78   See Section Three, interview with Dr. Jia-Sheng Wang, July 2006.
79   Herring, "Cancer and Chlorophyllin."
80   See Section Three, interview with Dr. Jia-Sheng Wang, July 2006.
81   WHO/IARC, "Volume 56 Some Naturally Occurring Substances: Food Items and Constituents, Heterocyclic Aromatic Amines and Mycotoxins."
82   WHO/IARC, "Volume 56 Some Naturally Occurring Substances: Food Items and Constituents, Heterocyclic Aromatic Amines and Mycotoxins."
83   See Section Three, interview with Dr. Jia-Sheng Wang, July 2006.
84   See Section Three, interview with Dr. Richard Lipsey, April 2006

CHAPTER THREE:

FOOD, FUNGUS, AND MYCOTOXINS

In order to understand what is medically and scientifically known so far about the level of health risks from ingestion of fungi and mycotoxins inherently in human food and animal feed, let's go back some 40 years to when aflatoxins were discovered. As briefly mentioned earlier, "In the 1960[s], more than 100,000 young turkeys on poultry farms in England died in the course of a few months from an apparently new disease that was termed 'turkey X disease.' It was soon found that the difficulty was not limited to turkeys. Ducklings and young pheasants were also affected, and heavy mortality was experienced. A careful survey of the early outbreaks showed that they were all associated with feeds, namely, Brazilian peanut meal. An intensive investigation of the suspect peanut meal was undertaken, and it was quickly found that this peanut meal was highly toxic to poultry and ducklings with symptoms typical of turkey X disease. Speculations made during 1960 regarding the nature of the toxin suggested that it might be of fungal origin. In fact, the toxin-producing fungus was identified as *Aspergillus flavus* (1961) and the toxin was given the name aflatoxin by virtue of its origin (*A. flavis*→Afla)."[1] This massive case of aflatoxicosis awakened scientific and medical professionals worldwide to the existence of this new mycotoxin and the potential deadly consequences from its ingestion.

The first recognized case of acute mycotoxin intoxication is identified as having occurred in France in 945 AD when a large number of people became ill from ergotism, which is a potentially fatal disease caused by metabolites of the fungi ergot.[2] However, not until the aflatoxin discovery was global interest ignited in toxigenic mycotoxins. According to the Department of Animal Science at Cornell University, the turkey X incident ". . . led to a growing awareness of the potential hazards of these substances as contaminants of food and feed causing illness and even death in humans and other mammals."[3] In fact, so many scientists joined the well-funded search for these toxigenic agents that "[t]he period between 1960 and 1975 has been termed the mycotoxin gold rush."[4]

However, as research funding dried up, so did scientific studies. Dr. Wang explains, "It's not as hot an issue as in the early 1970s and 1980s when a lot of academic institutes, companies, and government agencies in Washington were doing mycotoxin studies. In recent years, the USDA has a group whose work involves mycotoxins and the FDA has only a few people working on mycotoxins. There is really not too much funding for mycotoxin studies except for a few aflatoxin studies, which are mainly being done overseas as is my research project."[5] We will look at some of the possible reasons for this lack of mycotoxin research funding over the last 15 years or so, but first let's take a look at what we know so far from the research data, albeit limited, generated during the mycotoxin gold rush and subsequent years.

Just as certain species of fungi will grow on substrates indoors under favorable conditions (e.g., damp building materials), aflatoxin-producing fungi and other mold species will also grow on food crops when conditions such as temperature and humidity are conducive to fungal growth in the field, post harvest, during transportation, and/or during storage.[6] Fungal growth on crops results not only in fungal contamination but also in mycotoxin contamination if the fungus is toxigenic. When these contaminated agricultural products are processed for consumption and used as ingredients in manufactured products, various levels (depending on the manufacturing process) of fungi and mycotoxins will remain in the end product.[7] Therefore, not only can we inhale fungi and mycotoxins in damp, indoor environments, but we also ingest a certain amount of them daily in our food. In fact, Dr. Wang states that mycotoxin ingestion is unavoidable, as mycotoxins are naturally occurring toxins.[8]

Species of fungi inherently grow on predisposed food crops in the field, but not all food crops are prone to fungal growth of the same species. However, the possibility always exists that a species of fungi not normally seen infesting a particular crop in the field could contaminate it in storage if the product (i.e., grain, etc.) is not sufficiently dried after harvest and properly prepared for storage or if the product is improperly stored (i.e., too much humidity, not aerated, etc.). For example, aflatoxins have been found to grow on corn in the field, but not on wheat, reports to Dr. Bacon.[9] However, aflatoxin contamination of wheat *can* occur in storage if conditions support fungal growth of an aflatoxin-producing species. Garnett Wood, PhD, a chemist who performs mycotoxin testing for the FDA, clarifies, "Wheat and small grains do not appear to be an important source of aflatoxin contamination unless abused in storage."[10]

Although wheat is not known to support the growth of aflatoxin-producing fungi in the field, given the right conditions, it will support the growth of other species. For example, fungal species that produce other mycotoxins, such as the trichothecene deoxynivalenol (DON), do tend to grow on wheat in the field as well as on barley and oats, states Dr. Wang.[11] So we see that not all crops are prone to fungal growth of all species, just certain ones. Also important to understand is that one type of crop can be susceptible to fungal growth of multiple species. For example, corn (a large grain) is not only at risk from infestation of aflatoxin-producing fungi but also from growth of fungi that produce other mycotoxins, such as fumonisin.[12] This, of course, can result in multi-mycotoxin contamination. Furthermore, contamination of multiple mycotoxins can occur from infestation from a single species as many fungi produce numerous mycotoxins. In fact, published reports state, "Food and feed are usually contaminated by more than one mycotoxin because a certain strain of moulds may produce different mycotoxins."[13] However, just because a crop is predisposed to fungal growth of a specific species does not necessarily mean that fungal and mycotoxin contamination is always present.

In addition to the inherent, predisposition of a crop, climate will also have an effect on fungal growth and the accompanying levels of mycotoxin contamination. According to Dr. Wang, "Mycotoxin levels in grains also depend on the environment in which they are grown. For example, in southern states like Texas, the weather is usually hot and humid; there is a higher level of contamination of mold than in the northern states. That's true because the local products may have higher levels of aflatoxins and fumonisins. Northern states have different problems like the trichothecene toxins such as deoxynivalenol (DON), which tend to grow on wheat, barley, and oats. Corn and peanuts mainly have problems with aflatoxins and fumonisins."[14]

Dr. Wood expounds on the climate factor. He states that a crisis or severe outbreak of mycotoxin contamination resulting from increased fungal growth in a particular food crop can be caused by environmental factors such as temperature, humidity, drought stress, and the extent of rainfall. Whenever a severe outbreak occurs, "FDA will increase its monitoring efforts by obtaining a larger than planned number of samples from the affected areas," states Dr. Wood.[15]

Ingestion of mycotoxins can occur by eating the raw commodity (such as corn), by eating a manufactured product that contains the food crop (such as corn chips), or by eating a product (such as milk) that came from an animal that ate the contaminated grain (such as field corn). Toxicosis (poisoning) can occur in both humans and animals via consumption of food and feed, respectively, contaminated with fungal growth and the associated mycotoxin by-products. Toxicosis can be either acute (sudden onset) or chronic (developing over a longer time period).

Let's take aflatoxicosis, for example. According to the FDA, it is ". . . poisoning that results from ingestion of aflatoxins in contaminated food or feed."[16] The FDA states, "Acute aflatoxicosis is produced when moderate to high levels of aflatoxins are consumed. Specific, acute episodes of disease [that] ensue may include hemorrhage, acute liver damage, edema, alteration in digestion, absorption, and/or metabolism of nutrients, and possibly death. Chronic aflatoxicosis results from ingestion of low to moderate levels of aflatoxins."[17] Published research has shown that chronic ingestion of these levels of aflatoxin ". . . results in cancer, immune suppression, and other 'slow' pathological conditions."[18]

"Evidence of acute aflatoxicosis in humans has been reported from many parts of the world, namely the Third World countries like Taiwan, Ouganda[19] [Uganda], India, and many others. The syndrome is characterized by vomiting, abdominal pain, pulmonary edema, convulsions, coma, and death, with cerebral edema and fatty involvement of the liver, kidneys, and heart," reports Cornell University.[20] An outbreak of acute aflatoxicosis that occurred in 2004 demonstrates the possible deadly consequences of aflatoxin ingestion. Dr. Wang reports, "In Kenya, Africa, people ate aflatoxin-contaminated food. There were 317 people that got acute aflatoxicosis, and 125 folks died."[21]

Third World countries, such as those in Africa, have conditions that increase the likelihood of acute aflatoxicosis in humans. Some contributing factors include ". . . limited availability of food, environmental conditions that favor fungal development in crops and commodities, and lack of regulatory systems for aflatoxin monitoring and control."[22] Therefore, strict limitation of aflatoxin-contaminated food in these countries could lead to increased lack of food and excessive prices.[23] In other words, people in Third World countries face the dilemma of either eating contaminated food or starving.

Cornell University reports, "Aflatoxicosis is primarily a hepatic disease."[24] This means that the aflatoxins predominantly attack the liver, a fact proven by animal studies. For example, researchers report, "The liver is the primary target organ with liver damage occurring when poultry, fish, rodents, and nonhuman primates are fed aflatoxin B1."[25] The FDA further details, "Aflatoxins produce acute necrosis [death of living tissue], cirrhosis, and carcinoma of the liver in a number of animal species; no animal species is resistant to the acute toxic effects of aflatoxins; hence it is logical to assume that humans may be similarly affected."[26]

The association between aflatoxin ingestion and human liver deterioration and carcinogenicity is confirmed by the high rate of liver cancer in Third World countries. A cellular biologist states, "Liver cancer is the fifth most prevalent cancer in the world, and 80% of the cases are in the developing world. The primary causes of liver cancer in the developing world are the hepatitis B virus and aflatoxin and, most ferociously, the two combined."[27]

Dr. Santella states that aflatoxin and mycotoxin contamination in food in the U.S. is not as large of a problem as what occurs in Third World countries. She reports, "In the U.S. it does not seem to be as big a problem, probably because also in the U.S. we do not have high hepatitis B viral infection rates. All of our studies in Taiwan and the studies of Dr. [John] Groopman and other groups have shown that the people at highest risk for liver cancer are those who have both the aflatoxin exposure and are also carriers of either hepatitis B or C virus. That is probably because the viruses cause cell death in the liver that leads to cell proliferation. So, the cells in the liver keep dividing, but if you have aflatoxin DNA adducts formed in the liver in a dividing cell and the adducts are there when the cell divides, that is when you get the genetic alterations."[28] Replication of genetically altered cells can lead to cancer.

Dr. Santella continues, "So, the biggest risk is with the combination of hepatitis virus and the aflatoxins. That said, the U.S. population does have about a 2 percent carrier rate for hepatitis C virus. So, it is a problem."[29] Two percent doesn't sound like much, but based on current U.S. population statistics,[30] that equates to well over 6 million people at increased risk of liver cancer from aflatoxin exposure. According to a 2004 published CDC report, approximately 1.2 million people have chronic hepatitis B virus.[31] This subgroup of population also has a higher risk of liver cancer from aflatoxin ingestion.

Chronic aflatoxin ingestion is a risk to which we are all subject and is inevitable, to some degree. According to Cornell University, "Such exposure is difficult to avoid because fungal growth in foods is not easy to prevent."[32] In fact, "Aflatoxins are considered unavoidable contaminants of food and feed, even where good manufacturing practices have been followed," states Cornell University.[33]

Although aflatoxin contamination of food is considered a major problem in the developed world,[34] incidences of acute levels of aflatoxin poisoning are uncommon in developed countries because of the screening mechanisms and manufacturing processes that reduce aflatoxin content, keeping heavily contaminated food out of the marketplace.[35] According to the FDA, "The chances of [a person ingesting] acute levels of aflatoxin is remote in well-developed countries."[36] However, ". . . concern still remains for the possible adverse effects resulting from long-term exposure to low levels of aflatoxins in the food supply," according to Cornell University.[37]

Chronic ingestion of these aflatoxin levels will affect people's health differently. Just as "[a]nimal species respond differently in their susceptibility to the chronic and acute toxicity of aflatoxins,"[38] so do humans. The FDA reports, "The toxicity can be influenced by environmental factors, exposure level, and duration of exposure, age, health, and nutritional status of diet."[39] These same factors, and possibly others, also determine the degree to which aflatoxin ingestion affects human health—and were life and death determinants for the people in Kenya, Africa. These influencing factors, likewise, determine a person's level of susceptibility to chronic and acute toxicity from other mycotoxins as well.[40]

Since we all consume certain levels of aflatoxins in food even when "good manufacturing practices" are implemented, let's take a closer look at the potential health effects of aflatoxin ingestion and the foods possibly affected.

According to Cornell University, "Aflatoxins are produced primarily by some strains of *A. flavus* and by most, if not all, strains of *A. parasiticus* plus related species *A. nomius* and *A. niger*. Moreover, these studies also revealed that there are four major aflatoxins—B1, B2, G1, G2—plus two additional metabolic products, M1 and M2, that are of significance as direct contaminants of foods and feeds. The aflatoxins M1 and M2 were first isolated from milk of lactating animals fed aflatoxin preparations; hence, the M designation."[41] This means that when animals—including dairy cattle—are fed aflatoxin-contaminated grain, aflatoxin M1 is produced in their livers and may be excreted in their milk, depending on the level of contamination in the feed, according to Dr. Wood.[42] This potential for contamination is why monitoring of aflatoxin M1 is included in the FDA's compliance program for mycotoxins, states Dr. Wood.[43] Stricter guidelines apply to grains fed to dairy animals, which we will shortly discuss.

Published research shows, "Many substrates [crops/foods] support growth and aflatoxin production by aflatoxigenic molds."[44] Therefore, many different types of food can be contaminated with aflatoxigenic molds and aflatoxins. In order to get as complete a list as possible of foods potentially contaminated with aflatoxins, let's review lists of identified foods compiled by multiple sources.

One report on aflatoxins states, "Natural contamination of cereals, figs, nuts, tobacco, and a long list of other commodities is a common occurrence."[45]

The FDA states, "In the United States, aflatoxins have been identified in corn and corn products, peanuts and peanut products, cottonseed, milk, and tree nuts such as Brazil nuts, pecans, pistachio nuts, and walnuts. Other grains and nuts are susceptible but less prone to contamination. . . . The most pronounced contamination has been encountered in tree nuts, peanuts, and other oilseeds, including corn and cottonseed."[46]

Cornell University states, "Aflatoxins are detected occasionally in milk, cheese, corn, peanuts, cottonseed, nuts, almonds, figs, spices, and a variety of other foods and feeds. Milk, eggs, and meat products are sometimes contaminated because of the animal consumption of aflatoxin-contaminated feed. However, the commodities with the highest risk of aflatoxin contamination are corn, peanuts, and cottonseed. . . . Aflatoxin-contaminated corn and cottonseed meal in dairy rations have resulted in aflatoxin M1 contaminated milk and milk products, including non-fat dry milk, cheese, and yogurt."[47]

Although there are many mycotoxins that occur naturally in foods,[48] "[a]flatoxins have received greater attention than any other mycotoxin because of their demonstrated potent carcinogenic effect in susceptible laboratory animals and their acute toxicological effect in humans."[49] One research paper states, "Aflatoxin is a mycotoxin of global significance. . . [and] has attracted worldwide attention because it is a powerful toxin that damages genes."[50] This ability to alter genetic structure is the characteristic that makes aflatoxins potent carcinogens.

Dr. Santella states that although our bodies are designed to eliminate toxic substances—be they aflatoxins, arsenic, formaldehyde, or others—some toxins, such as aflatoxins, through the body's metabolizing process actually become carcinogenic. She explains, "One of the main purposes of the metabolism in the body is to rid the body of these chemicals. There are specific enzyme systems that are designed to metabolize these compounds and excrete them. Most of these chemicals are excreted. Volatile compounds come back out in your breath; some come out in urine; some are excreted in feces. However, for chemicals that are carcinogenic, sometimes the metabolism, which is designed to generate metabolites that can then be excreted easily, actually forms metabolites that are highly reactive. For example, aflatoxin in its original form is not carcinogenic. It cannot damage DNA, which, as we know, is the critical step in cancer development."[51]

She continues, "Aflatoxin is metabolized in the body to aid in excretion. When the body is exposed to aflatoxin, the body tries to metabolize it to a water-soluble form so it can be excreted in the urine. One of those metabolites is the reactive compound that damages DNA. So, when we are exposed to chemicals, whether they are carcinogenic or not, the major process of the body is to get rid of them. Generally, that works quite efficiently, but sometimes there is DNA damage, protein damage, and RNA damage as a result of the various metabolites that are formed during the process of trying to get rid of the carcinogen."[52] It is this DNA damage that then can lead to cancer—but does DNA damage always lead to cancer?

According to Dr. Santella, "Our DNA is constantly being damaged, even if we are not exposed to exogenous chemical carcinogens. There are normal cellular processes like metabolism of various hormones that generate, for example, reactive oxygen species that damage DNA. If you go out in the sunlight, your skin forms thymine dimers, a type of damaged DNA that leads to skin cancer. DNA damage is happening constantly. Since DNA is so critical to the cell, there are a number of enzyme systems in the body that can repair DNA. We have multiple DNA repair pathways, so even if you are exposed to a carcinogen and some small portion of it binds to your DNA, your body has the ability to fix that DNA, and then you are fine."[53]

She explains further, "Cancer develops if you have damaged DNA in a cell that then replicates, or divides, before the damage can be removed. That is how the genetic changes occur that are responsible for cancer development. So, if your DNA is damaged and your cell fixes it, no problem. However, if the DNA damage is there when the cell decided it was time to divide, that is when you have the problem. Studies that we, and others, have been doing are showing that DNA repair capacity—the ability to repair damage to your DNA, which we know varies from one person to another—is a major factor in understanding who gets cancer."[54]

The toxigenic and carcinogenic properties of aflatoxins are why they are ". . . probably the best known and most intensively researched mycotoxins in the world [and] . . . have been associated with various diseases, such as aflatoxicosis, in livestock, domestic animals, and humans throughout the world."[55] Research has revealed that other mycotoxins also have the similar ability as aflatoxins to alter DNA. According to Dr. Santella, fusarium C is another mycotoxin that forms reactive intermediates when metabolized in the body, and thus can alter DNA and form cancerous cells.[56]

The fact that toxic compounds inherently exist in food crops and can cause cancer and other possible negative health effects (depending on the mycotoxin, individual genetics, and other influencing factors earlier mentioned) gives us each reason to learn more. We should seek answers to the following questions: Which food crops are susceptible to mycotoxin-producing fungal growth? What are the health risks from eating food that comes from mycotoxin contaminated crops? Which mycotoxins are regulated? Are said regulations enforced? Is sample testing required? To what degree is compliance to mycotoxin regulations self-imposed by suppliers and/or manufacturers? And what is the penalty for noncompliance?

These questions should not be taken lightly because, as mentioned earlier, research shows, "Depending on the quantities produced and consumed, mycotoxins can cause acute or chronic toxicity in animals and humans."[57] Furthermore, depending on the answers to these questions, consumer health is either being protected—or shortchanged.

As we set out to answer these questions and more, the inherent balancing act of the often dueling priorities—health versus economics—in the complex world of "Food Safety" becomes apparent. Let's review our findings and see if the safeguards in place, indeed, keep us safe.

According to the FDA/Center for Food Safety & Applied Nutrition (CFSAN), "The occurrence of mycotoxins in foods and feeds is not entirely avoidable; therefore, small amounts of these toxins may be in foods and feeds."[58] However, the FDA further states, ". . . [F]ood is deemed adulterated if it contains poisonous or deleterious substance, such as mycotoxins, which may render it injurious to health. Mycotoxins can be considered added poisonous or deleterious substances because their presence in human food and/or animal feeds can be avoided in part by good agronomic and manufacturing practices. Strategies used by the Food and Drug Administration (FDA) to minimize mycotoxins in the U.S. food supply include establishing guidelines (e.g., action levels, guidance levels), monitoring the food supply through formal compliance programs (domestic and import), and taking regulatory action against product that exceeds action levels, where action levels have been established."[59]

Out of the estimated 300 to 400 mycotoxins,[60] two have action levels—aflatoxin and patulin—and two have guidance levels—fumonisin and vomitoxin (deoxynivalenol or DON).[61] FDA documents state that in 1969 the FDA set an action level for aflatoxins,[62] and in 2001 the FDA established an action level for patulin.[63] Action levels require compliance. On the other hand, guidance levels that regulate the other two mycotoxins do not require compliance. Guidance levels, which are sometimes referred to as "tolerances," are *just* recommendations. Since aflatoxin was the first mycotoxin to warrant an action level, let's first review what is known about the health effects of its exposure. Dr. Wang points out,

"Right now, only aflatoxin is listed as a human carcinogen. It is certainly a human carcinogen. In 1993 the International Agency for Research on Cancer [IARC] listed aflatoxin as human category 1 (confirmed human carcinogen); then in 2002, they reconfirmed the conclusion."[64]

The IARC documents, "Naturally occurring aflatoxins are carcinogenic to humans (Group 1)."[65] It also states, "Aflatoxin M1 is possibly carcinogenic to humans (Group 2B)."[66] The IARC reports, "Exposure to aflatoxin M1 occurs mainly through consumption of milk, including mother's milk. Lifetime exposure to aflatoxins in some parts of the world, commencing in utero, has been confirmed by biomonitoring."[67] According to the CDC, biomonitoring is ". . . the direct measurement of people's exposure to toxic substances in the environment by measuring the substances or their metabolites in human specimens, such as blood or urine. Biomonitoring measurements are the most health-relevant assessments of exposure because they indicate the amount of the chemical that actually gets into people from all environmental sources (e.g., air, soil, water, dust, [and] food) combined, rather than the amount that *may* get into them."[68]

Regarding aflatoxin regulations, Cornell University further explains, "The FDA has established specific guidelines on the acceptable levels of aflatoxins in human food and animal feed by establishing action levels that allow for the removal of violative lots from commerce. The action level for human food is 20 ppb [parts per billion] total aflatoxins, with the exception of milk which has an action level of 0.5 ppb for aflatoxin M1. The action level for most feeds is also 20 ppb. However, it is very difficult to accurately estimate aflatoxin concentration in a large quantity of material because of the variability associated with testing procedures; hence the true aflatoxin concentration in a lot cannot be determined with 100% certainty."[69]

Dr. Wood states that FDA regulations limit aflatoxin content in feed for dairy animals to 20 ppb so the resulting milk by-product will contain no more than 0.5 ppb aflatoxin content. He points out, however, that feed intended for beef cattle, swine, or poultry can contain up to 300 ppb of aflatoxin. According to Dr. Wood, "Aflatoxin M1 is the only mycotoxin that is examined in milk."[70]

The FDA states, "Action levels and tolerances are established based on the unavoidability of the poisonous or deleterious substances and do not represent permissible levels of contamination where it is avoidable."[71] In other words, even under the best conditions with good agricultural and manufacturing practices, a certain level of these poisonous substances will still taint susceptible food crops and resultant products.

The FDA further states, "Action levels and tolerances represent limits at or above which FDA will take legal action to remove products from the market."[72] However, according to an FDA spokesperson who asked to remain nameless, "The FDA does not require mycotoxin testing to be done at any location (farm, grain elevator, manufacturing plant), even for aflatoxins, which have an action level. This includes feed/food for both animal and human consumption."[73]

Furthermore, the FDA spokesperson reports, "As per a court ruling, action levels and/or guidance levels are not binding."[74] This is confirmed by an FDA compliance document that states, "Action levels are not binding on the courts, the regulated industry, or the agency (see: 55 FR 20782, May 21, 1990)."[75]

Richard Sellers, vice president of feed control and nutrition for the American Feed Industry Association (AFIA), explains that although the FDA doesn't require testing, if it does find an elevated level of aflatoxin in a manufacturer's food, perhaps that particular company should have been testing because the FDA can seek condemnation, recall, etc.[76]

Although condemnation, recall, or other enforcement action can result when aflatoxin (or patulin) levels exceed their respective action levels, how can regulators determine that aflatoxin (or patulin) levels in the food supply are at or below the designated action level when testing is not required and action levels are "not binding on the courts, the regulated industry, or the agency"? Sadly, the aflatoxin (and patulin) action levels appear to be regulations with no "teeth" and afford little to no enforceability!

According to the aforementioned FDA spokesperson, "Regarding grain and feed for animals, the FDA field collects and analyzes ~200 samples/year for the Center for Veterinary Medicine (CVM). These ~200 feed samples are tested for one to five mycotoxins (aflatoxins, fumonisins, zearalenone, ochratoxin A, and vomitoxin). Because the FDA has not established any guidance/tolerance/action levels, etc., for these two mycotoxins [zearalenone and ochratoxin A], we would handle any feed contaminated with zearalenone and/or ochratoxin A on a case-by-case basis."[77]

The FDA spokesperson adds, "Additional testing for mycotoxins on grain and food intended for human consumption is done at another center of the FDA (Center for Food Safety and Applied Nutrition or CFSAN)."[78]

We made multiple inquires via telephone and in writing to ascertain how many samples of food for human consumption are collected each year and analyzed for aflatoxin, patulin, and other mycotoxins. The FDA and several other governmental agencies would not cite any specific numbers. The only facsimile of an answer was provided by Dr. Wood. He said, "The compliance programs are flexible in nature; it is not a difficult task for [the] FDA to shift its monitoring efforts and resources so that a larger than planned number of samples can be collected and analyzed for a particular mycotoxin during a crisis year."[79] (An example of a crisis year would be one in which weather conditions increased fungal growth in crops.) Although we appreciate Dr. Wood's diplomacy, the inability to secure any "numerical" assurance regarding the quantity of samples tested each year of crops and foods intended for human consumption was less than reassuring. We couldn't help but wonder, "Is the minimum number of samples tested each year so low that U.S. citizens would be alarmed if the government disclosed this type of information?"

According to Dr. Wang, the monitoring for mycotoxin contamination in food for human consumption ". . . is done through spot-checking, which is mainly done at the farming level. Currently, there are several companies, like VICAM in Massachusetts and the Neogen Corporation in Michigan, that have developed testing methods to detect mycotoxin levels in grain or feed. They sell their products to farmers, grain elevators, and the government. With these types of tests, farmers can test their product themselves to see if it's over the limit or within the limit of government regulatory action and guidance levels. Also, grain elevators can test product before purchasing. . . . [T]esting is done randomly with scant (very, very few) levels from each site. Testing is not forced; it's just like a recommendation."[80]

Dr. Bacon, who is the research leader and supervisory microbiologist for the USDA Toxicology and Mycotoxin Research Unit, concurs. He points out that unless a mycotoxin has an action level, testing is likely not to occur. He states, "People do what they have to do. Those tests are expensive. That's my gut opinion. That's an expensive undertaking . . . It's just sort of like—well, why should we? They're just a guidance. The thing with fumonisin is that they're just guidance [levels]. We don't have to really do it. And then, when it becomes a part of the law, they'll do it. [But] if something is not mandated, it probably won't ever come to be. It's just human nature."[81] Therefore, unless mycotoxin testing is bottom-line driven, we can be fairly accurate in assuming that manufacturers in the food supply chain are not testing raw or finished products for mycotoxins not regulated by an action level.

Also not reassuring is the fact that we couldn't find any governmental agency responsible for the testing of mycotoxins in raw wheat. This lack of monitoring was especially of concern to us since most bread is made from wheat and bread is a major food staple for most families. When asked why no agencies were routinely testing samples of raw wheat, Dr. Bacon explained, "That's because the aflatoxins are not found on wheat."[82] Other mycotoxins (such as DON) are known to grow on wheat. However, raw wheat is not being tested for levels of DON (or any other non-aflatoxin mycotoxins). Dr. Bacon points out the reason why DON is not tested for in raw wheat either, "That's because that's a guidance."[83] Remember, guidance levels are just recommendations. Furthermore, the guidance level for DON only applies to finished wheat products, not raw wheat.[84]

Even when mycotoxins are regulated by an action level, as is the case with aflatoxin and patulin, compliance is voluntary. This policy leads us to ask the obvious question: What assurances do consumers have that the content of regulated mycotoxins in food does not exceed the action or guidance levels if 1) testing is not mandatory, and 2) the FDA is not even willing to assure consumers—like us—that at least a minimum number of spot samples are routinely tested each year for regulated mycotoxins? The fact that some level of mycotoxin testing is not mandatory at the manufacturing level concerns us immensely because many of the food groups susceptible to fungal growth in the field and/or in storage are copiously consumed by children: peanuts (raw peanuts, peanut butter, peanut snacks, etc.), corn (cereals, chips, etc.), wheat (cereals, chips, bread, etc.), and apples (apple juice, apple juice-containing products, etc.). What we all perceive as healthy additions to a wholesome diet may instead be servings of low doses of toxic poisons, which may be detoxified by our bodies or may be stored in our bodies, depending on genetics and other factors.

The other action-level-regulated mycotoxin is patulin. According to FDA/CFSAN, "In 2001, FDA established an action level of 50 micrograms per kilogram (50 ppb) for patulin in apple juice and in the apple juice component of a food that contains apple juice as an ingredient."[85] To help put this limit into perspective, one rotten apple can contain more than 10,000 ppb of patulin. The FDA/CFSAN states, "In fact, if one rotten apple (containing >10,000 parts per billion (ppb) patulin) is used along with 200 sound apples to make juice, the resulting patulin level in the juice could exceed FDA's action level for patulin."[86]

According to the FDA, "Patulin is a toxic substance produced by molds that may grow on apples. In the past, patulin has been found to occur at high levels in some apple juice products offered for sale in or import[ed] into the U.S."[87] Although the mycotoxin patulin is reported to be destroyed by fermentation, of concern is the fact that it is not destroyed during the pasteurization process.[88] According to the FDA, "Thermal processing appears to cause only moderate reductions in patulin levels, thus patulin present in apple juice will survive the pasteurization processes."[89]

The FDA further reports, "Patulin is a mycotoxin that is produced by certain species of *Penicillium*, *Aspergillus*, and *Byssochylamys* molds that may grow on a variety of foods including fruit, grains, and cheese. Patulin has been found to occur in a number of foods including apple juice, apples and pears with brown rot, flour, and malt feed. However, given the nature of the food, the manufacturing processes, or consumption practices for many foods, patulin does not appear to pose a safety concern—with the exception of apple juice."[90] The term *apple juice* includes ". . . single strength apple juice, reconstituted single strength apple juice (if the food is an apple juice concentrate), or the single strength apple juice component of the food (if the food contains apple juice as an ingredient)."[91]

Furthermore, ". . . [The] FDA believes that if processors do not implement controls for patulin, apple juice consumers may not be optimally protected from potential adverse effects due to long-term exposure to patulin from the consumption of apple juice. [The] FDA thus believes that it is appropriate that apple juice processors *voluntarily* establish controls for patulin."[92] Again, note the reference to "voluntary controls," even in regard to an action-level regulated mycotoxin. Voluntary controls are put in place by industries to reduce liability risks, build consumer confidence, and facilitate securing higher quality raw material from supply sources—all of which manufacturers hope translate into an increased bottom line. Or course, cooperative self-regulation also reduces further governmental involvement into private enterprise.

The FDA/CFSAN confirm, "Exposure over time to high levels of patulin may pose a health hazard."[93] The FDA/CFSAN also confirm that high levels of patulin can occur in apple juice, which is why the action level specifically regulates apple juice.[94] The main concern, of course, lies in the fact that apple juice is a longtime favorite of children that is oftentimes consumed on a daily basis. Due to children's smaller size in combination with frequent, long-term apple juice consumption, children could be at increased risk for the health effects associated with patulin ingestion. Even with high levels of patulin kept in check because of self-regulation "encouraged" by a regulatory action level, what (if any) are the health effects in children from chronic ingestion of low levels of patulin?

According to Aerotech P & K, a laboratory that performs fungal and mycotoxin analysis, patulin inhibits potassium uptake and is a possible carcinogen.[95] Potassium is an essential mineral, which means that the body is unable to manufacture it on its own.[96] Potassium is important in regulating pH (acidity/alkalinity) and water balance and plays an important role in nerve function and cellular integrity.[97] Furthermore, according to the European Mycotoxin Awareness Network, patulin has been shown to be immunotoxic and neurotoxic.[98] The organization also states, "In short-term studies, patulin causes gastrointestinal hyperaemia[99] [hyperemia], distension haemorrhage[100] [hemorrhage], and ulceration."[101]

Concerning is the lack of widespread knowledge regarding the existence of patulin in apple juice and apple juice products and the associated adverse health risks. Why isn't this information being disclosed to consumers, especially people with children? Are we again looking at the issue of economics versus health? Quite possibly, if the existence and potential adverse health effects of apple juice were widely known, apple juice might not be perceived as the healthy juice we have long thought it to be. We must consider the possibility that the grocery store shelves are lined with bottles of pasteurized apple juice laden not only with sugar but also with potentially health-affecting levels of the mycotoxin patulin, depending on how stringent manufacturers have adhered to the action level. As consumers, we must adhere to the motto *Buyer Beware*.

Next let's take a look at the prevalence of the two guidance-level-regulated mycotoxins—fumonisin and vomitoxin (deoxynivalenol or DON)—and the associated health risks from their ingestion.

Fumonisin is produced by various species of *Fusarium* and is a common natural contaminant of corn.[102] According to the FDA/CFSAN, "Fumonisins have been linked to fatalities in horses and swine. Recent studies have demonstrated the presence of fumonisins in human foods, including corn meal and breakfast cereals. Epidemiological investigations demonstrating a possible association of *F. verticillioides* with esophageal cancer and recent animal studies indicating the carcinogenicity of fumonisin FB1 . . . [has] highlighted the need to ensure that foods do not contain excessive amounts of fumonisins."[103] The FDA has established various guidance levels for fumonisins (FB1, FB2, and FB3) in certain foods.[104]

According to Dr. Wang, "Fumonisin is a big issue right now due to very high levels of contamination in corn and corn products in the United States. Before '97 in the U.S., there were no regulatory levels for fumonisin. Then in '97, the FDA and the USDA set up a guidance level of 0.2 ppm for fumonisin. Then they found that levels in almost 95 percent of U.S. corn products were over the point of 0.2 ppm. There was no way to follow the guidance level. For example, if they kept that guidance level, no U.S. corn products could be exported. In 2000, the USDA and the FDA lifted the guidance level to 2 ppm, raising it ten times the original guidance level."[105]

The FDA states, "The purpose of the guidance is to identify for the industry fumonisin levels that FDA considers *adequate* to protect human and animal health and that are *achievable* in human foods and animal feeds with the use of good agricultural and good manufacturing practices. FDA considers this guidance to be a prudent public health measure during the development of a long-term risk management policy and program by the agency for the control of fumonisins in human foods and animal feeds."[106] In other words, guidance levels that would *more than adequately* protect human and animal health may not be "achievable." Thus, lower guidance levels are not established—and a blind eye is turned to the remaining "unavoidable" level of health consequences.

It is true, we can't break the financial backs of the businesses that make up the food supply chain, but the question to ask is: Are regulatory levels of mycotoxins being set as low "as reasonably achievable"?

With guidance levels—testing for which may, or may not, be performed—designed only to provide "adequate" health protection, we must look for other ways to lower our mycotoxin consumption. For example, levels of fumonisins and other mycotoxins can be decreased during manufacturing processes. That means that the more refined the product, the less the mycotoxin content. The FDA/CFSAN gives an example using corn. Joint documentation states, "Because fumonisins are concentrated in the germ and the hull of the whole corn kernel, dry milling results in fractions with different concentrations of fumonisins. For example, dry milled fractions (except for the bran fraction) obtained from degermed corn contain lower levels of fumonisins than dry milled fractions obtained from non-degermed or partially degermed corn. Industry information indicates that dry milling results in fumonisin-containing fractions in descending order of highest to lowest fumonisin levels: bran, flour, meal, grits, and flaking grits."[107]

It is important to minimize levels of fumonisin ingestion because animal studies prove, "Fumonisin causes cancers of [the] liver or kidney, along with blood disorders and pulmonary edema in farm and experimental animals."[108] Published research states, "The major species of economic importance is *Fusarium verticillioides* . . . [It] is present in virtually all corn samples. Most strains do not produce the toxin, so the presence of the fungus does not necessarily mean that fumonisin is also present."[109] However, for a mold-sensitized person, the existence of fungi alone, even without mycotoxin contamination, can cause a hypersensitive reaction.[110]

In summary, the FDA states, "Although human epidemiological studies are inconclusive at this time, based on a wide variety of significant adverse animal health effects, the association between fumonisins and human disease is possible."[111] The IARC has designated fumonisin as a group 2B carcinogen—a possible human carcinogen.[112]

In regard to vomitoxin (deoxynivalenol or DON), it is a trichothecene produced by the *Fusarium* species. According to FDA/CFSAN, "Several adverse weather-related DON contamination episodes in the U.S. have motivated the FDA to issue guidance levels for food (wheat) and feed in 1982 and updated levels in 1993."[113] As mentioned earlier, this guidance level applies only to *finished* wheat products, not raw wheat. The FDA/CFSAN state, "FDA is continuing to study the scope and toxicological significance of the DON problem to determine if further regulatory measures are needed to control DON in food and feed products."[114]

Scientists report that DON is ". . . most prevalent and is commonly found in barley, corn, rye, safflower seeds, wheat, and mixed feeds."[115] Researchers state, "Deoxynivalenol (vomitoxin) causes anorexia at low levels and vomiting at higher levels and also damages the immune system."[116] This symptom is the reason deoxynivalenol is also referred to as vomitoxin or the food refusal factor.[117] Based on our personal experiences, vomiting is also one of the flu-like symptoms that people can experience from fungal re-exposure via food contamination. We have to wonder if this "food refusal factor" occurs because the triggering food was contaminated with deoxynivalenol!

People sensitized to fungi may find themselves especially sensitive to wheat. The FDA/CFSAN explain why, "It is not possible to completely avoid the presence of DON in wheat. DON is sometimes found in wheat grown under normal conditions. However, the fungus thrives in the cool, wet conditions that occurred in the Midwest in the spring and summer of 1993. When DON occurs in wheat, the levels are reduced by the processing of wheat into wheat products, like flour, but processing does not totally eliminate DON."[118] The inability to totally eliminate

DON, of course, may lead to higher incidences of negative health effects from chronic ingestion of deoxynivalenol for a wheat-staple nation such as the United States.

Because of the potential for adverse health effects from ingestion of DON, a guidance advisory level has been set that applies to wheat products. However, no guidance level is in place for testing the raw wheat itself, as earlier mentioned.[119] The FDA/CFSAN state, "The FDA has established a guidance level of 1 microgram per gram (1 ppm) for deoxynivalenol in finished wheat products that may be consumed by humans. No guidance level has been established for raw wheat intended for milling into human food products."[120] The FDA/CFSAN support the no-raw-wheat testing regulation. Joint documentation claims, "Because there is significant variability in manufacturing processes, an advisory level for raw wheat is not practical."[121]

Also of a concern is the level of effectiveness of the guidance level that is set for manufactured wheat products. It may not afford consumers the added protection it implies because, as discussed earlier, guidance levels are just recommendations. They are not enforced. Remember, monitoring of DON and the other guidance-level-regulated mycotoxin (fumonisin) is voluntary, which means testing of samples may—or may not—be taking place at the farming, manufacturing, and/or government levels.

On the other hand, raw grains susceptible to aflatoxin-producing fungi are sampled and tested because aflatoxins are regulated by an action level—not a guidance level. For this reason, Dr. Wood states, "Shelled corn designated for human use in the preparation of corn products is collected and analyzed by the FDA under our compliance programs."[122] Dr. Bacon explains the reason the FDA actively monitors shelled corn intended for human consumption is because aflatoxins are known to grow on corn, and aflatoxins (as just mentioned) have an established action level, not a guidance level.[123]

In summary, the FDA/CFSAN confirm, ". . . [DON] is a common contaminant of several grains including wheat, corn, barley, and rye. DON has been associated with a number of adverse health effects in humans and animals."[124] The IARC has designated deoxynivalenol as a Group 2B carcinogen—a possible human carcinogen.[125]

We, as consumers, must not be naïve. We must keep in the forefront of our minds the reality Dr. Bacon stripped to the bare truth: If mycotoxin testing is not mandated, human nature is not to do it. Why, you say? Money. Plain and simple—testing costs money. Money is, undoubtedly, also one of the reasons the FDA finds voluntary establishment of controls for monitoring mycotoxin levels appropriate.[126] Government intervention and testing both cost money.

Interestingly, reliance on manufacturers' voluntary testing of finished products is a continuous theme when it comes to mycotoxins, even in regard to aflatoxin—the one that has a set action level. For example, "Peanuts containing over 25 ppb are allowed to be shipped to a processor for manufacture into finished products. However, the processor must give *assurance* that the finished products produced will comply with the action level of 20 ppb total aflatoxins," states Dr. Wood.[127] Basically, government mycotoxin regulations give permission for the fox to guard the hen house! In other words, manufacturers are being allowed to regulate themselves! What level of *assurance* does this "honor" system give U.S. consumers that U.S. foods do not exceed the set action levels for aflatoxin and patulin and the recommended guidance levels for fumonisin and deoxynivalenol?

In addition to the mycotoxins regulated by action and guidance levels, scientists and various groups within the FDA and CFSAN also study and/or monitor a few mycotoxins that are not regulated. For example, the above-mentioned FDA spokesperson reports that the FDA analyzes feed content for two additional mycotoxins: ochratoxin A and zearalenone.[128]

According to a published report, "Ochratoxin, a toxin produced in *Penicillium* and *Aspergillus*, is mainly found in grain, nuts, and dried fruits and [is] usually associated with storage of such foods. The toxin damages the kidney, [and] causes cancer and immune suppression. . . . Ochratoxin has also been found in the milk of cows consuming contaminated grain."[129] Researchers further state, "Of the *Aspergillus* toxins, only ochratoxin is potentially as important as the aflatoxins. The kidney is the primary target organ. Ochratoxin A is a nephrotoxin [destructive to kidney cells] to all animal species studied to date and is most likely toxic to humans, who have the longest half-life for its elimination of any of the species examined."[130]

Another published article further explains, "When ingested as a food contaminant, OTA [ochratoxin A] is very persistent in human beings with a blood half-life of 35 days after a single oral dosage due to unfavourable[131] [unfavorable] elimination toxicokinetics. [Toxicokinetics is a subfield of toxicology that studies how toxins are absorbed by, metabolized by, and eliminated from the bodies of living things.[132]] This renders the toxin among the most frequent mycotoxin contaminants in human blood in the EU, the U.S., Canada, and elsewhere, where it has been investigated. OTA is neither stored nor deposited in the body, but heterogeneous body distribution may impose serious damage to the kidneys."[133] In other words, ochratoxin A does not evenly distribute in the body, and therefore concentrated levels could do the body damage during the elimination process.

Scientists report, "In addition to being a nephrotoxin, animal studies indicate that ochratoxin A is a liver toxin, an immune suppressant, a potent teratogen [causing developmental malformations], and a carcinogen. . . . Ochratoxin has been detected in blood and other animal tissues and in milk, including human milk. It is frequently found in pork intended for human consumption . . . How common is human exposure to ochratoxin? Studies from Canada, Sweden, West Germany, and Yugoslavia detected ochratoxin in human blood and serum. Analyses of urine from children in Sierra Leone detected both ochratoxin and aflatoxin throughout the year."[134]

Furthermore, the FDA/CFSAN state, "Ochratoxin A is a naturally occurring nephrotoxic fungal metabolite produced by certain species of the genera *Aspergillus* and *Penicillium*. It is mainly a contaminant of cereals (corn, barley, wheat, and oats) and has been found in edible animal tissues as well as in human blood sera and milk. Studies indicate that this toxin is carcinogenic in mice and rats. It is not completely destroyed during the processing and cooking of food; therefore, the implication of risk to human health and safety must be considered. [The] FDA needs current, up-to-date information on the incidence and levels of occurrence of this toxin in the U.S. for use in considering any necessary regulatory control measures for this substance."[135]

For this reason, the FDA/CFSAN has included ochratoxin A in a surveillance program along with aflatoxin, patulin, DON, and fumonisin FB1, FB2, and FB3.[136] In addition, Dr. Bacon states that his group at the USDA Toxicology and Mycotoxin Research Unit has recently assumed research responsibilities on ochratoxin A.[137]

Interestingly, although ochratoxin A is among the strongest carcinogenic compounds in rats and mice, and its toxicological profile includes teratogenesis, nephrotoxicity, and immunotoxicity,[138] it was classified by the IARC as only a 2B cancer compound, which designates it as possibly carcinogenic for humans.[139] However, a European report published in 2000 states that legislation authorities were discussing maximal residue levels for ochratoxin A in various foodstuffs.[140] Some 10 years later, at the second printing of this book, ochratoxin A remains—a mycotoxin not regulated in the United States.

Both ochratoxin A and zearalenone have been a focus of worldwide surveillance, along with other mycotoxins. The Codex Committee on Food Additives and Contaminants (CCFAC) lists a "Code of Practice for the Prevention and Reduction of Mycotoxin Contamination in Cereals, including Annexes on Ochratoxin A, Zearalenon[e], Fumonisins, and Trichothecenes."[141] Codex Alimentarius states, "The Codex Alimentarius Commission was created in 1963 by FAO [Food and Agriculture Organization of the United Nations] and WHO [World Health Organization] to develop food standards, guidelines, and related texts such as codes of practice under the Joint FAO/WHO Food Standards Programme."[142] The United States, along with many other countries, is a member of the Codex Commission.

This worldwide attention is merited. According to a published article on the nutritional health aspects of mycotoxins, the most frequently observed human health effects from zearalenone are developmental abnormalities and harmful effects to reproduction.[143] Literally, the development of future generations and the quality of their lives could depend on mycotoxin research.

The reality of this statement is further illustrated when we consider the fact that foods can contain multiple mycotoxins, resulting in simultaneous multiple exposures. According to Dr. Young, "Corn contains twenty-five different mycotoxin-producing fungi, including recognized carcinogens! Peanuts contain twenty-six."[144] How many different mycotoxins are in *one* spoonful of corn breakfast cereal that so many children sit down to eat every morning?

This potential for a mouthful of multiple mycotoxins in every bite brings up the concern that the existence and synergistic effects of multiple mycotoxins are not addressed within the current system of set screening levels: the action levels for aflatoxin and patulin or the two guidance levels—which may be no more than a façade—for fumonisin and deoxynivalenol. Furthermore, with the existence of 300 to 400 identified mycotoxins—plus an untold number of mycotoxins not yet identified—how many mycotoxins are wreaking havoc with our health from daily ingestion of commonly eaten foods that are contaminated with adverse-health-impacting levels of mycotoxins? Nutrition should be our goal, not chronic toxicity.

(ENDNOTES)

1 The Department of Animal Science, Cornell University, "Aflatoxins: Occurrence and Health Risks." Available at http://www.ansci.cornell.edu/plants/toxicagents/aflatoxin/aflatoxin.html (accessed February 2006).

2   Maja Peraica et al., "Contamination of Food with Mycotoxins and Human Health," *Arh Hig Rada Toksikol* 2001; 52: 23-35. Copy of published article provided by Dr. Maja Peraica, February 2006.

3   The Department of Animal Science, Cornell University, "Aflatoxins: Occurrence and Health Risks."

4   Bennett et al., "Mycotoxins," p. 3.

5   See Section Three, interview with Dr. Jia-Sheng Wang, July 2006.

6   FDA/CFSAN, "Aflatoxins." Available at http://www.cfsan.fda.gov/~mow/chap41.html (accessed February 2006); and The Department of Animal Science, Cornell University, "Aflatoxins: Occurrence and Health Risks."

7   FDA, "Aflatoxins," and the Department of Animal Science, Cornell University, "Aflatoxins: Occurrence and Health Risks."

8   See Section Three, interview with Dr. Jia-Sheng Wang, July 2006.

9   Interview with Dr. Charles Bacon, August 2006.

10   Interview with Dr. Garnett Wood, July 2006.

11   See Section Three, interview with Dr. Jia-Sheng Wang, July 2006.

12   FDA/CFSAN, "Mycotoxins in Domestic Foods." Available at http://www.cfsan.fda.gov/~comm/cp07001.html (accessed August 2006).

13   Maja Peraica et al., "Prevention of Exposure to Mycotoxins from Food and Feed," *Arth Hig Rada Toksikol* 2002; 53: 229-237. Copy of published article provided by Dr. Maja Peraica, February 2006. "Moulds" is the European spelling of the word *mold*.

14   See Section Three, interview with Dr. Jia-Sheng Wang, July 2006.

15   Interview with Dr. Garnett Wood, July 2006.

16   FDA, "Aflatoxins."

17   Ibid.

18   Bennett et al., "Mycotoxins," p. 7.

19   "Ouganda" is the French spelling of the word *Uganda*.

20   The Department of Animal Science, Cornell University, "Aflatoxins: Occurrence and Health Risks."

21   See Section Three, interview with Dr. Jia-Sheng Wang, July 2006.

22   The Department of Animal Science, Cornell University, "Aflatoxins: Occurrence and Health Risks."

23   Bennett et al., "Mycotoxins," p. 9.

24   The Department of Animal Science, Cornell University, "Aflatoxins: Occurrence and Health Risks."

25   Bennett et al., "Mycotoxins," p. 7.

26   FDA, "Aflatoxins."

27   Joe Cummins, "Increased Mycotoxins in Organic Produce?" Available at http://www.i-sis. org.uk/IMIOF.php (accessed February 2006).

28   See Section Three, interview with Dr. Regina Santella, July 2006.

29   Ibid.

30   The U.S. population was 309,858,430 as of 03:02 UTC (EST +5) July 30, 2010. From U.S. Census Bureau, "U.S. and World Population Clocks—POPClocks," which additionally states, "Coordinated Universal Time (UTC) is the equivalent of Eastern Standard Time (EST) plus 5 hours or Eastern Daylight Saving Time (EDT) plus 4 hours." Available at http://www.census.gov/main/www/ popclock.html (accessed July 2010).

31   CDC, "Incidence of Acute Hepatitis B—United States, 1990—2002," *MMWR*, January 2, 2004/52(51 & 52); 1252-1254. Available at http://www.cdc.gov/mmwR/preview/mmwrhtml/mm5251a3. htm (accessed October 2006).

32   The Department of Animal Science, Cornell University, "Aflatoxins: Occurrence and Health Risks."

33   Ibid.

34   Cummins, "Increased Mycotoxins in Organic Produce?"

35   The Department of Animal Science, Cornell University, "Aflatoxins: Occurrence and Health Risks."

36   FDA, "Aflatoxins."

37   The Department of Animal Science, Cornell University, "Aflatoxins: Occurrence and Health Risks."

38   Peraica et al., "Toxic Effects of Mycotoxins in Humans."

39   FDA, "Aflatoxins."

40   Peraica et al., "Toxic Effects of Mycotoxins in Humans."

41   The Department of Animal Science, Cornell University, "Aflatoxins: Occurrence and Health Risks."

42   Interview with Dr. Garnett Wood, July 2006.

43   Ibid.

44   Bennett et al., "Mycotoxins," p. 6.

45   Ibid.

46   FDA, "Aflatoxins."

47   The Department of Animal Science, Cornell University, "Aflatoxins: Occurrence and Health Risks."

48   Peraica et al., "Toxic Effects of Mycotoxins in Humans," and Cummins, "Increased Mycotoxins in Organic Produce?"

49   The Department of Animal Science, Cornell University, "Aflatoxins: Occurrence and Health Risks."

50   Cummins, "Increased Mycotoxins in Organic Produce?"
51   See Section Three, interview with Dr. Regina Santella, July 2006.
52   Ibid.
53   Ibid.
54   Ibid.
55   The Department of Animal Science, Cornell University, "Aflatoxins: Occurrence and Health Risks."
56   See Section Three, interview with Dr. Regina Santella, July 2006.
57   Peraica et al., "Toxic Effects of Mycotoxins in Humans."
58   FDA/CFSAN, "Mycotoxins in Domestic Foods."
59   Ibid.
60   Sorenson, "Fungal Spores: Hazardous to Health?" and Bennett et al., "Mycotoxins," p. 3.
61   FDA/CFSAN, "Mycotoxins in Domestic Foods."
62   FDA, "Sec. 683.100 Action Levels for Aflatoxins in Animal Feeds (CPG 7126.33)." Available at http://www.fda.gov/ora/compliance_ref/cpg/cpgvet/cpg683-100.html (accessed August 2006).
63   FDA/CFSAN, "Mycotoxins in Domestic Foods."
64   See Section Three, interview with Dr. Jia-Sheng Wang, July 2006.
65   WHO/IARC, "Volume 56 Some Naturally Occurring Substances: Food Items and Constituents, Heterocyclic Aromatic Amines and Mycotoxins."
66   Ibid.
67   Ibid.
68   CDC, "National Biomonitoring Program." Available at http://www.cdc.gov/biomonitoring/ (accessed August 2006). (Italics added.)
69   The Department of Animal Science, Cornell University, "Aflatoxins: Occurrence and Health Risks."
70   Interview with Dr. Garnett Wood, July 2006.
71   FDA/CFSAN, "Action Levels for Poisonous or Deleterious Substances in Human Food and Animal Feed," August 2000. Available at http://www.cfsan.fda.gov/~lrd/fdaact.html (August 2006).
72   Ibid.
73   Interview with Veterinary Medical Officer, FDA/Center for Veterinary Medicine, Office of Surveillance and Compliance, July 2006 with written confirmation of verbal interview via e-mail, July 2006.
74   Ibid.
75   FDA, "Sec. 683.100 Action Levels for Aflatoxins in Animal Feeds (CPG 7126.33)."
76   Interview with Richard Sellers, August 2006.
77   Interview with Veterinary Medical Officer, FDA/Center for Veterinary Medicine, Office of Surveillance and Compliance, July 2006 with written confirmation of verbal interview via e-mail, July 2006.
78   Ibid.
79   Interview with Dr. Garnett Wood, July 2006.
80   See Section Three, interview with Dr. Jia-Sheng Wang, July 2006.
81   Interview with Dr. Charles Bacon, August 2006.
82   Ibid.
83   Ibid.
84   FDA/CFSAN, "Mycotoxins in Domestic Foods."
85   Ibid.
86   FDA/CFSAN, "Juice HACCP Hazards and Controls Guidance." Available at http://www. cfsan.fda.gov/~dms/juicgu10.html (accessed August 2006).
87   FDA, "Sec. 510.150 Apple Juice, Apple Juice Concentrates, and Apple Juice Products– Adulteration with Patulin." Available at www.fda.gov/ora/compliance_ref/cpg/cpgfod/cpg510-150.htm (accessed August 2006).
88   FDA/CFSAN, "Patulin in Apple Juice, Apple Juice Concentrates and Apple Juice Products," September 2001. Available at http://www.cfsan.fda.gov/~dms/patubck2.html (accessed August 2006).
89   Ibid.
90   Ibid.
91   FDA, "Sec. 510.150 Apple Juice, Apple Juice Concentrates, and Apple Juice Products–

Adulteration with Patulin."

92   FDA/CFSAN, "Patulin in Apple Juice, Apple Juice Concentrates and Apple Juice Products."

93   FDA/CFSAN, "Juice HACCP Hazards and Controls Guidance."

94   Ibid.

95   Aerotech P & K, "Potential Mycotoxins in Indoor Environments." January 1, 2001. Available at http://www.aerotechpk.com/Resources/TechtipDetails.aspx?i=35&t=IAQ (accessed September 2006).

96   Goldberg, *Alternative Medicine*, p. 395.

97   Ibid., p. 398.

98   European Mycotoxin Awareness Network, "Fact Sheet 6, Patulin." Available at http://www.lfra.co.uk/eman2/fsheet6.asp (accessed September 2006).

99   "Hyperaemia" is the British spelling of the word *hyperemia*.

100   "Haemorrhage" is the British spelling of the word *hemorrhage*.

101   European Mycotoxin Awareness Network, "Fact Sheet 6, Patulin.".

102   FDA/CFSAN, "Mycotoxins in Domestic Foods."

103   Ibid.

104   Ibid.

105   See Section Three, interview with Dr. Jia-Sheng Wang, July 2006.

106   FDA, "CVM [Center for Veterinary Medicine] Update. Final Guidance Availability on Fumonisin Levels in Human Food and Animal Feeds," November 15, 2001. Available at http://www.fda.gov/cvm/CVM_Updates/fumongl.htm and http://www.cfsan.fda.gov/~dms/fumongu2.html (accessed August 2006). (Italics added.)

107   FDA/CFSAN, "Mycotoxins in Domestic Foods."

108   Cummins, "Increased Mycotoxins in Organic Produce?"

109   Bennett et al., "Mycotoxins," p. 12.

110   Marinkovich, "Fungal Hypersensitivity Pathophysiology."

111   FDA, "CVM [Center for Veterinary Medicine] and Fumonisins." Available at http://www.fda.gov/cvm/fumonisin.htm (accessed August 2006).

112   WHO/IARC, "Volume 56 Some Naturally Occurring Substances: Food Items and Constituents, Heterocyclic Aromatic Amines and Mycotoxins."

113   FDA/CFSAN, "Mycotoxins in Domestic Foods."

114   Ibid.

115   Bennett et al., "Mycotoxins," p. 16.

116   Cummins, "Increased Mycotoxins in Organic Produce?"

117   Bennett et al., "Mycotoxins," p. 16.

118   FDA/CFSAN, "Guidance for Industry and FDA/Letter to State Agricultural Directors, State Feed Control Officials, and Food, Feed, and Grain Trade Organizations: Final Guidance," September 16, 1993. Available at http://www.cfsan.fda.gov/~dms/graingui.html (accessed August 2006).

119   FDA/CFSAN, "Mycotoxins in Domestic Foods."

120   Ibid.

121   Ibid.

122   Interview with Dr. Garnett Wood, July 2006.

123   Interview with Dr. Charles Bacon, August 2006.

124   FDA/CFSAN, "Mycotoxins in Domestic Foods."

125   WHO/IARC, "Volume 56 Some Naturally Occurring Substances: Food Items and Constituents, Heterocyclic Aromatic Amines and Mycotoxins."

126   FDA/CFSAN, "Patulin in Apple Juice, Apple Juice Concentrates and Apple Juice Products," September 2001. Available at http://www.cfsan.fda.gov/~dms/patubck2.html (accessed August 2006).

127   Interview with Dr. Garnett Wood, July 2006. (Italics added.)

128   Interview with Veterinary Medical Officer, FDA/Center for Veterinary Medicine, Office of Surveillance and Compliance, July 2006 with written confirmation of verbal interview via e-mail, July 2006.

129   Cummins, "Increased Mycotoxins in Organic Produce?"

130   Bennett et al., "Mycotoxins," p. 13-14.

131   "Unfavourable" is the British spelling of the word *unfavorable*.

132   From CancerWEB's on-line medical dictionary, copyright 1997-2005. Available at http://cancerweb.ncl.ac.uk/cgi-bin/omd?query=toxicokinetics&action=Search+OMD (accessed August 2006).

133   Petzinger et al., "Ochratoxin A from a toxicological perspective," 2000, *Journal of Veterinary Pharmacology and Therapeutics*, 23 (2), 91-98. Available at http://www.blackwell-synergy.com/doi/abs/10.1046/j.1365-2885.2000.00244.x?cookieSet=1&journalCode=jvp (accessed August 2006).

134   Bennett et al., "Mycotoxins," p. 13-14.

135   FDA/CFSAN, "Mycotoxins in Domestic Foods."

136   Ibid.

137   Interview with Dr. Charles Bacon, August 2006.

138   Petzinger et al., "Ochratoxin A from a toxicological perspective."

139   WHO/IARC, "Volume 56 Some Naturally Occurring Substances: Food Items and Constituents, Heterocyclic Aromatic Amines and Mycotoxins."

140   Petzinger et al., "Ochratoxin A from a toxicological perspective."

141   FDA/CFSAN, "CFSAN 2003 Program Priorities Report Card," FDA, November 24, 2003. Available at http://www.cfsan.fda.gov/~dms/cfsann03.html (accessed August 2006).

142   Codex Alimentarius, "FAO/WHO Food Standards." Available at http://www.codexalimentarius.net/web/index_en.jsp (August 2006). "Programme" is the British spelling of the word *program*.

143   Kovacs, M., "Nutritional Health Aspects of Mycotoxins," *Orv Hetil*, 2004 August 22:145(34): 1739-46. Available at http://www.ncbi.nim.nih.gov/entrez/query.fcgi?cmd=Retrieve&db=PubM ed&list_uids=15493122&dopt=Abstract (accessed August 2006).

144   Young et al., *The pH Miracle*, p. 88.

CHAPTER FOUR:

STARVE THE FUNGI, NOT YOURSELF

Fungal and mycotoxin sensitivities can make eating a miserable task, creating dread and apprehension at each meal. In fact, for those of us who have fungal and mycotoxin sensitivities, the challenge of identifying foods that do not trigger antigenic responses can seem insurmountable. When we first started to adjust our diet seeking foods as free of fungi and mycotoxins as possible, the number of foods filled with fungal antigens seemed almost endless, given that so many of our everyday foods contain fungal-derived additives. No wonder food was making us sick! We were eating what we thought were healthy, whole foods—protein-rich almonds, whole wheat breads, sugar-cane sweetened sunflower butter as a substitute for peanut butter, corn, whole grain cereals—the list goes on. In the beginning, we didn't understand why we kept reacting, reacting, and reacting!

Although food sensitivities can be individually specific, we found (through a process of trial and error) that if one of us got sick from a food, usually we all got sick. Sometimes, one or two of us would react, while the others didn't. However, within a short amount of time, the particular food would begin to cause reactions in all of us. In time, we noticed this pattern was happening primarily with foods such as nuts, grains, and canned tomato products or with processed products made from crops inherently harvested and shipped in large quantities (i.e., peanuts). These finished products would often contain "hot pockets" of fungi and/or mycotoxins, despite manufacturing efforts to homogeneously blend the raw materials. A prime example of this hot pocket phenomenon is peanut butter. We would get halfway through a jar, gleeful not to be having any reactions, only to have our short-term success dashed with a severe reaction after eating the next serving.

How do we begin to eliminate fungal toxins from our diets when they naturally occur in so many foods? A list of top-ten mycotoxic foods, compiled by David A. Holland, MD, and coauthor Doug Kaufmann, gives some guidance on "foods to avoid:

1. Alcoholic beverages—Alcohol is the mycotoxin of the *Saccharomyces* yeast—brewer's yeast. Other mycotoxins besides alcohol can also be introduced into these beverages through the use of mold-contaminated grains and fruits. Producers often use grains that are too contaminated with fungi and mycotoxins to be used for table foods, so the risk is higher that you are consuming more than just alcohol in your beverage.

2. Corn—Corn is 'universally contaminated' with fumonisin and other fungal toxins such as aflatoxin, zearalenone and ochratoxin. Fumonisin and aflatoxin are known for their cancer-causing effects, while zearalenone and ochratoxin cause estrogenic and kidney-related problems, respectively. Just as corn is universally contaminated with mycotoxins, our food supply seems to be universally contaminated with corn—it's everywhere! A typical chicken nugget at a fast food restaurant consists of a nugget of corn-fed chicken that is covered by a corn-based batter that is sweetened with corn syrup!

3. Wheat—Not only is wheat often contaminated with mycotoxins, but so are the products made from wheat like breads, cereals, pasta, etc. Pasta may be the least-'offensive' form of grains since certain water-soluble mycotoxins, such as deoxynivalenol (vomitoxin), are partially removed and discarded when you toss out the boiling water that you cooked the pasta in. Unfortunately, traces of the more harmful, heat-stable and fat-soluble mycotoxins, such as aflatoxin, remain in the grain. Regarding breads—it probably doesn't matter if it's organic, inorganic, sprouted, blessed or not—if it came from a grain that has been stored for months in a silo, it stands the chance of being contaminated with fungi and mycotoxins.

4. Barley—Similar to other grains that can be damaged by drought, floods, and harvesting and storage processes, barley is equally susceptible to contamination by mycotoxin-producing fungi. Barley is used in the production of various cereals and alcoholic beverages.

5. Sugar (sugar cane and sugar beets)—Not only are sugar cane and sugar beets often contaminated with fungi and their associated mycotoxins, but they, like other grains, fuel the growth of fungi. Fungi need carbohydrates—sugars—to thrive.

6. Sorghum—Sorghum is used in a variety of grain-based products intended for both humans and animals. It is also used in the production of alcoholic beverages.

7. Peanuts—A 1993 study demonstrated 24 different types of fungi that colonized the inside of the peanuts used in the report. And this was after the exterior of the peanut was sterilized! So, when you choose to eat peanuts, not only are you potentially eating these molds, but also their mycotoxins. Incidentally, in the same study the examiners found 23 different fungi on the inside of corn kernels. That said, if you choose to plant your own garden in an attempt to avoid mycotoxin contamination of corn or peanuts, it does you no good if the seed (kernel) used to plant your garden is already riddled with mold.

8. Rye—The same goes for rye as for wheat and other grains. In addition, when we use wheat and rye to make bread, we add two other products that compound our fungal concerns: sugar and yeast!

9. Cottonseed—Cottonseed is typically found in the oil form (cottonseed oil), but is also used in the grain form for many animal foods. Many

studies show that cottonseed is highly and often contaminated with mycotoxins.

10. Hard Cheeses—Here's a hint: if you see mold growing throughout your cheese, no matter what you paid for it, there's a pretty good chance that there's a mycotoxin not far from the mold. It is estimated that each fungus on Earth produces up to three different mycotoxins. The total number of mycotoxins known to date numbers in the thousands. On the other hand, some cheeses, such as Gouda cheese, are made with yogurt-type cultures, like *Lactobacillus*, and not fungi. These cheeses are a much healthier alternative, fungally speaking."[1]

Our family experienced allergic-type reactions to the above listed foods so severe at times that we got violently ill. The most severe reaction came from sugar cane, which is used in many health food products. We quickly realized that we had to be just as diligent label readers in a health food store as in a regular grocery store. After reviewing the list of top-ten mycotoxin-contaminated food groups, we began to check labels specifically for those ingredients and found that when we avoided them, our food reactions greatly diminished.

There were no quick and easy answers to designing a diet free of fungi and mycotoxins. However, with effort, it was possible to reduce fungal and mycotoxin ingestion by carefully selecting food from groups less prone to fungal growth and the ensuing mycotoxin contaminants. It took conscious effort, especially in the beginning, but the results—increased levels of health and energy in each of us— made the effort all worth it in the end.

According to Dr. Straus, mold sensitized people can react to foods that contain fungal antigens.[2] So what exactly are fungal antigens? According to CRISP (Computer Retrieval of Information on Scientific Projects), fungal antigens are ". . . substances of fungal origin that have antigenic activity."[3] An antigen is ". . . any substance to which the body reacts by producing antibodies."[4] In other words, fungal antigens would elicit an immune response from the body, creating the production of antibodies.

Substances that induce antigenic activity are often referred to as allergens. "Allergens usually enter the body through breathing, absorption through the skin, by eating or drinking foods, or by injection, such as insect bites or vaccinations. Because the body judges the substances to be dangerous to its health, the immune system identifies them as antigens. The mobilized immune system then releases specific antibodies to deactivate the allergenic antigens, setting in motion a complex series of events involving many biochemicals. These chemicals then produce the inflammation or other typical symptoms of an allergy response."[5] This antigenic response explains why we and other mold-sensitized individuals react to foods that contain substances our bodies recognize as antigenic because of prior fungal exposure.

Important to understand is the fact that two different types of fungal contamination can exist in food—fungal antigens and/or mycotoxins. Some foods will contain just the fungal antigens (since not all fungi produce mycotoxins), while other foods will contain both because the fungal antigens present came from a mycotoxin-producing species. So is it fungal antigens (for example, possible fungal proteins or traces of fungal proteins in the food) or mycotoxins that cause reactions to certain foods?

Dr. Erasmus states, "It could be allergic reactions, or it could be another of the body's reactions to poisons. Every mold, for instance, has proteins in it, and these proteins are foreign to our bodies. If we eat food containing the mold and digest those proteins completely, those proteins will not cause a problem. If we don't digest them completely, our immune systems could trigger an allergy in response to the foreign proteins. Inhaled or on the skin, these proteins could trigger allergic reactions that lead to breathing difficulties or rashes. In addition, molds can contain toxic non-protein molecules [for example, mycotoxins] that don't belong in the body. Our immune systems can react to these toxins because the immune system is responsible for making sure that our bodies are intact from the inside and are not under attack. We're always under attack, but under normal circumstances, our immune systems ensure that the foreign proteins, foreign organisms, and foreign molecules do not cause harm to our bodies."[6]

According to Dr. Wang, "Reactions to mold and exposure to mycotoxins are different. Some people may be hypersensitive (allergic) to mold-contaminated environments, which may be caused by inhalation of or skin contact with the mold or mold spores. Human exposure to high levels of mycotoxins through food usually leads to toxicosis, such as liver toxicity for aflatoxins and hematologic toxicity for trichothecene toxins."[7]

Interestingly, although mycotoxins are capable of being acutely toxic in high doses and may be chronically toxic over time in lower doses, alone they will not elicit an immune response from the body, according to Dr. Wang.[8] The reason is because by themselves, mycotoxins are too small for the body to recognize. In other words, if a mycotoxin is not attached to a protein cell, it does not elicit an antibody response, yet when attached to a protein, an antibody response can be detected.[9]

Anthony Lupo, manager of technical service at Neogen Corporation, a company that develops, manufactures, and markets diagnostic test kits, including test kits for mycotoxin detection in foods, explains, "Generally mycotoxins are considered very small on a molecular scale (~300 molecular weight). While they are responsible for a myriad of toxicological effects, they are not viewed as allergenic. Allergenic reactions are the product of a person's immune response, and to have an immune response, the foreign substance must be large enough for the immune system to recognize it as a foreign invader. This is generally termed immunogenicity. Proteins are typically tied to allergenic responses in sensitive individuals as they are sufficiently large (5,000 molecular weight[10] and larger) to be recognized as foreign.

Neogen technology utilizes antibodies generated against the six major mycotoxins; however, when these antibodies were first developed, the mycotoxins themselves needed to be chemically modified and linked to carrier proteins in order to trigger the immune response outlined above."[11]

So does this potential lack of immunogenicity mean that the body will not have an allergic reaction to a mycotoxin? Or is there a certain threshold level of mycotoxins in the body that when met the body does recognize the mycotoxins as foreign substances and then creates antibodies?

Dr. Wang says, "Yes and no. Mycotoxins are usually small molecules and do not cause immune reactions alone. However, mycotoxins in human and animal bodies can bind to proteins; inhibit DNA, RNA, and protein synthesis; or regulate gene expressions, which may cause immune-toxic effects, including immunostimulation, hypersensitivity, and immunosuppression. On the other hand, mycotoxin-containing mold spores may play a certain role in causing individual hypersensitivity (allergic reaction)."[12] Therefore, the food reactions that mold-sensitized people experience can be a reaction to the fungal antigens, the mycotoxins that are attached to proteins, or both. Remember, if mycotoxins are present in a food, fungal proteins will not be far behind! Whichever the case may be, the solution remains the same: Avoid ingestion of foods containing fungi, fungal antigens, and mycotoxins. If there's no fungus present, there should be no fungal-produced mycotoxins present either.

No matter how hard we try, mycotoxin ingestion is inevitable. Some of us eat more, some less, depending on our diet. The FDA points out, ". . . [E]pidemiological studies in Africa and Southeast Asia, where there is a high incidence of hepatoma [tumor of the liver], have revealed an association between cancer incidence and the aflatoxin content of the diet."[13] These overseas studies lead us to ask the logical question: Is consumption of low levels of aflatoxins and other mycotoxins in U.S. foods (such as wheat, corn, and peanuts) silently contributing to the myriad of degenerative diseases and cancers that "naturally" occur in the United States?

Let's think about it for a minute. Since fungal antigens and mycotoxins attached to proteins can create an allergic immune response and allergies are known to cause a host of other ailments, the possibility exists that chronic, low level ingestion of mycotoxins may very well lead to degenerative diseases and cancers—even if it has not yet been proven by and documented in epidemiological studies by scientific and medical researchers. In fact, *Alternative Medicine: The Definitive Guide*, a leading health book, states, "Allergies can cause or contribute to asthma, bronchitis, rheumatoid arthritis, diabetes, ear infections, eczema, hives, migraines or cluster headaches, chronic fatigue syndrome, gastrointestinal disorders, glaucoma, kidney problems, weight gain, seizures, heart palpitations, depression, and even cerebral palsy and multiple sclerosis, among other conditions."[14] The next question to ask: Do these health conditions, stemming from the body's reaction to allergic components, lead to degenerative diseases and cancers?

Some people may not realize that they are having an allergic reaction to certain foods or specific ingredients. In fact, they may crave or become addicted to the very food to which they are allergic. This fact was discovered during the 1940s by environmental medicine pioneer Theron Randolph, MD.[15] "According to Dr. Randolph, an allergic reaction to food can last for up to three or more days, making the addiction to the food difficult to discover. This is because the symptoms that a person normally experiences when undergoing withdrawal from an addictive substance can actually improve or be suppressed if the person eats more of the addictive food."[16] This phenomenon—the suppression of symptoms caused by addictive/allergenic foods—has been labeled "masking."[17] This addictive type of response could be why some people become alcoholics, chocoholics, and/or even heavy coffee drinkers.

In addition to predisposed sensitivities from prior fungal exposure, Dr. Wang explains that genetic differences are the major reason people have different reactions when equally exposed to the same levels of toxins.[18] Physical, as well as psychological, reactions can occur. Dr. Rapp states, "Specialists in environmental medicine believe it is possible that *any* area of the body can be affected by an allergy or a food or chemical sensitivity. Substances called chemical mediators are released during allergic reactions and travel all over the body, not just to 'accepted' areas such as the lungs or nose."[19] Any area of the body includes the central nervous system and the brain, which, of course, can affect psychological behavior. Changes in behavior can indicate an allergic response to a food and/or environmental contaminant.

Identifying ingredients to which a person is reactive can be very difficult. For example, fungal-derived ingredients are in food products containing malt such as malted milk drinks, cereals, malted milk balls, and other candies.[20] So if a person has a reaction to a cereal, for example, is he/she having a reaction to the fungal-derived additive, a different additive, fungal contamination that occurred in the field or in storage, and/or the accompanying mycotoxin(s)? The existence of fungal and mycotoxin contamination in cereals, which are frequently consumed by children, warrants asking the following question: What levels of fungi and mycotoxins are our children eating in their breakfast cereals day in and day out, year after year?

According to Dr. Bacon, the mycotoxin fumonisin is primarily on field corn, which is used in the processing of corn meal, corn flakes, corn chips, and other processed corn products.[21] Dr. Wang states, "The fungi that produce fumonisin regularly can contaminate in the field products like corn, and almost always they produce the toxin."[22]

As Dr. Wang noted earlier, "Fumonisin is a big issue right now due to very high levels of contamination in corn and corn products in the U.S." He illustrates the severity of Fumonisin in U.S. foods with the following story, "We . . . collected some corn samples in developing countries through the collaboration studies. As a control, we used cornmeal purchased from a [U.S.] national grocery store chain. I wanted students to see how food from developing countries is more contaminated with fumonisins than U.S. food. However, the measurements didn't prove that. We found higher levels in the U.S. cornmeal than in the corn samples from developing countries!"[23]

If the U.S. cornmeal contaminated with higher levels of fumonisins as compared to the corn samples from developing countries is any indication as to the level of toxicity in U.S. corn products, consumers may very well have cause for concern. Significant to note is the fact that these extremely elevated levels of fumonisins existed in a U.S. *processed* product. That the product tested was a processed product is relevant because, according to Dr. Bacon, fumonisin levels are said to be greatly reduced during manufacturing by a washing process as fumonisin is very water soluble.[24] Why then, with the ability to greatly reduce fumonisin levels in a processed product via a simple washing process, were higher levels of fumonisin found in a U.S. manufactured corn product than levels found in raw corn products from developing countries?

Additionally, Dr. Wang states, ". . . [D]eoxynivalenol sometimes is in a very high level in certain foods, especially in cereal. In England, a study was done in which they collected normal people's urine. What they found was that when these people ate cereal, deoxynivalenol was found in their urine."[25] Does the prevalence of deoxynivalenol in cereal mean there is a good possibility that every time we sit down to eat a bowl of cereal in the U.S., it contains deoxynivalenol?

"Yes," states Dr. Wang. But, of course, "It depends on how much you eat and the contamination level of deoxynivalenol in the original raw cereal." He reminds, "You know it is also a co-contaminant issue, not only fumonisins (fumonisin B1, B2, and B3), but also fumonisin and aflatoxin at the same time."[26] So again, we are talking about multiple mycotoxin exposure via ingestion of everyday foods, such as cereal.

Different crops are susceptible to the growth of specific fungi. In other words, not all fungi grow on every type of crop. For example, fungi that produce both fumonisin and aflatoxin will grow on corn in the field, resulting in possible crop contamination with either fumonisin or aflatoxin—or both. However, wheat does not tend to get contaminated with aflatoxin-producing fungi in the field, but rather in storage. Dr. Wang states that fungal growth that produces aflatoxins predominantly occurs in storage on wheat and is not an inherent risk in the field.[27] However, he clarifies that other fungi do tend to grow on wheat in the field, such as those that produce deoxynivalenol (DON).[28] Since many cereals, breads, pastas, and tortillas are made with either corn or wheat and most of the grain in the U.S.

is stored in huge silos, the risk of aflatoxin and other mycotoxin contamination in these types of products could be quite high. According to Dr. Holland et al., ". . . [I]f it came from a grain that has been stored for months in a silo, it stands the chance of being contaminated with fungi and mycotoxins."[29]

Dr. Thrasher concurs, ". . . [I]n this country we store our grains in silos where mold then begins to grow, leaving behind mycotoxins and other mold by-products. In Mexico the grains are consumed as they are harvested. The Mexicans do not have the morbidity [disease] that we have, even though they may be slightly undernourished. A physician in S. California who sees illegal immigrants has come to the conclusion that these people are better off than we are with respect to morbidity."[30]

The issue of mycotoxins in U.S. grown food—especially grains—could be a taboo subject to broach, based on the response we received when researching this topic. We maneuvered a governmental maze of agencies as we searched for the appropriate people to whom questions could be addressed. Multiple times we asked for clarification of studies and reports that indicate further study is needed in regard to chronic, low level ingestion of mycotoxins. Multiple times we were given "slivers" of information, rather than complete answers that would directly answer our questions. Could it be that the powers that be don't want full interpretations, without qualifiers, of these studies out to the general public? Are the health risks (from mycotoxin ingestion to the U.S. population) that could be surmised from the ample studies in Third World countries being watered down?

Dr. Thrasher states, "You have surmised correctly."[31]

The possibility of a government and industry collusion should heighten concerns regarding fungal and mycotoxin contamination in all foods. In addition to contamination concerns in cereals, what about all that wholesome whole wheat bread? Are we being beguiled as to the healthy nature of grain consumption? Is the health of our country's greatest asset—our nation's children—being placed second to economics? If strings are being pulled by industry and political puppeteers to orchestrate a public façade of safe mycotoxin levels in U.S. foods, then we, as consumers, must demand more investigation and accountability in regard to the safety of our foods.

Remember, cereal and bread are just *two* of the food products out of many that we all eat that can be affected by fungal and mycotoxin contamination! Could this continual consumption of low levels of fungi and fungal by-products—aflatoxins and other mycotoxins—lead to serious, long-term health repercussions? These presumed "low" levels may not necessarily be low either. Remember Dr. Wang's testing revealed higher levels of fumonisin in U.S. cornmeal than in corn samples from Third World countries! At what point will chronic exposure to aflatoxins, other mycotoxins, and other fungal contaminants in U.S. foods ignite ill health effects? No one knows, as there are many contributing factors. According to Cornell University, ". . . [T]he expression of aflatoxin-related diseases in humans may be influenced by factors such as age, sex, nutritional status, and/or concurrent exposure to other causative agents such as viral hepatitis (HBV) or parasite infestation."[32] These same factors undoubtedly affect the health outcome from exposure to other mycotoxins as well.

The long-term promotion of a whole-grain-based diet full of bread, cereal, rice, and pasta in the U.S. has no doubt increased the health risks associated with fungal and mycotoxin exposure via ingestion. Just take a look at the original U.S. Department of Agriculture (USDA) food pyramid, which was developed in 1992, based on nutrition recommendations from the *Dietary Guidelines for Americans* and the Recommended Dietary Allowances (RDAs) published in 1989.[33] The original food pyramid that recommended 6 to 11 servings per day of bread, cereal, rice, and pasta at the base of the pyramid was prominently displayed on most packages of cereals, rice, and pastas—acting as an indirect product endorsement.

This grain and bread food group, of course, is the one most likely to contain the largest levels of fungi, aflatoxins, and other mycotoxins. Essentially, by following the USDA guidelines, people have been consuming a diet rich in fungi, aflatoxins, and other fungal metabolites! To make matters worse, these fungal contaminants were fueled by the recommended daily two to four servings from the fruit group. The food group with the highest detoxification potential, vegetables,[34] was recommended at three to five servings per day. The milk, yogurt, and cheese food group and the meat, poultry, fish, dry beans, eggs, and nuts group were both recommended at two to three servings per day. Dependent on the selections made from these categories, fungal components again could have been consumed. The last food group—fats, oils, and sweets—was recommended to be used sparingly, with no differentiation of good fats versus bad fats—or, in other words, healing fats versus killing fats.[35]

The success, or rather the failure, of this food program can be judged by the current state of health of our nation. Health literature reports, "It is no secret that our contemporary conventional medical system is in a state of terrible disarray. Though conventional medicine excels in the management of medical emergencies, certain bacterial infections, trauma care, and many often heroically complex surgical techniques, it has failed miserably in the areas of disease prevention and the management of the myriad [of] new and chronic illnesses presently filling our hospitals and physicians' offices."[36] It is reasonable to ask: How much of a catalyst was the original food pyramid to the "myriad" of degenerative diseases that plague our country today?

The USDA acknowledged the failure of the original food program by replacing it in 2005 with a new pyramid called MyPyramid.[37] According to the USDA, the design of the pyramid reflected both the then-current science and the *2005 Dietary*

Guidelines for Americans, which is published by the USDA and Health and Human Services (HHS).[38]

The USDA acknowledges, "The American diet is not in balance. On average, Americans don't eat enough dark greens, orange vegetables and legumes, fruits, whole grains, and low-fat milk products. They eat more fats and added sugars. To bring the diet into balance, MyPyramid recommends eating more of the under-consumed foods and less solid fats, added sugars, and caloric sweeteners, and foods rich in these."[39] However, it appears to be still heavy on grains, light on vegetables, and does not address which fats and oils are healing versus degenerative.[40] A more health-focused model for people, especially those battling fungal-related illnesses, would be one emphasizing vegetables.[41]

The message of the MyPyramid program may have proven too complex because a mere six years later in June of 2011, the US Department of Agriculture replaced MyPryamid with MyPlate. The new program is visually represented by a place setting that illustrates the five food groups,[42] which is easier to understand and emulate, but is it any better for health?

Helping to exposure the potential hidden agenda behind the original food pyramid is Harvard Professor Walter Willett, MD, DrPH, who describes it as ". . . one major source of false intelligence."[43] Dr. Willett makes a valid point: Since the USDA's mission is to promote American agriculture, not public health, is it "'. . . the best agency to be giving dietary advice? They have so many conflicts of interest.'"[44]

The issues relating to fungal and mycotoxin contamination of U.S. food—and the related health effects—are akin to a multi-industry, political hot button.

Governmental agencies along with industry and political powers have a vested interest in controlling and suppressing the subject of fungi and mycotoxins in food. Industries economically tied to commodities prone and/or susceptible to fungal contamination, such as wheat, corn, peanuts, dairy—just to name a few—do not want the inspection spotlight focused on their crops and/or resultant products. The fear, of course, is of a ripple effect felt all the way from the raw material supplier (i.e., the farmer) to the manufacturer of products containing the raw materials (i.e., whole wheat bread, corn chips, etc.) to the distributor (i.e., grocery stores).

Opinions vary about the degree to which fungal and mycotoxin contamination is a problem in the U.S. food supply. For example, Dr. Wang adamantly states, "In the U.S., actually, mycotoxin contamination is still a big problem."[45]

Dr. Santella has concerns specifically in regard to aflatoxin ingestion. As she earlier explained, "It is a problem [because] the U.S. population does have about a 2 percent carrier rate for hepatitis C virus," which is the subpopulation at greatest risk from aflatoxin exposure.[46]

However, David Eaton, PhD, who is a professor of environmental and occupational health sciences in the Toxicology Program at the University of Washington, believes, "It is an economic concern and a potential public health concern, although all of the evidence to date suggests that the efforts in regulating aflatoxins in the food supply in the U.S. are pretty effective in keeping the levels below those which cause any measurable effects on public health."[47] Scientists and doctors assess measurable effects of aflatoxin ingestion based on population rates of liver cancer because the liver is the organ primarily affected by aflatoxins.[48] [See interview with Dr. Eaton in Section Three.]

Although there may not be "measurable effects" on public health in regard to aflatoxin ingestion in the U.S. food supply, are there measurable effects on public health from the ingestion of *other* mycotoxins, individually and/or collectively, in the U.S. food supply? In other words, is there evidence that mycotoxins in U.S. food—even guidance-level-regulated mycotoxins for which testing may, or may not, be performed—are contributing to the levels of various cancers and other degenerative diseases that we see in our society today?

Dr. Bacon, who is the research leader and supervisory microbiologist for the USDA Toxicology and Mycotoxin Research Unit, states, "They would be a contributor, not necessarily *the* contributor. I would not be surprised if they are not contributing factors . . . but I don't believe they are a major contributor."[49]

A definitive answer may not be forthcoming any time soon, as the funding floodgates may very well be politically and economically controlled. In fact, funding U.S. scientists and medical doctors to study and document the health effects of aflatoxins, fumonisins, and other mycotoxins ingested in *U.S. grown food* could be viewed as politically and economically "incorrect." Furthermore, to continue to limit research that addresses the health effects of human ingestion of mycotoxins to U.S. industry-safe areas such as Third World countries creates an industry-welcomed façade that the U.S. food supply does not have levels of mycotoxin contamination that merit field studies. Scientific studies in Third World countries—that show certain cancers stem from ingestion of mycotoxin-contaminated foods—do not raise the same alarm bells in U.S. citizens as would similar results from studies conducted with U.S. grown food. Although this strategy may effectively control and limit public awareness regarding the potential health effects of mycotoxin ingestion in the U.S. food supply, it will not minimize the potential health effects. Health risks of which we are not aware—or refuse to acknowledge—can certainly still hurt us.

Dr. Wang believes that eventually mycotoxins will be studied more thoroughly in U.S. foods. "I think so. [Because] when you eat corn, you will get aflatoxin, fumonisin, and other mycotoxins. No one knows how those toxins together affect the human body. We only know the effects based on one toxin—aflatoxin or fumonisin alone. That is why our studies really focus on development of markers for multiple toxin exposures, for studying how corn or other food really affects human cancer risks."[50]

The health effects of multi-mycotoxin exposure have already been a focus of European countries, which have led the mycotoxin research crusade. According to Dr. Thrasher, "Current research is coming out of Europe, not the United States."[51] In particular, research is coming out of the Scandinavian countries, according Dr. Thrasher. He states, "They are far ahead of us—far, far ahead of us."[52]

Because of these research advancements, European countries have set lower limits of mycotoxin acceptability in their food. Dr. Santella concurs, "Our regulations are somewhat higher in terms of what is allowed as compared with other countries, like England."[53] In other words, U.S. food is more likely to contain higher levels of mycotoxins than food in European countries.

In fact, countries that purchase commodities from the U.S. require testing and certification of mycotoxin levels. If levels are above the maximum acceptable limits set by these other countries—levels that are lower than U.S. tolerance levels— then these countries will not accept the product for purchase, according to Lynn Polston, mycotoxins program manager with GIPSA (Grain Inspection, Packers, and Stockyards Administration). GIPSA is the U.S. organization that conducts, for a fee, analyses for mycotoxins in grains and other commodities intended for export.[54]

Unfortunately, higher mycotoxin tolerance levels may make the United States an exporter's haven, providing a market for foreign food products with mycotoxin contents higher than the levels that other countries will accept. For example, prior to the FDA establishing an action level for patulin, many countries, including Austria, Belgium, France, Norway, Sweden, and Switzerland had already established a maximum acceptable level of 50 ppb for patulin in apple juice and products containing apple juice from concentrate.[55] At that time, it was speculated by National Food Processors Association (NFPA) members that products with patulin levels exceeding the 50 ppb limit were being diverted to the U.S. from other countries that already had established the 50 ppb maximum acceptable level.[56] Do we want the United States to be a disposal site for foods with levels of mycotoxin contamination unacceptable to other countries? Do we, as U.S. citizens, want to be eating foods that other countries deem unacceptable, and thus have rejected?

If the cream of our crops—those testing with the lowest levels of mycotoxins—are being sold and exported to other countries, what quality of U.S. grown raw commodities is left for U.S. citizen consumption and product manufacturing? Again, economics is outweighing health. The fact that the U.S. government and industries leave the remaining food crops—which contain the higher levels of mycotoxins—for U.S. citizens to consume is of concern. Although one scientist we interviewed claimed that the lower European mycotoxin levels are without merit, others just as firmly believe they are justified. At what level of health deterioration in our country will these economic-based decisions be exposed?

A 2004 Hungarian publication states, "Mycotoxins occur in small amount[s] in the foods; however, their continuous intake even in microdoses can result in accumulation in the organism [humans and animals]. Synergic effects of the mycotoxins, as well as their possible additive multi-toxic effects, seem to be especially dangerous."[57] Thus, chronic ingestion of low levels of multi-mycotoxins may result in long-term health repercussions.

The European publication further states, "Public health risks of the toxins accumulating in the human and animal bodies during the long-term consumption of the mycotoxin-containing foods—even in small doses—have not been evaluated yet as thoroughly as their importance would require."[58] The importance of this, of course, lies in determining the degree to which multiple mycotoxins in food and feed contribute to cancer, other degenerative disease, and illness. Although these issues certainly warrant research in the U.S.—with U.S. grown commodities— studies may not be forthcoming as decisions affecting funding may be industry, politically, and economically driven, not scientifically based.

(ENDNOTES)

1 David A. Holland, MD, and Doug Kaufmann, "The Top-10 Myco-Toxic Foods." Available at http://www.mercola.com/2003/nov/5/toxic_foods.htm (accessed March 2006). Reprinted with permission from Mr. Kaufmann. Dr. Holland and Mr. Kaufmann are coauthors of *The Fungus Link* and *The Fungus Link, vol. 2*, which are available at http://www.knowthecause.com/ (accessed August 2006).

2   E-mail from Dr. David Straus, May 2006.

3   CRISP (Computer Retrieval of Information on Scientific Projects), "Fungal Antigen," NIH (National Institutes of Health). Available at http://crisp.cit.nih.gov/Thesaurus/00003136.html (accessed May 2006). As noted on the CRISP website, CRISP is a searchable database of federally funded biomedical research projects conducted at universities, hospitals, and other research institutions that is maintained by the Office of Extramural Research at the National Institutes of Health.

4   *Webster's New World Dictionary*, p. 27 (New York, NY, Pocket Books, 2003).

5   Goldberg, *Alternative Medicine*, p. 508.

6   Interview with Dr. Udo Erasmus, May 2006.

7   See Section Three, interview with Dr. Jia-Sheng Wang, July 2006.

8   Ibid.

9   Interview with Anthony Lupo, July 2006, and e-mail from Anthony Lupo, July 2006. For more information on Neogen Corporation see http://www.neogen.com/ (accessed August 2006).

10   According to Anthony Lupo, molecular weight is also sometimes referred to as atomic mass units (AMU).

11   Interview with Anthony Lupo, July 2006, and e-mail from Anthony Lupo, July 2006.

12   See Section Three, interview with Dr. Jia-Sheng Wang, July 2006.
13   FDA, "Aflatoxins."
14   Goldberg, *Alternative Medicine*, p. 507.
15   Ibid., p. 509.
16   Ibid.
17   Ibid., p. 510.
18   See Section Three, interview with Dr. Jia-Sheng Wang, July 2006.
19   Rapp, *Is This Your Child?* p. 35.
20   FDA, "Partial List of Microorganisms and Microbial-Derived Ingredients That Are Used in Foods."
21   Interview with Dr. Charles Bacon, August 2006.
22   See Section Three, interview with Dr. Jia-Sheng Wang, July 2006.
23   Ibid.
24   Interview with Dr. Charles Bacon, August 2006.
25   See Section Three, interview with Dr. Jia-Sheng Wang, July 2006.
26   Ibid.
27   Ibid.
28   Ibid.
29   Holland et al., "The Top-10 Myco-Toxic Foods."
30   E-mail from Dr. Jack Thrasher, August 2006.
31   Ibid.
32   The Department of Animal Science, Cornell University, "Aflatoxins: Occurrence and Health Risks."
33   USDA, "MyPyramid." Available at http://wwwmypyramid.gov (accessed March 2006).
34   Kris Novak, PhD, "Cancer Susceptibility: Role of Genes, Environment, and Diet." Available at http://www.medscape.com/viewarticle/448881_print (accessed March 2006).
35   Erasmus, *Fats That Heal Fats that Kill*, p. xxii.
36   Goldberg, *Alternative Medicine*, p. 8.
37   USDA, "MyPyramid," USDA.
38   Ibid.
39   Ibid.
40   Interview with Dr. Udo Erasmus, May 2006.
41   Ibid.
42   USDA, MyPlate, http://www.cnpp.usda.gov/myplate.htm (accessed May 2013).
43   Craig Lambert, "Demolish the Food Pyramid," *Harvard Magazine*, November-December 2001. Available at http://www.harvardmagazine.com/on-line/110156.html (accessed March 2006)
44   Ibid.
45   See Section Three, interview with Dr. Jia-Sheng Wang, July 2006.
46   See Section Three, interview with Dr. Regina Santella, July 2006.
47   See Section Three, interview with Dr. David Eaton, July 2006.
48   See Section Three, interviews with drs. Jia-Sheng Wang, July 2006, and David Eaton, July 2006.
49   Interview with Dr. Charles Bacon, August 2006. (Italics added.)
50   See Section Three, interview with Dr. Jia-Sheng Wang, July 2006.
51   Interview with Dr. Jack Thrasher, May 2006.
52   Ibid.
53   See Section Three, interview with Dr. Regina Santella, July 2006.
54   Interview with Lynn Polston, July and August 2006.
55   From correspondence to Dr. Terry Troxell, Division of Enforcement and Programs Policy, CFSAN/FDA from Allen W. Matthys, PhD, vice president, Technical Regulatory Affairs with the National Food Processors Association, November 1, 21996. Available at http://www.fda.gov/ohrms/dockets/dailys/00/Aug00/081600/c000002.pdf (accessed August 2006).
56   Ibid.
57   Kovacs, "Nutritional Health Aspects of Mycotoxins."
58   Ibid.

CHAPTER FIVE:

MORE TRIAL AND ERROR

After several weeks of no progress and then eventual setbacks from the homeopathic treatment plan, our ND gave us the name and phone number of an infectious disease specialist with whom we could consult. She also discussed sauna therapy, which is a documented method of detoxification.[1] According to Dr. Erasmus, "Saunas, increased fluid intake, and sweating from exercise help remove toxic materials from . . . [the] body."[2] The concept is easy to understand when it comes to chemical detoxification—you work up a sweat and the toxins exit via the pores in your skin. But how would the heat and moisture from a sauna affect fungal growth?

After discussions with our naturopath, we decided that Kurt would see the specialist, and our daughter and I would start sauna therapy. After Kurt's initial misdiagnosis and mistreatment, we wanted to proceed with caution. We felt there was no reason to set appointments for each member of our family to see the specialist until a course of treatment had proven effective. If it turned out that the chosen treatment method was not effective, there was no reason to subject all of us to potentially harmful side effects. We were just as cautious regarding alternative treatment options. If there was a risk of sauna therapy worsening our conditions, we didn't want Kurt trying it since he was already more fragile than the rest of us. In hindsight, exerting this level of caution was a good decision.

Kurt's continual deterioration, and the rest of the family's lack of progress, is why our ND recommended that we consult with an infectious disease specialist who claimed to have prior experience in treating fungal-related illnesses. We soon found out just how small a city with a population of 100,000 can be as word mysteriously spread through town that Kurt had seen an infectious disease specialist. People we didn't even know began asking if what we had was contagious! The small town rumor mill created unfounded fear in people, which caused a socially imposed quarantine on us. So, not only were we displaced, sick, and unemployed, but we now had everyone giving us a wide berth because they were afraid of catching what they were convinced was infectious!

Kurt arrived at his doctor appointment covered in hives from his neck to his toes, struggling with severe breathing difficulties, an unstable thyroid, and several other mold-related symptoms. The doctor's communication style was dictatorial, an extreme contrast to the educational philosophy of naturopathic medicine. Ignoring the doctor's arrogant attitude, Kurt patiently waited to learn the most up-to-date treatments for mold exposure. He listened while he labored to breathe, which often happened when he went anywhere because a whiff of mold could trigger a severe asthma attack. Kurt was skeptical when the doctor told him that unless he was immunocompromised from a disease such as cancer, AIDS (Acquired Immune Deficiency Syndrome), and/or leukemia, mold could not hurt him. Kurt, the doctor said, was suffering from just a bad allergic reaction to mold.

The doctor breezed over the fact that Kurt had been treated with two different types of steroids, both pre- and post-hurricane. The specialist apparently did not understand that steroids are immunosuppressive medications that can render a person immunosuppressed and even immunocompromised. Fred Lopez, MD, an infectious disease specialist at Louisiana State University's School of Medicine, explained in a PBS interview post-Katrina that people taking immunosuppressive medications like steroids are more likely to suffer complications from being exposed to mold.[3]

The infectious disease specialist treating Kurt did confirm that Kurt could have mold spores in his lungs, but he was adamant that mold spores can't grow inside human lungs or tissue. He said Kurt was having an acute allergic reaction to mold in his sinuses and lungs, but he was adamant that mold couldn't possibly affect Kurt's thyroid, because mold can't affect the endocrine system at all. When Kurt questioned him in regard to these statements, the infectious disease specialist became confrontational and told Kurt to prove him wrong. He challenged Kurt to find scientific and medical documentation that proved mold could affect the thyroid. This challenge remained in the back of our minds as we later began researching— searching for solutions to our still-remaining fungal-induced health problems.

Dr. Gray also has encountered infectious disease professionals who minimize the extent to which fungal exposure can negatively affect health. He states, "The real issue that is debated is: 'Are they pathogens?' Many of the members of the infectious disease communities will say, 'Well, we're not concerned about that. You found *Aspergillus* or *Penicillium* in the sputum or you found it in the stool, or you found *Stachybotrys* DNA in either the secretions from the lungs, the sputum, or you grew it out of the lungs—but it's not a pathogen.' Well, they're saying that because it is unlikely, but not impossible. We have seen *Aspergillus* nodules and other illnesses associated with it, such as hypersensitivity pneumonitis and a variety of other things that are clear pathologic processes in the presence of these molds and fungi; but, generally, they don't cause the kind of infection that we usually associate with sores and lesions in the skin or pneumonic processes that can be present and not causing a pneumonia. But the question is: If they are secreting toxins that are terribly destructive to multiple organ systems, is it reasonable to not consider them pathogenic? I believe that it is not. I believe that we have to pay attention to the toxic burden that is posed by the growth, colonization, or infection (or call it what you may) but the presence of these organisms in the system (generally in the lungs, but it can also be in the gut and it can also be in other

242 - MOLD: THE WAR WITHIN

tissues) and the generation of illness because the toxins are so toxic and so damaging. I think that that needs much more attention. We need to recognize that they are causing disease."[4]

To learn, after the fact, that other specialists of the infectious disease community also adhere the same illogical beliefs as those held by Kurt's infectious disease specialist confirmed the existence of a much larger problem: lack of up-to-date medical training and education at medical teaching facilities and/or lack of adequate continuing education. As we discuss the treatment choices prescribed by Kurt's infectious disease specialist, the extent of the doctor's illogicalness is further demonstrated. He prescribed Kurt two antifungals, Lamisil and nystatin. He gave Kurt several sample boxes of Lamisil that had the instructions and patient warnings removed from the inside of the boxes. A prescription was written for the nystatin.

Kurt questioned the doctor about which species of mold(s) these prescription antifungals were supposed to kill and whether there were tests that could identify the mold(s) to which he (Kurt) had been exposed. The doctor said this type of testing was too expensive. He also claimed it was too late to run these tests and asserted they don't always show which fungi are present. The MD stated he was giving Kurt these two antifungals because, together, they would kill whatever fungus to which Kurt had been exposed. He then claimed that Lamisil kills them all! He said he could give Kurt a stronger antifungal, but he would have to check Kurt into the hospital to administer it intravenously. However, he said he didn't think a stronger antifungal was necessary.

The doctor assured Kurt that after taking these two antifungals for one week, he would be as good as new! He said if that didn't work, he would run liver tests, wait a week, and then have Kurt take two more weeks of the same medications. The doctor again stated Kurt's thyroid irregularities could not possibly be mold induced. He believed these thyroid fluctuations were due to stress.

While filling the nystatin prescription, we asked the pharmacist for a printout of the warning literature for Lamisil. When we told the pharmacist that Kurt was being treated for an acute sinus reaction from mold exposure, she had the most unguarded reaction we have ever seen in a pharmacist. She looked bewildered and said, "Why would he be treating you with Lamisil? It is for toe fungus! And nystatin is used for *Candida*—overgrowth of yeast." She adamantly stated that neither of these antifungals was designed to treat a sinus fungal infection. When she asked if the doctor ran liver tests to get a baseline reading, we started to understand the toxicity level of antifungals. The pharmacist explained that baseline liver tests are recommended with antifungals so they can be used as a point of comparison after a month of treatment. She said that the doctor was probably getting around these liver tests by prescribing the Lamisil in two week intervals with one week off in between. We spoke with the other pharmacist on duty, who concurred that liver testing is usually conducted. He also stated that Lamisil usually is prescribed for a three-month period. We filled the nystatin prescription and went home to read the clinical pharmacological literature on both drugs.

The nystatin literature provided by the pharmacist stated, "Regular nystatin is not used to treat systemic fungal infections because of its negligible absorption from the gastrointestinal tract . . . Although nystatin is used orally, it is poorly absorbed from the GI tract. Following oral administration, it is almost entirely excreted in feces as [an] unchanged drug."[5] Additionally, it stated, ". . . [Nystatin] is effective only against *Candida* . . ."[6] To the best of our understanding up to this point, the literature told us that nystatin would be effective only in treating the *Candida* that had become overgrown in the intestinal tract. However, there was other information listed on the pharmaceutical printout regarding nystatin that we did not understand.

We asked the pharmacist and our ND about this hard-to-understand information, but both of them claimed to not understand it either. Regardless, both assured us that nystatin was safe and harmless. Furthermore, our ND reassured us that the naturopathic clinic routinely prescribed nystatin for *Candida* infections and that no harmful effects had been experienced by other patients. Based on these assurances and because the yet-to-be deciphered side effects listed on the clinical pharmacology printout were not listed on the regular patient handout, we decided—without fully understanding all the warnings on the pharmacological report—that Kurt should take the nystatin.

We thought—best-case scenario—the nystatin would eliminate the portion of Kurt's symptoms caused by yeast overgrowth and—worst-case scenario—the nystatin would do nothing. However, unknown to us at the time, the then-indecipherable drug warning had far reaching health implications. Furthermore, we later learned from a pharmacist at a national chain pharmacy that not all side effects of a drug are listed on the pharmacological report. For example, Avelox has 75 known side effects, yet only those required *by law* by the FDA will be included on the pharmacological report. Thus, patients are not being fully informed of all the potential health risks by the patient information sheets that accompany filled prescriptions or by the pharmacological reports, should they ask for them.

We did our best to carefully evaluate the Lamisil literature as well. It stated, "Laboratory and/or medical tests such as liver function tests may be done to monitor your progress or to check for side effects. This medicine may cause liver problems."[7] This printed forewarning echoed the warnings from the pharmacists and explained why the doctor had said he would intersperse treatment with two weeks on and one week off. The literature also confirmed what the pharmacist had told us regarding Lamisil. It stated, terbinafine (the generic name for Lamisil) ". . . is highly effective for treating onychomycosis [nail fungus] due to its fungicidal activity and ability to concentrate within the nail . . ."[8] Essentially, Kurt was risking liver damage to get rid of just nail fungus when he had fungi growing everywhere!

The doctor had prescribed a liver-toxic drug that was documented as being effective only in treating nail fungus—not the more widespread, fungal-related illnesses of the sinuses, lungs, and skin and imbalances in the intestinal tract. Not only did his actions not make sense to us, but they made us wonder if this particular doctor was grossly incompetent, had carelessly prescribed a drug about which he was uneducated, or had some alternative agenda of his own. For example, pharmaceutical companies often work hand in hand with doctors, researching new applications for existing drugs. Perhaps that was the situation in this case, as the pharmacological literature stated, "Current studies are evaluating the use of terbinafine in other mycotic infections."[9] Needless to say, we saw no scientifically or medically based evidence to support Kurt's taking the Lamisil.

During this same time frame, Kurt's thyroid mysteriously reverted back to a hypothyroid (low thyroid) state. He had started seeing another endocrinologist in hopes of finding a doctor knowledgeable about the effects of fungal exposure on the thyroid. She was skeptical about the thyroid problems being related to fungal exposure but was willing to take a more conservative, non-surgical approach. She said she would welcome reading any documentation that Kurt could provide regarding fungal exposure and thyroid function.

Shortly thereafter, a nurse practitioner prescribed an antibiotic for me because of a painful ear infection. When it did not respond to the antibiotic treatment, we suspected it was a fungal-related condition, a fact later confirmed when the ear pain went away after successful treatment of the fungi. At the same time as when the antibiotic was prescribed, the nurse practitioner (at the request of our ND) also prescribed nystatin for me to prevent further overgrowth of yeast, which could be caused by use of the antibiotic. Antibiotics destroy good intestinal organisms that normally keep harmful levels of other organisms in check. With less population of good organisms, an overgrowth of yeast and other harmful organisms can occur. Our ND believes that if you take an antifungal along with an antibiotic, the damage to the intestinal flora is minimized, because the antifungal will keep the undesirable overgrowth of yeast under control.

After a week on the nystatin, I noticed that the yeast covering much of my skin was disappearing. We discussed this discovery with our ND, which led to another prescription of nystatin—one for each member of our family. We all wondered: Would the nystatin be effective also in treating the other types of fungi that were flourishing in our bodies? Maybe the infectious disease specialist knew more than what was documented on the nystatin pharmacological printout. Maybe nystatin was effective against forms of fungi other than *Candida*.

We had no qualms about having the whole family on nystatin because we had repeatedly been told that it was harmless—so harmless that it is used even on infants for treatment of thrush. In fact, William Crook, MD, reports that the *Physician's Desk Reference* (PDR) states, "Nystatin is virtually nontoxic and nonsensitizing and is well tolerated by all age groups, including debilitated infants, even on prolonged administration."[10]

Furthermore, when we checked with a second pharmacy, we were told by their head pharmacist that the doctors from the naturopathic clinic at which we were receiving treatment leave patients on nystatin for years. This set off alarm bells deep inside of us, but in our desperate desire to regain our health, we listened to the professional word-of-mouth and written propaganda espousing the harmlessness of nystatin— misinformation for which our health later paid a price. We had unknowingly given our entire family a toxic substance.

After being on nystatin for a month, we found the health of our family spiraling downward at an unbelievably rapid rate. Our health bottomed out and remained there for months. We poured over the nystatin clinical pharmacology printout again and again, trying to decipher the medical terminology that, so far, no medical or pharmaceutical professional could explain. The cause of our health dropout was right in front of our eyes on a piece of paper, yet the information was not ascertainable without interpretation.

The terminology we were trying to understand reads as follows: "Nystatin . . . is nearly identical to amphotericin B in structure; both agents are polyene antifungals. . . . Like amphotericin B, nystatin binds to sterols in the cell membranes of both fungal and human cells. Nystatin is usually fungistatic *in vivo* but may have fungicidal activity at high concentrations or against extremely susceptible organisms. Nystatin has greater affinity for ergosterol, the sterol found in fungal cell membranes, than for cholesterol, the sterol found in human cell membranes; however, nystatin is too toxic to be used systemically. As a result of this binding, membrane integrity of both fungal cells and human cells is impaired, causing the loss of intracellular potassium and other cellular contents. . . ."[11]

Not until we spoke with Dr. Straus, a cell biologist, did we understand the meaning of the above warning. He explained, from a biological standpoint, what actually takes place in the human body when taking certain antifungals, such as nystatin, and why they have the potential to be toxic to the human body. He states, "Let me give you a good example of what is and isn't toxic. Penicillin is a wonderful antibiotic that we use against bacteria. The reason penicillin works so well against bacteria and doesn't harm humans is because penicillin attacks the cell walls of bacteria, which are entirely different from our cells. Human cells and bacteria cells are quite different."[12]

He further explains, "The reason that fungal antibiotics are so toxic is that fungal cells and human cells are more similar than bacterial and human cells. Some fungal antibiotics attack the cell membranes, and when they do, they also do damage to human cells. . . . Some of these antifungals attack the fungal cell membrane, and because the fungal cell membranes and our cell membranes are similar (we have cholesterol in our cell membrane and fungi have ergosterol, which are both sterols) when you attack one with an antibiotic, you have the potential to attack the other as well. That's why you have to be very careful when you take antifungals and have to be closely monitored because they have the potential to damage human tissue."[13]

It is quite disconcerting to realize that an MD, an ND, and a total of four pharmacists from two different pharmacies gave us incorrect information regarding a drug. To test our theory that these pharmacists were parroting pharmaceutical rhetoric instead of reading the actual clinical pharmacology printouts, we again called one of the pharmacies—a nationally known chain pharmacy that everyone has probably used at one time or another—and inquired about the side effects and associated risks of nystatin. We were told, again, that nystatin is so safe—harmless—that it is used for treatment of thrush even in newborn babies! We asked for the clinical pharmacology printout, went and picked it up, and showed it to the pharmacist, pointing out what Dr. Straus had explained. The pharmacist was stunned. He stammered, "Well, how can that be? We—uh—we—use this on babies for thrush." He then stated that he had never read the clinical pharmacology printout. He said, "*They* always say it is harmless." The "they" were not identified—possibly, the pharmaceutical representatives who sell nystatin or the doctors who prescribe nystatin.

Because of this experience, we realize that it is imperative to not just check with a pharmacist regarding a drug's toxicity, but also request a printout of the clinical pharmacological documentation to confirm the information provided by pharmacists and prescribing doctors. Furthermore, we must keep seeking clarification of pharmacological literature until we fully understand all the written warnings! This experience reinforces that no matter how many doctors and pharmacists give assurances, we must be diligent and verify verbal information with printed documentation: patient information sheets, pharmacological reports, and product insert literature from the pharmaceutical manufacturers, which includes disclosures designed to legally protect the pharmaceutical companies. Make this level of research a routine step when filling a prescription and safeguard the health of you and your family from the cost of medical and pharmaceutical misinformation.

Sadly, many of us spend hours shopping for just the right outfit or the perfect vacation destination, yet we spend little to no time researching the medical impact of prescription drugs on our health; we just pop them in our mouths, go about our life, and wait to "automatically" get better. We each, with our families, ultimately live the repercussions of doctors' decisions regarding our health. We'd better make sure we agree with the chosen treatment plan by researching not only the drugs that have been prescribed and the procedures that have been recommended, but also the possible associated side effects. This level of thoroughness must also be implemented with alternative treatment options. We all want to believe that doctors are infallible because, at times, they literally hold our lives in their very hands. We must, however, realize that we're all capable of making mistakes and misjudgments—even doctors and even pharmacists.

Researching the effectiveness and safety of treatment options can help avoid costly health choices. Before taking any prescribed drug or undergoing any medical procedure, we should inquire about the effectiveness of the pharmaceutical being prescribed and/or the procedure being recommended. Specifically, we should ask each prescribing physician: Based on your own clinical experience, what is the effectiveness of this drug and/or procedure? You may be surprised as the answers may not always be the pharmaceutical rhetoric you expect to hear.

For example, many doctors had recommended a course of treatments with broad spectrum antifungals, which are much stronger than the antifungal nystatin and carry warnings of liver toxicity. Because of potential damage to the liver, these antifungals are often prescribed with intermittent use and/or liver monitoring via blood tests. To collect enough data with which to make an informed decision for our family, we asked multiple doctors the same question: What is the percentage of effectiveness of these broad spectrum antifungals based on *your* clinical observations? It was because of the answers to this question that we determined the risks outweighed any potential benefits. We were repeatedly told by multiple medical doctors that, although more powerful, these broad spectrum antifungals do not always work. We were reassured, however, by the same doctors that if the first broad spectrum antifungal did not work, other broad spectrum antifungals would be prescribed until one did work!

The following medical case summary exemplifies this medical philosophy and illustrates the potential resistance of fungi to antifungal pharmaceuticals. The treating physicians reported, "Phaeohyphomycosis is a clinical entity caused by dematiaceous fungi. We describe a clinical case of phaeohyphomycosis due to *Cladosporium cladosporioides* in a 45-year-old white male, apparently healthy, human immunodeficiency virus-negative. The patient was treated with terbinafine [an antifungal drug] for nine months, with regression of a skin lesion. Three months after discontinuation of the therapy, there was a clinical and mycological relapse [fungal re-infection]. After progression of the disease with inadequate treatment, there was no response to amphotericin B [another antifungal drug] and flucytosine [yet another antifungal drug]. Finally, we obtained a clinical response with itraconazole oral [a fourth antifungal drug] solution at 600 mg day (-1) for a six-month period."[14]

It took 15 months of treatment with two antifungal drugs and an untold number of months of treatment with yet two more antifungal drugs before this man's infection from *Cladosporium cladosporioides* was successfully treated. It should be noted that all of these antifungal pharmaceuticals come with warnings of possible liver damage.[15] After taking these liver-toxic antifungals for more than a year, we must ask the degree to which (if any) these drugs compromised the man's liver.

It should be noted that *Cladosporium cladosporioides* was one of the species cultured from air samples collected in the New Orleans area post-Katrina.[16]

Many people, like us, are willing to share information that would possibly help others avoid costly health decisions, thus enabling the recipients of the information to make more health-enhancing choices. Even treatment plans that appear harmless can be deceiving, for example, sauna therapy. The naturopathic clinic where we were treated had an in-house sauna with a cold-plunge tub. The naturopaths advised patients to take a 10–15 minute sauna followed by a cold plunge; the process is repeated three times. The theory behind taking a cold plunge after a hot sauna is that it will shock your system, thus stimulating circulation.

The saunas, piggybacked with cold plunges, were invigorating at first, but after five or six sessions, my strength and health began to go downhill. We checked with the sauna attendant and our ND and were told that deteriorating health sometimes happens when a person has a lot of toxins. They explained that the heat from the sauna draws stored toxins out of the body's cells and fat tissues, and once disturbed, these toxins have to work their way out of the body via eliminatory channels, such as the skin, liver, and kidneys. In fact, some detoxification centers, such as the Environmental Health Center in Dallas, Texas, conduct laboratory testing in order to monitor the toxin levels in the blood and/or fat tissue of patients undergoing sauna and detoxification programs.[17]

Not all clinics using sauna therapy as a method for detoxification monitor patients via laboratory evaluation. Ours did not offer this service to us. In fact, at the time we were undergoing sauna therapy, we were not even aware that sauna detoxification could draw toxins out to such a degree that the increased level of toxins could cause damage to eliminatory organs such as the liver and kidneys.

Dr. Callaghan, who also uses laboratory testing at COEM, clarifies that sauna detoxification will not damage organs to the point of being fatal, but it can make patients feel "sluggish and crummy." For this reason, he states that he and Allan Lieberman, MD, medical director of COEM, use vitamin and mineral therapy to bind toxins for a safer and more comfortable elimination from the body. Dr. Callaghan explains that at COEM, testing is done both before and after patients undergo a four-week detoxification program in order to see if toxin levels have increased or decreased. He states that most oftentimes toxin levels in the urine increase. He says this increase in elimination of toxins is because the body is similar to a sponge, soaking up toxins. During a period of detoxification, the sponge (cells in the body) is "squeezed," which releases the toxins.[18] Unless these released toxins are bound with vitamins and minerals to ensure removal from the body, these toxins can recirculate through the body, defeating the purpose of detoxifying.

Possibly with extensive vitamin and mineral support, we might have had a different response to sauna therapy. However, without adequate vitamin and mineral support, after two more sauna sessions, both our daughter and I became extremely ill for months. Also, during the same two-week time period, our entire family had been taking the nystatin, so it is difficult to ascertain the degree to which each contributed to the decline of our health. Regardless, the result was that our entire family spent the first long, cold Montana winter constantly sick, battling one infection after another. We were in a continual state of sickness. We didn't get sick, begin to feel better, and then catch some other type of bug. It was a debilitating sickness, compounded by flus, throat infections, lung infections, and a list of other ailments. We weren't able to shake an existing infection before another one would crawl onto the back of our fragile health.

The two weeks of sauna detox, compounded by the nystatin, in retrospect, proved to be detrimental to an unquestionable degree. These two treatment methods greatly undermined what was left of our health. Not only did we have to start the battle all over again, but we were advancing forward with our health struggling at a new, low ebb. We were frustrated and discouraged—but not beaten.

The human spirit has an innate, underlying strength that can rally in moments we least expect and give us direction and inspiration from within. For this God-given ability, we are grateful. What could have been a moment of resolved failure instead was a source of motivation and drive. Deep down inside, we had the unshakable gut feeling that there were solutions to our health problems; we just had not yet found them. At this point, however, our naturopath told us that the current state of our health may be just how it was going to be—that we were not going to be able to recover and that we would have to just "adjust" our lives. She said this comment to me specifically in regard to Kurt, who had not regained any of the ground lost from the health-undermining nystatin we had stopped taking several weeks earlier.

Maddened and motivated by this defeatist attitude (and still under the care of our naturopath), we started to research alternative solutions. We would often sit and think, pressing into the recesses of our minds, trying to come up with any possible neutralizing agent for mold and mycotoxins. We scoured books, searched the internet, and eventually networked across the continent of North America, interviewing PhDs, MDs, NDs, MPHs, and other health and environmental specialists. We were on a quest for information that would again open the doors of health to us—a quest that led to the reams of research that became the catalyst for this book, the subsequent interviews, and the restoration of our family's health.

One by one, we began to implement changes, watching for effective results. We slowly improved as we built a healing regime, adding each new effective component until, to our joy and relief, our health not only returned but improved beyond where it had been before the hurricane!

We re-examined our diet yet again and cut everything except water, vegetables, and lean protein, scrutinizing every label for additives. (For example, many frozen vegetables now have added seasonings, sauces, and broths.) Once we were non-reactive on this even more restricted diet, we slowly began adding one food at a time to see if we reacted. We wanted to make sure we hadn't earlier introduced a food prone to fungi to which we might have been reacting, unwittingly re-exposing ourselves to fungal contaminants that could have been causing ongoing symptoms. We eventually found a few products that we were able to tolerate without a reaction of any kind, from which we built meals, which over time allowed us to expand our food selections.

Gone were the days of freely tossing items into the grocery cart without several minutes of label reading, deliberation, and family consensus. There is strength in numbers. When one is weak, the rest must be strong. We realized at this point that if we were to get better, we would have to keep our diet completely free of foods that didn't foster health. We focused on therapeutic whole foods that would give our bodies the nutrients to rebuild a strong immune system and eliminated foods containing caustic agents such as sugars, additives, antibiotics, hormones, fungi, and mycotoxins.

During this same time period, we had a phone consult with a medical environmental specialist who shared that garlic and oregano possess antifungal properties. Interestingly, we had already been researching for several weeks the healing properties of garlic in relation to fungi. For years, we had used a concoction of garlic and onions to ward off both viral and bacterial infections. For years, it worked successfully—until Hurricane Katrina. We had tried our trusted concoction while in the shelter the first two weeks after evacuation, but the home remedy that used to keep us sniffle-free had been completely ineffective.

Encouraged, however, by the newly corroborated information from the environmental specialist, we started experimenting again with garlic. We thought maybe the previously tried water-based garlic and onion concoction hadn't been strong enough, so we tried making a batch of garlic oil. In the past, garlic oil had proven stronger when it successfully treated two of our family members who had become ill after two weeks of mold exposure years earlier. I made an especially strong batch with extra-virgin olive oil. I sealed it in a glass jar and left it on the counter. In a few days a bubbling mass of oil, garlic, and who-knows-what-else was flowing from the jar all over our counter. In the past, I would have simply drained the oil off and stored it in a brown glass bottle, but the intensity of this bubbling action was different; it had me concerned. I researched the matter on several food industry and governmental websites and found that multiple cases of botulism had been reported with this type of homemade garlic preparation.

We had wanted to try treatment with garlic oil because a study we had read stated that garlic extract in oil may be more effective in fighting mold as a remarkable increase in the activity of garlic extracts was noted when mixed with oil.[19] But, not wanting to risk possible poisoning from botulism, we tossed out the batch of garlic oil and moved on to chewing raw garlic cloves, swallowing spoonfuls of chopped fresh garlic, and experimenting with garlic freshly juiced in a commercial-strength juicer at a local juice shop, The Funkey Monkey. We were laughing at ourselves the whole while, but also wondering if we would get any results.

As part of our search, we spoke with David Christopher, master herbalist and director of The School of Natural Healing.[20] The school offers many correspondence courses on herbal medicine and books with herbal remedy recipes, three of which had been instrumental in Kurt's avoiding knee surgery from a torn ACL (anterior cruciate ligament) several years earlier. In regard to our current health dilemma, Mr. Christopher suggested essential oils, garlic, and onions. We were already experimenting with onion and copious amounts of garlic; we also had already been using an essential oil nose ointment to help reduce sinus swelling, which made it easier for us to breathe.

Health literature states, "Essential oils are highly concentrated extracts—typically obtained either by steam distillation or cold pressing—from the flowers, leaves, roots, berries, stems, seeds, gums, needles, bark, or resins of numerous plants. They contain natural hormones, vitamins, antibiotics, and antiseptics. Known also as *volatile oils* because they evaporate easily in air, essential oils are soluble in vegetable oil, partially soluble in alcohol, and not soluble in water. Because they are so concentrated, they are likely to irritate mucous membranes and the stomach lining if taken internally. It is, therefore, best to use essential oils externally only, such as in poultices, inhalants, bath water, or on the skin (a few drops)."[21] We eventually found two essential oils to be quite beneficial in soothing and healing our hives and rashes, specifics of which we will share in the next chapter.

We have been proponents of vitamin and herbal supplements and have used them as a means of preventative health care for nearly a decade before Hurricane Katrina. We have always viewed vitamins and herbs as a big shield that helps protect our health and reduce the toxic toll on our bodies from exposure to the multiple contaminants in the world. For years we had researched vitamin and herbal remedies, a background which proved very helpful at this point. We began to search reference books we had used for years, such as the *School of Natural Healing* and *Prescription for Nutritional Healing*,[22] the latter of which is an in-store reference book at many health food stores. We also found helpful books with which we were previously unfamiliar, such as the aforementioned *Alternative Medicine: The Definitive Guide*,[23] which gives thorough, easy-to-understand explanations of how various systems work within the body and supplies specific information on natural treatment options for various health conditions.

Over the years we have also appreciated the availability of David Christopher to answer questions regarding natural healing. For those interested, every Monday through Thursday from 1 to 2 p.m. Mountain Time he takes phone calls—free of charge—from the public regarding natural remedies. To speak with Mr. Christopher during this allotted time period, call The School of Natural Healing at 800-372-8255. For more information on available correspondence curriculums, see www. schoolofnaturalhealing.com. Once on the school's website, you can also click through to the website of Christopher Publications, which features many herbal healing books. The one we have found to be an excellent resource, often containing information not found in other books, is the earlier mentioned *School of Natural Healing* by John R. Christopher, MH, ND,[24] who was David Christopher's father. For readers wanting to explore natural healing options, The School of Natural Healing is extending a 20 percent discount on the book *School of Natural Healing* and on all other publications by Dr. Christopher to those who identify themselves as readers of *Mold: The War Within* when placing their order via the toll-free number.

Having been encouraged by multiple sources about the natural healing properties of certain foods, vitamins, and herbs, we began a systematic process of trial and error. We continued to scour the aforementioned books, as well as many, many others, looking for possible answers. Additionally, we searched scientific and medical data banks and conducted many first-hand interviews with scientific and medical professionals. We share with you the documentation from these multiple sources that address the individual healing components that ultimately became our treatment plan: vitamins, herbs, and other health supplements along with a diet free of sugar, fungi, antibiotics, hormones, and chemicals as much as possible. We share not only what worked, but why. Please be aware that most of these vitamins, herbs, and nutritional supplements have multiple properties and applications. We focus on the properties and treatments that were applicable to our particular health situation.

As already mentioned, we experimented extensively with garlic. Untold numbers of books, articles, and sections within books exist that address the healing properties of garlic. In summary, garlic is an antifungal,[25] a stimulant for the immune system, a natural antibiotic,[26] a powerful detoxifier,[27] and an antiviral agent.[28] It helps clear congested lungs, coughs, bronchitis, and sinus congestion, is a preventative measure for colds and flu,[29] and kills parasites.[30]

In our search for a super powerful garlic supplement, we tried a brand that documents as an ingredient 180 mg of a patented, extracted allicin powder, which is believed to be the substance responsible for garlic's antibacterial properties.[31] Because of this belief, the manufacturer of this particular product formulated and patented an extraction process that is said to extract the allicin and stabilize it. It sounds good, but it didn't work for us—zip, zilch, nada.

The ineffectiveness of the allicin-extracted powder may be because there are several other active properties present in garlic, including 33 isolated sulphur-containing compounds.[32] The synergistic effects between compounds is said to also contribute to the healing properties of garlic. Thus, when one active compound is isolated from the others, the integrity of the healing properties of the garlic may diminish or be lost.

We also tried a well-known and highly advertised deodorized garlic supplement but did not find it effective either. Again, it is possible that the integrity of the healing properties of garlic is diminished or lost when the odor is removed. Interestingly, health literature states that the odorous factor in garlic is the allicin.[33] Eventually, we did find a garlic supplement that proved extremely effective, which we will share as our story unfolds. We now use this supplement as needed and use raw garlic in cooking and juicing. Gone are the days of chewing whole garlic cloves!

We also experimented with ginger. Since adding external moisture and heat had worsened our condition, which we discovered from the saunas, we wondered if a drying agent taken internally would help deprive the fungi of the moisture necessary to grow. Fresh ginger root can be purchased at almost any grocery or health food store. Health books state ginger is helpful to the respiratory and digestive tracts in dislodging congestion;[34] is good for colds, chronic bronchitis, and coughs;[35] and is an antifungal and antibacterial agent.[36] While we were sick, we used ginger daily. Now we cook with it, juice with it, and take supplements as needed. We use both fresh and encapsulated ginger powder.

Another herb we found helpful was cayenne. It acts as a catalyst for other herbs;[37] helps ward off colds, sinus infections, and sore throats;[38] and is an antibacterial agent.[39] We used cayenne daily when we were sick to heighten the properties of the ginger and garlic.

We started using goldenseal as an immune support. We had used it in the past for colds, sore throats, and other infections. We did not expect this herb to do anything more than it had done in the past—support the immune system. However, the drying properties of this herb turned out to be very beneficial. Literature states that goldenseal is an excellent drying, mucus-reducing remedy for the upper respiratory tract, nasal inflammations, or ear infections.[40] It fights infection; strengthens the immune, lymphatic, and respiratory systems;[41] is a remedy for colds and upper respiratory tract infections; and has antibiotic and antimicrobial properties.[42] Goldenseal contains alkaloids, such a berberine, which has a broad spectrum of activity against a variety of microorganisms, including bacteria, protozoa, and fungi, specifically *Candida albicans*.[43] Because of this ability to inhibit the growth of *Candida*, literature states that goldenseal can be used to prevent the overgrowth of yeast that often accompanies the use of antibiotics.[44] Furthermore, berberine has been shown to ". . . activate macrophages (cells that digest cellular debris and other waste matter in the blood)." [45]

We used goldenseal in combination with ginger, cayenne, and garlic. This combination worked so well that it enabled Kurt to breathe after incidences of fungal re-exposure (either from fungal contaminants in the air or food) would trigger asthmatic attacks. In our opinion, goldenseal is an amazing herb. We still use it as needed, but sparingly. It is important to note that documentation states not to use golden seal for prolonged periods.[46] For this reason, we reserved the use of goldenseal for times of respiratory distress. Had unrestricted use been possible, goldenseal might have brought long-term health improvements, but as it was, it served as an excellent crisis intervention herb.

Long before Hurricane Katrina and our fungal exposure, the herb yarrow had proven helpful to our family. We had used yarrow-infused baths when our children had chicken pox to open the pores in their skin, which allowed the poisons from the virus to be drawn out. This drawing herb sped up the scabbing and healing process. Our thought was, if yarrow could work to draw out the toxins from chicken pox, why wouldn't it work in the same manner to draw out fungal toxins through our bleeding hives? Yarrow is documented to open pores, allowing the release of toxins. [47] We limited the yarrow baths to no more than five to ten minutes, once or twice a week, in order to not draw out too many toxins at once.

Up to this point, we were using the garlic, ginger, cayenne, and goldenseal (with periodic breaks) in combination with yarrow-infused baths, as needed. This regime eventually became an important support remedy, but alone it was no more than management of symptoms. We had not yet found a treatment plan that supplied our bodies with adequate "ammunition" with which to conquer the fungi and enable us to march to victory, reclaiming full restoration of our health.

We continued to search for answers. We still had a deep, overriding gut feeling that there were attainable solutions. We just had not come across them yet. We were determined to unearth what could possibly even be an "easy" solution to such a myriad of miseries brought on by fungal and chemical exposures. As we watched our children, sick from respiratory-related illnesses, full of rashes and bleeding hives, hitting bottom right along with us after taking the nystatin, we refused to believe that our current health was how it was going to be for the long haul, as our naturopath had suggested. We were also aware at this point that poisoning from mycotoxins could possibly lead to cancers and degenerative diseases. These bleaker realities instilled in us an intense level of motivation and drive. We also thought of the untold number of people battling similar health conditions from fungal and chemical exposures, knowing full well the toxic medical options that doctors would be offering them. We wanted not only to identify solutions that would restore the health of our family, but also to document for others suffering from mold and chemical exposures the information that we had found to be of value.

As we researched the mold issue, we came to understand that fungal exposure can silently affect people's health in normal, everyday living conditions. We started to read about sick building syndrome, building related illnesses, and SIDS and how these conditions may be caused by fungal exposure. The overwhelming magnitude of the extent to which mold can affect human health fueled our drive to restore our health to pre-exposure vitality and then share the information with others so they could learn from our research, draw their own conclusions, and potentially benefit as well. This mission was the catalyst behind this book.

Much of what ultimately ended up being effective for us began through meeting certain individuals. These "chance" meetings could be viewed as fate, manifest destiny, God-directed, pure coincidence of being in the right place at the right time, or whatever you want to call it; regardless, meeting people who cared enough to share their knowledge, and in some instances products, enabled us to formulate an effective treatment plan that restored our family's health. It is an amazing story to look back on in the sense that what if we hadn't made a particular phone call or gone into a certain store that ultimately led to the discovery of an effective component of our treatment plan? We might not be where we are today, writing about what effectively worked for our family, because we might still be sick.

One of these caring, knowledgeable individuals is Julie Gobin, the owner of The Funky Monkey (juice shop) in Billings, Montana. She shared some information that led us to juice a combination of vegetables and herbs, which proved powerful and effective and restored approximately 50 percent of our health after just several weeks. "All from juicing?" you might ask. We wouldn't have thought this alternative solution would have been so effective either; we are grateful to Julie for sharing her knowledge.

Juicing is often thought of as something reserved for people who are super health conscious. Prior to personally experiencing the benefits of juicing, we thought, "Why juice? We'll eat the entire vegetable, which includes the fiber. There's got to be more nutritional value in eating the whole vegetable, right?" In fact, the only reason we started juicing was because our naturopath had suggested we experiment with fresh-juiced cabbage as a treatment option for Kurt's hyperthyroid condition. (Kurt's thyroid had reverted back, yet again, to a hyper state.)

One night we were in The Funky Monkey talking with Julie, asking her how to dilute the fresh garlic juice that one of her employees had juiced for us a few nights earlier. We were trying to formulate a potent herbal remedy of fresh-juiced garlic, onion, and other herbs to support our immune systems, as we were constantly getting sick. After a short visit with Julie, we left the juice shop with a new understanding of juicing. First, fresh-juiced vegetables, fruits, and herbs should be consumed within the first 15 minutes after juicing. It is during this short time period that the live enzymes are still "alive."[48] If you let a fresh-juiced drink sit or store it in the refrigerator, it nutritionally becomes a "dead" food. Second, there is no need to juice a large amount of fresh garlic in a commercial juicing machine to use in an herbal remedy because the garlic will be so strong you won't be able to use it! Third, just throw a palatable amount of fresh garlic, onion, and other vegetables into the juicer each time for a freshly juiced drink. These concepts may seem elementary, but for people, like us, not familiar with the practice of juicing, these beginner points were necessary.

Another night we stopped in at The Funkey Monkey and were most intrigued by a matter-of-fact comment Julie made. She said that fungi don't grow in an alkaline environment, for the most part. Figuratively speaking, we were hit by a concept we had read about years earlier: Alkalize or die. The theory is to raise the pH of the body to an alkaline state through consumption of alkaline foods or alkaline-based supplements. The belief is that viruses and bacteria can't flourish in an alkaline biological terrain. What a theory—raise the level of alkalinity and pH in the body by consuming an alkaline diet, and thus kill internal fungi!

Julie shared a story about a man sick from leukemia who consumed a diet solely of carrots to increase his body's alkalinity as a part of his treatment plan. As the story was told to us, the man turned orange from eating so many carrots but beat his leukemia! We found this story very intriguing because mycotoxins have the ability to alter DNA, which is essentially what cancer cells are—cell mutations. The fact that carrot juice reportedly aided in curing a form of cancer was not lost on us. Since certain mycotoxins are carcinogenic, we thought, "Maybe there's something to this carrot juicing."

Our primary focus up to this point had been to "starve" the mold by withholding sugar and carbohydrates. This, to a large degree, had turned our diet into a base of vegetables and lean proteins, which was a combination of alkalinity from the vegetables and acidity from the proteins. Although this nutritional approach may have limited the spread of the fungi, it was not effective by itself in ridding the body of existing fungi and mycotoxins. We were eager to see if a more alkaline-based diet would make any noticeable changes. Additionally, the alkalinity theory made sense, as most fungi do grow in an acidic environment, according to Dr. Chew.[49] It is of no surprise then that anti-*Candida* diets are primarily alkaline. As for the looming possibility of cancer, if fungi and mycotoxins are successfully treated and removed from the body using diet and food supplements, then the catalyst to cell mutations is also gone.

We bought a juicing machine and began juicing. We originally chose a Champion juicer because of its durability. An acquaintance who is an ND had once told us that the Champion juicer would withstand the daily rigors of juicing carrots. He said that daily juicing of carrots would literally tear the basket-style juicers to shreds. Since our recovery, we have switched to an Omega Juicer, as we have found that it produces a much cleaner juice. For more information on Omega juicers, see www.OmegaJuicers.com.

Armed with our juicer, we started juicing fresh carrots—a very alkaline vegetable —with fresh garlic and ginger. We juiced morning, noon, and night, drinking anywhere from 8 to 12 ounces of our carrot concoction at each juicing. Needless to say, we juiced a lot of carrots, because it takes several carrots to make just one glass of juice! We found Costco to be a great source for our increasing produce needs. The ability to buy in bulk at Costco helped make our fungal-free diet affordable. In fact, we were such frequent shoppers at Costco, shopping every three to four days for carrots (and other fresh produce), that Costco employees eventually asked us if we had horses! For Costco locations, see www.costco.com/warehouse/locator.aspx.

It is interesting that even though carrots contain sugar, we did not experience any kind of "sugar rush" from the freshly juiced carrot juice, which was quite surprising since we had been on a diet void of sugar and most carbohydrates. In fact, at the time Kurt was still having trouble eating and keeping food down. He was not only able to tolerate the carrot juice, but consuming the alkaline, liquid nutrients was a major turning point for his health—for everyone's health.

After three months we slowly reduced our carrot juice intake from 8 to 12 ounces three times a day down to 8 ounces twice a day and eventually down to 4 to 8 ounces once a day. The sign that indicated a need for these reductions was slight sugar headaches. Once we would reduce our consumption of carrot juice, the sugar headaches disappeared. When the sugar headaches reappeared, we reduced our consumption again. Although we can't explain physiologically what caused the change in the amount of carrot juice we could tolerate, it appears in the beginning our bodies needed and used all the nutrients in the carrot juice, whereas when we began to experience the slight sugar headaches, our bodies no longer needed as much. These types of adjustments require "listening" to your body. Now that we are healthy, we continue to juice at least three times a week. It is a big favorite of our entire family. In fact, we were so amazed by the results we experienced from juicing carrots that we did some research to find out why juicing, specifically carrots, was so effective given our health condition. First, let's take a look at what we found in regard to the benefits of juicing in general.

Juicing is a fascinating medicinal aid. "Fresh fruit and vegetable juices contain a broad array of vitamins, minerals, enzymes, and various co-factors that both enhance and complement individual nutrients, so your body gets the most good from them," according to Dr. Bernard Jensen of Escondido, California.[50] Important to understand is that eating the same quantity of raw carrots, garlic, and ginger would not have delivered the same results. This difference is ". . . because juices are assimilated with very little effort on the part of the digestive system, [and] their nutrients have a health-building impact at a relatively low cost in energy."[51] In other words, the carrot juice delivered valuable nutrients to our bodies without our bodies having to work for them!

Furthermore, "Fresh juice is more than an excellent source of vitamins, minerals, enzymes, purified water, proteins, carbohydrates, and chlorophyll. Because it is in liquid form, fresh juice supplies nutrition that is not wasted to fuel its own digestion as it is with whole fruits, vegetables, and grasses. As a result, the body can quickly and easily make maximum use of all the nutrition that fresh juice offers."[52]

For this reason, juicing can be a vital source of nutrients for people who are too ill to eat and/or unable to keep their food down, as was the case with Kurt. According to Dr. Christopher, "Fresh herbal juices expressed from leaves or roots are probably the best medicinal aid when available. . . . Juices are far superior to plain water in supplying needed liquid to the body, because the appetite diminishes. Juices do not wear out and tire the body in digestion; they are in composition similar to a transfusion. They go almost directly into the blood stream, while furnishing rapid nutrition. Juices are especially valuable because the growing herb filters out all the inorganic poisons (such as the chlorine and fluorine) in the water."[53]

This ease of digestion of fresh juices is echoed by Dr. Erasmus, "[Fresh juices] are mineral-, vitamin-, and enzyme-rich and require little energy to digest; our cells can use energy not needed for digestive functions—which normally consume a great deal of it—for healing and housecleaning our bod[ies]."[54]

The health benefits of juicing are widely documented and should not be underestimated. "Juicing is an excellent means of adding fruits and vegetables to your diet. Since juice contains the whole fruit or vegetable—except for the fiber, which is the indigestible part of the plant—it contains virtually all of the plants' health-promoting components. Because fresh juices are made from *raw* fruits and vegetables, all of the components remain intact. Vitamin C and other water-soluble vitamins can be damaged by overprocessing or overcooking. Enzymes, which are proteins needed for digestion and other important functions, can also be damaged by cooking. Fresh juice, however, provides all of the plants' healthful ingredients in a form that is easy to digest and absorb. In fact, it has been estimated that fruit and vegetable juices can be assimilated in twenty to thirty minutes."[55]

Now let's take a look at what we found in regard to the health benefits of carrots and carrot juicing.

"Carrot juice is the king of the vegetable juices. Extremely high in pro-vitamin A, which the body converts to vitamin A, it also contains vitamins B, C, D, E, and K, as well as the minerals calcium, phosphorous, potassium, sodium, and trace minerals. The alkaline minerals, especially calcium and magnesium, contained in carrot juice help soothe and tone the intestinal walls. At the same time, these minerals help to strengthen bones and teeth. Skin, hair, and nails benefit from its high protein and mineral content. Fresh carrot juice stimulates digestion and has a mild diuretic effect. But perhaps its most important contribution to body health is in its tonic and cleansing effect on the liver. Through regular use, carrot juice helps the liver to release stale bile and excess fats. When fat levels are reduced, cholesterol levels are reduced."[56]

Most importantly, "[w]hen eaten raw, carrots are efficient colon-cleansers, which tone the bowel, reduce the re-absorption of estrogen, and lower cholesterol."[57] These benefits could be significant contributing factors as to why fresh-juiced carrots revitalized our health by an estimated 50 percent, as ". . . some of the fungal mycotoxins are very estrogenic," according to Dr. Thrasher.[58] The importance of this nugget of information bears repeating. Carrots, when eaten or juiced raw, reduce the re-absorption of estrogen. Therefore, carrots, when eaten or juiced raw, could reduce the re-absorption of estrogenic compounds such as the estrogenic mycotoxins.

Although freshly juiced carrots will not contain the complete fiber of the vegetable, the juice will still contain some fiber. Additionally, if more fiber is desired, some of the pulp can always be stirred back into the juice, making it thicker. Fiber from carrots ties up bile acids, cholesterol, and toxins and carries them out of the body, lowering cholesterol levels and reducing risk of cardiovascular disease, according to Dr. Erasmus.[59]

Juicing is an easy way to increase antioxidants. "An antioxidant is a natural biochemical substance that protects living cells against harmful free radicals. Antioxidants readily react with oxygen-breakdown products and neutralize them before oxidative damage occurs."[60] In other words, "[a]ntioxidants are natural compounds that help protect the body from harmful free radicals. These are atoms or groups of atoms that can cause damage to cells, impairing the immune system and leading to infections and various degenerative diseases such as heart disease and cancer. Antioxidants, therefore, play a beneficial role in the prevention of disease. Free radical damage is thought by scientists to be the basis for the aging process as well."[61]

Many of us may not realize that "[c]ooking destroys a part of the vitamins in foods and leaches minerals."[62] So if we want to obtain the maximum nutritional value possible, it is vital to eat vegetables raw. Dr. Erasmus, a strong proponent of eating foods in the form that is as close to nature as possible, states, "Remember that the more vegetables are eaten raw, the better. . . . This is because raw foods contain enzymes that help digest these foods and whatever else we eat. Our digestive glands need to secrete fewer enzymes to do the work of digestion. When foods are cooked, these enzymes—being protein—are destroyed (denatured). Our digestive system must then produce more digestive enzymes, increasing its workload and wearing it out faster."[63]

How many of us throw vegetables into a pan, turn on the burner, and heat them until they are boiling hot? What we are doing is destroying much of the valuable nutrients—the vitamins and enzymes—of the vegetables. This nutrient-reducing process helps explain the positive results we experienced from drinking freshly juiced raw carrots—all the vitamins and enzymes were intact. Dr. Erasmus states, "It is surprising how much better we start to feel when we eat more raw foods or their fresh juices. Carrots and dark green leafy vegetables are especially good for us."[64] Both are also high in beta-carotene.[65]

Dr. Erasmus continues, "Much of nature's supply of antioxidants is removed from foods by processing practices geared to convenience, taste, shelf life, and profit rather than health. Having lost antioxidants from foods, we must return them to our diet by replacing processed foods with whole foods, drinking freshly pressed juices, using foods super-rich in essential nutrients, or taking concentrates of these essential nutrients. . . . Antioxidants do prevent free radicals from getting out of hand. They do slow down aging processes and neutralize toxins."[66]

As mentioned earlier, carrot juice is rich in pro-vitamin A, which the body converts to vitamin A as needed. "Any leftover beta-carotene then acts as an antioxidant, breaking free radical chain reactions and preventing the oxidation of cholesterol."[67] Leading health literature states, "Vitamin A acts as an antioxidant, helping to protect the cells against cancer and other diseases, and is necessary for new cell growth. It guards against heart disease and stroke and lowers cholesterol levels. This important vitamin also slows the aging process. *Protein cannot be utilized by the body without vitamin A.*"[68] Since excess protein can be present in the body from the fungal antigens and the antibodies that are created by the body's immune response to these foreign substances,[69] drinking freshly juiced raw carrots may have helped our bodies use any additional proteins, freeing up the energy of our bodies to work in other areas.

By understanding how certain foods help our bodies work more efficiently and strengthen our immune systems, we can take a more proactive role in supporting and preserving our health. For example, carrots are rich in beta-carotene,[70] which is said to increase the number of T cells.[71] T cells are an important part of the body's immune response. They ". . . react to and destroy specific invading antigens, cancerous cells, or infectious agents."[72] By providing our bodies with an increased source of *natural* beta-carotene, we may have increased the number of T cells in our bodies, helping to destroy the foreign fungal cells.

T cell production is especially important when battling health repercussions from fungal exposure because mycotoxins can suppress the immune system by interfering with the production of T cells. For example, compounds produced by *Stachybotrys* cause immunosuppression, as Dr. Straus explained earlier, "These compounds, for example, cyclosporin A, cause inhibition of the production of interleukin 2 by activated T cells. Interleukin 2 is necessary for the development of T cell immunologic memory. This stops expansion of the number and function of antigen-selected T cell clones, resulting in immunosuppression."[73]

Obviously, T cells are just one component of the immune system, and the immune system is just one of the many complex systems that intrinsically function together to keep us healthy. As we researched to find out why fresh juiced carrots had helped strengthen our immune systems, we began to understand that the body, to some degree, if given the nutrients it needs, has the capacity to heal itself. By increasing our understanding of how the body uses various nutrients to derail the disease process, we began to base food selections on nutrient content—not taste buds or convenience. With this new thought process in place, we began to focus on raw vegetable consumption even more. We must all realize, whether we want to or not, that the body cannot squeeze nutrients out of junk food!

In order to understand more about how the body prevents the process of disease formation, we refer you to a book mentioned earlier, *Alternative Medicine: The Definitive Guide.*[74] It is one of the many books we read on the subject; it is comprehensive, yet easy to understand.

After experiencing the healing power of freshly juiced carrots, we began to understand more than ever that food choices do affect health, vitality, and even life expectancy, and if not given the sufficient nutrients needed, our bodies will not be able to function as they were intended. When drinking freshly juiced raw carrots improved our condition by an estimated 50 percent, it strengthened our confidence in our deep conviction that there were attainable answers—possibly more simple answers—that would fully restore our health. We continued to focus on nutrient content and nutritional value—one bite at a time—and forged ahead in our research.

There were a few vitamin supplements our ND prescribed that were quite helpful. For example, she prescribed 200 mg of magnesium per day for both Kurt and me to help regulate intestinal "plumbing." Magnesium is also an essential mineral, which means that our bodies cannot manufacture it, and it must be derived from diet and/or supplements.[75]

Our naturopath also prescribed a combination of vitamins to treat a hypothyroid (low thyroid) condition that I have had for many years. Hypothyroidism is a common condition for many people. "According to the American Association of Clinical Endocrinologists (AACE), at least six million Americans suffer from hypothyroidism, yet only half of them have been properly diagnosed."[76] Many hypothyroid people, like me, have battled the frustration of blood tests that reveal normal levels of thyroid hormones while still suffering many of the clinical symptoms of hypothyroidism. This lack of correspondence between laboratory and clinical data frequently occurs with patients, according to medical reports.[77] My subthyroid clinical symptoms went away after I started taking the vitamin supplements prescribed by our naturopath.

The supplements were as follows: 400 IU of vitamin E, 10,000 IU of vitamin A, 30 mg of zinc, 500 mg of L-tyrosine, and 200 mcg of yeast-free selenium. These supplements, and others, are noted in *Alternative Medicine: The Definitive Guide* as nutrients that help enhance thyroid function.[78]

Interestingly, the health effects of vitamin E have been cited in at least one study pertaining to fungi. It was found that ". . . supplementation of the culture medium with this antioxidant also had a negative effect on the sporulation"[79] of the studied strains. The fact that vitamin E negatively affected the ability of more than one species of fungi to sporulate was noteworthy to us.

In regard to vitamin A, it is documented that oftentimes the livers of people who are hypothyroid (low thyroid) are unable to convert beta-carotene into vitamin A.[80] As earlier mentioned, if the conversion process does not convert beta-carotene to vitamin A, the beta-carotene will still act as an antioxidant.[81] However, if the conversion process is not occurring, a vitamin A supplement may be needed.

Our naturopath prescribed fresh-juiced cabbage for Kurt's hyperthyroid (elevated thyroid) condition. When this cruciferous vegetable proved ineffective, she prescribed a broccoli supplement. Certain vegetables will naturally suppress thyroid hormone production.[82] These thyroid-suppressing foods can be taken in supplement form or eaten as whole foods: broccoli, Brussels sprouts, cabbage, cauliflower, kale, mustard greens, rutabagas, spinach, turnips, soybeans, peaches, and pears.[83]

Prior to the hurricane, Kurt had a slight hyperthyroid (elevated thyroid) condition that was kept under control by taking the same broccoli supplement. This hyperthyroid (elevated thyroid) condition required no prescription medication. After exposure to multiple fungi and chemicals post-hurricane, his thyroid levels crashed to an extreme hypothyroid (low thyroid) level. When he was improperly treated with steroids, his thyroid skyrocketed to an extreme hyperthyroid (elevated thyroid) condition. Then-inexplicably, his thyroid later (yet again) flipped back to a hypothyroid (low thyroid) condition. His thyroid remained unstable for several months until the fungi were effectively treated, after which his thyroid gland began to regulate and gradually return to its pre-hurricane state.

The importance of sharing so much information regarding the effects of fungal exposure on the thyroid gland is that the two endocrinologists we saw were not familiar with the fact that fungal exposure could negatively impact thyroid function. Medical documentation shows otherwise. According to the Department of Pathology at Virginia Commonwealth University, fungi are one of the causes of thyroiditis, which is inflammation of the thyroid gland.[84] The Bobby R. Alford Department of Otolaryngology-Head and Neck Surgery at Baylor College of Medicine also cites fungi, as well as parasites, as agents that can cause thyroiditis.[85] As we all understand at this point, some fungi are parasitic.

Dr. Callaghan states that doctors who are mold literate know to test mold-exposed patients for thyroiditis.[86]

According to the New York Thyroid Center at Columbia University Medical Center, hypothyroidism (low thyroid) can result from thyroiditis. It states, "In the case of thyroiditis, hypothyroidism is caused by destruction of the thyroid gland by an inflammatory process. When thyroid cells are attacked by the inflammation, these cells die. Without thyroid cells, the thyroid is no longer able to produce enough thyroid hormone to maintain the body's normal metabolism. Hypothyroidism or an underactive thyroid gland results." It further states, "Occasionally . . . the thyroid has been so destroyed that it can never produce normal quantities of thyroid hormone. In this case, permanent hypothyroidism results and medication is necessary."[87]

Kurt's second endocrinologist noted that his thyroid was inflamed, although she did not draw the correlation between an inflamed thyroid gland and fungal exposure. However, with adequate information, as documented above, it appears to be quite simple: 1) Kurt got exposed to fungi. 2) His thyroid gland became inflamed (from the fungi). 3) This inflammation caused death of thyroid cells. 4) With fewer functioning thyroid cells than normal, his thyroid function plummeted, which resulted in the initial, extreme hypothyroidism (low thyroid). 5) The steroids caused his thyroid gland to resume functioning, causing an extreme hyperthyroid (elevated thyroid) condition. 6) A hypothyroid (low thyroid) condition returned because the fungi were still causing thyroid cell death, thus reducing the thyroid gland's ability to produce adequate hormone levels. 7) When the fungi were effectively treated, his thyroid gland function gradually returned to its pre-hurricane state. Based on medical documentation, in our opinion, this sequence of events is what caused the irregular, rollercoaster functioning of Kurt's thyroid.

In addition to fungi being able to cause inflammation of the thyroid gland, some can also produce substances that are documented as endocrine disruptors. The thyroid gland is a part of the endocrine system, which is comprised of the ". . . pineal, pituitary, hypothalamus, thyroid, parathyroid, adrenals, pancreas, gonads or sex glands, and other glandular tissue located in the intestines, kidneys, lungs, heart, and blood vessels."[88]

The Chlorine Chemistry Council, which is a business council of the American Chemistry Council, a national trade association representing manufacturers and users of chlorine and chlorine-related products, documents that naturally occurring endocrine-active chemicals (phytoestrogens and mycoestrogens) can be produced by plants or fungi. It further states, "Before a chemical substance can be defined as an 'endocrine disruptor,' it has to be scientifically proven that it has the ability to produce adverse health effects in humans or wildlife."[89]

According to the Center for Bioenvironmental Research at Tulane University, endocrine disruptors are ". . . synthetic chemicals and natural plant compounds that may affect the endocrine system (the communication system of glands, hormones, and cellular receptors that control the body's internal functions). Many of these substances have been associated with developmental, reproductive, and other health problems in wildlife and laboratory animals. Some experts suggest these compounds may affect humans in similar ways."[90] At the time of this writing, we were unable to locate studies regarding the health effects resulting from human exposure to fungal endocrine disruptors.

The Center for Bioenvironmental Research confirms that fungi are a source of endocrine disruptors. It states, "Environmental estrogens are the most studied of all the endocrine disruptors. Natural compounds capable of producing estrogenic responses, such as the phytoestrogens, occur in a variety of plants and fungi." [91]

Not surprising, the EPA is interested in this special class of chemicals—naturally occurring non-steroidal estrogens (NONEs).[92] The EPA states, "These are natural products derived from plants (phytoestrogens) and fungi (mycotoxins). These chemicals occur widely in foods and have the potential to act in an additive, synergistic, or antagonist fashion with other hormonally active chemicals."[93] In other words, mycotoxins have the potential to act in an additive, synergistic, or antagonist fashion with the hormonally active agents in our bodies.

In addition to an inflamed thyroid gland, Kurt's thyroid was also enlarged, which is often called a goiter. MedlinePlus states that this condition, ". . . is associated with hyperthyroidism usually together with toxic symptoms . . ."[94] Toxic symptoms refer to what professionals in the medical field call toxic nodules, which have the potential to grow uncontrollably, sometimes requiring surgical removal.[95] This better-understood medical condition is why Kurt's first endocrinologist recommended surgery, as he did not understand that Kurt's thyroid condition could have been caused by fungal exposure, which once treated would enable his thyroid to return to its pre-hurricane state. As mentioned earlier, Kurt's second endocrinologist was willing to take a slower "wait-and-see" approach, for which we were thankful, because after effective fungal treatment, Kurt's thyroid function progressively returned to its pre-hurricane, slightly elevated state, which is controllable with broccoli and/or broccoli supplements.

Interestingly, Dr. Thrasher states that he and his colleagues see people in their clinic with severe thyroid problems. He explains, "We don't know whether they had them before exposure or after exposure, but we do see a pretty good number. I don't know the percentage. I am never surprised to see an abnormally functioning thyroid gland. If we take a look at what the thyroid gland is susceptible to and what causes abnormal thyroid gland function, we see chlorinated hydrocarbons, particularly PCBs (poly-chlorinated biphenyls). The molds produce chlorinated mycotoxins. Nobody's looked at that."[96] Therefore, the possibility exists that thyroid disfunction may be a common health effect from mycotoxin-producing fungal exposure.

Another supplement we used was lipoic acid. As we researched, we came across a peer-reviewed medical study indicating that lipoic acid had ". . . been shown to protect against, or reverse, the adverse health effects of mycotoxins."[97] We felt a noticeably positive effect on our health from taking the lipoic acid supplement, although once healthy, we discontinued use.

The information we have shared about supplements is for educational purposes; it should not be interpreted to mean the supplements discussed are safe for anyone else's use. The supplements and dosage levels prescribed for us by our treating physician should not be used for self-treatment. Use of any of the supplements should be discussed with each person's treating, licensed physician. Side effects are possible from supplements, and certain supplements may not be appropriate for everyone. For example, although lipoic acid is an antioxidant, in a study in which mice were induced with diabetes mellitus, lipoic acid was documented to have deleterious effects on postimplantation embryos.[98] A further example of a supplement with potential negative side effects is vitamin A, which can increase the risk of birth defects, such as heart abnormalities, cleft lip, or cleft palate with supplementation of levels as low as 15,000 IU.[99] "However, no evidence exists linking similar or higher doses of vitamin A obtained through the diet to such birth defects," according to health literature.[100] Again, check with your physician before increasing dosage levels of existing supplements and/or adding any new supplements.

Although the carrot juice and aforementioned thyroid supplements were health enhancing, we were still in search of a natural substance that would kill the fungi, neutralize the mycotoxins, and restore our immune systems, thus enabling full restoration of our health. We continued our quest.

(ENDNOTES)
1   Goldberg, *Alternative Medicine*, p. 293.
2   Erasmus, *Fats That Heal Fats that Kill*, p. 410.
3   PBS, "Is It Safe to Return?"
4   Interview with Dr. Michael Gray, March 2007.
5   "Clinical Pharmacology–Customized Monograph: Nystatin," *Clinical Pharmacology*, 2005 Gold Standard.
6   Ibid.
7   Lamisil, 12/13/05, patient information sheet, Wal-Mart Pharmacy, information expires 1/19/06.
8   "Clinical Pharmacology–Customized Monograph: Terbinafine," *Clinical Pharmacology*, 2005 Gold Standard.
9   Ibid.
10   Crook, *The Yeast Connection Handbook*, p. 120.
11   "Clinical Pharmacology–Customized Monograph: Nystatin."
12   Interview with Dr. David Straus April 2006.
13   Ibid.
14   Vieira MR et al., "Phaeohyphomycosis due to *Cladosporium cladosporioides*," *Med Mycol*. 2001 Feb; 39(1): 135-7. Available at http://www.ncbi.nlm.nih.gov/entrez/query.fcgi?cmd=Retrieve&db=PubMed&list_uids=11270401&dopt=Abstract (accessed March 2006).
15   Dr. Fungus, "Amphotericin B deoxycholate." Available at http://www.doctorfungus.

org/thedrugs/Ampho_Deoxycholate.htm; Dr. Fungus, "Ketoconazole." Available at http://www.
doctorfungus.org/thedrugs/Ketoconazole.htm (accessed February 2006).

16   Chew, "Mold and Endotoxin Levels in the Aftermath of Hurricane Katrina: A Pilot Project of Homes in New Orleans Undergoing Renovation."

17   According to information on the website of the Environmental Health Center in Dallas, Texas. Available at http://www.ehcd.com/services/treatments.htm (accessed September 2006).

18   Interview with and e-mails from Dr. Timothy Callaghan, January 2007. For more information on COEM, see: www.coem.com (accessed January, 2007).

19   Obagwu J. and Korsten L., "Control of Citrus Green and Blue Molds with Garlic Extracts," *European Journal of Plant Pathology*, vol. 109, no. 3, March 2003, 221-225. Available at http://www.ingentaconnect.com/content/klu/ejpp/2003/00000109/00000003/05111329 (accessed February 2006).

20   For more information on The School of Natural Healing see: www.schoolofnaturalhealing.com or call 800-372-8255.

21   Balch, *Prescription for Nutritional Healing*, p. 86.

22   Phyllis A. Balch, CNC, *Prescription for Nutritional Healing*, Fourth edition (New York, Avery Publishing Group, 2000).

23   Burton Goldberg, *Alternative Medicine*, (Berkeley, California, Celestial Arts, 2002).

24   John R. Christopher, *School of Natural Healing* (Springville, Utah, Christopher Publications, 1976).

25   Balch, *Prescription for Nutritional Healing*, p. 71; Goldberg, *Alternative Medicine*, p. 261-262; and Stefan Chmelik, *Chinese Herbal Secrets*, p. 138 (Garden City Park, NY, Avery Publishing Group, 1999).

26   Penelope Ody, *The Complete Medicinal Herbal*, p. 33 (New York, NY, DK Publishing, Inc. 1993), and Goldberg, *Alternative Medicine*, p. 261-262.

27   Paavo Airola, PhD, *The Miracle of Garlic*, p. 20-21 (Sherwood, OR, Health Plus Publishers,1980).

28   Goldberg, *Alternative Medicine*, p. 261-262.

29   Ibid.

30   Chmelik, *Chinese Herbal Secrets*, p. 138.

31   Airola, *The Miracle of Garlic*, p. 32.

32   Ibid.

33   Airola, *The Miracle of Garlic*, p. 32, and Goldberg, *Alternative Medicine*, p. 677.

34   Amanda Ursell, *The Complete Guide to Healing Foods*, p. 114 (New York, NY, Dorling Kindersley Publishing, Inc. 2000).

35   Jethro Kloss, *Back to Eden*, p. 130 (Loma Linda, CA, Back to Eden Publishing Co., 1995).

36   Goldberg, *Alternative Medicine*, p. 627.

37   Balch, *Prescription for Nutritional Healing*, p. 92, and Christopher, *School of Natural Healing*, p. 447.

38   Balch, *Prescription for Nutritional Healing*, p. 92.

39   Ody, *The Complete Medicinal Herbal*, p. 46.

40   Ibid., p. 67.

41   Balch, *Prescription for Nutritional Healing*, p. 98.

42   Goldberg, *Alternative Medicine*, p. 263.

43   Ibid., and Crook, *The Yeast Connection Handbook*, p. 132.

44   Goldberg, *Alternative Medicine*, p. 263.

45   Ibid.

46   Balch, *Prescription for Nutritional Healing*, p. 98.

47   Kloss, *Back to Eden,* p. 202.

48   Balch, *Prescription for Nutritional Healing*, p. 717.

49   E-mail from Dr. Ginger Chew, July 2006.

50   Stephen Blauer, *The Juicing Book*, p. vii. (Garden City Park, NY, Avery, 1989).

51   Ibid.

52   Ibid., p. ix.

53   Christopher, *School of Natural Healing*, p. 537-538.

54   Erasmus, *Fats That Heal Fats that Kill*, p. 409.

55   Balch, *Prescription for Nutritional Healing*, p. 717.

56   Blauer, *The Juicing Book*, p. 45-46.
57   Goldberg, *Alternative Medicine*, p. 191.
58   See Section Three, interview with Dr. Jack Thrasher, April 2006.
59   Erasmus, *Fats That Heal Fats that Kill*, p. 401.
60   Goldberg, *Alternative Medicine*, p. 23.
61   Balch, *Prescription for Nutritional Healing*, p. 53.
62   Erasmus, *Fats That Heal Fats that Kill*, p. 306.
63   Ibid.
64   Ibid., p. 318.
65   Goldberg, *Alternative Medicine*, p. 589.
66   Erasmus, *Fats That Heal Fats that Kill*, p. 192.
67   Balch, *Prescription for Nutritional Healing*, p. 57.
68   Ibid., p. 15. (Italics added.)
69   Interview with Dr. Udo Erasmus, May 2006.
70   Goldberg, *Alternative Medicine*, p. 589.
71   Ibid., p. 627.
72   Ibid., p. 21.
73   See Section Three, interview with Dr. David Straus, April 2006.
74   Burton Goldberg, *Alternative Medicine*, (Berkeley, California, Celestial Arts, 2002).
75   Goldberg, *Alternative Medicine*, p. 395.
76   Ibid., p. 366.
77   Ibid.
78   Ibid., p. 367.
79   Tamás Emri et al., "Effect of Vitamin E on Autolysis and Sporulation of *Aspergillus nidulan*," July 2004, *Applied Biochemistry and Biotechnology*, vol. 118, issue 1-3, pp. 337-348.
80   Balch, *Prescription for Nutritional Healing*, p. 15.
81   Ibid., p. 57.
82   Goldberg, *Alternative Medicine*, p. 1030.
83   Ibid.
84   Virginia Commonwealth University (VCU) Department of Pathology, "Thyroid Gland," VCU. Available at http://www.pathology.vcu.edu/education/endocrine/endocrine/newthyroid/ (accessed September 2006).
85   Rance W. Raney, MD, "Inflammatory Diseases of the Thyroid," January 6, 1994. Available at http://www.bcm.edu/oto/grand/1694.html (accessed September 2006).
86   Interview with and e-mails from Dr. Timothy Callaghan, January 2007.
87   The New York Thyroid Center at Columbia University Medical Center, "Thyroiditis," The New York Thyroid Center at Columbia University Medical Center. Available at http://cpmcnet.columbia.edu/dept/thyroid/thyroiditis.html (accessed September 2006).
88   Goldberg, *Alternative Medicine*, p. 25.
89   Chlorine Chemistry Council, "Endocrine Disruption," February 19, 1998. Available at http://c3.org/chlorine_issues/health/enddisrupt.html (September 2006).
90   E-hormone, Center for Bioenvironmental Research at Tulane University, "Endocrine Disrupting Chemicals." Available at http://e.hormone.tulane.edu/edc/html (September 2006).
91   Ibid.
92   EPA, "Endocrine Disruptor Screening Program; Proposed Statement of Policy," Federal Register Environmental Documents. Available at http://www.epa.gov/fedrgstr/EPA-PEST/1998/December/Day-28/p34298.htm (accessed September 2006).
93   Ibid.
94   MedlinePlus Medical Dictionary. Available at http://www2.merriam-webster.com/cgi-bin/mwmednlm (accessed March 2006).
95   Mayo Clinic, "Thyroid nodules."
96   See Section Three, interview with Dr. Jack Thrasher, April 2006.
97   Rogers SA, "Lipoic acid as a potential first agent for protection from mycotoxins and treatment of mycotoxicosis," *Arch Environ Health*. 2003; 58(8): 528-32. Available at http://wwmedscape.com/medline/abstract/15259433?prt=true (accessed January 2006).
98   Padmanabhan R. et al., "Beneficial effect of supplemental lipoic acid on diabetes-induced

pregnancy loss in the mouse," *Ann N Y Acad Sci*, 2006 Nov; 1084: 118-31. Available at http://www.ncbi.nlm.nih.gov/entrez/query/fcgi?db=pubmed&cmd=Retrieve&dopt=AbstractPlus&list_uids=17151296&query_hl=1&itool=pubmed_docsum (accessed September 2006).

99   Goldberg, *Alternative Medicine*, p. 865.

100   Ibid.

CHAPTER SIX:

HELP FROM THE KAHUNAS

We were encouraged by the positive response we had to the antioxidants in the alkalizing carrot juice and to the thyroid supplements. However, we were continuously getting sick, which meant our immune systems were still impaired. We continued to research, searching for the weakness of our formidable foe (fungi) and the reason it was able to make us so sick. When we read that certain fungi produce poisons called VOCs and mycotoxins, we realized we were battling a multi-dimensional enemy. Our bodies were having to contend with not only the fungi, but also the poisons from the VOCs and mycotoxins, the latter of which are proven immune suppressors.[1] We started to understand why our immune systems were so fragile and why we were constantly sick.

The concept of mycotoxins is easy to understand. They are weapons certain fungi use to ensure their survival against other organisms. When fungi colonize internally, they essentially try to take over what they view as their territory (our bodies), emitting their poisons to ensure their survival.[2] As the human body responds to these foreign stimuli, *The Battle Within* begins.

We began to search for a natural way to neutralize the poisonous toxins—the mycotoxins. We began researching a supply source for a product that we had used for detoxification years earlier—noni juice. Traditionally, it has been used to neutralize toxins from fish poisoning in the South Pacific.[3] We wondered if it might also neutralize fungal toxins.

So what is noni juice? According to the *Honolulu Star-Bulletin*, "The noni plant was used in traditional healing throughout Polynesia and is being promoted worldwide for all kinds of health problems and diseases."[4] In fact, the health claims of noni are so prevalent that Brian Issell, MD, (then-internist, oncologist, and clinical sciences program director at the Cancer Research Center of Hawaii, University of Hawaii), began Phase I of a study to evaluate the medicinal value of noni. This study, which began in 2001, was funded by a $170,000 two-year grant from the National Institutes of Health (NIH).[5] When the NIH grant was not renewed in July of 2004, the Phase I study continued with support from the Hawaii Community Foundation.[6] Before we discuss the study in more detail, let's first take a closer look at what is known about noni.

Historically, ". . . noni has been cultivated throughout communities in the South Pacific for hundreds of years."[7] Because of the widespread prevalence of the noni fruit in Polynesia, noni juice and products made from the fruit are usually identified by the region in which the fruit was grown, for example, Tahitian noni or Hawaiian noni. The scientific name for the fruit is *Morinda citrifolia*. Published literature states, "*Morinda citrifolia* is technically an evergreen shrub or bush, which can grow to heights of fifteen to twenty feet. It has rigid, coarse branches which bear dark, oval, glossy leaves. Small white fragrant flowers bloom out of cluster-like pods which bear creamy-white colored fruit. The fruit is fleshy and gel-like when ripened, resembling a small breadfruit. The flesh of the fruit is characteristically bitter and when completely ripe produces a rancid and very distinctive odor."[8]

There are several different methods in which noni juice can be processed. Check out the processing method before making a purchasing decision because the manufacturing process *will* affect the quality of the end product. For example, "Traditionally, the fruit was picked before it was fully ripe and placed in the sunlight. After being allowed to ripen, it was typically mashed and its juice extracted through a cloth."[9] The traditional preparation of noni juice is still available through select suppliers who process the juice via natural fermentation. However, the mass production of noni juice and other noni products has commercialized many production operations.

In order to facilitate increased production, many noni manufacturers produce a fresh-squeezed noni juice versus the traditional, fermented noni juice, which requires three to six weeks of aging.

The *fresh-squeezed* designation means that ". . . the juice is pressed directly from ripe fruits using a mechanical device and bottled directly into glass or plastic containers and is not allowed to ferment. These products are either pasteurized or refrigerated to preserve their integrity."[10] During the pasteurization process, the noni juice is heated to a high temperature to kill harmful bacteria. Unfortunately, this elevated temperature can also destroy important plant enzymes, which are more heat-sensitive than vitamins.[11] In fact, plant enzymes are destroyed by being heated above 118 degrees Fahrenheit and are deactivated or destroyed by pasteurizing, canning, and microwaving.[12] Therefore, a raw, unpasteurized noni juice product will retain more of the health-revitalizing live plant enzymes.

Another important point to consider when evaluating a noni juice product is the possibility of added contaminants from a water source any time a noni juice has been reconstituted from dehydrated fruit. Dehydrated fruit can be made from either green or ripe noni.[13] The obvious problem with this type of preparation is that the manufacturer may be reconstituting the dehydrated noni fruit with non-purified municipal water, full of chlorine, fluoride, and other contaminants. So claims of "100% organic" do not ensure a healthy product if it has been reconstituted. The same is true of noni juice reconstituted from noni concentrates.

There are several additional factors to consider before making a noni purchase. However, before we get into a more in-depth discussion regarding different processing methods, let's first address the most important factor to consider: the geographic region in which the noni fruit (from which the product is made) was grown. The quality of raw material will affect the quality and pureness of the manufactured product. As with any consumable product, negative health effects from environmentally contaminated ingredients are always a possibility. The two main geographical regions supplying the noni market are Tahiti and Hawaii. Both are utopias of paradise, or are they?

Tahiti and other French Polynesian islands are colonies of France under colonial rule of the French government.[14] It is sad, but true, that from 1966 to 1974[15] France conducted 44 nuclear tests in the atmosphere and from 1975 to 1991 conducted 115 underground tests on two tiny South Pacific atolls, Moruroa and Fangataufa.[16] (Atolls are ring-shaped coral reefs that enclose lagoons.)

The fact that the nuclear testing took place is not in question. However, the effects, on both the environment and human and animal health, have been a subject of dispute. Let's take a closer look at available documentation. A report released on February 2, 2006, by a French Polynesia Assembly (local parliament) Inquiry Commission that reviewed the "secret" documents of the Defense Ministry dated from 1965 to 1967 shows ". . . incontestable and precise proof of lying by the authorities who conducted the nuclear testing. Although the authorities maintained that the tests were clean and that the radioactive fallout did not affect the population, the report shows the contraire [opposite]; each one of the tests conducted between 1966 and 1967 caused radioactive fallout on the islands in French Polynesia."[17]

Since all of the atmospheric tests conducted through 1974 were detonated from the same location, conclusions have been drawn that each of the incidences of nuclear atmospheric testing caused fallout on the Polynesian islands, including Tahiti. *Pacific Magazine* cited a Tahiti newspaper, the *Tahitipresse,* that reported, "The island of Tahiti was subjected to fallout from each of France's atmospheric nuclear tests 1,200 kms (720 miles) away more than 30 years ago . . ."[18] The *Pacific Magazine* further states, "The news media's extracts of the inquiry committee's report and final conclusions claim, 'it is not exaggerating to think that the (radioactive) fallout occurred on (the island of) Tahiti after each atmospheric test.'"[19]

According to *Pacific Magazine*, "The inquiry committee's report required nearly six months of investigation . . ."[20] The Centre for Research and Information on Peace and Conflict reports, "After having auditioned government ministers, medical experts, and health workers, the Inquiry Commission expresses its strong conviction that the aerial nuclear tests had severe consequences for health, not only for those who worked on the test sites, but for the entire population of French Polynesia."[21] The approximate population of the entire French Polynesian region was 250,000 in 2006.[22] The Centre for Research and Information on Peace and Conflict further reports, "The visits made by the Inquiry Commission, accompanied by experts in radiological analysis (CRIIRAD), to the islands of Mangareva,

Tureia, and Hao have confirmed that . . . the fallout of past nuclear testing [is] still measurable today . . ." Their findings also confirm ". . . the extremely poor condition in which the population and these islands were left by the military once the testing [was] completed."[23]

According to *Pacific Magazine*, "The inquiry committee report claims a link between the tests and number of cancer cases in French Polynesia, noting the study under way by the French National Institute of Health and Medical Research (INSERM) . . ."[24] The effects of nuclear fallout on the environment were not specifically addressed in published documentation. For example, the environmental effects were not discussed in the English synopsis of the Inquiry Commission's Report. The effects from the underground nuclear tests were also not addressed. However, further investigation may be forthcoming. According to *Pacific Magazine*, "The report recommends the creation of a new inquiry committee to investigate the underground French nuclear tests . . ."[25] that were conducted from 1975 to 1991. In addition, the Inquiry Commission recommends that this new council ". . . follow-up on the consequences of nuclear tests on the health of people and on the environment."[26]

Greenpeace has also called for an independent scientific investigation of the Moruroa test site, one of the two atolls on which the 115 underground nuclear tests were conducted.[27] "'After years of secrecy and denials about the environmental impacts of testing, the French Government has had to concede there has been damage to coral,' said Greenpeace Australia campaigns manager Benedict Southworth."[28] He further points out, "'Last year the French revealed that plutonium had leaked into the lagoon at Moruroa, and one year later they admit to cracks in the coral of the atoll . . ."[29]

The news about the fractures in the coral at Moruroa and the plutonium leaks raise serious concerns about radioactive leaks into the environment.[30] However, at this point in time, without a scientific study revealing the long-term environmental effects of 25 years of nuclear testing (both atmospheric and underground) to inform us of the possible level of *agricultural contamination*, we must rely on our common sense. We must ask ourselves: Which is more beneficial for our health—a noni product produced from fruit grown in Tahiti, which lies 720 miles from 25 years of nuclear testing or a noni product produced from fruit grown in Hawaii, which lies 2,700 miles north of the French Polynesia testing ground?

For our family's long-term use, we decided to err on the side of caution and began to research Hawaiian manufacturers of noni products. According to Scott Nelson, PhD, at the College of Tropical Agriculture and Human Resources, University of Hawaii, "There are a number of significant quality control issues that may impact the good reputation of Hawaii noni juice products, including product dilution (or amendments), product contamination, chemical residues, lack of standardization for biologically active ingredients, and poor product consistency (e.g., significant variability in juice attributes)."[31]

He further explains, "Some juice producers choose to add sugar to their raw juice products without informing their clients. This practice may lead to a significant amount of alcoholic fermentation if contaminating yeasts are present in the juice. Other producers water down their juice products, an unfortunate practice which affects the levels of biologically active compounds in noni juice. Some noni juicing facilities in Hawaii have relatively unsanitary conditions or poor worker hygiene. This can result in juice products that are contaminated with unwanted microorganisms. Fruits that are not washed or plants and fruits that are sprayed with pesticides may result in juice that has chemical residues within. Many noni juice producers are not aware of which biologically active ingredients they can or should be considering when striving to produce a standardized, consistent juice product."[32]

In addition to the aforementioned factors that can cause variations in the quality and consistency of noni juice, the chosen processing method will also have an impact. "Noni fruit juice and juice products are processed and prepared in Hawaii by a variety of methods. For example, noni juice may be fermented versus unfermented or fresh-squeezed versus drip-extracted. The 'traditional' juice is both drip-extracted and fermented/aged for at least two months. The 'non-traditional' method of juice extraction is by pressing or squeezing the juice from ripe fruits. Noni juice may be amended with other additives or diluted or bottled in its pure state. It may be bottled with or without pasteurization. . . . Because fermented noni juice usually has a low pH (approx. 3.5), pasteurization may not always be necessary," reports Dr. Nelson.[33] The North Dakota State University Extension Service confirms, "A pH of 4.6 or lower is required for safe canning without the use of pressure processing."[34] Pressure processing is a form of pasteurization.

Our goal was to locate the purest, most unprocessed form of noni juice available. Unfortunately, the information needed to evaluate a product, such as the specifics of the processing method, is not always on the product label or manufacturer's website. In order to find out this information, we had to call the manufacturers themselves to inquire about of their respective processing methods.

After extensive research, we selected a product line manufactured in Hawaii. We were especially interested in the certified organic noni juice, because it would not have the pesticide and/or chemical residue that is likely to be present in/on nonorganic noni fruit. Health literature states, "Eating organically grown foods is extremely important, as recent studies have shown that organic foods are not only far more free from carcinogenic pesticide contaminants than conventionally grown foods, they are also richer in the essential nutrients and trace elements necessary for cancer prevention, including beta carotene, vitamin E, and selenium."[35] Even though not every piece of food we ate was organic, we felt that if we were going to use a food in a medicinal manner, it should be organic.

We were also interested in the organic noni juice from this company because it was processed using the traditional method (aged three to six weeks, drip-extracted, and then squeezed into juice) versus the freshly squeezed method (aged a certain number of days to soften and then squeezed into juice).

Essentially, the traditional manner is a drip-extraction process. The noni fruit is picked at its peak of ripeness and allowed to soften for one to three days and then fermented in a closed container. Airtight lids ensure an anaerobic (no air) environment necessary for processing. Containers that do not have a pressure relief valve must be periodically "burped" to release pressure that builds from the fermentation process. As the fruit continues to ripen (naturally fermenting), the flesh of the fruit softens. The fruit is allowed to naturally ferment (similar to how a wine is aged) for three to six weeks.[36]

Dr. Nelson further describes the traditional method, "During this time the noni juice separates (drips) gradually from the pulp. . . . The noni juice collects inside the containers and ferments as it gradually seeps and sweats from the fruits."[37] By the end of this slow fermentation process, the fruit has essentially become 70 percent liquid; it is then squeezed with a press and bottled. A traditionally fermented noni juice will have nothing added to the fruit juice—no water, sugar, filler juices, yeasts, or nutrients.[38]

It is important to understand that fermented juice has very little sugar, since most of the sugar ferments out in the aging process. Literature states, "Through fermentation, the sugars in noni juice are transformed to organic acids, causing the pH to be reduced and acidity to increase."[39] Since sugar feeds fungi, it was critical for us to use a fermented noni juice that had been aged to a state in which it contained little to no sugar.

Unfermented or ". . . fresh juice, of course, contains the sugars that are inherent in the fruit fresh from the plant."[40] Since fresh fruit is a simple sugar, which is not beneficial for fungal-related conditions, a nonfermented noni (fruit) juice containing natural sugars could have been counterproductive to our health. Remember, no matter how "healthy" a source from which the sugar comes, it is still sugar and will still feed the fungi. For this reason, we avoided any fresh-squeezed, nonfermented noni juice products, as they would contain the same sugar as the raw fruit. Fruit was not allowed in our antifungal diet.

Processing methods will also affect the chemical profile of the end product. For example, if a noni juice is fermented, it will have a different chemical profile, according to Dr. Issell.[41] Likewise, adding grape, blueberry, or raspberry juice to noni juice will also change the chemical composition, points out Dr. Issell.[42]

Adding "natural" flavoring agents to noni juice is a practice that is often used by manufacturers to enhance the taste of their products and boost profits.

Because of the commercialization of noni products, many noni manufacturers have "modernized" and "streamlined" their processing operations. Deviations from the traditional method of processing, of course, will not produce a noni juice with the same chemical profile as that which healing folklore originated. The chemical profile will be different. Specialty suppliers do exist that remain true to the traditional method of processing. They invest time into their products and allow the juice to naturally ferment. The end product is a traditionally fermented noni juice, free of chemical-altering additives. We sought to use a noni juice that would be as close as possible to the traditional toxin-neutralizing noni juice.

The products we initially tried from the company in Hawaii included: two 32-ounce bottles of organic pasteurized noni juice, two 32-ounce bottles of organic unpasteurized noni juice, one 4-ounce bar of noni soap, one 1-ounce bottle of noni tincture, and one bottle of 90 capsules of noni powdered extract. We were careful to try one product at a time, so if we had a reaction, we would know which product had caused it. Likewise, if we saw improvement, we wanted to know which product was responsible for the improvement.

We started by taking the unpasteurized noni juice rather than the pasteurized, because we were looking for minimal processing that would not denature the natural healing properties of the fruit. The pasteurized noni juice from the original company was raised to 160 degrees for 5 seconds.[43] The temperature of the unpasteurized noni juice, of course, was not raised, which enabled the juice to retain all its vitamins, live enzymes, good yeasts, and other beneficial organisms. We figured the juice with the most live properties would be the most effective.

The results we experienced were astounding. Liquid began to come up from our lungs. One of the problems we had continually battled was the accumulation of fluid in our lungs. When we took the unpasteurized noni juice, it was as if this fluid was being squeezed out of our lungs, which enabled us to cough up all sorts of ugly yuck. The feeling of compression on our chests that we had from water in the lungs began to subside. Only then did we realize how much discomfort we had been tolerating without really realizing or consciously acknowledging it.

We also experienced some gentle die-off headaches, which was evidence that the fungi were dying.[44] This die-off reaction is addressed in *The Yeast Connection Handbook*, "Many people who take nystatin or other prescription or nonprescription antiyeast medications feel worse for several days after they start taking them. Symptoms include fatigue, depression, aching, irritability, and abdominal pain. Yet, when they continue the medication, the symptoms disappear.

Such symptoms are due to 'die-off reactions. . . . [W]hen you kill many yeasts in your digestive tract with nystatin or other antifungal drugs, it may take a few days (or longer) to get rid of the dead yeast products."[45] The fact that our die-off headaches were gentle—not the debilitating need-to-be-in-a-dark-room headache that we experienced when initially taking the nystatin—leads us to believe that the unpasteurized noni juice not only kills fungi but also helps the body to more effectively eliminate the dead fungal products.

Another noticeable effect of the unpasteurized noni juice was the clearing of what we called *brain fog*. The toxins from fungi can affect your ability to think, reason, and process. They can affect your memory, even short-term memory. For people who have not experienced the effects of mold toxins, it may be hard to understand the extent to which these particular poisons can reduce the level of a person's functionality, even when performing simple day-to-day tasks. For example, multiple times, I turned on the stove burner to cook food, got distracted, and then completely forgot that I had even begun to cook something! I became very adept at cleaning the under-the-burner area of our stove because I kept boiling food over and burning it. Other times, I would stir the food (periodically) for 20 minutes, waiting for it to get hot and then realize I had forgotten to turn on the burner. Just as annoying, sometimes I would turn on a burner but then put the pan on a different burner. One time, when Kurt did the same thing, he placed a plastic cutting board on what should have been a cold burner. Obviously, as the burner became increasingly hot, it melted the plastic. We now have the shape of the round coil from the stove burner imprinted on one side of our cutting board as a daily reminder of how far we have come.

We are not alone in experiencing these types of frustrations. In fact, while researching this book, we were repeatedly told stories of behavioral changes in people who had suffered mold exposure. Although both males and females can be affected in this manner, symptoms have often been described as raging PMS (premenstrual syndrome) symptoms that mysteriously develop after mold exposure. Dr. Thrasher explains why mood and personality changes would make sense after exposure to mold: Many of the mycotoxins are estrogenic compounds, which can create an excess of estrogen in the body. This excess of estrogen can wreak havoc particularly on the female system.[46]

Another indicator that these PMS-type symptoms are mold-related is that they do not respond to treatments that are medically known to alleviate hormonal imbalances. For example, in my particular situation, our ND prescribed homeopathic, herbal, and nutritional supplements to treat these hormonal fluctuations—none of which worked. In fact, some products designed to regulate hormonal imbalances actually exacerbated the problem. Finally, after taking the unpasteurized noni juice for several weeks, these wild hormonal fluctuations subsided and "PMS" symptoms went away.

Traditionally the topic of menstrual cycles is generally of a private nature, but we felt it important to point out that the exposure to the estrogenic compounds from the toxigenic fungi has the ability to alter a woman's menstrual cycle. For this reason, we will simply say, if a woman is experiencing unexplained PMS-type symptoms and/or irregular menstrual cycles, it may be a good idea to look for a possible source of exposure to hidden toxigenic (mycotoxin-producing) fungi and discuss fungal exposure as a possible cause with a doctor. Based on "our" personal experience, once the fungal exposure was eliminated, both in the external environment and internal (inside the body) environment, all systems (including female) went back to normal. If this candidness helps just one person, it was worth our sharing. Hormonal fluctuations can be a very complicated side effect of fungal exposure that can easily be incorrectly attributed to other factors.

We also saw a noticeable improvement in our children's health from the use of the unpasteurized noni juice. Although their old spark and pre-hurricane level of energy had not yet returned, they were starting to become more active, running, jumping, and playing more than they had been able to in several months.

The dose levels of noni that we started with were 3 ounces twice a day for Kurt and me, 2 ounces twice a day for our eight-year-old daughter, and 1 ounce twice a day for our two-year-old son. With just the first 32-ounce bottle, we could see major improvements. According to Dr. Nelson, refrigeration is not necessary. He states, "Fermented juice (when uncontaminated and with low pH, e.g., approximately 3.5–4.0) will store well at room temperature without pasteurization."[47] Regardless, we stored our unpasteurized noni juice in the refrigerator. We prefer to drink our noni juice cold, as it tastes better when chilled. In fact, "Noni juice, when aged well, will actually be reminiscent of a good wine. It will be strong, but will have a very mellow, nicely aged taste."[48] We found this analogy to be fairly accurate. Although, for clarification purposes, "There is not a significant amount of alcohol produced by fermentation of noni juice."[49]

Although we were extremely pleased with the results we experienced from the organic unpasteurized noni juice, we tried the organic pasteurized noni juice to see if we would notice any difference. Therefore, after consuming the first 32-ounce bottle of unpasteurized juice, we switched to the pasteurized. The taste and smell of the pasteurized juice was somewhat stronger than that of the unpasteurized, which made us wonder if it was going to be more effective. We quickly found that not to be the case. We immediately began to regress: Fluid started to accumulate in our lungs again; we began to have symptoms akin to stomach and intestinal flu; our bleeding hives worsened, and we developed sore throats, colds, and general overall malaise.

We increased our dose of the pasteurized noni juice and found that we had to take approximately three times as much of the pasteurized to even start to come close to the effects we received from the unpasteurized. Clearly, the unpasteurized noni juice was the one for us to use long-term, but why had the unpasteurized juice been so much more effective than the pasteurized? The increased potency of unpasteurized noni juice is best explained by Dr. Young who states, "Pasteurizing juice—and almost all of it, even in the health food stores, is pasteurized—evaporates the enzymes and destroys the life force." In fact, Dr. Young instructs to ". . . avoid any pasteurized noni products."[50]

We ordered more unpasteurized noni juice and once it arrived, resumed our initial dosages. We used the unpasteurized noni juice to combat the fungi and mycotoxins, evaluating our results by monitoring the reduction of visible fungi and other symptoms. We quickly regained the ground we had lost while taking the pasteurized noni juice and then made additional progress.

Another product that proved quite helpful was the Rosemary Lavender Noni Soap. The ingredients of the noni soap are saponified vegetable oils (olive, soy, coconut, and palm), lavender essential oil, rosemary essential oil, and noni powder. The noni soap immediately soothed our skin, helped ease the itching, and, ultimately, along with the unpasteurized noni juice, helped heal our hives. After five months of using both the noni soap and the unpasteurized noni juice, the scars on our skin from hives were barely noticeable.

Prior to using the noni soap, we had tried, to no avail, an oatmeal soap that was supposed to ease itching and heal scabs, a glycerin soap, a baby soap, and a few others—none of which helped. We were amazed and pleased when the Rosemary Lavender Noni Soap gave our itchy, bleeding, dry, scaly skin relief.

Why did the noni soap work such wonders? We researched some of the ingredients and found, interestingly, that rosemary is identified as having antifungal properties.[51] In addition, certain varieties of rosemary activate the metabolism in the outer layer of skin and improve cell regeneration.[52] Lavender, one of the other ingredients, is reported to have therapeutic properties for the skin as well. Health literature states of lavender, "The classic oil of aromatherapy has the broadest spectrum of benefits—it can be used undiluted on burns, small injuries, skin ulcers, eczema, and insect bites. Lavender's high ester content gives it a calming, almost sedative quality."[53] Then, of course, there are the skin healing properties of noni. Literature states, "One of the most prevalent historical uses of noni was in poultice form for cuts, wounds, abrasions, burns, and bruises."[54] We also have used noni juice in combination with aloe vera for pain relief from a burn when aloe vera alone did not provide sufficient relief.

The noni tincture did not help alleviate our skin condition, but it was quite effective in reducing swelling, itching, and pain from insect bites (both yellow jackets and mosquitoes).

As an alternative to juice, we also tried the noni capsules for a short time period.

They were produced from certified organic or wildcrafted Grade A ripe noni fruit and contained the entire fruit—including pulp, peel, and seeds.[55] Interestingly, the noni capsules appeared to be more effective against the fungal growth in our sinus cavities than was the unpasteurized noni juice. The noni in the capsules acted as a strong drying agent. The more our sinuses dried out, the more they healed. However, because the noni capsules gave us a slight "upperish" feeling, we discontinued their use and resumed our dosages of unpasteurized noni juice.

Why did we experience a different response from the noni capsules versus the liquid unpasteurized noni? One possible reason is that the capsules we took contained the whole fruit, including seeds, pulp, and peel, all of which may have additional properties that would not be present in the juice from fermented, pressed fruit.

After ten days on the initial dose of unpasteurized noni juice, we were making solid inroads in the battle against the fungi that had taken hold in our bodies. We observed consistent progress, measured by diminishing symptoms. Gone was the congestion and feeling of incredible pressure on our chests from built-up fluid in our lungs. In addition, we had seen about a 50 percent reduction in hives.

We were eventually able to formulate some theories based on our personal observations of our own healing of the effects of the unpasteurized noni juice. It appeared to draw fungi from the inside to the outside of our bodies, through the pores of our skin and other eliminatory channels. This process went on for months, sometimes in cycles. For example, the fungi on the bottoms of our feet would begin to clear, and then another round would come out from more deeply infected tissues. The fungi on other parts of our skin acted in a similar fashion; sometimes it would be denser in certain areas, coming out in patches. We also had a continual weeping from the pores in our skin. We could see this pattern in each of us; just when we thought the fungi were all gone, another round would appear. We were making progress, but we could see that the process of the fungi working out of our bodies was going to take time.

At our next appointment, our naturopath saw the progress we were making but felt adamant that we each needed to be taking a higher dose. Astute observation of changes in a patient's physical and mental condition is the value of working closely with a licensed physician. We would not have known to increase our dosage levels at this time. We are grateful to our naturopath for recognizing this necessity. It was a critical turning point in our health recovery.

Our naturopath increased the number of doses per day from twice a day to four times a day, with Kurt and me each taking 3 ounces, our daughter taking 2 ounces, and our son taking 1 ounce. After we began taking the new dosage levels, we saw an increase in fungal demise, with fungi consistently moving from the inner tissue to the outer layer of skin. No longer was the fungi coming out in rounds, interspersed with days at a time void of signs of fungal exodus. At this point, we could see that this particular brand of unpasteurized noni juice was making solid, consistent headway against the fungi (possibly multiple species) that had set up camp in our tissues and other areas of our bodies. It also appeared to be neutralizing the mycotoxins. We continued on this dosage and frequency level through 36 bottles (32 ounces each), after which we began a maintenance dose of 1 to 3 ounces each per day, as needed.

After we saw the visual effects of the unpasteurized noni juice successfully combat fungi and mycotoxins, we began to research scientific and medical studies to see if the antifungal and toxin neutralizing properties of noni had been scientifically studied. There are several significant published, peer-reviewed research papers regarding scientific studies of noni.[56] Let's take a look at a few.

One animal study evaluated the effects of noni on tumors in mice. It was done by the Department of Pharmacology at the John A. Burns School of Medicine at the University of Hawaii at Manoa. The doctors report, "We have confirmed the anti-tumor potential at animal level that noni fruit juice could inhibit murine tumor growth with a definite curative potential. . . . Noni fruit juice is not cytotoxic in cell cultures . . . but the juice can indirectly kill the cancer cells via activation of the cellular immune system involving macrophages, natural killer cells, and T cells. Therefore, noni fruit juice is [a] powerful antitumor immunostimulator of plant food origin without having toxicity."[57] That the juice was not cytotoxic in cell cultures means that the juice did not kill cells in the cultures.

Another study, led by the Department of Physiology at the Louisiana State University Health Sciences Center, tested the effects of noni juice using ". . . human placental vein and human breast tumor explants [living tissue transferred to culture form] as sources for angiogenic [the formation and differentiation of blood vessels] vessel development. The published report states, "The results obtained in these studies demonstrate that juice from *Morinda citrifolia* is effective in:

(1) inhibiting new angiogenic growth in human placental vein explants
(2) reducing the rate of capillary proliferation and the development of vascular networks in vein discs in which angiogenesis does occur
(3) inducing apoptosis [programmed cell death] in newly formed angiogenic networks
(4) suppressing both angiogenic incidence and vessel development in human breast cancer explants."[58]

Interestingly, the report states (in contrast to the earlier study involving the anti-cancer activity of noni fruit juice against tumors in mice), ". . . the effects of noni on angiogenesis do not appear to be mediated by the immune system, as no leukocytes [white blood cells] are available in the human vein disk culture."[59] In other words, no immune response is possible when working with a culture versus a live specimen. This is noteworthy because, since the stimulation of the immune system was not responsible for the tumor-inhibiting effects of noni juice on the tumor explants, other properties or activations—individual or combined—resulting from the noni juice used in this study must be responsible.

Another study performed at the University of Hawaii at Manoa focused on ". . . the cytotoxic effects of water and ethanol extracts of whole noni fruit, pulp, peel, and seed" on breast cancer cell lines. Somewhat surprising, the initial data indicates that the noni extract preparation used in this particular study demonstrates general cytotoxicity on both normal breast cells and on non-invasive and invasive breast carcinoma cells.[60] The significance, of course, is that the extracts of whole noni fruit—including pulp, peel, and seed—was toxic to normal breast cells as well as to the carcinogenic breast cells. The other two aforementioned studies used noni fruit juice, which of course would be void of the pulp, peel, and seeds (as long as the juice used was a pressed or squeezed form, not reconstituted from dehydrated or concentrate sources of noni that included the pulp, peel, and seed). The difference in the findings of these studies could lead one to hypothesize that the cytotoxicity was caused by one of the additional components—the pulp, peel, and/or seeds, a combination thereof, or some unidentified influencing factor in the study. However, a conclusion is undeterminable based on the data available from the study.

Does the cytotoxicity of extracts of whole noni fruit under certain laboratory conditions mean that noni products that contain the pulp, peel, and/or seeds of the noni fruit may not be safe for human consumption? The answer to this question cannot be scientifically concluded based on culture and/or animal studies because direct correlations cannot be drawn from culture and/or animal studies to predict human outcome, according to Dr. Straus at Texas Tech University Health Sciences Center.[61] For as much as this study reveals, we still do not know which component was cytotoxic, nor do we know if it would be cytotoxic to cells in the human body. Further research and studies addressing possible toxicity of the noni seed, leaf, root, and bark need to be conducted, according to David Backstrom, ND.[62]

This line of study is important because if an ingredient is toxic at the cellular level even in cell cultures, it could be an indication that it would also be toxic to cells in the human body. Human toxicity may not necessarily occur because cells in cell cultures can react differently than living cells in the body, according to Dr. Straus.[63] However, if seed inclusion in a product is of a concern, noni juice that is reconstituted from noni powder purchased in bulk should be avoided, as it could include seeds. A report states, "Most of the drying and powdering operations are involved with whole fruit (slicing, drying, and grinding of whole fruits into powder, including seeds)."[64] Additionally, it is important to note that noni capsules that list noni fruit as the ingredient can contain the fruit, as well as the pulp, peel, and seeds, as they are all a part of the fruit. Eventually, scientific studies may prove whether noni seeds are cytotoxic to human cells, but at this time, it is still scientifically unknown.

Up to this point, all the cited studies have involved culture samples and/or animals. When asked about these previous studies, Dr. Issell, who began the first human study investigating the effects of noni in humans in 2001 at the Cancer Research Center of Hawaii, University of Hawaii,[65] states, "The studies you cite are very different and are most likely irrelevant to what we are able to do in human testing. These preclinical systems have yet to be shown to have any prediction for human results. Product ingredients and dosing are vastly different from what we can do in humans."[66]

Dr. Issell explains the human noni study, "We are conducting the earliest type of study (Phase I) to identify the tolerance and potential toxicities (adverse side effects) of different doses of noni in cancer patients for whom no known effective treatment is available."[67]

Phase I refers to the phases of clinical trials and human research as defined by the FDA in the Code of Federal Regulations.[68] Typically, "In Phase I clinical trials, researchers test a new drug or treatment in a small group of people (20–80) for the first time to evaluate its safety, determine a safe dosage range, and identify side effects."[69]

The broad long-range objective of Principal Investigator, Dr. Issell, and his research team is ". . . to define the usefulness of noni extracts for cancer patients. The hypothesis to be tested is that noni at a specified dosing provides cancer patients with a sufficient benefit to toxicity profile to be useful as a therapeutic. Specific aims of this study are [to]:
1. Determine the maximum tolerated dose of capsules containing 500 mg of freeze-dried noni fruit extract.[70]
2. Define toxicities associated with the ingestion of noni.
3. Collect preliminary information on the efficacy of noni in respect to anti-tumor and symptom control properties to help select specific patients for subsequent Phase II studies.
4. Identify chemical constituents of the extract that can be used to characterize the bioavailability and pharmacokinetics of noni food supplements."[71]

"So far, we have entered over 50 patients into the study at different dose levels. We have found no dose-limiting toxicities with doses up to 28 capsules (14 grams) daily. We are continuing to enter new patients, and in the absence of any dose-limiting effects, we are hoping to identify the best dose based on quality of life measures. We have found that some quality of life measures are better at certain doses when compared with the other doses and are hoping to identify a maximum quality of life sustaining dose that we will use for subsequent Phase II efficacy studies with placebo controls. Through this analysis, we also hope to identify patients most likely to be helped by noni as subjects for a Phase II study," states Dr. Issell.[72]

Since results could vary greatly based on the processing of a noni product, the obvious question to ask is: How has the noni product that is being used in the study been processed? "Initially, we had noni supplied in freeze-dried formulation by Innovative Nutriceuticals. They subsequently stopped production and helped us get supplied with a dehydrated formulation from another supply source," answered Dr. Issell.[73]

In the first product, according to Dr. Issell, "The whole fruit was freeze dried."[74] The designation of "whole fruit" means that the fruit, pulp, peel, and seed were all used. Noteworthy is the fact that the second product used in the study did not contain the seed and are dehydrated versus freeze-dried. When asked if any differences were observed in the health effects of the two different forms of noni capsules supplied and if the unforeseen product change elongated the study due to a need for product continuity over a specific duration, Dr. Issell replied, "No different effects with formulation change were observed. We were able to continue increasing dosage within the original study design since no different effects were observed. A minimum of five patients are observed for a minimum of four weeks before going to the next dose level."[75]

When asked if the exclusion of seeds in the dehydrated noni extract was a factor considered when selecting the replacement product or if fermentation versus non-fermentation issues played a part in the final product selection, Dr. Issell replied, "Issues about seeds and fermentation are speculative at best. We chose the product we did for our study because it was the best defined and most consistent one we could get."[76]

Since human scientific and medical noni studies are still in their infancy, much is still scientifically and medically undetermined at this time, such as the medicinal properties of traditionally fermented noni (aged three to six weeks) as compared with non-fermented noni (aged for only a few days to allow the fruit to soften). Dr. Issell believes, "Fermenting noni will likely give a different profile of chemical ingredients. However, we have no knowledge as to whether this is a good or bad thing for noni's medicinal properties. It is speculative at best to state that fermentation is important for its medicinal properties."[77] However, as discussed earlier, a fermented noni juice will not have the sugar content of that of nonfermented noni juice.

Assisting Dr. Issell in this study are researchers Carolyn Gotay, PhD, a psychologist; Faith Inoshita, RN, MS, a clinical research nurse; and other research team members. The study has been detailed in local media. "They are not only looking at anti-tumor properties, but monitoring participants' quality of life to see whether noni relieves fatigue, pain, and other cancer symptoms," reports the

Honolulu Star-Bulletin.[78] "Patients fill out questionnaires about how they're feeling, how their pain is, their fatigue, and how they function," continues the *Honolulu Star-Bulletin.*[79] It further reports, "Dr. Adrian Franke, associate specialist, and Laurie Custer, research associate, are analyzing the ingredients of noni, as well as blood and urine samples from patients, to see what chemicals may have anti-cancer activity."[80] In addition, measurements of tumors are taken and monitored.[81] Thus, the results of Phase I will provide the first clinically observed, human health effects of noni with measurable components.

Dr. Issell adds, "We are studying the chemical profile of the product we are using and the degree to which different ingredients get absorbed and excreted in the patients participating in our study so that noni products can be standardized for chemical consistency."[82] He points out, "Scopoletin is the ingredient we are using to standardize noni. It is a biologically active chemical that we have been able to measure in the blood and urine of patients after noni is taken."[83]

Published studies report, "Scopoletin is scientifically proven to have antibacterial properties."[84] However, isolating one component for standardization may not deliver the full healing potential of noni, as there are many active ingredients, some of which may also synergistically interact. For example, noni fruit and fruit juice contain alkaloids, polysaccharides, vitamins, and minerals in addition to scopoletin.[85] In our opinion, attempting to isolate one component for standardization is a bit like messing with the goose that laid the golden egg, cliché or not.

The Phase I study was completed June 2006.[86] Before completion, Dr. Issell reported that they had seen increased improvement in the quality of life in patients and had gotten improvement at higher dose levels compared with lower dose levels.[87] An abstract of early Phase I study findings was published in the *Official Journal of the International Society for Quality of Life Research*, which contains abstracts accepted for presentation at their annual conference. The abstract concluded, "Quality of life measures may be useful for selecting dosing for subsequent phase II/III efficacy trials, particularly for complementary and alternative medicine agents that are commonly used but of unproven value."[88]

Another abstract of early Phase I findings was published in the *Journal of Clinical Oncology*, which publishes abstracts presented at or in conjunction with the American Society of Clinical Oncology (ASCO) Annual Meeting Proceedings. The ASCO abstract concluded, "Quality of life measures may be useful for selecting dosing (maximum quality of life sustaining dose) for phase II efficacy trials of non toxic dietary supplements."[89] Both abstracts reported that no adverse toxic events were attributable to noni and no measured tumor regressions were noted.[90]

The data collected in Phase I was carefully analyzed to select a dose of noni for ". . . subsequent Phase II studies with scientific confidence," explains Dr. Issell.[91]

In 2009, the following conclusions drawn from the Phase I data were published in the Journal of Dietary Supplements[92]:

- A noni dose of 4 capsules four times daily (8 grams) is recommended for Phase II testing where controlling fatigue and maintaining physical function is the efficacy of interest.

- These results suggest that an optimal quality of life sustaining dose focused on maintenance of physical activity and control of fatigue appears to be four capsules four times daily (8 grams daily).

Typically, "In Phase II clinical trials, the study drug or treatment is given to a larger group of people (100–300) to see if it is effective and to further evaluate its safety. In Phase III studies, the study drug or treatment is given to large groups of people (1,000–3,000) to confirm its effectiveness, monitor side effects, compare it to commonly used treatments, and collect information that will allow the drug or treatment to be used safely."[93] Whether the Phase II study will take place and provide further scientific evidence to support Polynesian folklore and the widespread anecdotal reports of the healing properties of noni is uncertain at this time.

Personal testimonies like ours will continue to speak for the healing potential of noni. We can only attest to our experience with the traditionally fermented raw (unpasteurized) noni juice from which we observed our family's healing from fungal- and mycotoxin-related illnesses. Since we were not using any other form of treatment at the time aside from our sugar- and carbohydrate-restricted diet and the combination of fresh juiced ginger, garlic, and carrots, it is our opinion that the unpasteurized noni juice was responsible for the progression of our healing, drawing the fungi out of our bodies and neutralizing the mycotoxins.

Noni's ability to neutralize toxins is further supported by historical reports that South Pacific natives used noni to neutralize toxins from bites, stings, and poisonous fish.[94] In addition, noni is historically reported to have been used by island healers to kill intestinal parasites,[95] a term that describes parasitic fungi colonizing in the intestines. These toxin-neutralizing and anti-parasitic properties further explain noni's effectiveness in combating fungal-related conditions.

Further proof that the unpasteurized noni juice was the component in our healing regimen that effectively combated the fungi and mycotoxins is evidenced by the fact that when we stopped taking the noni juice for a day or two (testing to see if we could go on a maintenance dose), we would visually see a fungal relapse. Eventually, we were able to reduce our noni consumption to a maintenance dose without a fungal relapse.

The noni juice was not the only supplement that was a part of our treatment plan. After several weeks of fungal treatment with the unpasteurized noni juice, we began to slowly add other alternative healing aids to further support our healing process, boost our immune systems, and support the detoxification process. However, even with these added components, we feel the noni juice still handled the "lion's share" of the elimination of the fungi.

People often ask if other brands of noni juice carried at grocery store chains or health food stores will be as effective as the unpasteurized noni juice we used during our recovery. The answer to that question depends on the following factors, which can alter the chemical makeup of the end product and directly affect product effectiveness:

1. Is it raw (unpasteurized) noni juice or pasteurized?
2. Is it sourced from certified organic noni fruit or non-organic?
3. Is it fermented or fresh-squeezed?
4. Is the juice diluted with water or an alternate juice, such as grape or guava?
5. Is it additive-free or spiked with additives, such as sugar?

During our initial recovery, we selected a brand of noni juice with the following attributes:

1. A Hawaiian noni juice
2. A raw (unpasteurized) noni fruit juice
3. Sourced from certified organic noni fruit
4. Processed traditionally with fermentation
5. No dilution
6. No additives

The brand of noni juice we used during our recovery is no longer available, but we were able to locate another company, Healing Noni Inc., that manufactures noni juice with these same specifications. Our family has continued to use unpasteurized noni juice as a food supplement due to its overall health benefits. We have found the quality of the certified organic raw (unpasteurized) noni juice from Healing Noni Inc., to be of equal or higher quality as our initial supply source.

How can you purchase noni like a local? The Hawaiian word Kama'aina (ka-ma-eye-nah) is used to refer to someone who has lived in Hawaii for a long time. In Hawaii it is tradition to give a Kama' aina discount to locals. The owners of Healing Noni would like to embrace you as a local by extending a one-time 10% Kama' aina discount for you to use when placing your first order. To receive this discount, identify yourself as a reader of *MOLD: The War Within* when ordering via the toll-free number 1-877-662-4610 or enter the promo code "MoldMentor" in the coupon code field when ordering online at www.HealingNoni.com/raw-organic-noni-fruit-juice-32oz/ . Substantial discounts are also offered on multiple bottle orders.

Please note, when we used noni juice in medicinal doses, we were under the care of our medical doctor who was monitoring our health and progress. The dose levels that our doctor prescribed for us should not be used to determine dose levels for anyone else unless prescribed by their own doctor. People need to consult with their own physicians to determine the appropriate dosage levels for themselves based on their own health conditions before using noni juice as a form of medicinal treatment. Our results may or may not be typical, even given similar diet restrictions and supplement regimens.

For more information on noni juice, download our digital book *Why Noni?* It is available on Amazon and through other digital retailers.

Disclaimer: The information and research presented in this book is for educational purposes only. It is not intended to be medical advice or replace the services of a licensed health care professional. The supplements referenced in this book have not been evaluated by the Food and Drug Administration and are not approved drug products. The information and products presented in this book are not intended to diagnose, treat, cure, or prevent any disease. All physical and mental conditions should be diagnosed, treated, and monitored by a licensed, health care provider.

(ENDNOTES)

1 Interview with Dr. Charles Bacon, August 2006; See Section Three, interview with Dr. Jia-Sheng Wang, July 2006; See Section Three, interview with Dr. David Straus, April 2006; and Ammann, "Is Indoor Mold Contamination a Threat to Health?"

2   See Section Three, interview with Dr. Jack Thrasher, April 2006; Interview with Dr. Michael Gray, March 2007.

3   Rita Elkins, M.H., *Noni (Morinda citrifolia)*, p. 12 (Pleasant Grove, Utah, Woodland Publishing, 1997).

4   Helen Altonn, "Noni Shows Cancer Promise," July 24, 2005, *Honolulu Star-Bulletin*. Available at http://starbulletin.com/2005/07/24/news/story5.html (accessed July 2006).

5   Altonn, "Noni Shows Cancer Promise."

6   Ibid.

7   Elkins, *Noni (Morinda citrifolia)* p. 13.

8   Ibid., p 7.

9   Ibid., p 13.

10   Scott C. Nelson, PhD, "Noni Cultivation and Production in Hawaii." Available at http://www.ctahr.hawaii.edu/noni/downloads/noni33_50.pdf (accessed July 2006).

11   Goldberg, *Alternative Medicine*, p. 230.

12   Ibid., p. 230.

13   Nelson, "Noni Cultivation and Production in Hawaii."

14   Bengt Danielsson, "Poisoned Pacific: The Legacy of French Nuclear Testing," March 1990, pp. 22-31, vol. 46, no. 02, *Bulletin of the Atomic Scientists*. Available at http://www.thebulletin.org/article.php?art_ofn=mar90danielsson (accessed March 2006).

15   Derek Woolner, "Current Issues Brief No. 47 1994/95 Raison d' Etat and popular response: The resumption of French nuclear testing in the South Pacific," Parliamentary Research Service, June 22, 1995. Available at http://www.aph.gov.au/library/pubs/cib/cib94-95.htm (accessed March 2006).

16   Danielsson, "Poisoned Pacific: The Legacy of French Nuclear Testing," and Woolner, "Current Issues Brief No. 47 1994/95 Raison d' Etat and popular response: The resumption of French nuclear testing in the South Pacific."

17   Centre for Research and Information on Peace and Conflict (CDRPC), "Synopsis of the Inquiry Commission's Report," Lyon France. Available at http://nautilus.rmit.edu.au/forum-reports/Synopsis_of_Hirshon.pdf. (accessed July 2006).

18   *Pacific Magazine*, "French Polynesia: Tahiti Subjected to Nuclear Test Fallout, Says Leaked Report," January 26, 2006. Available at http://www.pacificislands.cc/pina/pinadefault2.php?urlpinaid=19864 (accessed July 2006).

19   Ibid.
20   Ibid.
21   CDRPC, "Synopsis of the Inquiry Commission's Report."
22   *Pacific Magazine*, "French Polynesia: Tahiti Subjected to Nuclear Test Fallout, Says Leaked Report."
23   CDRPC, "Synopsis of the Inquiry Commission's Report."
24   *Pacific Magazine*, "French Polynesia: Tahiti Subjected to Nuclear Test Fallout, Says Leaked Report," and CDRPC, "Synopsis of the Inquiry Commission's Report."
25   Ibid.
26   CDRPC, "Synopsis of the Inquiry Commission's Report."
27   Greenpeace, "Cracks in Moruroa Atoll Confirm Dangers of Global Nuclear Industry," May 5, 1999. Available at http://www.greenpeace.org/australia/news-and-events/media/releases/nuclear-power/cracks-in-moruroa-atoll-confir (accessed February 2006).
28   Ibid.
29   Ibid.
30   Ibid.
31   Nelson, "Noni Cultivation and Production in Hawaii."
32   Ibid.
33   Ibid.
34   NDSU Extension Service, "Why Add Lemon Juice to Tomatoes and Salsa Before Canning?" Available at http://www.ext.nodak.edu/food/lemnjuic.pdf (accessed August 2006).
35   Goldberg, *Alternative Medicine*, p. 589.
36   Interview with David Marcus, president of Hawaiian Herbal Blessings, July 2006.
37   Nelson, "Noni Cultivation and Production in Hawaii."
38   Interview with David Marcus, president of Hawaiian Herbal Blessings, July 2006.
39   Jay Ram, "Noni Processing and Quality Control: Protecting the Image of Hawaiian Products," from: Proceedings of the 2002 Hawaii Noni Conference, S.C. Nelson (ed.), University of Hawaii at Manoa, College of Tropical Agriculture and Human Resources, 2003. Available at http://www.ctahr.hawaii.edu/noni/Downloads/noni25_28.pdf. (accessed July 2006).
40   Ram, "Noni Processing and Quality Control: Protecting the Image of Hawaiian Products."
41   E-mail from Dr. Brian Issell, July 2006.
42   Ibid.
43   Interview with David Marcus, president of Hawaiian Herbal Blessings, July 2006.
44   Crook, *The Yeast Connection Handbook*, p. 123.
45   Ibid.
46   Interview with Dr. Jack Thrasher, May 2006.
47   Scott C. Nelson, PhD, "*Morinda citrifolia* Species Profiles for Pacific Island Agroforestry," April 2006. Available at http://www.traditionaltree.org/Morinda(noni).pdf (accessed July 2006).
48   Ram, "Noni Processing and Quality Control: Protecting the Image of Hawaiian Products."
49   Ibid.
50   Young et al., *The pH Miracle*, p. 97 and 165.
51   Goldberg, *Alternative Medicine*, p. 625.
52   Ibid., p. 82.
53   Ibid.
54   Elkins, *Noni (Morinda citrifolia)*, p. 25.
55   Interview with David Marcus, president of Hawaiian Herbal Blessings, July 2006.
56   Significant noni medical research papers as listed in "Cytotoxicity of Water and Ethanol Extracts of *Morinda citrifolia (L.)* Against Normal Epithelial and Breast Cancer Cell Lines," by A. Johnson et al.: Indian, "Screening of indigenous plants for antihelmintic action against human *Ascaris lumbrocoides*: Part II," *J Physiol Pharmacol* 1975 Jan-Mar 19; Younos C. et al., "Analgesic and behavioral effects of *Morinda citrifolia*," *Planta Med* 1990 Oct; 56(5): 430-4; Hiramatsu T. et al., "Induction of normal pehnotypes in ras-transformed cells by damnacanthan from *Morinda citrifolia*," *Cancer Lett* 1993 Sep 30; 73(2-3); 161-6; Hirazumi A. et al., "Anticancer activity of *Morinda citrifolia* (noni) on intraperitoneally implanted Lewis lung carcinoma in syngenic mice," *Proc West Pharmacol Soc* 1994; 37: 145-6; Hirazumi A. et al., "Immunomodulation contributes to the anticancer activity of *Morinda citrifolia* (noni) juice," *Proc West Pharmacol Soc* 1996; 39: 7-9; Hiwasa et al., "Stimulation

of uv induced apoptosis of human fibroblast UVr-1 cells by tyrosine kinase inhibitors," *FEBS Lett* 1999 Feb 12; 444 (2-3): 173-6; Hirazumi A. et al., "An immunomodulatory polysaccharide-rich substance from the juice of *Morinda citrifolia* (noni) with antitumouractivity," *Phytother Res* 1999 Aug; 13 (5) 380-7; Wang et al., "Cancer preventative effect of *Morinda citrifolia* (Noni)." *Ann NY Acad Sci* 2001 Dec; 952: 161-8.

57   Eiichi Furusawa, MD, PhD, "Anti-cancer Activity of Noni Fruit Juice Against Tumors in Mice," Proceedings of the 2002 Hawaii Noni Conference, S.C. Nelson (ed.), University of Hawaii at Manoa, College of Tropical Agriculture and Human Resources, 2003.

58   Conrad A. Hornick et al., "Inhibition of angiogenic initiation and disruption of newly established human vascular networks by juice from *Morinda citrifolia* (noni)," *Angiogenesis* 6: 143-149 (Printed in the Netherlands, Kluwer Academic Publishers, 2003).

59   Ibid.

60   A. Johnson et al., "Cytotoxicity of Water and Ethanol Extracts of *Morinda citrifolia (L.)* Against Normal Epithelial and Breast Cancer Cell Lines," Proceedings of the 2002 Hawaii Noni Conference, S.C. Nelson (ed.), University of Hawaii at Manoa, College of Tropical Agriculture and Human Resources, 2003.

61   See Section Three, interview with Dr. David Straus, April 2006.

62   Interview with Dr. David Backstrom, president and CEO of Noni Maui, July 2006.

63   See Section Three, interview with Dr. David Straus, April 2006.

64   Ram, "Noni Processing and Quality Control: Protecting the Image of Hawaiian Products."

65   Helen Altonn, "Noni's medical worth being tested," January 8, 2004, *Honolulu Star-Bulletin*. Available at http://starbulletin.com/2004/01/08/news/story8.html (accessed July 2006).

66   E-mail from Dr. Brian Issell, July 2006.

67   Ibid.

68   NIH, "Information on Clinical Trials and Human Research." Available at http://www. clinicaltrials.gov/ct/info/phase?style=nohdr (accessed July 2006).

69   Ibid.

70   According to an e-mail from Dr. Brian Issell, July 2006, the initial noni supplied was a freeze-dried formulation by Innovative Nutriceuticals. When this company subsequently stopped production, it helped Dr. Issell secure a new supply source of noni for the noni study, which was a dehydrated formulation from Noni Maui.

71   Cancer Research Center of Hawaii, University of Hawaii, "The Noni Study." Available at http://www.crch.org/CenStudyNoni.htm (accessed July 2006).

72   E-mail from Dr. Brian Issell, July 2006.

73   Ibid.

74   Ibid.

75   Ibid.

76   Ibid.

77   Ibid.

78   Altonn, "Noni's medical worth being tested."

79   Altonn, "Noni Shows Cancer Promise."

80   Ibid.

81   Helen Altonn, "Noni Research will begin to test the plant against cancer and its symptoms," July 19, 2001, *Honolulu Star-Bulletin*. Available at http://starbulletin.com/2001/07/19/news/story11.html (accessed July 2006).

82   E-mail from Dr. Brian Issell, July 2006.

83   Ibid and Brian F. Issell et al., "Pharmacokinetic Study of Noni Fruit Extract" Journal Of Dietary Supplements, Volume 5, Issue 4 December 2008 , pages 373 - 382 at http://www.informaworld. com/smpp/content~db=all~content=a906947009.

84   Kayser O et al., "Antibacterial activity of extracts and constituents of *Pelargonium sidoides* and *Pelargonium reniforme*," December 1997, 63 (6): 508-10, *Planta Med*. Available at http://www. ncbi.nlm.nih.gov/entrez/query.fcgi?cmd=Retrieve&db=PubMed&list_uids=9434601&dopt=Abstract (accessed July 2006).

85   College of Tropical Agriculture and Human Resources, University of Hawaii, "The Noni Website." Available at http://www.ctahr.hawaii.edu/noni/chemical_constituents.asp (accessed July 2006).

86   Interview with Faith Inoshita, January 2007, and http://clinicaltrials.gov/ct2/show/
NCT0033878
87   Altonn, "Noni Shows Cancer Promise," and Altonn, "Noni's medical worth being tested."
88   Brian F. Issell et al., "Quality of Life (QOL) Assessment in a Phase I Trial of Noni" *Official Journal of the International Society for Quality of Life Research* Volume 14 Number 9 November 2005 at http://www.isoqol.org/2005ConfAbstracts.pdf.
89   B.F. Issell et al., "Quality of Life Measures in a Phase I Trial of Noni" *Journal of Clinical Oncology*, 2005 ASCO Annual Meeting Proceedings. Vol 23, No. 16S, Part I of II (June 1 Supplement), 2005: 8217 at http://meeting.ascopubs.org/cgi/content/abstract/23/16_suppl/8217.
90   See endnotes 88 and 89.
91   E-mail from Dr. Brian Issell, July 2006.
92   Issell, B., "Using Quality of Life Measures in a Phase I Clinical Trial of Noni in Patients
          with Advanced Cancer to Select a Phase II Dose", *J Diet Suppl 2009: 6(4): 347–359.*
93   NIH. "Information on Clinical Trials and Human Research."
94   Elkins, *Noni* (*Morinda citrifolia*), p. 12. Ibid., p. 16.
95   Ibid., p. 7.

CHAPTER SEVEN:

WHAT'S IN A BRAND NAME?

After a month of using the unpasteurized noni, we were all making consistent, visible progress against the fungi that foraged within our bodies, yet we were still getting sick—sore throats, lingering lung congestion, ear aches, general overall fatigue, and continuous exhaustion. Our immune systems were still in an obviously weakened state from the fungi, the effects of the mycotoxins, and the medical mistreatment.

Since our evacuation by the Red Cross to our new home, we had not been healthy enough to accept the few social invitations that had been extended to us. Most people kept their distance. After all, we were sick from exposure to *something* following Hurricane Katrina! The unknown is scary. It was for us, too.

Some particularly friendly people kept inviting us to social gatherings, but we had been too sick to do anything more than send them an e-mail of decline. Finally, after a month of treatment with the unpasteurized noni, I felt well enough to call to thank them for the many invitations. A pleasant conversation ensued during which concern was expressed for our family, because we had been constantly sick since arriving from Louisiana. Permission was asked to bring over a care package for us. We were surprised at this outpouring of generosity from complete strangers. The unexpected act of kindness was welcome after months of forced isolation from being so sick. It is true that being sick can be a very lonely time, even if there are four of you sick together! You tend to feel that everybody else in the world is going on with their normal everyday lives, while yours is not normal at all!

The next day these benevolent people arrived with six boxes of fresh vegetables and fruits, canned foods, and various vitamin supplements. It was a level of generosity not often seen. What is noteworthy about this story is that the combinations of supplements ended up being the final link to our health restoration. To share what worked and why, we will discuss our family's personal experiences in regard to these supplements and present applicable medical and scientific documentation. We will not identify or discuss specific brands or manufacturers, but instead we will discuss the results of health studies that pertain to these specific supplements and examine ingredients that may be instrumental in increasing the effectiveness of certain supplements. After all, it's not the brand name that's important, it's the list of ingredients, quality of ingredients, manufacturing processes, and preservation ability of the packaging. Furthermore, manufacturers often revamp formulas or change supply sources of raw materials, both of which can alter product efficacy. Regular monitoring of ingredients and levels of product effectiveness is necessary. For this reason, we do not recommend or endorse the use of any supplement, any particular brand of supplement, or any particular manufacturer of supplements in this chapter or any other chapter in this book.[1] We are just sharing what our family found to be effective for us.

Prior to receiving the care package supplements, we had already been experimenting with various vitamins, as we knew our immune systems needed more support. We were still constantly battling one sickness after another and were lacking strength and stamina. We had been searching scientific and medical research for answers on how to fully overcome the powerful immune-suppressive effects of mycotoxins, to no avail. However, to our surprise, when we started taking this particular group of vitamin supplements, we stopped getting sick. We were shocked. No other combination of vitamins had noticeably boosted our immune systems, let alone to a level that we stopped getting sick. Why?

To answer that question, we started researching. We wanted answers based on current scientific and medical research, if possible, to give you, the readers of this book, answers right along with us. We wanted to be able to offer you more than just our personal testimonies. In the end, what we found gave us logical scientifically and medically based explanations as to why this combination of supplements was effective.

First, let's take a look at the particular vitamins and herbs that collectively were a critical component of our healing program and boosted our immune systems out of months of misery and sickness. We'll begin by discussing the adult supplements that were the final components to the treatment plan that restored Kurt's and my health, and then we'll address the specific supplements for kids that completed a successful treatment plan for our children.

Kurt and I each took a daily combination, three times a day, of the following: four vitamin C, two echinacea, and two garlic supplements. This combination effectively boosted our immune systems to the point that our sore throats went away and our ear aches cleared up; we could breathe pain free for the first time in months. We saw noticeable results just days after starting this combination of supplements. After having these initial products have such a positive and health-enhancing effect, we slowly added additional supplements, which we will later review. We were careful to try these additional vitamins, herbs, and other supplements one at a time so we would know which ones were adding noticeable health benefits and effectively combating our mold- and chemical-related illnesses. Before we move on to these other supplements, let's review scientific and medical findings relating to what we call the *immune-boosting trio*: vitamin C, echinacea, and garlic supplements.

Scientific and medical literature and research studies clearly document the health-enhancing properties of these three supplements. Let's review some findings.

The first component of the immune-boosting trio, vitamin C (also called ascorbic acid) is a powerful antioxidant, which means it protects cells against damage by free radicals. Free radicals are reactive by-products of normal cell activities that can damage cells, proteins, and DNA by altering their chemical structures.[2] Human and animal studies confirm that diets supplemented with high levels of vitamin C stimulate the immune system.[3] Furthermore, the results of one animal study reveal that high levels of dietary vitamin C supplementation have a beneficial effect on aflatoxin B1-induced immunosuppression.[4] High levels of dietary vitamin C were fed to fish that had been treated with aflatoxin B1 to induce an immunocompromised state. The high levels of vitamin C supplementation increased particular parameters of the nonspecific immunity of the treated fish to levels similar to those found in the control fish that had not been poisoned with aflatoxin B1.[5] The results of these scientific studies lead us to surmise that high intake of vitamin C may help counteract the immune-suppressive effects of mycotoxins, particularly aflatoxins, in humans.

Health literature further states that vitamin C is ". . . essential for tissue growth, wound healing, and absorption of calcium and iron, and utilization of folic acid."[6] It must be consumed and replenished daily because it is a water-soluble vitamin, which means it cannot be stored by the body except in insignificant amounts.[7] Scientists at UC Davis report, "Population studies show that individuals with high intakes of vitamin C have lower risks of a number of chronic diseases, including heart disease, cancer, eye diseases, and neurodegenerative conditions."[8] The researchers note as a possible contributing factor (influencing the lower risk of these particular chronic diseases in people who consume foods high in vitamin C and/or augment their diets with vitamin C supplements) the possibility that these people may embrace more healthful diets and/or lifestyles.[9] We also focused on healthy lifestyle modifications, some of which we have already shared and some of which we will soon discuss. Adhering to a healthy way of life undoubtedly added to our successful results, as vitamins, herbs, and other food supplements alone cannot undo the damage from a destructive lifestyle.

The second component of the immune-boosting trio is the echinacea. Echinacea is documented and widely used as an effective treatment for upper respiratory tract infections associated with colds and flu.[10] Several research studies confirm echinacea's viral-fighting abilities. However, results from other studies are less conclusive. What we do know is that people continue to use echinacea to treat viral colds and flus. It is because of this widespread use of echinacea as well as the public health burden of the common cold that the effects of echinacea use continue to be evaluated and documented in multiple research studies.[11] However, the data generated from these various research studies are often difficult to collectively summarize because the findings are subject to many variables, depending on the design of the study. For example, echinacea used in the various studies are prepared with different methods, with different parts of the herb, and sometimes in combination with other herbs, to treat yet a variety of viral illnesses.

In an attempt to draw a scientifically and medically based consensus from all these various studies, a meta-analysis was conducted by Sachin A. Shah, Doctor of Pharmacy, C. Michael White, Doctor of Pharmacy, and colleagues at the School of Pharmacy at the University of Connecticut, Storrs Campus.[12] A meta-analysis is a ". . . quantitative statistical analysis that is applied to separate but similar experiments of different and usually independent researchers and that involves pooling the data and using the pooled data to test the effectiveness of the results."[13] The conclusions drawn from the meta-analysis confirm the immune-stimulating effects of echinacea and its effectiveness in fighting viral infections.

Dr. Shah reported at the 2006 annual meeting of the American College of Clinical Pharmacology that use of echinacea before the onset of full-blown symptoms of the common cold reduces the incidence by more than a half and the duration by almost two full days.[14] The results from the echinacea meta-analysis were published in the peer-reviewed journal *Lancet Infectious Diseases* in 2007. Based on study findings, the authors concluded that echinacea decreased the odds of developing the common cold by 58% and decreased the duration of a cold by 1.25 days.[15] Dr. Shah, Dr. White, and colleagues are not affiliated with any company that manufactures and/or sells vitamins or herbs. Likewise, the study was not funded by any company that manufactures and/or sells vitamins or herbs.[16]

Dr. Christopher explains the effectiveness of echinacea, "Echinacea is a very effective blood purifier, and it is a powerful and stimulating antiseptic and antiputrefactive agent. It is very valuable for correcting autoinfection where a person has not been eliminating toxins well and the tissues and fluids have become septic or putrefactive with resultant weakness."[17] These additional properties help explain why our family has consistently experienced a viral-fighting response from the use of a high quality grade of echinacea.

The third component of the immune-boosting trio is the garlic supplement. We consider it the "super" supplement of the trio. By the time we are done reviewing our research, you will understand why. Technically, garlic is considered an herb. Its health-promoting properties have long been established and documented in health literature. As mentioned in an earlier chapter, an untold numbers of books, articles, and sections within books address the healing properties of garlic, documenting it as an antifungal,[18] a stimulant for the immune system, a natural antibiotic,[19] a powerful detoxifier,[20] an antiviral agent,[21] and an enhancer of antibody production.[22] Furthermore, garlic helps clear congested lungs, coughs, bronchitis, and sinus congestion; is a preventative measure for colds and flu;[23] and kills parasites.[24] As we know from earlier discussions, some species of fungi are parasitic.

The garlic was the supplement about which we were most curious, for obvious reasons. As we shared earlier, we had experimented with every conceivable form of garlic until we had it coming out of our pores, literally. Chewing multiple raw cloves of garlic had done nothing—except nearly make us sick. A highly advertised, celebrity endorsed garlic supplement had provided no noticeable health improvements. An allicin-extracted garlic supplement manufactured with purported innovative technology had proven completely ineffective. So why did the garlic that arrived in our care package, when two tablets were taken three times a day in combination with four vitamin C and two echinacea, effectively restore our immune systems?

In order to answer this question, we took a close look at the ingredients listed on the label of the bottle of garlic, searching for any "special" component that would explain the "super" effectiveness of this particular brand of garlic. Our first discovery was that the tablets were enteric coated. The enteric coating is designed to protect the garlic tablet from dissolving until it reaches the intestinal tract, thus protecting the garlic from stomach acids. The beneficial properties of garlic when taken in the form of raw garlic or garlic supplements without an enteric coating will not reach the intestinal tract with the same intensity and integrity as will enteric coated garlic tablets. When an enteric coated garlic tablet reaches the intestinal tract and begins to dissolve, it is like a bomb filled with antifungal and antibacterial dynamite that explodes right on target to kill the enemy!

Our second discovery was that the garlic we were using contained extract from peppermint leaves. Hmmm . . . We had never tried peppermint—in any form—with our garlic. We began to research the healing properties of peppermint. We found that according to the University of Maryland Medical Center (UMMC), peppermint oil has antiviral properties against a number of infectious agents, and it is an effective decongestant.[25] Then finally, after months of scouring many research books we finally found what we were looking for: An additional property of peppermint that could be responsible for the increased potency of the garlic supplement. *Back to Eden*, a trusted source of herbal information since 1939, states, "Peppermint is a general stimulant. A strong cup of peppermint tea will act more powerfully on the system than any liquor stimulant, quickly diffusing itself through the system and bringing back to the body its natural warmth and glow without the usual tendency to relapse."[26] Hmmm . . . What an ingenious pairing—increase the effectiveness of garlic by propelling it with peppermint, a natural, healthy stimulant. The concept made sense to us. We were off to the grocery store for garlic and—what else? Peppermint, of course! We figured if a commercial supplement manufacturer could make a potent garlic supplement propelled by peppermint, so could we. How hard could it be?

We tried peppermint tea with garlic powder. We tried peppermint leaves with garlic in our carrot juice. We tried peppermint and garlic blended, crushed, and chopped together. Before long, our kids were waving the white flag. Not only were we not getting a comparable potency to that which had been provided in the commercial garlic formula, but we were barely able to choke these concoctions down!

We continued to research. We found a study that documented increased effectiveness of garlic against certain species of fungi when the garlic was mixed with oil. The study stated, "A remarkable increase in the activity of garlic extracts was observed when extracts were mixed with oil. Consequently, the treatment comprising 1% extract plus oil was as effective (100% control) as the fungicide treatment in controlling both green and blue molds on Valencia oranges."[27] Hmmm . . . If our hunch was right, the peppermint extract in the garlic supplement must be prepared in an oil base.

We called the manufacturer's product information line and confirmed that, yes, oil *is* used in the processing of this particular brand's garlic supplement and that, yes, the peppermint extract is in the form of an oil extract. The confirmation of this additional component—oil—further explained why we, suffering from fungal exposure, had such effective results from this brand's enteric coated garlic supplement. The oil boosts the effectiveness of the garlic, which when released in the intestines is propelled by the peppermint, thus making it much more effective than natural forms of garlic and competitive brands of garlic that are not enteric coated or made from the same formula. Not only was this particular garlic formula boosting our immune systems when taken with the vitamin C and the echinacea, but it was also aiding in the battle against the fungi! For the first time, we began to understand why alternative healing books document garlic as an agent that kills fungi and parasites (i.e., parasitic fungi).[28] Furthermore, by taking this garlic supplement in combination with vitamin C and echinacea, the peppermint in the garlic formula acts as a multi-supplement stimulant and enhances the healing properties of the other two supplements as well.

Among other supplements, we now keep enteric coated garlic, vitamin C, echinacea, noni, and peppermint tea as staples in our medicinal cabinet to fight fungal re-exposures and oncoming illnesses. We take this combination of supplements to boost our immune systems when we feel an illness coming on. We have experimented with other brands of enteric coated garlic and have found them to be effective as well, as long as we take unpasteurized noni about one half hour after having taken the enteric coated garlic. We have found that taking the noni afterward is necessary because when the enteric coated garlic kills yeasts and fungi, the organisms release mycotoxins, which the noni neutralizes, in our opinion and observation. We feel this combination powerfully combats fungi.

Another supplement Kurt and I found essential in helping us ward off colds, flus, and other everyday illnesses was an adult supplement of fruits and vegetables concentrates. As long as we took two of the fruits and vegetables concentrates each day along with the other supplements discussed, we remained healthy. However, if we stopped taking any one of the supplements out of this combination during our healing period, our immune systems immediately began to falter. To "test" the effectiveness of the supplement of fruits and vegetables concentrates, several times we stopped taking just that one particular supplement and instead increased our daily food intake of fruits and vegetables. We would eat four to five servings of fruits and vegetables (mostly vegetables) at every meal with vegetable snacks in between. Our thinking was that we would be able to obtain comparable stability in our health from increased consumption of fruits and vegetables as we received from taking two of the supplements of fruits and vegetables concentrates each day. We were truly surprised when fresh produce was not tantamount to supplements of fruits and vegetables concentrates.

Before we delve into exploring why lightly steamed, table-served vegetables did not provide the same nutritional support, in our opinion, as the supplement of fruits and vegetables concentrates, let's take a look at published, peer-reviewed scientific and medical studies regarding the health effects of supplementation with concentrates of fruits and vegetables.

Our research reveals that the associated health benefits of eating fruits and vegetables have been a focus of scientific studies for some time. In fact, a published report states, "Epidemiological studies have shown that *low* plasma levels of antioxidant micronutrients, which are commonly found in fruits and vegetables, are associated with increased risk for diseases such as heart disease, cancer, metabolic disorders, and the like."[29] In other words, if people don't eat enough fruits and vegetables, they may have low levels of antioxidant micronutrients in their blood and be more susceptible to heart disease, cancer, metabolic disorders, and other conditions.

Because of the importance of consuming adequate quantities of fruits and vegetables, researchers have begun to study the effectiveness of supplements of fruits and vegetables concentrates. The results of a double-blind, randomized, placebo-controlled study conducted by the University of Florida published in 2006 reveals enhanced immunity and antioxidant capacity in humans by consumption of a dried, encapsulated fruits and vegetables juice concentrates.[30] Researchers investigated the effects of 59 healthy law students who consumed either encapsulated fruits and vegetables juice concentrates or placebo capsules for 77 days. The results revealed that the students who consumed the fruits and vegetables juice concentrates exhibited fewer total symptoms of illness during the study period than the controls. Blood tests of the students who supplemented with the fruits and vegetables juice concentrates confirmed increased levels of nutrients in plasma, resulting in increased antioxidant capacity, reduced DNA damage in lymphocytes, and increased circulating gammadelta T cells.[31] Clearly, health-enhancing benefits were proven from daily supplementation of fruits and vegetables juice concentrates.

Another double-blind study published in 2004, designed to monitor the dietary habits of a group of healthy, middle-aged men and women to assess the effects of supplementation with a natural phytonutrient preparation from fruits and vegetables concentrates, measured blood levels of various antioxidant micronutrients— beta-carotene, vitamin C, vitamin E, selenium, and folate.[32] The results state,

"Significant increases in blood nutrient levels after active supplementation were observed for beta-carotene, vitamin C, vitamin E, selenium, and folate. Ranges measured, after supplementation, often fell into those associated with a reduced risk for disease."[33] In other words, by taking a daily supplement of fruits and vegetables concentrates, study participants were able to increase antioxidants in their blood often to a level considered to reduce risk for disease.

The researchers point out that the data suggests, ". . . although generally health conscious, participants still fell short of the recommended five portions of fruits and vegetables per day."[34] As the first study noted, fruits and vegetables are a primary source of antioxidant micronutrients. Without adequate levels of antioxidant micronutrients in their blood, people are at increased risk for heart disease, cancer, metabolic disorders, and other conditions. Therefore, without increased consumption of fruits and vegetables and/or supplementation with fruits and vegetables concentrates, these "generally health conscious" people are at increased risk for the aforementioned diseases.

The 2004 study confirms, "Supplementation with mixed fruit and vegetable juice concentrates effectively increased plasma levels of important antioxidant nutrients and folate."[35] Therefore, supplementing daily with a concentrated form of fruits and vegetables can keep us healthier, because the concentrates can increase the levels of antioxidant nutrients and folate in our blood to levels that have been associated scientifically with the reduction of disease!

An earlier all-male study published in 2003 found similar results when supplementation with dehydrated juice concentrates from mixed fruits and vegetables was studied to determine the effects on selected vitamins and antioxidants in the blood of participants.[36] Study researchers report, "A mixed fruit and vegetable concentrate increases plasma antioxidant vitamins and folate and lowers plasma homocysteine in men."[37] The study concludes, "In the absence of dietary modification, supplementation with a fruit and vegetable concentrate produced responses consistent with a reduction in CHD [coronary heart disease] risk."[38] So by taking a daily supplement of fruits and vegetables concentrates, scientific studies indicate we can reduce our risk of developing heart disease!

Although some published sources caution that phytochemicals in the form of supplements will provide only selected nutrients, not deliver the diversity of compounds found naturally in foods,[39] the results from the just-reviewed three studies clearly demonstrate the health-preserving effects of daily consumption of fruits and vegetables concentrates in supplement form. Our family's personal experience from taking supplements of fruits and vegetables concentrates, which contributed to the nutrients our bodies needed for healing and restoring optimal health, exemplifies the health-preserving effects of fruits and vegetables concentrates supplementation. However, remember that consistent quality results are dependent on consistent quality raw materials. In order to find out from where each supplement manufacturer obtains its raw materials and how the raw products are cultivated, fertilized, and harvested, it is necessary to make inquires at the manufacturing level in regard to each product. Check to see if the product is designated as organic. If the product is not labeled as organic, check to see if the fruits and vegetables concentrates come from whole plants that are certified as organic or at least derived from plants grown with sustainable, chemical-free methods. A supplement of fruits and vegetables concentrates may not carry the certified organic label even if the fruits and vegetables come from certified organic plants, because other added ingredients may not meet the organic certification standards. Always verify that no artificial colors, flavors, or preservatives are added.

We feel it particularly important to confirm a manufacturer's policy in regard to the use of sludge fertilizer. [See Section One, Chapter Four for an earlier discussion on sludge fertilizer.] According to the EPA, "Organically produced food cannot be produced using excluded methods, sewage sludge, or ionizing radiation."[40] However, the term *organic* has been used in direct connection with fertilizer processed from sewage sludge. For example, the New York *Organic* Fertilizer Company (NYOFCo), which was formed in 1993 in Hunts Point, New York, and closed in 2010, due to lawsuits prompted by the stench emitting from the facility, bore the designation *organic* in name alone.[41] According to the Office of the Attorney General, "NYOFCo accepts sewage sludge from New York City's sewage treatment plants and processes it into fertilizer pellets for sale to out-of-state agricultural operations."[42] The NRDC further explains, "The Hunts Point sewage plant treats sewage from more than a half million people in New York City. The NYOFCo sludge facility has an exclusive contract with the city to process up to 825 wet tons of sludge per day, which it processes into pelletized fertilizer for out-of-state export."[43] The question is: Although the use of sludge fertilizer is not allowed under organic certification, where does fertilizer processed by an *organic* fertilizer company end up?

Another factor to consider that can increase or decrease the nutrient content of fruits and vegetables concentrates in supplement form is the amount of time that elapses from harvest to processing. Oftentimes manufacturers of fruits and vegetables concentrates have information about product processing methods available on their websites. Look for specific information regarding the amount of time that lapses from harvest to product processing. With the availability of today's high-tech processing equipment, manufacturers should be touting the benefits—the ability to process plants within hours of harvest, which ensures maximum nutrient retention in the supplement. If a manufacturer is not heralding this level of quality processing, it is most likely not using sophisticated processing methods that capture the highest possible level of natural nutrients. Remember, the level of quality control of both the raw material production and the product processing directly affects the end nutritional value provided in the fruits and vegetables concentrates.

Even after we realized the advanced processing abilities to capture natural nutrients almost fresh off the vine, we were still surprised when an increased diet of fresh fruits and vegetables didn't come close to providing the same nutritional value as did daily consumption of our supplements of fruits and vegetables concentrates. But just think about it for a moment. How much nutritional value is lost from the time fruits and vegetables are picked, packed, stored, transported, and delivered to the grocery store? We must also consider the nutrient value lost during shelf time before purchase and the time that passes in our own refrigerators. Furthermore, unless we purchase all organic produce that is locally grown, chemical ripening agents are frequently used to artificially ripen produce during transport as fruits and vegetables are often picked before they are fully ripe. When fruits and vegetables are picked before they are fully ripe, what happens to the nutrient content? What's more, the chemical ripening agents add just that—chemicals—that end up either being detoxified by or stored in our bodies! For these reasons, we should all strive to eat locally (or possibly regionally) grown organic produce.

Although we all may not eat five or more servings of fruits and vegetables a day, based on the results from the aforementioned studies, we can still obtain the equivalent detoxifying and health-sustaining nutrients by supplementing with concentrates of fruits and vegetables. Of course, supplements of fruits and vegetables concentrates should not be used as a substitute for daily consumption of fruits and vegetables, but rather to ensure that adequate levels of the necessary nutrients are met or exceeded. Based on research conducted by manufacturers of fruits and vegetables concentrates, a daily supplement can deliver key phytonutrients equivalent to those found in multiple servings of fruits and vegetables. Essentially, the supplement provides a daily nutrient boost, which otherwise couldn't be achieved unless we sat down and ate 5 to 10 servings of fresh-off-the-vine fruits and vegetables all in one sitting. Our bodies reap the benefits of the concentrated nutrients, which are delivered all at once, without having to use energy to digest the whole fruits and vegetables. Our bodies can then utilize the unexpended energy to make the best use of the increased level of phytonutrients, which are bio-available since they were derived from whole plant concentrates. Remember, it is especially important that fruits and vegetables concentrates come from plants grown with chemical-free methods that have no added artificial colors, flavors, or preservatives, because you want to receive the added nutrients without added contaminants!

Regardless of a manufacturer's nutrient claims, proof of the quality and bioavailability of the nutrients provided in a fruits and vegetables concentrate should be self-evident. We experienced a reduction in energy, concentration, strength, stamina, and immunity when we were not supplementing with the fruits and vegetables concentrates; we would then begin to succumb to signs of illness. The symptoms of oncoming illness would quickly disappear when we added the fruits and vegetables concentrates back into our daily health regime, which also included unpasteurized noni and the immune-boosting trio: vitamin C, echinacea, and garlic supplements.

Before we talk more specifically about the children's fruits and vegetables concentrates, let's take a look at some of the circumstances in which we found the use of the immune-boosting trio (vitamin C, echinacea, and garlic) and the fruits and vegetables concentrates health-preserving. First of all, we have found that if we take this combination of supplements at the first sign of an oncoming illness, we usually quickly rebound and don't even get sick. For this reason, we now strive to keep this combination of supplements on hand to combat colds, flus, and other ailments.

We also use this combination as a preventative measure when our bodies are under added stress, such as when work requires us to burn the candle at both ends or when we react to contaminants in the air or food. For example, when a nearby forest fire blew smoke and fine particulate into town, we initially experienced respiratory distress and allergic symptoms. Inhalation of the airborne contaminants had put extra stress on our respiratory systems, which taxed our immune systems. We were able to negate the initial negative impact and avoid any long-term detrimental health effects by taking supplements of vitamin C, echinacea, garlic, and fruits and vegetables concentrates.

Another scenario in which we use this combination to keep our immune systems strong is when we are re-exposed to mold, for example, by unwittingly walking into a building filled with airborne mold contaminants. Limiting re-exposure was especially crucial immediately post-healing when even limited exposure to this type of environment could still cause a severe reaction and sometimes make us extremely ill because of pre-exposure sensitivities. Fortunately, by immediately taking the combination of vitamin C, echinacea, garlic, and fruits and vegetables concentrates, we can avoid or greatly minimize backlash from mold re-exposure. The combination of supplements boosts our immune systems, which enables our bodies to quickly destroy inhaled foreign particles.

This combination of supplements is also effective in nullifying some, if not all, of the effects that can occur from unwittingly eating mold-contaminated food or antigenic ingredients, both of which can trigger an immune response. Violent reactions from food allergies, unfortunately, have happened more times than we care to admit. We would get to feeling better and then think, "We can eat this or that," and then our bodies would rebel. As we now know, a vast array of foods can contain fungal contaminants and/or antigens. Food sensitivities have been a frustrating part of our recovery process. No longer can we eat foods normally considered "healthy." In fact, at the time of this writing, we continue to focus on a fungal-free diet.

Kurt and I realize that our family very well may never have completely regained our pre-hurricane health without the nutritional support from supplements to boost our immune systems. In fact, because of the lifestyle changes we implemented, including the diet modifications and supplements discussed in this book, Kurt and

I are now in better health than we were before the hurricane! By one year post-hurricane, I had lost 50 pounds and Kurt had lost 30 pounds. We are now 100 percent recovered from the initial post-hurricane exposure, aside from the fact that we still experience fungal and chemical sensitivities in certain situations. Over time, the intensity of these sensitivities has slowly diminished. It is our hope that they will eventually lessen to the point of not occurring at all. Of course, in the meantime, we will continue to use unpasteurized noni and the immune-boosting trio: vitamin C, echinacea, and garlic supplements, along with the fruits and vegetables concentrates to avoid ill health effects from developing from inhalation and ingestion of fungal contaminants.

Kurt and I were not able to reduce our daily usage of the vitamin C, echinacea, and garlic supplements until our bodies were completely free of the fungi and related mycotoxins that were responsible for the suppression of our immune systems. Our immune systems remained in a weakened state, requiring additional support from these supplements until we fully completed the unpasteurized noni treatment discussed in the prior chapter. Eventually, after we completed our fungal treatment, we were able to reduce our usage of the vitamin C, echinacea, and garlic supplements without experiencing a relapse. This was truly a happy day when we realized our recovery had reached this point!

Our children, likewise, experienced immune-supporting effects from the following children's supplements: a chewable multivitamin, chewable fruits and vegetables concentrates, chewable vitamin C, and liquid echinacea. In fact, the multivitamin and chewable fruits and vegetables concentrates, when used together, were a critical component of their treatment plan. Just as the earlier discussed combination of adult supplements boosted Kurt's and my immune systems, these two children's supplements boosted our children's immune systems.

Our children regained their health quickly after starting to supplement with the chewable multivitamins and chewable fruits and vegetables concentrates. We were curious as to which supplement was responsible for the increased health and vitality of our children or if it was attributable to both; we wanted to know if the combination of both supplements would be required to sustain their newly regained health. For this reason, we experimented by having the children take one supplement at a time. We found that as long as they continued to take daily these two supplements—the chewable multivitamins and chewable fruits and vegetables concentrates—in combination with our family's earlier-described fungal-free diet, they remained healthy. However, if they missed taking either of these two supplements for even a few days (before having successfully completed the fungal treatment), they would experience noticeable deterioration in their health, such as throat, sinus, and lung infections. When the children resumed taking both the chewable multivitamins and the chewable fruits and vegetables concentrates, their symptoms of re-emerging mold-related illnesses would again disappear usually within two to three days.

We were extremely thankful for the effectiveness of the children's supplements and amazed with the results. In fact, seeing our children vibrantly smiling, laughing, running around, and showing their old spark for the first time in many months was such a stress reducer for Kurt and me that this combination of children's supplements also contributed to our health recovery, albeit indirectly! The less stress we experienced, the more our bodies were able to focus on healing.

We added use of the chewable vitamin C and liquid echinacea when the children showed the first signs of a potential oncoming illness, which wasn't often as long as they continued to use daily the chewable multivitamins and chewable fruits and vegetables concentrates. In fact, one bottle of liquid echinacea, which we primarily reserved for the children's use, lasted over a full year! During this time period, when our children did get sick, it was when their daily doses of chewable multivitamins and chewable fruits and vegetables concentrates were not taken for a few days. Clearly, the use of these two health-sustaining children's supplements could not be irregular during our children's healing period. Otherwise, the children would succumb to illness with physical symptoms presenting within a day or two. Even after our family completed its fungal treatment and the children's immune systems became stronger, our children's health continued to require the added nutritional support from the combination of chewable multivitamins and chewable fruits and vegetables concentrates in order to stay optimally healthy. The reason, most likely, is because children's immune systems are still developing. We find it imperative that our children continue take these two supplements daily because without them, their health is just not as robust.

We feel strongly, as parents, that children's diets be supplemented with both a high quality multivitamin and fruits and vegetables concentrates. According to the *Journal of Allergy and Clinical Immunology*, "Lack of adequate macronutrients or selected micronutrients, especially zinc, selenium, iron, and the antioxidant vitamins, can lead to clinically significant immune deficiency and infections in children."[44] To ensure our children's daily health, we know we can count on quality brands of chewable multivitamins and chewable fruits and vegetables concentrates, because we found quality supplements effectively combated the immune-suppressive effects that resulted from exposure to mold and chemicals post-Katrina. Additionally, we know we can count on chewable vitamin C and liquid echinacea should our kids begin to show the slightest signs of an oncoming illness, because the use of these additional immune-boosters usually stops any illness from fully taking hold, and thus the children don't get sick.

To find a combination of supplements that stabilized our family's health and provided us with effective, predictable results we could count on was a welcome and empowering change. With stronger immune systems in place, we started accomplishing more than just sleeping, eating, and merely existing in a state of constant exhaustion and sickness. It was at this point that we realized the burden of searching for answers to our health problems had been lifted as our strength and stamina began to grow. We began to smile, laugh a little, and regain our sense of humors, which in itself shocked us because until then, we hadn't realized how dreary our lives had become filled with sickness. Reaching this point in our recovery may not seem like much when you read about it, but to begin to have more control over our daily lives rather than have our health dictate our every decision was a monumental milestone in our journey to health.

Other supplements that proved health-restoring for our family were the following: lecithin, an adult and children's fish oil supplement, which provides the omega-3 oils (EPA and DHA), and a probiotic. Let's take a look at why these particular supplements would have been noticeably effective against mold- and chemical-related illnesses.

Regarding lecithin, health literature states, "Lecithin is a type of lipid that is needed by every living cell in the human body. Cell membranes, which regulate the passage of nutrients into and out of the cells, are largely composed of lecithin. The protective sheaths surrounding the brain are composed of lecithin, and the muscles and nerve cells also contain this essential fatty substance."[45] It is easy to see why taking a good grade of lecithin would help restore and maintain health for both adults and children. Since many manufacturers of lecithin encapsulate the lecithin in rather large gel capsules, which are not easily swallowed by most children, look for a manufacturer that makes a chewable or granular form of lecithin.

The fish oil supplement was a gel capsule of refined fish oils that provides both EPA (elcosapentaenoic acid) and DHA (docosahexaenoic acid), which are two of the three major forms of omega-3 oils. The other major type of omega-3 fatty acid is ALA (alpha-linolenic acid).[46] According to the University of Maryland Medical Center (UMMC), "Omega-3 fatty acids are considered essential fatty acids, which means that they are essential to human health but cannot be manufactured by the body."[47] For this reason, omega-3 fatty acids, which are polyunsaturated fatty acids, must be obtained from the diet either in the form of food and/or food supplements.[48] According to the American Heart Association (AHA), omega-3 oils are plant-derived (ALA) or marine-derived (EPA and DHA).[49] ALA is found in ". . . flaxseeds, flaxseed oil, canola (rapeseed) oil, soybeans, soybean oil, pumpkin seeds, pumpkin seed oil, purslane, perilla seed oil, walnuts, and walnut oil."[50] Good sources of EPA and DHA are cold water fish, such as halibut, mackerel, herring, salmon, and sardines.[51]

The Mayo Clinic reports, "There is evidence from multiple large-scale population (epidemiologic) studies and randomized controlled trials that intake of recommended amounts of DHA and EPA in the form of dietary fish or fish oil supplements lowers triglycerides, reduces the risk of death, heart attack, dangerous abnormal heart rhythms, and strokes in people with known cardiovascular disease, slows the buildup of atherosclerotic plaques ('hardening of the arteries'), and lowers blood pressure slightly."[52] A 2000 report from Northwestern University notes additional health benefits. It states that consumption of DHA, a marine-derived omega-3 oil, is vital to the healthy structure and function of the human brain.[53] Therefore, consumption of a marine-derived source of omega-3 oils (DHA and EPA) can be beneficial to both heart and brain functions.

A 2006 published, peer-reviewed review article states, "Evidence suggests that increased consumption of n-3 fatty acids from fish or fish-oil supplements [EPA and DHA], but not of alpha-linolenic acid [plant-derived ALA], reduces the rates of all-cause mortality, cardiac and sudden death, and possibly stroke."[54] In other words, in order to best ensure receiving the aforementioned cardiac-related health benefits, the results from this scientific review indicate that direct consumption of dietary fish and/or fish oil (EPA- and DHA-containing) supplements is necessary. These findings are significant because scientific studies do not indicate that the body converts the plant-derived omega-3 oil (ALA) to EPA or DHA in significant amounts. According to the University of Maryland Medical Center, "Once eaten, the body converts [plant-derived omega-3 oil] ALA to EPA and DHA, the two types of omega-3 fatty acids more readily used by the body."[55] However, the American Heart Association clarifies, "Although some alpha-linolenic acid is converted to longer-chain omega-3 fatty acids, the extent of this conversion is modest and controversial."[56] EPA and DHA are longer-chain omega-3 fatty acids.[57]

Modest conversion rates are noted in other studies as well. A 2007 research study indicates that consumption of increased levels of plant-derived omega-3 oil (ALA) increases levels of EPA but not levels of DHA.[58] A 2006 study reports an estimated net fractional ALA inter-conversion from consumption of 700 mg of plant-derived omega-3 oil (ALA) supplementation to 21% EPA and 9% DHA.[59] Earlier studies in 1998 document poor conversion of plant-derived omega-3 oil (ALA) to the longer-chain polyunsaturated omega-3 fatty acids, EPA and DHA.[60]

In summary, research attributes heart-related benefits to consumption of marine-derived omega-3 oils (EPA and DHA). Both EPA and DHA are two types of essential omega-3 oils, which means that the body cannot manufacture the compounds on its own. The American Heart Association and scientific studies indicate "modest" conversion from plant-derived omega-3 oil (ALA) to EPA and even less conversion to DHA. Therefore, both EPA and DHA must be obtained from eating cold-water fish and/or supplementing with a marine-derived source of omega-3 oils (EPA and DHA), such as a fish oil supplement. Based on results from scientific studies, it is prudent for people seeking cardiovascular benefits to discuss with a licensed physician supplementation with a marine-derived source of omega-3 oils (EPA and DHA). Of course, only an EPA and DHA product manufactured with strict quality controls should be used. Beware of rancid, mercury contaminated, or otherwise damaged fish oils. Dr. Young states, "Make sure the EFAs [essential fatty acids] you get are fresh to ensure good results. (Break open a capsule—there should be no 'fishy' odor.)"[61]

In addition to the scientifically proven cardiovascular benefits of marine-derived omega-3 oils (EPA and DHA), the University of Maryland Medical Center (UMMC) discusses additional health-promoting effects of omega-3 oil consumption. It reports, "Extensive research indicates that omega-3 fatty acids reduce inflammation and help prevent certain chronic diseases such as heart disease and arthritis. These essential fatty acids are highly concentrated in the brain and appear to be particularly important for cognitive and behavioral function." In fact, UMMC states, ". . . omega-3 and omega-6 fatty acids play a crucial role in brain function as well as normal growth and development."[62]

Before supplementing with the fish oil supplement, Kurt and I knew very little about the benefits of consuming marine-derived omega-3 oils (EPA and DHA). Thus, we were surprised when we experienced improved memory a few days after beginning to supplement our diet with the fish oil. It also appeared, from our personal observations, that our brain neurons were firing more quickly when supplementing with the fish oil as compared to when we were not taking the supplement. It was as if our brains were waking up and becoming more alive! In other words, our mental acuity was heightened when supplementing with the fish oil. An increased level of mental sharpness could lead to a variety of positive life changes, such as new or improved personal relationships, employment opportunities, promotions and raises—the list goes on.

The American Heart Association (AHA) promotes consumption of both plant- and marine-derived omega-3 oils. In 2002 the AHA documents, "Large-scale epidemiological studies suggest that individuals at risk for CHD [coronary heart disease] benefit from the consumption of plant- and marine-derived omega-3 fatty acids, although the ideal intakes presently are unclear."[63] In 2007 the AHA reports, "Omega-3 fatty acids benefit the heart of healthy people and those at high risk of—or who have—cardiovascular disease."[64]

Thus, heart benefits are not limited to the consumption of just marine-derived omega-3 oils (EPA and DHA). Consumption of plant-derived omega-3 oil (ALA) has heart-enhancing effects as well. For example, the American Heart Association reports findings of a 16-year observational study led by Christine M. Albert, MD, MPH, who is an assistant professor of medicine at Harvard University Medical School, "Women who reported eating diets rich in oils containing alpha-linolenic acid (ALA) seemed to have a lower risk of dying from heart disease and sudden cardiac death than women whose diets are low in the plant-derived fatty acid . . ."[65]

Seeking to maximize health benefits, we eventually added to our family's health regime a plant-derived omega-3, omega-6, and omega-9 oil blend supplement, the results of which we discuss in the next chapter. However, since the body has to work to convert the plant-derived omega-3 (ALA) to EPA and DHA, Dr. Young states that it ". . . may be better to take the preformed (animal) omega-3s."[66]

The children's fish oil supplement was a favorite of our entire family, including Kurt and me. They are a tasty source of omega-3 oils (EPA and DHA) that our children love! A published report states, "DHA is the predominant structural fatty acid in the central nervous system and retina and its availability is crucial for brain development. . . . Intake of EFAs and DHA during preschool years may also have a beneficial role in the prevention of attention deficit hyperactivity disorder (ADHD) and enhancing learning capability and academic performance."[67] Supplementation with the children's fish oil appeared to improve school performance of our grade-school-age daughter.

Probiotic were also a critical component of our healing program. Probiotics are supplements of friendly bacteria designed to restore proper intestinal flora.[68] It may seem like a strange concept, but "[i]nside each of us live vast numbers of bacteria without which we could not remain in good health. There are several thousand billion in each person (more than all the cells in the body) divided into over 400 species, most of them living in the digestive tract. Certain of these bacteria help to maintain good health, while others have a definite value in helping us regain health once it has been upset."[69]

Less than optimal intestinal flora can occur for a variety of reasons. "Certain drugs, especially antibiotics, can severely upset this delicate balance—penicillin will kill friendly bacteria just as efficiently as it will kill disease-causing bacteria. Steroids (cortisone, ACTH, prednisone, and birth control pills) also cause great damage to the bowel flora."[70] The intestinal flora of our family had been assaulted on multiple fronts. Kurt had taken prescribed steroids and antibiotics, our children had taken prescribed steroids, and I had taken prescribed antibiotics. In addition, we all had been exposed to the initial flora-disrupting fungi, mycotoxins, and chemicals post-hurricane. Because of all these destructive forces, there was no question whether the integrity of our intestinal floras had been massively disrupted.

Furthermore, literature states, "Every antifungal treatment for candidiasis will reduce the numbers of 'friendly' bacteria that inhabit the intestines."[71] Since we had all taken a prescribed antifungal, we had yet another reason to find a probiotic that would effectively re-establish our intestinal flora. The antifungal, nystatin, undoubtedly contributed to the destruction of our friendly bacteria, which further explains our rapid decline in health after having taken it for several weeks. Improper bowel flora can interfere with the body's ability to utilize nutrients and eliminate unwanted toxins.[72]

Restoring a sufficient level of health-enhancing microbes, or probiotics, is critical to proper bowel function.[73] Health sources state, "Probiotics maintain the health of the cells lining the small and large intestines, which boosts the digestive tract's ability to absorb nutrients and fight unwanted toxins."[74]

Other substantial documented benefits from using probiotics ". . . include:

- The manufacture of certain B vitamins, including niacin (B3), pyridoxine (B6), folic acid, and biotin.
- Enhanced immune system activity. . . .
- Production of antibacterial substances that kill or deactivate hostile disease-causing bacteria. Friendly bacteria do this by changing the local levels of acidity, by depriving pathogenic bacteria of their nutrients, or by actually producing their own antibiotic substances.
- Anticarcinogenic effect since probiotics are active against certain tumors.
- Improved efficiency of the digestive tract . . ."[75]

Health literature documents some important points to understand regarding probiotics, "The difficulty with probiotics is getting them into your intestines in an active form. *L. acidophilus* and other bacteria can easily be destroyed during the manufacturing process or by heat and light while being stored. They can also be dissolved by acids in the stomach when probiotics are taken orally. When selecting a supplement, therefore, it is important to find one that will deliver viable (live) bacterial cultures to where they can do you some good."[76]

Understanding the fact that probiotics can actually be dissolved by stomach acids helps explain why we had not gotten noticeable results from the pharmaceutical grade probiotics prescribed by our naturopath until we started taking them with carrot juice, which is extremely alkaline. In our opinion, our stomach acids had been dissolving the probiotics until we added the alkalizing carrot juice, which then protected the probiotics from the full brunt of our stomach acids. However, even then, we did not notice a significant effect from taking the pharmaceutical grade probiotics.

We experimented with other brands of probiotics, including some that were enteric coated. Again, an enteric coating protects the probiotics from stomach acids and enables the tablet to dissolve in the intestinal tract. Although the enteric coated probiotics and some of the nonenteric coated ones initially worked effectively, there was a problem of diminishing potency once the seals on the bottles were broken. These probiotics would deliver "live" probiotics for the first ten days or so after the bottles had been opened, after which they noticeably lost potency, even though we kept them in the refrigerator. One solution was for Kurt and me to share a bottle of adult probiotics and the children to share a bottle of children probiotics rather than having a bottle open for each of us; by sharing a bottle, we used up the probiotics twice as fast and thus minimized the level of reduced potency. We also found through our trial and error with various brands that if we took too high of a dose of probiotics, we got intestinal cramps! According to health literature, "Other than mild gastrointestinal upset, the use of probiotics does not produce any side effects."[77]

If anyone is in doubt about the importance of probiotics, just consider the following: The small intestine is about 22 to 25 feet long and the large intestine is about 5 feet long.[78] Not many of us like to think of what can get lodged, wedged, trapped, or grow inside approximately 27 feet of intestinal space. Not exactly a dinner-time topic, but it is a topic debated in scientific and medical fields as to whether or not fungi can colonize in the intestines. Doctors, such as Dr. Gray, who routinely treat mold-exposed patients, are aware of this possibility. However, the majority of the medical community is not, at least at this writing. We believe future scientific and medical studies will eventually prove that fungal colonization and infestation of the intestines is not only possible, but probably quite common. If we were not sure of this medical probability before finding effective probiotics, we were thoroughly convinced afterward. Effective probiotics enabled intestinal putrification to void from our bodies, a process that went on for months, which was a sure sign probiotics were needed!

We have become keenly aware from our family's experiences that you have to prepare your health and the health of your family *today* for tomorrow's disasters. Most certainly, with stronger immune systems already in place, our bodies will be more equipped to sustain health during times of trauma and exposure to mold and chemicals that come our way from natural disasters, flooding, water leaks, or sick buildings. Furthermore, we all need strong immune systems in place to deal with everyday life stresses, which over time can also undermine good health. A leading health publication states, "Treatment of chronic disease currently accounts for 85% of the national health-care bill. This state of affairs is due to the fact that we spend almost nothing to treat the causes of chronic disease before major illness develops, according to a report from the American Association of Naturopathic Physicians."[79] Why wait for illness to develop from poor lifestyle habits and then spend huge sums on heroic lifesaving measures,[80] when we can prevent many of these chronic diseases from occurring? According to former Surgeon General C. Everett Koop, in his 1988 *Report on Nutrition and Health*, dietary imbalances are the leading preventable contributor to premature death in the U.S.[81] This conclusion is confirmed by the CDC. It states 54% of heart disease, 37% of cancer, 50% of cerebrovascular disease, and 49% of atherosclerosis (hardening of the arteries) is preventable through lifestyle modification.[82]

The selection of a high quality, effective vitamin line is a health- and life-enhancing step. According to Harvard doctors, "Inadequate intake of several vitamins has been linked to chronic diseases, including coronary heart disease, cancer, and osteoporosis."[83] Because "[m]ost people do not consume an optimal amount of all vitamins by diet alone,"[84] these Harvard doctors state, "We recommend that all adults take one multivitamin daily."[85]

Although Kurt and I did not take a multivitamin during our healing period because our healing regime included combining several individual vitamins, herbs, and other health supplements as discussed in this book, we each later began taking a multivitamin. We continue to find that as long as each member of our family takes a daily multivitamin along with a supplement of fruits and vegetables concentrates, we stay healthy. If we feel the rare sniffle, cough, or flu coming on, we immediately add into our health regime the combination of vitamin C, echinacea, and garlic supplements, which usually wards off the oncoming illness. So far, this immune-boosting trio has not let us down!

Of course, everyone should consult a licensed health care professional for diagnosis, for treatment options, and for consultation of natural alternatives such as vitamin, herb, and/or other supplement(s). Use of some alternative supplements, as well as diet changes, can affect the effectiveness of certain prescription drugs. For example, increases in the amount of omega-3 fatty acids in the diet (via increased consumption of foods rich in omega-3 oils and/or omega-3 oil supplements) may enable certain cholesterol-lowering medications to work more effectively.[86] In other words, with adequate consumption of omega-3 oils, less cholesterol-lowering medication may be required.

The members of our family remained under the care of our naturopath and endocrinologist for the duration of our healing from the post-Katrina mold and chemical exposures that caused multiple illnesses and systemic effects. We were fortunate to have been evacuated to a state that recognizes alternative physicians, such as naturopaths, as licensed, primary care physicians assigned with certain pharmaceutical-prescribing rights. Depending on your particular state's laws, your private health insurance may pay for alternative products prescribed by traditional and/or alternative physicians. Furthermore, medical treatment costs that are not covered by private and/or state insurances for illnesses incurred because of federally recognized natural disasters may be reimbursable by FEMA. These reimbursements are evaluated and processed on a case-by-case basis. We are thankful and grateful to all involved in our family's healing process. We look forward to continuing to enhance our health-focused diet with the vitamins, herbs, and other supplements that proved health restorative for our family.

(ENDNOTES)

1 The information and research presented in this book is for educational purposes only. It is not intended to be medical advice or replace the services of a licensed health care professional. The information in this book has not been evaluated by the Food and Drug Administration. The information and/or products presented in this book are not intended to diagnose, treat, cure, or prevent any disease. All physical, mental, and psychological conditions should be diagnosed, treated, and monitored by a licensed, knowledgeable health care provider. The publisher and authors shall not be liable and/or responsible for any loss, injury, and/or damage allegedly arising from the information and/or use of any products mentioned in this book.

2   The Merck Manuals, "Vitamin C (ascorbic acid)." Available at http://www.merck.com/mmhe/sec12/ch154/ch154l.html (accessed March 2006); MedlinePlus, "Medical Dictionary." Available at http://www2.merriam-webster.com/cgi-bin/mwmednlm?book=Medical&va=free%20 radical (accessed March 2006).

3   Sahoo et al., "Immunomodulation by dietary vitamin C in healthy and aflatoxin B1-induced immunocompromised rohu (Labeo rhyolite)," *Comp Immunol Microbiol Infect Dis.* 2003 Jan;26(1):65-76. Available at http://www.ncbi.nlm.nih.gov/entrez/query.fcgi?db=pubmed&cmd=Retrieve&dopt=Abst ractPlus&list_uids=12602688&query_hl=10&itool=pubmed_docsum (accessed March 2006); Goldberg, *Alternative Medicine*, p. 590.

4   Sahoo et al., "Immunomodulation by dietary vitamin C in healthy and aflatoxin B1-induced immunocompromised rohu (Labeo rhyolite)."

5   Ibid.

6   Goldberg, *Alternative Medicine*, p. 396.

7   MedicineNet.com, "Definition of ascorbic acid." Available at http://www.medterms.com/script/main/art.asp?articlekey=12536 (accessed March 2006).

8   Jacob RA et al., "Vitamin C function and status in chronic disease," *Nutr Clin Care*, 2002 Mar-Apr;5(2):66-74. Available at http://www.ncbi.nlm.nih.gov/entrez/query.fcgi?itool=abstractplus&db =pubmed&cmd=Retrieve&dopt=abstractplus&list_uids=12134712 (accessed March 2006).

9   Ibid.

10   Goldberg, *Alternative Medicine*, p. 260.

11   National Center for Complementary and Alternative Medicine (NCCAM), "Echinacea for the Prevention and Treatment of Colds in Adults: Research Results and Implications for Future Studies," October 2005. Available at http://nccam.nih.gov/research/results/echinacea_rr.htm (accessed March 2006).

12   Martha Kerr, "Echinacea Cuts Cold Incidence," September 20, 2006, Reuters Health Information. Available at http://www.medscape.com/viewarticle/544846 (accessed October 2006).

13   Medline Plus Medical Dictionary. Available at http://www2.merriam-webster.com/cgi-bin/mwmednlm?book=Medical&va=meta-analysis (accessed March 2006).

14   Kerr, "Echinacea Cuts Cold Incidence."

15   Shah SA, et al., "Evaluation of Echinacea for the Prevention and Treatment of the Common Cold: A Meta-Analysis," *Lancet Infect Dis.* 2007 Jul; 7(7):473-80. Available at http://www.ncbi.nlm. nih.gov/pubmed/17597571 (accessed July 2010), and Shah SA, et al., Erratum in *Lancet Infectious Dis.*, Volume 7, Issue 9, Page 580, September 2007. Available at http://www.thelancet.com/journals/laninf/ article/PIIS1473-3099(07)70207-4/fulltext (accessed July 2010).

16   Interview with drs. Sachin A. Shah and C. Michael White, April 2007.

17   Christopher, *School of Natural Healing*, p. 98.

18   Balch, *Prescription for Nutritional Healing*, p. 71; Goldberg, *Alternative Medicine*, p. 261-262; and Stefan Chmelik, *Chinese Herbal Secrets*, p. 138 (Garden City Park, NY, Avery Publishing Group, 1999).

19   Penelope Ody, *The Complete Medicinal Herbal*, p. 33 (New York, NY, DK Publishing, Inc. 1993), and Goldberg, *Alternative Medicine*, p. 261-262.

20   Paavo Airola, PhD, *The Miracle of Garlic*, p. 20-21 (Sherwood, OR, Health Plus Publishers,1980).

21   Goldberg, *Alternative Medicine*, p. 261-262.

22   Ibid., p. 676.

23   Ibid.

24   Chmelik, *Chinese Herbal Secrets*, p. 138.

25   UMMC, "Peppermint." Available at http://www.umm.edu/altmed/ConsHerbs/Peppermintch. html (accessed July 2006).

26   Kloss, *Back to Eden*, p. 165.

27   Obagwu et al., "Control of Citrus Green and Blue Molds with Garlic Extracts."

28   Balch, *Prescription for Nutritional Healing*, p. 71; Goldberg, *Alternative Medicine*, p. 261-262; and Chmelik, *Chinese Herbal Secrets*, p. 138.

29   Kiefer et al., "Supplementation with mixed fruit and vegetable juice concentrates increased serum antioxidants and folate in healthy adults," *J Am Coll Nutr.* 2004;23(3):205-11. Available at http://www://www.jacn.org/cgi/content/abstract/23/3/205 (accessed July 2006). (Italics added.)

30   Nantz et al., "Immunity and antioxidant capacity in humans is enhanced by consumption of a dried, encapsulated fruit and vegetable juice concentrate," *J Nutr.* 2006 Oct;136(10):2606-10. Available at http://www.ncbi.nlm.nih.gov/entrez/query.fcgi?db=pubmed&cmd=Retrieve&dopt=AbstractPlus&list_uids=16988134&query_hl=18&itool=pubmed_DocSum (accessed November 2006).

31   Ibid.

32   Kiefer et al., "Supplementation with mixed fruit and vegetable juice concentrates increased serum antioxidants and folate in healthy adults."

33   Ibid.

34   Ibid.

35   Ibid.

36   Samman S et al., "A mixed fruit and vegetable concentrate increases plasma antioxidant vitamins and folate and lowers plasma homocysteine in men," *J Nutr.* 2003 Jul; 133(7): 2188-93. Available at http://www.ncbi.nlm.nih.gov/entrez/query.fcgi?db=pubmed&cmd=Retrieve&dopt=Abstract Plus&list_uids=12840177&query_hl=1&tool=pubmed_docsum (accessed March 2006).

37   Ibid.

38   Ibid.

39   Sereana Howard Dresbach et al., "Phytochemicals–Vitamins of the Future?" The Ohio State University Fact Sheet Extension. Available at http://ohioline.osu.edu/hyg-fact/5000/5050.html (accessed September 2006).

40   US EPA, Agriculture, "Organic Farming." Available at http://www.epa.gov/agriculture/torg. html (accessed July 2010).

41   Biosolids Success Stories, "Southeastern Colorado Cross-country Biosolids Hauling and Agricultural Application." Available at http://www.biosolids.org/docs/source/SECO.pdf (accessed on July 20, 2010); and Rocchio, Patrick "NYOFCo Plant Winds Down," July 14,2010. Available at http://www. nypost.com/p/news/local/bronx/nyofco_plant_winds_down_zNevJsxbL9qMihuerL8sgL (accessed on July 20, 2010).

42   Office of the Attorney General, Media Center, "Attorney General Cuomo Sues South Bronx Fertilizer Company to End Noxious Odors that Threaten Health and Well-Being of Hunts Point Residents," February 5, 2009. Available at http://www.ag.ny.gov/media_center/2009/feb/feb5a_09.html (accessed July 20, 2010).

43   NRDC Media Center, "Stench from South Bronx Sewage Plants Targeted in Lawsuit by NRDC, Community Group, and Residents," July 9, 2008. Available at http://www.nrdc.org/media/2008/080709.asp (accessed July 20, 2010).

44   Cunningham-Rundles et al., "Mechanisms of nutrient modulation of the immune response," *J Allergy Clin Immunol*, 2005; 115(6):1119-28; quiz 1129. Available at http://www.medscape.com/medline/abstract/15940121?prt=true (accessed March 2006).

45   Balch et al., *Prescription for Nutritional Healing*, p. 57.

46   UMMC, "Omega-3 fatty acids." Available at http://www.umm.edu/altmed/articles/omega-3-000316.htm (accessed June 2006).

47   Ibid.

48   Ibid.

49   Penny M. Kris-Etherton et al., "AHA Scientific Statement: Fish Consumption, Fish Oil, Omega-3 Fatty Acids, and Cardiovascular Disease," *Circulation*, 2002; 106:2747. Available at http://circ.ahajournals.org/cgi/content/full/106/21/2747 (accessed June 2006).

50   UMMC, "Omega-3 fatty acids."

51   Charles R. Harper et al., "Beyond the Mediterranean Diet: The Role of Omega-3 Fatty

Acids," *Preventative Cardiology* 6(3): 136-146, 2003. Available at http://www.medscape.com/viewarticle/458440_7 (accessed March 2006), and UMMC, "Omega-3 fatty acids."

52   Mayo Clinic, "Drugs & Supplements: Omega-3 fatty acids, fish oil, alpha-linolenic acid." Available at http://www.mayoclinic.com/health/fish-oil/NS_patient-fishoil (accessed March 2006).

53   David J. Lin, "All About DHA," 2000. Available at http://www.astro.northwestern.edu/~lin/DHA/ (accessed March 2006).

54   Chenchen Wang et al., "Review Article: n-3 Fatty acids from fish or fish-oil supplements, but not alpha-linolenic acid, benefit cardiovascular disease outcomes in primary- and secondary-prevention studies: a systematic review," *American Journal of Clinical Nutrition*, Vol. 84, No. 1, 5-17, July 2006. Available at http://www.ajcn.org/cgi/content/full/84/1/5 (accessed September 2006).

55   UMMC, "Omega-3 fatty acids."

56   Kris-Etherton et al., "AHA Scientific Statement: Fish Consumption, Fish Oil, Omega-3 Fatty Acids, and Cardiovascular Disease."

57   Bourre JM, "Dietary omega-3 fatty acids for women," *Biomed Pharmacother*, 2007 Feb-Apr;61(2-3):105-12. Available at http://www.ncbi.nlm.nih.gov/entrez/query.fcgi?db=pubmed&cmd=Retrieve&dopt=AbstractPlus&list_uids=17254747&query_hl=18&itool=pubmed_docsum (accessed March 2007).

58   Attar-Bashi NM et al., "Failure of conjugated linoleic acid supplementation to enhance biosynthesis of docosahexaenoic acid from alpha-linolenic acid in healthy human volunteers," *Prostaglandins Leukot Essen Fatty Acids*, 2007 Mar;76(3):121-30. Available at http://www.ncbi.nlm.nih.gov/entrez/query.fcgi?db=pubmed&cmd=Retrieve&dopt=AbstractPlus&list_uids=17275274&query_hl=14&itool=pubmed_docsum (accessed March 2007).

59   Burdge GC et al., "Conversion of alpha-linolenic acid to eicosapentaenoic, docosapentaenoic, and docosahexaenoic acids in young women," *Br J Nutr*. 2003 Nov; 90(5):993-4; discussion 994-5. Available at http://www.ncbi.nlm.nih.gov/entrez/query.fcgi?db=pubmed&cmd=Retrieve&dopt=AbstractPlus&list_uids=12323090&query_hl=29&itool=pubmed_docsum (accessed March 2006).

60   Brouwer DA et al., "Gamma-linolenic acid does not augment long-chain polyunsaturated fatty acid omega-3 status," *Prostaglandins Leukot Essent Fatty Acids*. 1998 Nov;59(5):329-34. Available at http://www.ncbi.nlm.nih.gov/sites/entrez?Db=pubmed&Cmd=ShowDetailView&TermToSearch=9888208 (accessed March 2006).

61   Young, *The pH Miracle*, p. 163.

62   UMMC, "Omega-3 fatty acids."

63   Kris-Etherton et al., "AHA Scientific Statement: Fish Consumption, Fish Oil, Omega-3 Fatty Acids, and Cardiovascular Disease."

64   American Heart Association, "Fish and Omega-3 Fatty Acids," April 27, 2007. Available at http://www.americanheart.org/presenter.jhtml?identifier=4632 (accessed April 2007).

65   American Heart Association, "Meeting Report," November 8, 2004. American Heart Association's Scientific Sessions, 2004. Available at http://www.americanheart.org/presenter.jhtml?identifier=3025993 (accessed March 2006).

66   Young, *The pH Miracle*, p. 163.

67   Singh M., "Essential fatty acids, DHA and human brain," *Indian J Pediatr*. 2005 Mar; 72(3):239-42. Available at http://www.ncbi.nlm.nih.gov/entrez/query.fcgi?db=pubmed&cmd=Retrieve&dopt=AbstractPlus&list_uids=15812120&query_hl=8&itool=pubmed_docsum (accessed March 2006).

68   Goldberg, *Alternative Medicine*, p. 35.

69   Ibid.

70   Ibid., p. 36.

71   Ibid., p. 624.

72   Alternative Medicine.com, "The Dirty Truth About Probiotics." Available at http://www.alternativemedicine.com/common/adam/DisplayMonograph.asp?storeID=02AD61F001A74B5887D3BD11F6C28169&name=ConsArticles_TheDirtyTruthAboutProbioticsca (accessed July 2006).

73   Ibid.

74   Ibid.

75   Goldberg, *Alternative Medicine*, p. 36.

76   Ibid.

77   Ibid., p. 37.

78   Encyclopedia Britannica. Available at http://www.britannica.com/EBchecked/topic/330544/ large-intestine (accessed August 6, 2010).
79   Goldberg, *Alternative Medicine*, p. 8.
80   Ibid.
81   Ibid., p. 8 and 9.
82   Ibid., p. 9.
83   Fairfield et al., "Vitamins for chronic disease prevention in adults: scientific review," *JAMA*, 2002, Jun 19;287(23):3116-26. Available at http://www.ncbi.nlm.nih.gov/entrez/query.fcgi?db=pubm ed&cmd=Retrieve&dopt=AbstractPlus&list_uids=12069675&query_hl=1&itool=pubmed_docsum (accessed March 2006).
84   Fairfield et al., "Vitamins for chronic disease prevention in adults: clinical applications," *JAMA*, 2002, Jun 19;287(23):3127-9. Available at http://www.ncbi.nlm.nih.gov/entrez/query.fcgi?db=p ubmed&cmd=Retrieve&dopt=AbstractPlus&list_uids=12069676&query_hl=1&itool=pubmed_docsum (accessed March 2006).
85   Ibid.
86   UMMC, "Omega-3 fatty acids."

CHAPTER EIGHT:

DETOXIFY WHILE REDUCING TOXIC EXPOSURES

Once we got the fungal invasions under control and restored our immune systems to the point we stopped constantly getting sick, we focused our attention on researching detoxification aids. We wanted to cleanse from our bodies the hurricane fungal and chemical toxins, as well as the toxic loads stored in our bodies from a lifetime of living in a dangerously polluted world. Unfortunately, the human body acts like a sponge, absorbing toxic contaminants with which we come into contact.[1] We all must look for health-enhancing methods of detoxification whether we have been exposed to hurricane-size contaminants or not!

We added an amino acid supplement to our diet as an experiment. Health literature states, "Amino acids are the building blocks of protein."[2] Eight amino acids are essential, which means that the body is unable to manufacture the compounds on its own.[3] Thus, these eight amino acids ". . . must be obtained from food sources or from dietary supplements."[4] Although we didn't notice a significant difference, we found that it provided a subtle reduction in allergic symptoms.

Noticing this slight improvement, we started researching specific properties of individual amino acids. Our interest in the possible detoxifying and healing properties of amino acids was heightened by a medical doctor who shared with us that intravenous glutathione treatments given to outpatients at a detoxification clinic made mold survivors feel "absolutely great." This improvement, however, was short-lived and dependent on the intravenous deliverance of the glutathione.

The positive effects of glutathione are easy to understand. Health literature states, "It is a powerful antioxidant that inhibits the formation of and protects against cellular damage from free radicals. . . . Glutathione protects not only individual cells but also the tissues of the arteries, brain, heart, immune cells, kidneys, lenses of the eyes, liver, lungs, and skin against oxidant damage."[5]

Since we were looking for non-invasive health solutions, intravenous glutathione treatments were not an option, regardless of the exorbitant cost. Furthermore, oral supplementation of glutathione was not promising either. According to Dr. Eaton, stomach acid will break down the glutathione, which may not resynthesize.[6] However, health literature documents that the production of glutathione by the body can be boosted by taking the two amino acids, NAC (N-acetyl cysteine) and L-methionine.[7] This information led us to supplement with 700 mg of L-methionine and 800 mg of NAC, which proved to be a strong detoxifying combination.

L-methionine is one of the essential amino acids and is a powerful antioxidant.[8] Health literature states, "As levels of toxic substances in the body increase, the need for methionine increases. The body can convert methionine into the amino acid cysteine, a precursor of glutathione. Methionine thus protects glutathione; it helps to prevent glutathione depletion if the body is overloaded with toxins. Since glutathione is a key neutralizer of toxins in the liver, this protects the liver from the damaging effects of toxic compounds."[9]

Health literature further documents, "The sulfur-containing amino acid cysteine is needed to produce the free radical fighter glutathione and to help maintain it at adequate levels in the cells. N-acetyl cysteine (NAC) is a more stable form of cysteine that can be taken in supplement form. NAC is used by the liver and the lymphocytes to detoxify chemicals and other poisons. It is a powerful detoxifier of alcohol, tobacco smoke, and environmental pollutants, all of which are immune suppressors."[10]

According to Dr. Eaton, NAC is used to treat certain diseases and acetaminophen (Tylenol) poisoning. He states, "It works extremely well because Tylenol overdoses basically deplete the glutathione in your liver, and the subsequent amounts of Tylenol that are in the body after the glutathione is depleted then damage your liver. You can prevent that damage by letting your liver resynthesize glutathione by giving it plenty of extra cysteine. N-acetyl cysteine crosses cell membranes much better than cysteine itself and is much more stable against oxidation."[11]

According to Dr. Young, "Supplementing with N-acetyl cysteine has been shown to increase glutathione levels in the kidneys, bone marrow, and particularly in the liver, which uses both compounds for protection against mycotoxins."[12]

He continues, "Research on N-acetyl cysteine focused on acetaldehyde (a primary mycotoxin of yeast and fungus also found in cigarette smoke) shows just how powerful it is. In a series of trials of many different nutrients, L-cysteine was shown to reduce the deadly effect of acetaldehyde by 29 percent. Glutathione brought it down 33 percent. But N-acetyl cysteine reduced it to zero! This potent antidote to the polluted modern environment is a normal component of the body, but to receive its maximum benefits requires supplementation."[13]

Furthermore, Dr. Young states, "Studies show that N-acetyl cysteine can also bond to toxic heavy metals such as lead, mercury, and cadmium and escort them out of the body."[14] Therefore, in environments laden with mold and heavy metals, such as the widespread Katrina-hit area, NAC could prove to be life and health enhancing.

According to Dr. Rapp, other forms of glutathione treatment are now available, such as a glutathione nasal spray with a doctor's prescription.[15]

Another surprising source of detoxification came from an oil blend of plant-derived omega-3, omega-6, and omega-9 oils. When we first tried the omega oil blend, we had no expectations whatsoever. We truly thought that maybe our bodies' needs for omega oils were already being met since our naturopath had prescribed diet supplementation with ground flaxseed, which provides plant-derived omega-3 oils. Also, we were taking a supplement that provided a marine-derived source of omega-3 oils (EPA and DHA) from which we had noticed positive effects.

After we started supplementing with the omega oil blend, we immediately noticed additional improvements in our health. However, before we detail our observations from taking the omega-3, -6, and -9 oil blend, let's review a few noteworthy scientific and medical studies that will help us understand why the body needs omega-3 oils from *both* marine-derived sources (EPA and DHA) in the diet or from taking fish oil supplements and a plant-derived source (ALA) in the diet or from supplements. We were under the initial misconception that our bodies needed only one or the other, not both.

Scientific research indicates that consumption of the three different types of omega-3 oils (marine-derived EPA and DHA and plant-derived ALA) do not provide the same health benefits. For example, a 2006 review article that was published and peer-reviewed states, "Evidence suggests that increased consumption of n-3 fatty acids from fish or fish-oil supplements [EPA and DHA], but not of alpha-linolenic acid [plant-derived ALA], reduces the rates of all-cause mortality, cardiac and sudden death, and possibly stroke."[16] Chalk up one scientific reason for a diet rich in marine-derived omega-3 oils (EPA and DHA) or the use of such a supplement.

The American Health Association (AHA) promotes consumption of both plant-derived omega-3 oil (ALA) and marine-derived omega-3 oils (EPA and DHA). In 2002 the AHA documents, "Large-scale epidemiological studies suggest that individuals at risk for CHD [coronary heart disease] benefit from the consumption of plant- and marine-derived omega-3 fatty acids, although the ideal intakes presently are unclear."[17] Chalk up one scientific reason for a diet rich in both plant-derived omega-3 oil (ALA) and marine-derived omega-3 oils (EPA and DHA) or the use of such supplements.

In 2007 the AHA reports, "Omega-3 fatty acids benefit the heart of healthy people and those at high risk of—or who have—cardiovascular disease."[18] However, the AHA explains, "Although some alpha-linolenic acid [plant-derived omega-3 oil] is converted to longer-chain omega-3 fatty acids [EPA and DHA], the extent of this conversion is modest and controversial."[19] Scientific studies confirm limited conversion from plant-derived omega-3 oil (ALA) to EPA and even less conversion to DHA.[20] Chalk up another scientific reason for a diet rich in marine-derived omega-3 oils (EPA and DHA) or the use of such a supplement.

Therefore, since research 1) attributes heart-related benefits to consumption of marine-derived omega-3 oils (EPA and DHA), and 2) the American Heart

Association and scientific studies reveal "modest" conversion from plant-derived omega-3 oil (ALA) to EPA and even less conversion to DHA, scientific evidence indicates that it is prudent to consume a diet rich in marine-derived omega-3 oils (EPA and DHA) and/or discuss with a licensed physician the use of such a supplement.

The question to ask at this point is: If you are consuming a diet rich in marine-derived omega-3 oils (EPA and DHA) and/or use such a supplement, does scientific research support additional diet augmentation or supplementation with plant-derived omega-3 (ALA) and omega-6 oils? Initially, we thought, "Oil is oil, right? Why would we need any additional oil supplementation if we were already supplementing our diet with an EPA- and DHA-providing fish oil supplement and consuming flax seeds, which are rich in omega-3 oil (ALA)?"

The answer is quite clear. Just as research indicates that consumption of the omega-3 oils (marine-derived EPA and DHA and plant-derived ALA) doesn't provide the same health benefits, neither does consumption of the two essential omega oils, omega-3 and omega-6, provide the same health benefits. For example, according to the University of Maryland Medical Center, "Omega-3 fatty acids help reduce inflammation and most omega-6 fatty acids tend to promote inflammation. An inappropriate balance of these essential fatty acids contributes to the development of disease while a proper balance helps maintain and even improve health."[21] The omega oil blend that we took contained a higher ratio of omega-3 oil to omega-6 oil, which likely helped reduce the inflammation in our bodies caused by our bodies' immune responses to the mold and chemical exposure. With less inflammation, we experienced reduced allergic responses to allergens.

The University of Maryland Medical Center further points out, "It is important to maintain an appropriate balance of omega-3 [marine-derived EPA and DHA and plant-derived ALA] and omega-6 (another essential fatty acid) in the diet as these two substances work together to promote health."[22] Linoleic acid (LA) is the main omega-6 oil, which is in most vegetable oils, such as safflower, corn, peanut, and sesame.[23] Chalk up one scientific reason for a diet rich in marine-derived omega-3 oils (EPA and DHA), plant-derived omega-3 oil (ALA), and plant-derived omega-6 oil (LA) and/or the use of such supplements.

It is important to note that the American Heart Association reports that long-term observational studies indicate additional benefits from consumption of plant-derived omega-3 oil (ALA).[24] Chalk up one scientific reason for a diet rich in plant-derived omega-3 oil (ALA) and/or the use of such a supplement.

These scientific and medical findings indicate that we should obtain adequate amounts of both marine-derived omega-3 oils (EPA and DHA) and plant-derived omega-3 oil (ALA) as well as omega-6 oil (LA) from either food sources and/or food supplements. Therefore, a diet rich in marine-derived omega-3 oils and/or the use of such supplements meet different essential fatty acid (EFA) needs of the body than does a diet rich in plant-derived omega-3 oil (ALA), omega-6 oil (LA), and omega-9 oil, and/or the use of such supplements. Not all oil is equal; not all oil serves the same purpose.

Now let's review the health effects that we experienced (in addition to the already mentioned reduction in inflammation) from supplementation with the omega oil blend. We believe it helped detoxify our bodies and increase liver function. The whites of our eyes became very white, probably whiter than they were before the hurricane. We noticed toxins coming out of our skin, yet not damaging our skin as this detoxification process had earlier, creating hives, rashes, pain, and extreme itching.

Why would have we experienced these types of benefits from taking an omega oil blend? According to Dr. Erasmus, "Good oils, especially those containing omega-3s, carry oil-soluble toxins out of the body through the skin even in the absence of sweating—sweating just speeds up the process. It is the natural movement of oils to the skin that is the reason for their effectiveness. Oils also increase the energy levels of our cells. This gives our cells improved capacity for whatever their functions are, including detoxification. They may help the liver do its detoxification jobs better."[25]

We were very surprised and pleased with the visible results. Over time, the scars that remained from the months of hives began to heal. Our skin became soft and supple. Our hair, much of which had fallen out directly post-hurricane, began to grow back. By the one-year anniversary of Katrina, Kurt's thinning hair and receding hairline was thickening and creeping back down his forehead, and I had a new growth of hair coming in that was about three inches long! We also could feel a boost in energy at the cellular level, which resulted in additional weight loss in both Kurt and me. In fact, because of the changes in our diet and the addition of the omega oil blend and the other supplements in our restorative health regime, Kurt had lost 35 pounds and I had lost 50 pounds.

The biggest unexpected result was the ability of the omega oil blend to clear our heads beyond the "brain-defogging" process that had already occurred from taking the unpasteurized noni juice. To our pleasant surprise, both the speed and clarity in our ability to think, remember, and reason further increased. From a physical standpoint, we could feel a drawing out of toxins from our sinus cavities and other areas within our heads as inflammation decreased. We surmise that the omega oil blend was able to reduced inflammation, which then allowed our body processes to draw oil-based toxins from our brains, sinus cavities, and toxin-filled tissues; with less toxic inflammation, we were able to think, remember, and reason more clearly.

According to Dr. Thrasher, everything we smell goes directly into our brains.[26] With this fact in mind, we hate to think of all the airborne, volatile toxins (mycotoxins,

VOCs, chemicals, and pulverized debris) that we, and others, inhaled post-Katrina that went straight into our brains. Unaware of the extent of potential effects from toxic inhalation, we were simply aware that our post-hurricane cognitive abilities were much slower than normal, a condition that we half-jokingly referred to as *mold brain damage*. At the time, we were not aware that exposure to certain mycotoxins produced by toxigenic species is suspected to actually cause encephalopathy, which is brain damage.

We are not alone in our perception of a brain-fog condition resulting from fungal exposure. We have spoken with several people who became sick after fungal exposure, and they too talk of the inability to think, remember, and reason as before their exposure to fungi. Although we had clearly made progress against the fungi, had more stamina, and had stopped constantly getting sick, we still had lingering effects of this brain fog. In fact, not until after the omega oil blend began to reduce inflammation, presumably drawing toxins from our brain tissues and other areas, and increase our clarity of thinking, did we realize the full extent to which our mental sharpness had been affected.

Interestingly, although difficult to describe, we noticed distinct differences in the health-enhancing effects from consuming a fish oil supplements as compared to consuming an omega oil blend. Consumption of a quality fish oil supplement was remarkable in how it increased brain function. It appeared to enable our brain neurons to fire more quickly, as if our brains were waking up and becoming more alive! We did not notice this effect from consumption of the omega oil blend. However, consumption of the omega oil blend created a feeling of calmness and contentment, which we attribute to the likely reduction of inflammation, removal of toxins, and improved hormonal efficiency.

It is difficult to tell specifically if one product or the other contributed more to the return of our pre-hurricane cognitive ability. We view the two supplements—one that provided a source of marine-derived omega-3 oils (EPA and DHA) and the other that provided a source of plant-derived omega-3 oil (ALA), omega-6 oil (LA), and omega-9 oil—as a package deal. Just as scientific research reflects, we observed that the use of both supplements provided different health benefits. We were elated to have found two products that enabled us to think, remember, and reason as we did before our fungal and chemical exposure. What is even more amazing is that the health improvements that we experienced from the use of both of these products (which were not taken simultaneously during our trial-and-error period) were apparent even after just several days of use. For example, the omega oil blend reduced inflammation, which in turn likely released toxins. In fact, it drew toxins out of the body with such speed and effectiveness that we were able to tolerate only 1 teaspoon a day in the beginning. Dr. Erasmus states that people who are sick will want to start slowly, as we did, because the oil will enable their bodies to detoxify more efficiently, which shouldn't be done too rapidly.[27] Of course, as with the use of any supplement, the use of an omega oil blend, which is available at health food stores, should be reviewed with a licensed physician.

Another supplement we researched and considered the use of was digestive enzymes. In theory, these enzymes (from either an animal or microorganism source) digest the proteins that your body is unable to, either because of lack of digestive enzymes in your intestinal tract or overproduction of antibody proteins that your body produces in an immune response to antigen proteins.[28] However, since digestive enzymes come from either an animal source (such as pancreatic enzymes) or from microorganisms (such as enzymes made by *Aspergillus* and/ or other fungi), it is possible to have an antigenic response, thus creating more antigen proteins!

According to Dr. Erasmus, an alternative solution is to eat a lot of vegetables since digestive enzymes are also supplied by eating raw food. He states that they can also be obtained from probiotics.[29] Since we were already consuming a diet largely comprised of raw food and taking probiotic supplements, we decided not to risk possible allergic and/or antigenic reactions to either type of digestive enzyme product. Additionally, our ND recommended that we not use digestive enzymes because our illnesses were fungal-related.

Another area we started looking at was current exposure to irritating, detrimental chemicals. Until we became sensitive post-hurricane to additives, antibiotics, hormones, and chemicals, we did not realize how many hidden chemicals we were exposed to on a daily basis. We started to identify chemical exposures that we could eliminate from our lives. We thought if we lowered the overall toxic burden of our bodies, then maybe they would be able to better detoxify the fungal and chemical toxins incurred from Hurricane Katrina.

One of the first changes we made was to store food in glass jars instead of using plastic containers to ". . . avoid bacteria growth and toxins that can seep into foods from plastics."[30] According to Dr. Erasmus, "Research has shown that plastic leaches into foods that come in contact with it and the more oil/fat that is in the food, the faster is the leaching of plastic into the food."[31] For this reason, we use glass as much as possible and chose products packaged in glass whenever available.

Another contaminant to which we became extremely sensitive post-hurricane was formaldehyde. It is ". . . normally present at low levels (usually less than 0.03 ppm) in both outdoor and indoor air."[32] However, formaldehyde-containing products and building materials can further increase indoor air levels. For example, we noticed an unpleasant odor after we brought home several new pieces of inexpensive pressed board furniture, which excessively off-gassed formaldehyde into our indoor air. Eventually, we had to replace this furniture with more expensive, all-wood furniture, which we selected from showroom floor samples. We purchased showroom floor samples in order to minimize further formaldehyde contamination in our home, because showroom samples, having sat out for eight or nine months, have already undergone the initial off-gassing of chemicals inherent in the manufacturing of even all-wood furniture. In other words, let the initial formaldehyde and other chemical off-gassing occur inside the showroom—not inside your home!

After replacing the formaldehyde-laden pressed wood furniture, we were able to notice a definite and significant increase in air quality in our home. The formaldehyde in the adhesives used to make the pressed wood furniture was no longer off-gassing and recirculating throughout our home via our central air system.

In indoor environments, formaldehyde can cause ill health effects to occupants, depending upon the level. According to Thad Godish, PhD, "Formaldehyde is a potent sensory irritant that causes irritation of the eyes, nose, throat, sinuses, as well as the lower respiratory system (cough, shortness of breath, chest pain). It can also apparently cause headaches, unusual fatigue, disturbed sleep, and diarrhea. It can also predispose an individual to infections of the eyes and sinuses. These symptoms can all be caused by formaldehyde at relatively low exposure levels (>0.05 ppm).[33]

"The odor of formaldehyde is distinctive but is often masked by the odors of other materials," according to Dr. Godish.[34] Its level of off-gassing is also affected by the level of heat and moisture. These climate-dependent fluctuations of formaldehyde off-gassing can make causative identification of symptoms more difficult, as does the fact that formaldehyde can affect people's health differently. The American Cancer Institute reports that some people are very sensitive, while others have no reaction to the same level of exposure.[35]

People in high-risks groups will likely have increased negative health effects from formaldehyde exposure. Pediatrician Scott Needle, MD, explains why children are at higher risk, "Children are more vulnerable than adults because they have developing immune systems." Dr. Needle cites other factors that increase health risks to children, "Children breathe faster than adults (inhale more times per minute). They are closer to the ground and to contaminants that settle to the ground. They put things in their mouths and play on the ground. Also, children are smaller so it doesn't take as much to produce effects on them."[36]

Dr. Thrasher, who has published several papers regarding the health effects of formaldehyde exposure, explains why formaldehyde can cause such an array of health problems, "First of all, formaldehyde is a great conjugator or adductor. What I mean by that is, it binds to every protein in your body and changes that protein so that the immune system recognizes it as foreign. By doing this, formaldehyde is very antigenic, meaning it stimulates the production of antibodies. When you breathe it in, it starts binding with all the proteins in the nasal cavity, which changes those proteins, and then antibodies to those changed proteins are created."[37] As we have already learned, excess proteins in the body can cause inflammation and other symptoms anywhere in the body.

Unfortunately, the use of formaldehyde is widespread. According to the National Cancer Institute (NCI), formaldehyde is used in ". . . pressed wood products such as particleboard, plywood, and fiberboard, glues and adhesives, permanent press fabrics, paper product coatings, and certain insulation materials."[38] Pressed wood, of course, is used in the construction of furniture, cabinets, and other building materials.[39] Furthermore, the NCI states, "Formaldehyde can also be released by burning wood, kerosene, natural gas, or cigarettes; through automobile emissions; or from natural processes."[40]

Formaldehyde is a hazardous substance that is monitored by the Agency for Toxic Substances and Disease Registry (ATSDR). The agency reports that formaldehyde is ". . . also used as a preservative in some foods and in many products used around the house, such as antiseptics, medicines, and cosmetics. It is also known as methanal, methylene oxide, oxymethyline, methylaldehyde, and oxomethane."[41] ATSDR states that household sources of exposure include fiberglass, carpets, permanent press fabrics, paper products, and some household cleaners.[42] Interestingly, the Consumer Product Safety Commission claims, "Products such as carpets or gypsum board do not contain significant amounts of formaldehyde when new. They may trap formaldehyde emitted from other sources and later release the formaldehyde into the indoor air when the temperature and humidity change."[43]

According to the EPA, "Formaldehyde, a colorless, pungent-smelling gas, can cause watery eyes, burning sensations in the eyes and throat, nausea, and difficulty in breathing in some humans exposed at elevated levels (above 0.1 parts per million). High concentrations may trigger attacks in people with asthma. There is evidence that some people can develop a sensitivity to formaldehyde. It has been shown to cause cancer in animals and may cause cancer in humans. Health effects include eye, nose, and throat irritation; wheezing and coughing; fatigue; skin rash; severe allergic reactions. [It] may cause cancer."[44]

The EPA further states, "In homes, the most significant sources of formaldehyde are likely to be pressed wood products made using adhesives that contain urea-formaldehyde (UF) resins. Pressed wood products made for indoor use include: particleboard (used as sub-flooring and shelving and in cabinetry and furniture); hardwood plywood paneling (used for decorative wall covering and used in cabinets and furniture); and medium density fiberboard (used for drawer fronts, cabinets, and furniture tops). Medium density fiberboard contains a higher resin-to-wood ratio than any other UF pressed wood product and is generally recognized as being the highest formaldehyde-emitting pressed wood product."[45]

The EPA explains, "Average concentrations in older homes without UFFI [urea-formaldehyde foam insulation] are generally well below 0.1 (ppm). In homes with

significant amounts of new pressed wood products, levels can be greater than
0.3 ppm."[46] Therefore, older homes may be healthier in terms of formaldehyde
exposure than newer homes because the formaldehyde-containing building
materials have had years in which to off-gas. According to the EPA, newer homes
could have levels higher than three times (> 0.3 ppm) the level the EPA designates
as elevated (0.1 ppm).

The serious health effects of formaldehyde exposure have been scientifically
studied. Because of the results of these studies, "In 1987, the U.S. Environmental
Protection Agency (EPA) classified formaldehyde as a probable human carcinogen
under conditions of unusually high or prolonged exposure. . . . In 1995, the
International Agency for Research on Cancer (IARC) concluded that formaldehyde
is a probable human carcinogen. However, in a reevaluation of existing data in June
2004, the IARC reclassified formaldehyde as a known human carcinogen," reports
the National Cancer Institute.[47] Additionally, ATSDR states, "The Department
of Health and Human Services (DHHS) has determined that formaldehyde may
reasonably be anticipated to be a carcinogen."[48]

Because of the potential negative health effects of formaldehyde, multiple agencies
have set permissible exposure limits for formaldehyde. As you will see, the
exposure limits set by the following agencies vary. The NCI reports, "In 1987,
OSHA passed a law that reduced the amount of formaldehyde to which workers
can be exposed over an 8-hour work day from 3 ppm to 1 ppm. In May 1992, the
law was amended, and the formaldehyde exposure limit was further reduced to
0.75."[49] OSHA limits are based on a 40-hour workweek.[50] According to ATSDR,
"The National Institute for Occupational Safety and Health (NIOSH) recommends
an exposure limit of 0.016 ppm."[51] According to the CDC, NIOSH levels are based
on an 8-hour time weighted average (TWA).[52] Furthermore, the Washington State
Department of Health reports that NIOSH recommends a 15-minute exposure
ceiling of 0.1 ppm.[53] The Consumer Product Safety Commission reports, "A 1985
HUD regulation covering the use of pressed wood products in manufactured
housing was designed to ensure that indoor levels are below 0.4 ppm."[54] The
EPA considers levels over 0.1 ppm ". . . unusually high."[55] Although these
government agencies have not set the same level for permissible exposure limits
of formaldehyde, they each have acknowledged that negative health repercussions
from formaldehyde exposure exist.

Due to the increased awareness of formaldehyde toxicity because of the Katrina
FEMA trailer debacle and pressure from industry groups, such as U.S. wood
manufacturers, Senator Mike Crapo sponsored a bill, the "Formaldehyde Standards
for Composite Wood Products Act", that was ultimately signed by President Barack
Obama on July 7, 2010. According to U.S. Congressional documentation, "The act
amends the Toxic Substances Control Act to reduce the emissions of formaldehyde
from composite wood products and for other purposes."[56] The text of the act states
that by January 1, 2013, the composite wood products sold in the United States

will have to meet formaldehyde emission standards of about 0.09 ppm.[57] The new national 0.09 ppm regulated level of formaldehyde sounds good on the surface. It is lower than the 0.1 ppm that the EPA considers "unusually high." But is 0.09 ppm a safe level of airborne formaldehyde? According to Dr. Godish, "Both the California Air Resources Board and [the] Department of Health have set an action guideline of 0.10 ppm for remediation and a target acceptable level of 0.05 ppm. However, studies by Australian scientists have shown adverse physiological reactions even at 0.05 ppm."[58] Because of the findings of these studies, Dr. Godish cites 0.03 ppm as a "feasibly safe level" of airborne formaldehyde.[59]

Undoubtedly, the enactment of the "Formaldehyde Standards for Composite Wood Products Act" was a step in the right direction. However, only with the passing of time will we be able to evaluate its effectiveness. Unfortunately, the act contains some language that is favorable to industry, not health. The act, which applies to finished goods, contains several exclusions, such as, "The term 'finished good' does not include— (i) any component part or other part used in the assembly of a finished good…"[60] The act does not define the terms *component part* or *other part used in the assembly*, which could give manufacturers the ability to include formaldehyde-emitting *component parts* or formaldehyde-laden *other parts used in the assembly* of their products.

The act also contains some exemptions. The new 0.09 ppm regulation does not apply to hardboard, structural plywood, structural panels, or structural composite lumber, which means homes and other structures can continue to be built with wood that contains higher levels of formaldehyde. Senator Crapo's measure establishes formaldehyde health standards in composite wood used in products such as shelving, cabinets, furniture and other goods, according to the Idaho Statesman.[61]

To have lower levels of formaldehyde emitting from shelving, cabinets, furniture, and other goods will reduce indoor levels of formaldehyde in homes, businesses, modular and mobile homes, and RV trailers. The question remains, however, will this initial effort to reduce indoor formaldehyde levels be enough given that structural composite woods can still contain higher levels of formaldehyde? How much formaldehyde will be emitting into the indoor air from the structural composite woods?

The HUD regulation of 0.4 ppm will remain in place for HUD housing, even though 0.4 ppm is four times the level that the EPA considers "unusually high". Indoor formaldehyde levels in mobile homes and RV trailers will likely continue to be concentrated at higher levels due to less indoor air volume. According to Dr. Godish, "Travel trailers, like mobile homes, have had a long history of being contaminated with formaldehyde." He explains this level of toxicity occurs because historically formaldehyde use in the manufacturing of RVs has not been subject to any regulation at all.[62]

Dr. Godish elaborates, "The construction of mobile homes is regulated by the Department of Housing and Urban Development. HUD regulates emissions of formaldehyde from particleboard and wood paneling and requires manufacturers to post a *health notice* in a prominent location inside the mobile home indicating the nature of health concerns associated with formaldehyde exposures in mobile homes."[63]

Although the same formaldehyde-laden materials are used in travel trailers and other RVs (exposure to which can cause the same negative health effects), HUD regulations and warnings do not apply to RVs because they are not subject to HUD regulation.[64] Therefore, if you buy a mobile home, you will be warned of the potential health hazards of your purchase. However, if you buy an RV, you will not. Interestingly, trailers and motor homes are often manufactured by the same companies that manufacture mobile homes, according to Dr. Godish.[65] Given that some of these manufacturers do indeed produce both products, how can these companies, in good conscience, warn one group of consumers of the potential health hazards of manufacturing components, but yet not another (i.e., purchasers of RVs)? Obviously, these particular companies place more value on their bottom lines than on the health of their customers.

Another element of concern is the importing of materials for use in RV manufacturing. Dr. Godish explains, "Trailer/RV manufacturers can use formaldehyde-emitting wood products manufactured in the U.S. or those imported from other nations. Pressed wood products produced in south Asia are not subject to any formaldehyde emission requirements and in fact are often potent emitters of formaldehyde."[66] Prior to the enactment of the "Formaldehyde Standards for Composite Wood Products Act", with no regulations or permissible exposure limits, there was nothing stopping RV manufacturers from purchasing and importing pressed wood materials highly laden with formaldehyde, which in turn elevated the levels of airborne formaldehyde inside RVs when these materials off-gassed. Manufacturers could choose to purchase the less expensive Asian wood products to boost their bottom lines, regardless of the cost to their customers' health.

The verbiage in the "Formaldehyde Standards for Composite Wood Products Act" does not indicate that the new formaldehyde regulation level would apply to the "structural" composite wood used in the manufacturing of RVs. Very little can be expected to change in regard to formaldehyde levels inside RVs unless RV manufacturers voluntarily seek structural materials that emit lower levels of formaldehyde.

At the time of this writing, it remains to be seen how the "Formaldehyde Standards for Composite Wood Products Act" will affect the price of imported Asian composite wood products. The act may cause an increase in price because Asian manufacturing practices will have to change in order to meet the lower regulated level of formaldehyde, which is required by the act to be verified through third-party testing. If a price increase occurs in Asian composite wood imports, Senator Crapo will have accomplished one of his goals. Evening the playing field between U.S. and Asian wood producers was one of the reasons Senator Crapo originally sponsored the bill. According to the Idaho Statesman, Crapo said he wanted to require foreign imports of composite wood, largely from China, to comply with standards he believes are already voluntarily followed by U.S. manufacturers.[67] This time the lobbying power exerted by U.S. industry groups, wood manufacturers in this case, worked in favor of the consumers' health.

Even with the enactment of the "Formaldehyde Standards for Composite Wood Products Act", because of the widespread use of formaldehyde and the yet-known effectiveness of the act, the caveat *Buyer Beware* is apropos. The National Cancer Institute recommends, "Before purchasing pressed wood products, including building materials, cabinetry, and furniture, buyers should ask about the formaldehyde content of these products."[68] Better yet, look for formaldehyde-free alternative products, which likely are better health-preserving options. Various companies have begun to serve this health-conscious niche in the market place by developing and designing formaldehyde-free products in their respective product categories. For example, Johns Manville offers a formaldehyde-free insulation, and Columbia Forest Products produces a formaldehyde-free hardwood plywood.

Another product that warrants mention is DensArmor, the paperless interior drywall that was recently developed by Georgia Pacific. According to the company's website, by substituting a glass mat facing for the paper facing found on regular sheetrock, design engineers at Georgia Pacific have created a product that is more mold resistant than traditional drywall. Paper facing is a significant source of food for fungi if adequate moisture accumulates.

Because of the research and development efforts of these companies, it is possible to create a more health-conscious living environment by minimizing the use of formaldehyde-containing and mold-susceptible building materials. However, just because a product is formaldehyde-free and/or mold resistant, doesn't mean that it doesn't contain other components that could be harmful to health. Consumers need to confirm that a product does not contain other potentially harmful chemicals or substances.

By supporting companies willing to invest in healthier product alternatives, we not only potentially protect our health and the health of our families but we encourage the continuation of this type of health-benefiting research and development. As consumers, we should request information about the levels of toxic off-gassing in product materials and demand healthier alternatives. If we, as consumers, refuse to buy products containing carcinogenic materials, companies will have no choice but to produce healthier product alternatives. We, personally, greatly appreciate companies that invest in the health aspects of product research, making it possible to purchase health-enhancing innovative products.

Interestingly, being sick and ultra-sensitive to fungi, chemicals, and other contaminants has enabled us to make long-term, health-enhancing changes in our life, because it quickly exposed to us products made of inferior and/or contaminated raw materials. In our search for nonreactive, high-quality products, we often found that the products we could tolerate were manufactured by companies that practice more earth-friendly methods of producing and manufacturing, which increases the quality of their products. One company is Stonyfield Farm, which produces yogurt and milk products while supporting an environmentally conscious philosophy of production and manufacturing. We found the Stonyfield Farm brand of plain yogurt to be a more powerful source of probiotics than other brands of yogurt. During our intense period of healing, we ate only the plain Stonyfield Farm yogurt because it has half the sugar content as compared to Stonyfield Farm's flavored yogurts. By consuming less sugar, we maximized the benefits from the probiotic cultures. Since our recovery, we now enjoy other flavors! The noticeable, added health benefit from the cultures in Stonyfield Farm yogurt may be in part because one of the six cultures is proprietary to Stonyfield Farm in the U.S., which means that the proprietary culture is not present in any other brand of yogurt in the U.S.[69]

When your body gets to a point it can no longer tolerate the additive-, hormone-, and toxic-filled "junk" that some manufacturers produce, it really sinks in that not all food is manufactured with health in mind. Quality raw materials are key to a nutritional end product. According to Stonyfield Farm president and CEO Gary Hirshberg, "Most business models follow this premise: Make the product as cheaply as you possibly can, so you have an enormous margin left over. Use this huge margin to buy advertising everywhere and try to convince the consumer to try your product."[70] We must not fall prey to the mind-bending advertising campaigns of food manufacturers that follow this type of business philosophy. Brands that have "healthy" advertising budgets promoting them do not necessarily contain healthy raw ingredients. Nor do they necessarily follow nutrient-preserving processing and manufacturing methods. We must not only become careful label readers, but we must also search for the manufacturers using the highest quality raw ingredients free of additives, antibiotics, hormones, and processing practices that destroy vital nutrients and vitamins.

Initially, this level of food research may feel like a lot of extra work that you don't have time for in an already busy life. It did to us, but frankly, having become ultra-sensitive to additives, antibiotics, and hormones, we had no choice but to take the time to locate a few "safe" products so we could eat something! Essentially, we all are faced with the same reality: If we don't make time today to plan for tomorrow's health—by making life-supporting food choices—tomorrow may never come.

We must use our time wisely to make choices that support our individual health needs, as well as the health of our planet. To locate and support companies that invest in our health by manufacturing products that are health-building—not health-destroying, filled with chemicals our bodies must detoxify—is a right we each have and must exercise. We must not let ourselves become living cesspools, diseases from which conditions new names are created—as the famous Greek philosopher, Plato, so stated: "And we have made of ourselves living cesspools and driven doctors to invent names for our diseases."[71]

We, as consumers, also have the power to shape the ecological future of our world by giving our consumer dollars to companies that not only invest in the development of life-enhancing products but also use manufacturing methods that do not destroy our environment. Even better is to support companies that have taken social responsibility to the next level by supporting and using health-enhancing and earth-preserving organic farming methods and invest their money in research and development of materials and products free of known toxic substances.

We, as a family, will continue to look for innovative, health-preserving and health-promoting companies that are investing in consumers' health—our health. With each purchase, we are essentially saying, "We accept and support the origination process of this product." Let's do our research as consumers and be careful which companies we support with our purchasing dollars, because in doing so, we determine which companies will still be in business to produce the products of tomorrow.

According to Mr. Hirshberg, "'. . . [I]f we don't recognize the power we have with our purchase, which is otherwise known as a vote, to vote for the kinds of products, services, and, ultimately, industries that will take toxins out of the biosphere or that will help give our grandchildren a break, then we're missing the boat. We have the power to encourage businesses to do the right thing. Otherwise, our grandchildren are going to be sitting there saying: What . . . were you people doing? How can you have blown this bountiful, gorgeous, incredible thing called the planet Earth?'"[72]

If we become passionate about protecting our health, our planet, and the future we are leaving for our children and generations to come, then we must refuse to financially support companies that pilfer and pollute our land. We must empower ourselves—especially in the wake of destruction—to make life-preserving choices, such as selecting healthy food equivalent to food therapy and/or locating a mold-free environment in which to live. We each ultimately live the repercussions of our daily choices—be them good or bad. Simply put, we need to maximize the good and minimize the bad in terms of what we eat, drink, breathe, and touch. We can't expect optimal health if we are living in a mold-contaminated house, a formaldehyde-filled trailer, or a neighborhood akin to a toxic waste dump. What we are exposed to—whether from food toxins or environmental contamination—will inevitably have an impact on our health.

Whether we are healthy today, suffering from mold-related illnesses, and/or battling a degenerative disease, we will have the best chance of preserving and even restoring our health in a multitoxic world if we follow the philosophy set forth in

400 B.C. by Hippocrates, another famous Greek philosopher, who believed: "Let food be thy medicine, and let thy medicine be food."[73]

If we use this philosophy to guide our every bite, we will be stronger and healthier—in mind, body, and spirit—to cope with life's inevitable challenges, be them widespread natural disasters or individual hurdles. We must realize: What we eat today, to some degree, is a self-manifesting destiny of tomorrow's health and can even determine which of us survives—and which of us doesn't—the next natural disaster or even the next act of terrorism.

The fact that we had success with natural remedies—vitamins, herbs, and other food supplements—in combination with a diet full of vegetables and void of as many additives, antibiotics, hormones, mycotoxins, fungal antigens, and chemicals as possible is really not surprising considering that leading universities have been researching components in natural foods as cures for disease for years. For example, researchers at Johns Hopkins and Oregon State University have found that ". . . prophylactic intervention with chlorophyllin or supplementation of diets with foods rich in chlorophylls may represent practical means to prevent the development of hepatocellular carcinoma [liver cancer] or other environmentally induced cancers."[74] Broccoli is a food high in chlorophyllin.

In addition, Dr. Wang states that he and other scientists are studying the effects of ". . . green tea polyphenols and green tea extracts to see if they can prevent the effects of aflatoxins. Structurally, they do not bind with the aflatoxin as . . . chlorophyllin [does], but if you drink green tea, the natural active components can increase your body's metabolizing enzymes. In this manner, the polyphenols and green tea extracts can detoxify aflatoxins."[75]

Treating and preventing diseases with nutrition has long been viewed as a viable concept. Thomas Edison stated, "The doctor of the future will no longer treat the human frame with drugs, but rather will cure and prevent disease with nutrition."[76] We now believe, because of our own health restoration, in the power of the human body to heal itself—if supplied with the proper nutrients before irreversible degeneration has occurred.

Dr. Erasmus further explains this concept, "Nutrients are the building blocks for making our bodies. The body is made out of food, water, air, and light. Our genetic program knows exactly what to do with those building blocks, provided we get them all and get enough of each. . . . If we don't get enough of any *one* essential nutrient, we cannot stay healthy. With inadequate intake of any one of the essential nutrients, our health will deteriorate. We will develop deficiency symptoms that are degenerative in nature—some of them look a lot like our degenerative diseases. The deficiency symptoms that develop from not getting enough of any one or more of the essential nutrients get worse with time. If we don't get enough for long enough, we die. That's how important they are."[77]

Dr. Erasmus continues, "If we bring back an adequate amount of the lacking essential nutrient(s) before we die, the symptoms of deficiency are reversed. That's because life, using our genetic program, knows exactly where to take each essential nutrient molecule, what to do with it, and how to use it to build a body where everything works. If we bring back enough of them while we are deteriorating on our way to dying, then all of the problems that come from not getting enough are reversed. If we want optimum health, optimum amounts of all of these essential nutrients that are building blocks for healthy bodies must be supplied."[78]

The important point to understand is that our treatment plan was a combination of vitamins, herbs, and other food supplements and a diet free of sugar, fungi, and chemicals as much as possible. We focused on these building blocks to restore our health. Each was interdependent on the other. For example, if we had been consuming a diet full of foods containing sugars, fungal antigens, mycotoxins, additives, antibiotics, hormones, and other chemicals, then the health-restoring products our family took would not have been as effective. The healing and detoxifying components would have been busy detoxifying our bodies from our contaminant-laden diet!

We had to aggressively fight for the restoration of our health. There were no quick fixes. It took work and perseverance on our part—and time. Ironically, because of the mold and chemical exposures we incurred post-Katrina and the subsequent changes that we had to make in our diet and health regime, we very well may enjoy a higher quality of health and life in the next 10 to 20 years than we otherwise would have had we not been forced—because of the lack of effective medical solutions—to research alternative health remedies in a quest to restore our health. The fact that our health was revitalized beyond its pre-hurricane state speaks volumes for the treatment regime we followed and the healing ability of alternative treatments. Should the information that we shared inspire you to begin your own journey to optimal health, do it wisely in consultation with a licensed doctor(s) as we did. Pay attention to warning signs of fungal re-exposure; and most of all, remain vigilant—don't get complacent. Be proactive and fight for *your* health, no matter what your condition or circumstances.

(Endnotes)

1   Interview with and e-mail from Dr. Timothy Callaghan, January 2007.

2   Goldberg, *Alternative Medicine*, p. 395.

3   Ibid.

4   Balch, *Prescription for Nutritional Healing*, p. 50.
5   Ibid., p. 55.
6   See Section Three, interview with Dr. David Eaton, July 2006.
7   Balch, *Prescription for Nutritional Healing*, p. 56.
8   Ibid., p. 50.
9   Ibid.
10   Ibid., p. 56.
11   See Section Three, interview with Dr. David Eaton, July 2006.
12   Young et al., *The pH Miracle*, p. 167.
13   Ibid.
14   Ibid., p. 166.
15   Interview with Dr. Doris J. Rapp, July 2010.
16   Chenchen Wang et al., "Review Article: n-3 Fatty acids from fish or fish-oil supplements, but not alpha-linolenic acid, benefit cardiovascular disease outcomes in primary- and secondary-prevention studies: a systematic review."
17   Kris-Etherton et al., "AHA Scientific Statement: Fish Consumption, Fish Oil, Omega-3 Fatty Acids, and Cardiovascular Disease."
18   American Heart Association, "Fish and Omega-3 Fatty Acids."
19   Kris-Etherton et al., "AHA Scientific Statement: Fish Consumption, Fish Oil, Omega-3 Fatty Acids, and Cardiovascular Disease."
20   Attar-Bashi et al., "Failure of conjugated linoleic acid supplementation to enhance biosynthesis of docosahexaenoic acid from alpha-linolenic acid in healthy human volunteers"; Burdge et al., "Conversion of alpha-linolenic acid to eicosapentaenoic, docosapentaenoic, and docosahexaenoic acids in young women"; and Brouwer et al., "Gamma-linolenic acid does not augment long-chain polyunsaturated fatty acid omega-3 status."
21   UMMC, "Omega-3 fatty acids."
22   Ibid.
23   Goldberg, *Alternative Medicine*, p. 193.
24   American Heart Association, "Meeting Report."
25   Interview with Dr. Udo Erasmus, May 2006.
26   See Section Three, interview with Dr. Jack Thrasher, April 2006.
27   Interview with Dr. Udo Erasmus, May 2006.
28   Ibid.
29   Ibid.
30   Jeanne Marie Martin with Zoltan P. Rona, MD, *Complete Candida Yeast Guidebook*, p. 260 (Roseville, California, Prima Health, 2000).
31   Interview with Dr. Udo Erasmus, May 2006.
32   Consumer Product Safety Commission, "An Update on Formaldehyde: 1997 Revision," Consumer Product Safety Commission. Available at http://www.cpsc.gov/CPSCPUB/PUBS/725.html (accessed August 2006).
33   Thad Godish, "Indoor Environment Notebook: formaldehyde and trailers." Available at http://web.bsu.edu/IEN/archives/082506.htm (accessed October 2006).
34   Ibid.
35   National Cancer Institute, "Formaldehyde and Cancer: Questions and Answers."
36   Interview with Scott Needle, MD, October 2006.
37   See Section Three, interview with Dr. Jack Thrasher, April 2006.
38   National Cancer Institute, "Formaldehyde and Cancer: Questions and Answers." Available at http://www.cancer.gov/cancertopics/factsheet/Risk/formaldehyde (accessed September 2006).
39   Godish, "Indoor Environment Notebook: formaldehyde and trailers."
40   National Cancer Institute, "Formaldehyde and Cancer: Questions and Answers."
41   ATSDR, "ToxFAQs for Formaldehyde," CDC. Available at http://www.atsdr.cdc.gov/tfacts111.html (accessed August 2006).
42   Ibid.
43   Consumer Product Safety Commission, "An Update on Formaldehyde: 1997 Revision."
44   EPA, "An Introduction to Indoor Air Quality: Formaldehyde." Available at http://www.epa.gov/iaq/formalde.html (accessed September 2006).

45   Ibid.

46   Ibid.

47   National Cancer Institute, "Formaldehyde and Cancer: Questions and Answers."

48   ATSDR, "ToxFAQs for Formaldehyde."

49   National Cancer Institute, "Formaldehyde and Cancer: Questions and Answers."

50   ATSDR, "ToxFAQs for Formaldehyde."

51   Ibid.

52   CDC, "Formaldehyde." Available at http://www.cdc.gov/niosh/idlh/50000.html (accessed August 2006).

53   The Washington State Department of Health, "Formaldehyde." Available at http://www.doh. wa.gov/ehp/ts/IAQ/Formaldehyde.HTM (accessed September 2006).

54   Consumer Product Safety Commission, "An Update on Formaldehyde: 1997 Revision."

55   EPA, "An Introduction to Indoor Air Quality: Formaldehyde."

56   U.S. Congress, "S.1660", Govtrack.us. Available at http://www.govtrack.us/congress/ billtext.xpd?bill=s111-1660 (accessed August 2010).

57   Dan Popkey, "Crapo says formaldehyde law will aid U.S. companies", The Idaho Statesman, July 9, 2010. Available at http://www.idahostatesman.com/2010/07/09/1261299/crapo-says-formaldehyde-law-will.html (accessed August 2010).

58   Godish, "Indoor Environment Notebook: formaldehyde and trailers."

59   Interview with Dr. Thad Godish, October 2006.

60   U.S. Congress, "S.1660", Govtrack.us. Available at http://www.govtrack.us/congress/ billtext.xpd?bill=s111-1660 (accessed August 2010).

61   Dan Popkey, "Crapo says formaldehyde law will aid U.S. companies", The Idaho Statesman, July 9, 2010. Available at http://www.idahostatesman.com/2010/07/09/1261299/crapo-says-formaldehyde-law-will.html

62   Godish, "Indoor Environment Notebook: formaldehyde and trailers."

63   Ibid. (Italics added.)

64   Ibid.

65   Ibid.

66   Ibid.

67   Dan Popkey, "Crapo says formaldehyde law will aid U.S. companies", The Idaho Statesman, July 9, 2010. Available at http://www.idahostatesman.com/2010/07/09/1261299/crapo-says-formaldehyde-law-will.html (accessed August 2010).

68   National Cancer Institute, "Formaldehyde and Cancer: Questions and Answers."

69   T. Foster Jones, "A Culture for Change, Stonyfield Farm Wants to Save the Planet, one business at a time," The Costco Connection, September 2006, vol. 21, no. 9, p. 16-18. Available at http:// www.costcoconnection.com/connection/200609/ (accessed September 2006).

70   Ibid.

71   Dr. Bob Martin, "Archived Famous Health Quotes." Available at http://www.doctorbob. com/famoushealthquotes.html (accessed December 2006).

72   Jones, "A Culture for Change, Stonyfield Farm Wants to Save the Planet, one business at a time."

73   Neena Ghandi, "Nutraceutical Industry News," Kelley School of Business, Indiana University, November 2005. Available at http://www.kelley.iu.edu/lifesc/News/nov05_Nutraceuticals. html (accessed August 2006).

74   Patricia A. Egner et al., "Chlorophyllin intervention reduces aflatoxin—DNA adducts in individuals at high risk for liver cancer." PNAS, December 4, 2001, vol. 98, no. 25. Available at www. pnas.org/cgi/doi/10.1073/pnas.251536898 (accessed August 2006).

75   See Section Three, interview with Dr. Jia-Sheng Wang, July 2006.

76   Ghandi, "Nutraceutical Industry News."

77   Interview with Dr. Udo Erasmus, May 2006.

78   Ibid.

SECTION THREE

PROFESSIONALLY SPEAKING

Gina M. Solomon, MD, MPH, serves as Deputy Secretary for Science and Health at the California EPA. Prior to joining Cal/EPA, she was a senior scientist in the health and environment program at the Natural Resources Defense Council (NRDC) in San Francisco, California, and Assistant Clinical Professor of Medicine at the University of California at San Francisco (UCSF) in San Francisco, California, where she was the Associate Director of the UCSF Pediatric Environmental Health Specialty Unit. Dr. Solomon is a specialist in adult internal medicine, preventive medicine, and occupational and environmental medicine. Her work has included research on asthma, diesel exhaust, pesticides, contaminants in breast milk, and threats to reproductive health and child development. In addition, she was the lead scientist on the NRDC team that collected air samples post-hurricane in New Orleans October and November of 2005.

Dr. Solomon received her undergraduate degree from Brown University, attended medical school at Yale, and did her residency and fellowship training at Harvard. Dr. Solomon serves on the U.S. EPA's Science Advisory Board Drinking Water Committee and is on the National Academy of Sciences Committee on Toxicity Testing and Assessment of Environmental Agents. She is Vice President of San Francisco Bay Area Physicians for Social Responsibility.

Dr. Solomon has authored numerous articles and reports and is coauthor of the book Generations at Risk: Reproductive Health and the Environment *and recipient of an award from the American Medical Writers Association. Dr. Solomon received The Breast Cancer Fund's Heroes Award in 2002, the Clean Air Award for Research from the American Lung Association of the Bay Area in 2004, and a 2004 award of recognition from California Safe Schools. The interview was conducted July 2006.*

LB: You were the lead scientist overseeing the NDRC's air quality study that was done in the New Orleans and surrounding areas after Hurricane Katrina. What factors did you consider when deciding to use the nonviable versus the viable methodology of air sampling for mold?

Dr. S: Let me give you some background on this project. My training is in occupational and environmental medicine. I have seen patients who are concerned about illnesses related to mold, but I had never sampled for mold before. I went to New Orleans to do an assessment of environmental issues at the request of some local community groups and some of the elected officials representing some of the flooded areas of New Orleans. I went there expecting to look at issues about sediment and contaminated floodwaters, but I was struck by the overwhelming smell and sight of mold.

We actually scrapped a lot of our original plans and decided that somebody needed to get out there and start measuring for mold, because our meetings with EPA and DEQ made it abundantly clear that they were not testing for mold and weren't planning to. So, over the course of literally two weeks, I attempted to become a mold expert. One of the main things I did was talk with a lot of mold experts around the country and folks who are good at sampling—industrial hygienists. On our team, the true mycologist is Dr. Mervi Hjelmroos-Koski, who is the second author on the paper written on the New Orleans testing, published in *Environmental Health Perspectives*.[1]

When I talked about testing for mold with scientists, it was clear that there is no consensus about the best way to measure for mold. The standard measurement in outdoor air is usually total spore count. The standard measurement, in as much as there is any standard that tends to be done on indoor air, is viable spores. We were interested in testing both indoors and outdoors, so we had a problem. We also wanted to be able to compare the indoor results with the outdoor results. We realized that we had to pick one or the other. As a physician, it seemed a little silly to me to look just at viable spores, because whether a spore is living or dead is irrelevant to how your body's immune system perceives it. In other words, if you're going to mount an allergic reaction or an immune response to the mold, your immune cells will not differentiate between a dead spore and a living spore. The immune cells will go nuts, whatever that spore is, if they're going to go nuts—if you're a person who is sensitive. So, from a clinical perspective, I just wanted to know how many spores were out there.

LB: Also, dead spores can still have mycotoxins on them.

Dr. S: Absolutely. The only reason the living spores matter is in the case, perhaps, of people who have very impaired immune systems who actually might get an infection from *Aspergillus* or some other invasive fungus. Although this is definitely a concern, the other health concerns are so much more widespread, so they are a much greater concern to me. Obviously, if the total spore count is high, you know that a fair number of those spores are viable, so you know there is a risk of all of the health effects related to mold.

LB: Without performing viable testing, it's impossible to know what the actual species are and sometimes even the genus. Without knowing the species, we don't know if mycotoxins are being produced since not all species produce mycotoxins. Since many of the negative health effects from mold exposure are caused by the

---

1   Gina M. Solomon et al., "Airborne Mold and Endotoxin Concentrations in New Orleans, Louisiana after Flooding, October–November 2005," *Environmental Health Perspectives*, online June 12, 2006. Available at www.ehponline.org/members/2006/9198/9198.pdf (accessed June 2006).

mycotoxins, can we get a full estimation of the risk to human health without this additional information?

Dr. S: I guess when I started this, I was more focused on allergy. I am aware of the mycotoxin problem and, frankly, had been skeptical about mycotoxins. Obviously, they cause health problems in animal studies. There's lots of conflicting stuff in human studies regarding mycotoxins. I wasn't quite convinced of the serious problems involved with mycotoxins; mostly, I was thinking of people with allergies and asthma. Now I'm becoming more concerned about the mycotoxin issue. I think that, over time, that issue will become more widely understood and will be taken with the seriousness that it deserves.

LB: Several of the air quality professionals that I have spoken with say it is best to do both types of testing—nonviable and viable—if there is the budget. Were budget limitations the reason why the NRDC didn't do both types of testing?

Dr. S: Yes, that was a major reason. We wanted to test in a lot of neighborhoods, both indoors and outdoors, and in areas that were flooded as well as comparison locations that were not flooded. We also wanted to test for hours rather than just minutes in order to gather good estimates of 24-hour exposures. With all of those priorities, a limited budget, and limited time with a team on the ground in New Orleans, we had to make some hard decisions very quickly. One of the things we sacrificed was viable sampling. It would still be useful to do that.

LB: Was there anything noteworthy that was not written in the paper published in *Environmental Health Perspectives*?

Dr. S: Yes, there are some other interesting results. I believe Dr. Hjelmroos-Koski is actually working on a manuscript right now that gets into some of the details of the microbiology and the mycology. She took some tape samples, which were able to give her a little more information about the genus and even species, in some cases, although she wasn't necessarily able to get down to the species level. The other thing that she looked at is the hourly counts. We did samples from 6 hours all the way up to 24 hours, and she was able to break it down to hour-long increments or even 30-minute increments. In some of these indoor environments, it was incredible what happened. There would be wild fluctuations. In one house, the *Stachybotrys* was undetectable for about four out of the six hours of sampling. Then, over a one-hour period, there was a spike in *Stachybotrys* that was incredible. Clearly, it had sporulated during that short period of time and our equipment caught it. When we averaged the results to create the 24-hour average, there was definitely detectable *Stachybotrys*, but the spore counts were not impressive because there was only a brief period of time they were high. But if anybody had been so unfortunate as to have been in there during that period of time (when the *Stachybotrys* was releasing its spores), that would have been terrible. What I think it also shows is that some of these methods that go in for a very short period of time and gather spores over a 15-minute period can really miss serious mold problems.

LB: That's why there's such value in the NRDC testing—because it was done over a longer period of time.

Dr. S: I think we picked up certain things from doing it over a longer period of time. It's a trade-off. You get certain things. I've also been in e-mail communication with a faculty member from Tulane University who's a mycologist. When she returned after the flooding, she just put out some petri dishes in her house and nearby. She said that there was so much growth in the petri dishes that it was really hard to identify stuff; it was all just growing on top of itself—total chaos. She was able to identify some species.

LB: There is a lot of debate in the medical community regarding the extent to which mold can negatively affect human health. What are your thoughts?

Dr. S: I certainly agree that there is a lot more to be learned about mold and that there are a lot of surprises still to come about ways that mold can affect health. I tend to be cautious about what can be said about mold health effects. I certainly think that we can clearly say that *Stachybotrys* is not good for you. It's a very, very bad idea for people to be in an environment where they're inhaling that mold and other molds for that matter. Some of that science is still very controversial, but what we do know is that this is bad stuff. The concentrations that people can be breathing down in New Orleans—even right now—are not healthy, from right after the hurricane ended continuing into the present—not healthy.

LB: The NRDC took samples in the New Orleans area in October and November of 2005. Do you know much about the state of mold contamination there right now?

Dr. S: Not a lot. All I know is what people are telling me, which is that with the warm weather, people are smelling and seeing mold again.

LB: That's what we're hearing, too.

Dr. S: I think it had been getting better. I was there in March and April, and it actually seemed better. A lot of houses had been gutted. There wasn't the same heavy smell of mold all over when you walked down the street. I was thinking—this is good; the air quality is improving. I didn't have the equipment to do testing at that point, but just from observation, it was getting better. Now I hear it seems to be getting worse, which is not surprising given the seasonality of mold. We have been trying to find funding to go back and do more testing.

LB: That would be great.

Dr. S: It's very expensive to do. We got some private funding right after Hurricane Katrina to go in and do some environmental testing, and that money is all gone. It's been tough to get people to realize that there is an ongoing need.

LB: How much is needed to fund another series of tests, if that's okay to ask?

Dr. S: With travel costs for the scientists, equipment rental, shipping, and the costs of the laboratory analysis itself, it would cost about $10,000 to do a decent overview survey of mold comparable to the one we did last fall.

LB: One of the things that was pointed out in your report is that some of the fully remediated houses still had high spore levels.

Dr. S: We even found *Stachybotrys* inside one fully remediated house. Although that was disturbing, the remediated houses clearly looked better. When comparing the fully remediated with the partially remediated or—Lord help us—the unremediated houses, there is clearly a trend toward lower spore concentration. The question is: Were those spores that we were measuring in the fully remediated houses blowing in from neighbors' houses, from outside, or were they indicating that there was still mold growth inside? We couldn't tell. We did indoor and outdoor samples simultaneously at most of these sites or, at worst, in one quick sequence. One day we went there and sampled indoors, and the next day we sampled outdoors to try to get a point of comparison. In the fully remediated places, the levels were generally lower indoors than outdoors, so that was good. It likely meant that the spores we had measured inside had come in from the outdoor air, but we couldn't be sure of that. We could go back and test some of those places that were fully gutted and remediated just to be sure that they're doing okay.

LB: I think what people need to take away from what you are saying is that once they've remediated their homes, they still need to do spot-checks later in time— maybe one month later, maybe three months later, etc. From talking with biologists, I know it's possible that the mold growth could start again from an outside source if enough indoor moisture is present. People need to be very careful because they could be living in a home where mold is growing, but they think they have completely remediated it. I was told that there is still a lot of debris piled around that is conducive to mold growth, and so until that situation gets under control, all these homes and buildings that have been remediated are at risk for contamination from that type of debris.

Dr. S: Absolutely. If the buildings do not have significant moisture left in them, the contamination might not really take hold. The humidity levels tend to be so high in New Orleans that it might actually let the general mold in the piles of debris come along and take hold again.

LB: What should people take away from reading your report in *Environmental Health Perspectives*?

Dr. S: One message is that the mold spore concentrations were stunningly high; the concentrations in the outdoor air were high enough to be a health concern, which is not something that people realized right away. People were, in many cases, wearing protective equipment when they went indoors, but outdoor concentrations of mold were off the charts in some areas. I think that was a very sobering finding. We interpreted the lower mold concentrations in the remediated homes as somewhat good news, although we were cautious about how happy to be. We felt there was some evidence that doing full remediation could have addressed problems. We thought that this was an urgent public health situation that required people be provided with good personal protective equipment, that they be warned and urged to wear personal protective equipment, and that people with conditions that could make them more sensitive should really think twice before going back into the city, especially those with children or elderly people. We wanted to get the word out so people could protect themselves. We also think that the government agencies need to be tracking this problem. They should be doing ongoing mold sampling in the city. At minimum, they should have rooftop monitors at a number of locations in the city and warn people when the mold levels are high. That's consistent with national practice. It is not being done.

LB: The National Weather Service has just recently started providing mold counts again for the New Orleans area.

Dr. S: Oh, they have started them again? Last I checked they hadn't. I'm glad they have restarted; we can keep an eye on them now.

LB: Your report summarizes the toxic chemicals that were detected in and around New Orleans that are in the sediment from the floodwaters—heavy metals, pesticides, banned pesticides, arsenic, industrial chemicals, and more.

Dr. S: And a heck of a lot of diesel fuel.

LB: Yes, and diesel fuel.

Dr. S: And other petroleum chemicals.

LB: At this point, do you know if the EPA is going to come in and clean or remove the contaminated topsoil?

Dr. S: There's been a lot of buck passing. The EPA says there might be funds to do this cleanup, but that the Corps of Engineers would be the agency that would need to do the cleanup. The Corps is saying that they are focusing on the levees, which of course is super important. The Corps also is saying that they need to hear exactly which areas need to be cleaned up, and the EPA says that they're still testing and evaluating the problem.

Then there's been this issue of sort of moving the benchmarks. This winter the DEQ abruptly came out with a new, revised safe level of arsenic, just after arsenic levels over the previous safe level were discovered all over New Orleans. It was just in a one-page memo. They revisited their arsenic standards and declared that a much higher level of arsenic is perfectly safe. That looked suspicious to many people. It seems that the DEQ's obvious solution was to just move the safe level and declare the higher levels that were being discovered in the city safe. There's been a fair amount of funny business.

Meanwhile, this winter, the rains did move that sediment around and washed some of it into the storm drains. In many cases, when the EPA went back to do repeat sampling, they said they couldn't find any sediment. They wouldn't do a second sample. The problem with that is that we know that the contaminants in that sediment were probably washing into the storm drains, and some were just soaking into the soil. That's the soil that kids may be playing on and that families may be planting vegetables in, so it wasn't very reassuring to know that it was just soaking into the dirt.

LB: In an NRDC press release, you are quoted as saying that if the contamination in the sediment is not cleaned up, ". . . residents could be faced with serious health risks, including cancer, neurological disease, and hormonal and reproductive system problems, over the long term."[2] What are the initial signs that an illness might be starting to brew?

Dr. S: The issues are about risks to young children who might be crawling around and in direct contact with the sediment in or with sediment-contaminated soil. The major concern here is specifically with the arsenic and petroleum compounds in the sediment. Arsenic is known to cause cancer, neurological problems, and reproductive problems with long-term exposure. This is not something that's likely to happen to adults or after relatively short-term exposure (i.e., over months). The initial signs of concern in a child might be irritant effects from the contaminants in the sediment, such as rashes, skin irritations, or a cough.

LB: Is anyone "immune" to the effects of the types of contaminants that are in the sediment, over the long-term?

Dr. S: There is no such thing as immunity to contaminants since immunity develops only toward microbes. However, some people seem to be much more resistant to any toxic effects than others. It is likely that the majority of people will be fine, but

2   NRDC Press Release, "New Testing Shows Widespread Toxic Contamination in New Orleans Soil, Neighborhoods." Available at http://www.nrdc.org/media/pressRelease/051201.asp (accessed June 2006).

some people will likely get sick. The factors influencing susceptibility may be genetic, nutritional, behavioral (referring primarily to children), and also connected to underlying diseases.

LB: Can you talk about the levels of gram-negative bacteria that were found and the levels of endotoxins? From reading your report, it sounded as if you found lower levels of endotoxins than expected.

Dr. S: Well, we actually expected to find a lot of endotoxins. Endotoxins are made by gram-negative bacteria, which we knew were in the floodwaters because the floodwaters were full of sewage and because tests showed a lot of coliform and *E. coli* bacteria, specifically in the floodwaters. We knew that the floodwaters had gone everywhere, including into people's homes. We knew that endotoxins would be released in that type of setting, so we put some effort into doing endotoxin sampling. We were surprised to find that the levels we were detecting were not higher indoors than out and that endotoxin levels were not high, consistently, in the flooded areas versus the non-flooded control sites.

Interestingly enough, the CDC reported that they had done a little bit of endotoxin testing and they could find high levels of endotoxins, so we are not actually sure whether to trust our results. Designing a method of testing that takes into consideration the different characteristics of organisms is the tough thing about designing this study. One of the things about endotoxins is they don't become airborne as easily as mold. Mold has incentive—the more airborne its spores become, the better their chance of spreading. It's pure, basic biology. The endotoxin is basically just cell wall fragments from bacteria; it's not something that the bacteria are trying to spread. It spreads when it's disturbed. The way we did our sampling was to try not to disturb the sampling environment. We actually would try to tiptoe into a house, try to not move anything around, carefully set up the equipment, and tiptoe out. The places we sampled were uninhabited. During the period of sampling, it would just be still and quiet.

If instead we had done a different sampling protocol and we had put on full protective gear (which we had anyway, in most cases), and if we had been protected from head to foot and had good respirators and had just gotten in there and started moving furniture, ripping and pulling up the wet carpet and trying to get it out to the curb—doing all the things that people would do if they were trying to get things cleaned up in their houses (sweep, vacuum, anything like that)—I'll bet we would have found a heck of a lot of endotoxins.

We did the same kind of thing outdoors when we sampled people's backyards. We tucked the samplers so that they would be sheltered from the wind, and in a couple cases, when it turned out that there was actual disturbance going on nearby, we decided not to sample because we didn't want to mess up our remote samples—if there was tree work or something like that going on. We probably substantially underestimated the real endotoxin risk, but we had only the data that we had to report. We reported what we had and explained the possible reasons why we may have found what we did. Endotoxins are potentially another risk that people may have been facing there. We just don't have the data to show it.

LB: How much risk are endotoxins to people's health?

Dr. S: Endotoxins are a tricky thing. At high concentrations they are known to decrease normal lung function. So if you take healthy, strapping young men and expose them to endotoxins, their lung function will diminish predictably and reliably; they are clearly not good for the lungs. They also create a very powerful inflammatory effect. Basically, our immune systems are primed to respond to endotoxins almost more powerfully than to any other substance. Endotoxins will produce major inflammation whenever they are in your body. The interesting and peculiar thing about endotoxins is that some studies indicate that low, low levels of endotoxins can actually sort of help in decreasing risk of asthma, especially in children. Nobody has really resolved that paradox yet. Then there are other studies that indicate endotoxins may increase risk of asthma. Endotoxins are a bit of an enigma at very, very low concentrations. There are definitely people who say they have a role to play in normal immune modulation in early development. That may be true, but high levels of endotoxins are never good. So if there were a lot of gram-negative bacteria in New Orleans, there's no way to frame that as good in any way, shape, or form.

LB: Are people at high risk for health implications from mold exposure, such as children, the elderly, and the immune compromised, also at high risk from endotoxin exposure?

Dr. S: Yes, in general. Anybody who has a respiratory problem is going to get sick more easily from either endotoxins or mold. Anybody whose body is less able to bounce back and recuperate will be at higher risk, which can include young children, the elderly, or people with underlying illnesses.

LB: Is there any way to judge which exposure is more dangerous for health—chemicals, molds, endotoxins, or mycotoxins? Since many of these are intermixed, has there been any research exploring the compounding negative effects on health?

Dr. S: The problem with environmental health is that people are never exposed to just one thing, and nobody has figured out the health effects of this whole combined mess of toxins to which people are exposed. There are a lot of good reasons to believe that they interact with each other and that the combined exposure may well be far worse than any of those exposures alone. That's never been studied. New Orleans was a vast experiment, I suppose, except that all the people who were involved in that experiment were rather unwilling and unwitting. It is probably the first time that there's ever been such a huge number of people exposed to this particular mix of contaminants at these kinds of high doses.

LB: If a data bank of medical case histories is compiled from those who became sick from Katrina, scientific and medical professionals will probably not be able to look back and get any sort of definitive picture as to the synergistic effects or even which organic or nonorganic toxin was responsible for someone's illness.

Dr. S: In addition, the evacuation means that people who were affected by Katrina are all over the country, so many of them are not being tracked. It's not clear that every potential case of Katrina-related illness is being reported to anybody.

LB: Given the current situation—that chemicals may not be properly cleaned up or removed and may just dilute and sink into the ground—do you think New Orleans, right now, is a safe environment for people in the high-risk groups—children, the elderly, and the immunocompromised—to be living?

Dr. S: That is a good question. The flooded areas in New Orleans are still, as far as I understand, a high-risk place. People should obviously make their own decisions. People should realize that if they are going back into flooded neighborhoods and they have an underlying illness or are bringing young children or elderly folks, there are some significant health risks.

LB: Pregnant women, too—I've seen pregnant women on TV in affected areas.

Dr. S: Oh, yeah.

LB: Is it an overstatement to say that people are at risk of shortening their life spans or their quality of life by not removing themselves from the types of contamination that remain in the flooded areas?

Dr. S: It's hard to say. That's a tough question. I think that it's clear there is a health risk. How significant the health risk is, is tough to quantify. All I can say is that if I had a young child, I would think twice before going back and spending time in one of the flooded neighborhoods. If I had an underlying illness—diabetes or a lung disease, for example—I would also think twice about doing it. But there are a lot of things that we all do every day that may or may not shorten our life spans. The thing that bothers me right now is that people don't necessarily understand that there is a risk. If people understand there is a risk and decide they want to go back anyway, that's as fine as deciding you want to go skiing. You might run into a tree, but you wanted to do it and that's fine. People need to know. From my perspective, I'm not going to tell people that they should or shouldn't go back. I just think it's important to tell them that there is a health risk.

LB: For people who have returned and are already experiencing negative health, if you were their doctor, what would you be advising them?

Dr. S: It depends on the person. When I was there, I worked with a woman who was having terrible sinus congestion that was probably from the mold, but she was doing very important work and very much wanted to stay. She found that she could take antihistamines and decongestants and wear a respirator in certain areas. She managed fine and is still doing well. There are other people who might need to get out of there and other people who could probably manage. It's a very individual decision. I don't think it makes sense to give blanket recommendations. I think people need to have the facts. As a physician seeing individual patients, I would talk with each person—each decision might be different.

LB: What do you most want people to take away from this interview?

Dr. S: I think people need to be careful, they need to be informed, and the government needs to help clean this problem up. The data do not suggest that New Orleans is uninhabitable or that it will remain toxic over the long term. But the data do show that there is a problem that needs to be addressed.

LB: Thank you and those at the NRDC for your advocacy on behalf of the people working and living in the New Orleans area.

Jack D. Thrasher, PhD, (1938-2017) was an Immunotoxicologist who had assisted hundreds of individuals injured by toxic chemicals and molds. His expertise included evaluation of and assistance to the environmentally injured, appropriate environmental testing, recommendations for medical evaluation, and researching and writing current scientific and medical literature. The scope of evaluations included molds, formaldehyde, pesticides, solvents, heavy metals, hydrogen sulfide, and cyanide.

Dr. Thrasher received his PhD in Human Anatomy from UCLA. His dissertation was in Cell Biology, which included toxicology and immunotoxicology. Since 1984, Dr. Thrasher had been a consultant and expert witness for cases involving environmental toxicology and immunotoxicology. He consulted for Progressive Healthcare Group, ImmunTox, LLC, and the Center for Immune and Toxic Disorders. Dr. Thrasher authored and coauthored numerous research papers, articles, and books. Much of his research is available on his website at www.drthrasher.org. The interview was conducted April 2006.

LB: There are two different methods of mold measurement. One is spores/m^3 and the other is cfu/m^3. What is the difference between these two types of mold measurements?

Dr. T: The first one is when microscopists look at mold spores; they are looking at viable and nonviable mold spores, just the total count. The second one is when they look at cfu's of the number of spores that were viable and grew when cultured. You have to remember that some of the things that are generated by molds are dead and some are alive.

LB: After Hurricane Katrina, the NRDC took air samples in New Orleans and surrounding areas. They found four toxigenic molds present: *Aspergillus*, *Penicillium*, *Cladosporium*, and *Stachybotrys*. Have you seen these NRDC test results?

Dr. T: I have not seen any of those results, but that's exactly what I would expect to be found in the air, plus others.

LB: The NRDC measurements were based on spores/m^3, so the test results measured both living and dead spores. How many spores/m^3 can people be exposed to without their health being negatively impacted?

Dr. T: There is absolutely no data to tell us what concentration is safe or not safe.

The thing that we have to consider is colonies of fungi. (I prefer to call them fungi rather than molds.) Colonies of fungi shed different types of particulate matter into the air. Let me explain what I mean by that: They shed spores, and the spores range from 2 microns in diameter up to 7–9 microns in diameter. They also shed fragments of their cell bodies that we call hyphae, so we've got hyphae fragments in the air as well.

Nobody has taken into consideration, until Dr. David Straus's group at Texas Tech University began studying it, that colonies also shed what we call very fine particulate matter [see interview with Dr. Straus in Section Three].[3] This was first studied by Dr. Górny, who characterized what was going on from the mold growth that goes on in the colony.[4] This particulate matter is smaller than the size of the spores; it goes from about less than 2 microns in diameter down to 0.2 microns in diameter. That fraction is some 300–400 times more concentrated than the spore count. That fraction also contains all of the toxic chemicals that are produced by the molds and the spores. We are looking at an environment that is not only mold spores but also mold spores plus the fine particulate matter. Furthermore, the fine particulate matter is shed by frequencies of 1–20 hertz, which are the frequencies associated with normal human activity, e.g., television, radio, conversation, walking, dancing, etc. Therefore, the shedding of this material is independent of mold spore release.

LB: Do all of these additional, unaccounted-for factors make the NRDC's measurements of mold spores/m³ gross underestimates of the actual toxicity of the air?

Dr. T: Yeah, well, they're meaningless. When we look at a report, we are also concerned not only about the genus but also about the species within the genus of the particular mold that is found. For example, there are six or seven species of *Aspergillus* that are potentially very dangerous. There are three or four species of *Penicillium* that are very dangerous, and if you do not speciate it, you have no idea what is in the air. So to rely strictly on mold spore counts is a bunch of bull; that's my opinion. Another example is *Stachybotrys*. If you look at *Stachybotrys*, there are two types of *Stachybotrys*. One is called chemo type A, which produces the very toxic trichothecenes (such as satratoxin H and G), and also stachylysin, which is a hemolytic agent [destroys red blood cells with subsequent release of hemoglobin]. The second is the non-chemo type A that produces a whole variety of other toxic compounds, such as spirocyclic drimanes, which are highly toxic both to the nervous system and to the immune system. So just to do a spore count is bull. I

3   T.L. Brasel et al., "Detection of Airborne Stachybotrys chartarum Macrocyclic Trichothecene Mycotoxins on Particulates Smaller than Conidia," *Appl Environ Microbiol*, 2005 January; 71(1): 114-122. Available at www.pubmedcentral.gov/articlerender. fcgi?tool=pmcentrez&rendertype=abstract&artid=544211 (accessed April 2006).

4   Górny RL, "Filamentous Microorganisms and Their Fragments in Indoor Air—a Review," *Ann Agric Environ Med*, 2004, 11:185-197. Available at http://www.aaem.pl/pdf/11185.htm (accessed April 2006).

look at these reports and shudder at what the so-called current industrial hygienists think and that they think what is going on is appropriate.

LB: The NRDC did not identify by species at all.

Dr. T: Then the report is meaningless. All it tells us is that there are *Aspergillus* and *Penicillium* in the air. You have to speciate each; you do not report *Aspergillus/Penicillium*. That is meaningless because we don't know what is in the air. Now the other thing that goes on simultaneously with the mold growth is the growth of both gram-negative and gram-positive bacteria. Many of the gram-negative bacteria produce what we call endotoxins. The gram-positives also can produce what we call bacterial cell wall toxins. These also get into the air. The current research, which we have to do in the test tube and with research animals as we can't do it with humans, shows that these compounds are synergistic with the mycotoxins.

LB: The NRDC found partially remediated buildings with anywhere from 79,000–638,000 spores/m^3. Even in fully remediated buildings, spore counts of 45,000–100,000 spores/m^3 were found.

Dr. T: Okay, we've got that count; now multiply that by 300.

LB: Why do we multiply by 300?

Dr. T: Because that is the effective particulate load of the fine particulate plus the spores.

LB: So what kind of health ramifications do you think will come from all this exposure to mold and mold particulate?

Dr. T: Let's take the fine particulate matter—we know that the spores can get into the nasal cavity and some down into the lungs. There's no question about that because they're small enough that they can get down into the lungs. They are from 2–7 microns in diameter. What do you think is happening with something much smaller than that, say down to 0.1–0.2 microns in diameter? It is getting deep into the airways and, therefore, is being taken up by the macrophages and absorbed into the blood and everything. [Macrophages contain granules or packets of chemicals and enzymes, which serve the purpose of ingesting and destroying microbes, antigens, and other foreign substances.] No one is even considering what is going on in that respect except Dr. Straus's group, Dr. Górny, me, and a few others who are really concerned about this health issue.

LB: The CDC published a report addressing mold prevention and health in the aftermath of hurricanes Katrina and Rita. The report states that if people are going into a mold-infested building just to retrieve personal property or to do a small amount of cleaning, they don't need to wear personal protective equipment.

Dr. T: The CDC is quite mistaken. The CDC has not considered the fine particulate matter that we were just discussing. Secondly, you cannot make that as a blanket statement. I'll give you the reason why: There are some people out there with genetic polymorphism. The one enzyme that I am concerned about is called glutathione S-transferase-1, which is an isoenzyme of the glutathione S-transferases. Some 50 percent of the population in the U.S. lacks that particular enzyme. We don't know why it's so high here. It's called GSTM1 null.

These people are highly susceptible to mycotoxins. The mycotoxin I can describe in detail for you that has been studied in this regard is aflatoxin B1. When people with this genetic polymorphism get exposed to aflatoxin B1, they have a higher risk of liver cancer and a higher risk of asthma. If they have a higher risk of asthma and cancer, what else do they have a higher risk for? They also have a higher risk for aflatoxin-albumin adducts. In addition, the epoxide of aflatoxin B1 adducts to the guanosine of DNA, with a predilection for mitochondrial DNA versus nuclear DNA. What this means is that the detoxification pathway that GSTM1 is involved with is not working at all. GSTM1 conjugates the epoxides of aflatoxin B1 to glutathione so that the epoxides can be excreted via the urine. Thus, if one has excess epoxides, which bind to sulfur and other groups on proteins, then the excess epoxides will bind to these proteins.

Albumin is the most concentrated protein in the blood and is, therefore, used as an experimental tool. One can test for albumin-epoxide conjugates to determine if there has been a significant exposure to a given epoxide. When the epoxide binds to such a protein (albumin) it denatures the protein; therefore, the immune system recognizes the denatured proteins (aflatoxin-albumin adduct) as foreign and makes antibodies against the aflatoxin-albumin adduct. Therefore, antibodies in the blood against this adduct means exposure, and such antibodies can be tested for in the body.

The potential danger regarding the adduct formation with guanosine in DNA is mutations. Such adducts can cause two types of mutations: nuclear DNA and mitochondrial DNA. This is probably the basis for aflatoxins causing cancer. If the mutations are sufficient in mitochondrial DNA, one would have abnormally low energy production. We have seen the latter in at least one or more infants.

LB: Is there any supplement that these people can take to replace this missing enzyme?

Dr. T: They can't replace the enzyme but they can take antioxidants, which will help. Regardless, if they are missing the enzyme, they are still missing the enzyme.

LB: Don't we all get exposed to a certain amount of aflatoxins in food, such as peanut butter, wheat products, and corn products?

Dr. T: Absolutely, we do. There's no question about it, but that's a baseline exposure that we're all exposed to. Now add to that the exposure you're getting from an infested building. You're adding to an already baseline-toxilogical environment, so you're just adding to it. The fact that we're being exposed through our food is meaningless. We're all being exposed.

LB: Now this 50 percent of the population, do they fall into the group that I have read so much about that is called the immunocompromised? [A note to the reader: As you can see, at this point we did not understand that each member of our family fell into a high-risk category and that Kurt was considered immunosuppressed/ immunocompromised because of his pre-hurricane treatment with prescription steroids for allergy management. This was revealed as we continued our research.]

Dr. T: They would not be immunocompromised. They are detoxification compromised, which would lead to immunocompromised because if they can't detoxify and get rid of these free radicals, these free radicals are going to bother and depress or affect the immune system.

LB: Would they be considered immunosuppressed?

Dr. T: What we can say is that they are immune disregulated. Sometimes they can show suppression, depending on what arm of the immune system you're looking at, while other portions of the immune system are enhanced. I'll give you an example. We see that some 40 percent of the people exposed to molds and toxic chemicals, in general, but also molds or fungi, have decreased natural killer cell activity, which means that they are immunocompromised. On the other hand, when we look at their T and B cells, they are going to have elevated T and B cells, which means that portion of the immune system is enhanced. Thus, their immune systems are disregulated.

LB: What about people with thyroid conditions? In our experience, the fungi or mycotoxins appear to be attacking the thyroid. Are people with hypothyroidism or hyperthyroidism considered immunocompromised, or would they fall into one of these other categories?

Dr. T: Here's the category we have to look at: Who are the most susceptible?—the very young, the very elderly, those with pre-existing illnesses, those with genetic defects, and pregnant women. I don't know how to answer the question about thyroid problems because in the clinic we see people with severe thyroid problems. We don't know whether they had them before exposure or after exposure, but we do see a pretty good number. I don't know the percentage. I am never surprised to see an abnormally functioning thyroid gland. If we take a look at what the thyroid gland is susceptible to and what causes abnormal thyroid gland function, we see chlorinated hydrocarbons, particularly PCBs (poly-chlorinated biphenyls). The molds produce chlorinated mycotoxins. Nobody's looked at that.

LB: Can antioxidants neutralize the effects of chlorinated mycotoxins?

Dr. T: They may or may not. We really don't completely understand yet how they are affecting the thyroid. They may or may not help an individual who has glutathione S-transferase-1 deficiency. We don't know.

LB: We read that some of these molds produce a form of estrogenic hormone.

Dr. T: Yes, some of those mycotoxins are very estrogenic.

LB: Do you have an opinion as to how widespread and severe the health implications will ultimately be from all the mold exposure in the hurricane-hit areas?

Dr. T: People say they have this "Katrina Cough" and try to laugh it off, but it's got to be a lot more serious than that.

LB: What actually is the Katrina Cough?

Dr. T: Inflammatory conditions of the lungs. No question about that.

LB: From mold exposure?

Dr. T: That is correct. Not only the molds but also all the other things I talked about—the bacteria and all the other things that are in the air. The same thing goes on inside the home. This goes back to the pulmonary hemosiderosis that Dr. Dorr Dearborn and the people from the CDC originally looked at; they found the black mold, identified it as *Stachybotrys,* and said this is causing the hemosiderosis. They did not look at all the other things that were in the indoor air at that time, so they are saying that black mold is causing the problem. It's not only the black mold and all the other molds; it's also the entire mixture to which people are being exposed.[5]

LB: After Hurricane Katrina, the air was just thick with particulate from airborne debris, including chemicals, smoke, and mold spores.

Dr. T: We don't know all the toxic chemicals to which people were exposed and that got into the soil and water.

LB: What do you think the probability is of this increased exposure to molds, particulates, and chemicals leading to an increase in cancer levels? Do you think there will be a direct correlation?

Dr. T: It's too early to say. When you get into cancer, cancer is the end stage of a disease process. My wife, for example, just died last July of kidney cancer. That's

5   For a review of the Ohio infant incidences, see Section One, Chapter Six.

an end stage. The question is: What led up to the end stage? She was an ex-smoker. Was it smoking? Was it the mycotoxins in the tobacco or what? We don't know, but her end stage was cancer. Up until she got to her end stage and death, she went through all the symptoms and health problems that you have. She went into olfactory sensitivity; she went into chronic fatigue; she went into fibromyalgia. She went into all these health problems before she succumbed to the cancer, so it's a continuous process. The answer to your question, I'll say, within a reasonable scientific probability, would be yes. We will see an increase, but what types of cancers, I don't know.

LB: A couple of medical studies researched lipoic acid and vitamin E and found these vitamins were effective in treating the effects of mold and mycotoxin exposure.

Dr. T: Vitamin C, vitamin E, and there's another one on the market no one knows about called Neprinol.[6] It's purchased only through the Internet. It's a mixture of Co-Q 10[7] and two or three enzymes that will also help from the antioxidant point of view. I take this on a daily basis, and I've never felt better.

LB: Are these enzymes the same enzymes 50 percent of the population are lacking that you talked about earlier?

Dr. T: No, these are enzymes directed at breaking up fiber from fibroids that are a result of an inflammatory process. For example, I have had arthritis in my fingers for years that has inhibited me from playing the piano, but I've been on this stuff for three months, and I now can play the piano.

LB: Does Neprinol help break fiber down so your body can metabolize it?

Dr. T: Yes.

LB: Will exposure and sensitization to one mold species create a sensitization to other mold species as well?

Dr. T: Yes, it may well do that because there is cross-reactivity. Molds make antigens, and the antigens consist of a variety of things but particularly a protein of some sort. The proteins share amino acid sequences, so, for example, if you get a protein from *Penicillium chrysogenum* and a protein from *Aspergillus versicolor* that are very similar in nature, you may well have cross-reactivity.

---

6   Neprinol is manufactured by Arthur Andrew Medical: http://www.enzymus.com (accessed April 2006).

7   "Coenzyme Q10 is an antioxidant similar to vitamin E. It also plays a crucial role in the generation of cellular energy, is a significant immunologic stimulant, increases circulation, has anti-aging effects, and is beneficial for the cardiovascular system," according to Balch et al, *Prescription for Nutritional Healing*.

LB: Would a person who is allergic to penicillin, since penicillin is derived from *Penicillium*, be predisposed to mold allergies and sensitivities?

Dr. T: It's possible. If people have existing allergies, it puts them in a more sensitive category, meaning their sensitivity threshold is lower.

LB: Are people allergic to penicillin good candidates for antifungals, or would this type of allergy put them at high risk for antifungal treatment?

Dr. T: There're a couple antifungals out on the market; Sporanox is one, and there's another new one out that's been approved by the FDA. I don't think they would cross-react with that if they have a mold infestation. I call it an infestation because we don't know whether it's an infectious process or a colonization process, but people are colonized with these molds.

LB: Are mycotoxins found in human tissue, organs, and blood serum because they are being produced inside the human body, because the mycotoxins were inhaled, or both?

Dr. T: Both. I'll give an example of people with aspergillosis caused by *Aspergillus fumigatus*. *Fumigatus* causes aspergillosis of the lungs. *Fumigatus* also produces a mycotoxin called gliotoxin. When researchers look at the blood of people with aspergillosis caused by this particular species of *Aspergillus*, they find gliotoxin in the blood. That tells me that in the infectious stage, that particular organism is producing a mycotoxin. There are several studies that point to this; unfortunately, we have not been looking at all the other mycotoxins as to what's going on.

Trichothecene mycotoxins, which are produced by *Stachybotrys*, were studied by Dr. Straus, Dr. Brasel, and others in their group.[8] They looked at people who were symptomatic in a building that contained *Stachybotrys* and found that they had trichothecene mycotoxins in their blood. Then they looked at some 20 people versus a control group in the study they did. That's telling me, yes, the mycotoxins get in the blood either by inhalation or, if the *Stachybotrys* does cause a colonization, by the production of the mycotoxins by the *Stachybotrys*. But what we know about *Stachybotrys* is that it probably does not colonize in humans because we do not find it in any type of colonization in humans, but *Aspergillus* does colonize.

LB: Do you think people in the high-risk categories should be in New Orleans or any other mold-affected area, or would that be too high a risk to their health?

Dr. T: People in the high-risk groups should stay away from any exposure.

---

8   Brasel TL et al., "Detection of Trichothecene Mycotoxins in Sera From Individuals Exposed to *Stachybotrys chartarum* in Indoor Environments," *Arch Environ Health*, 2004; 59(6); 317-23. Available at http://www.medscape.com/medline/abstract/16238166?prt=true (accessed August 2006).

LB: You were one of the expert witnesses in the *Gorman* case that was highly publicized in Los Angeles. A newspaper quoted one of the defense attorneys as saying the child had autism. Was that autism mold related? Can mold induce autism or symptoms of autism?

Dr. T: Well, that's what they are trying to tell the public, but it's not the truth.

LB: So this was not a case of true autism?

Dr. T: Let me describe it to you. Number one, the child was born with global developmental delay, which would indicate to me—as a scientist, cell biologist, and toxicologist—that he had mitochondrial damage. We see infants with mitochondrial damage with global developmental delay. In other words, everything is delayed—the neurological system and the rest of the organ systems. Number two, MRIs were done on that child at the age of 12 months and he had normal-looking brain structure. Number three, at two years of age the MRIs were repeated and he had cystic structures in the myelin that surround all the ventricles of his brain. Autistic children do not have cystic structures in the brain.

LB: How do you think these cystic structures were formed?

Dr. T: We surmised that the cystic structures resulted from damage to the myelin. We surmised, as experts, that that child had damage to the myelin of the brain. The myelin is that lipoprotein that goes around and insulates the neurons. If you interfere with that myelin process, you're going to interfere with all functions of the brain.

LB: Can that condition be induced from fungal exposure?

Dr. T: Yeah, it can be; that's my opinion because we see demyelinization in both adults and children who have been exposed. We see it mainly in the peripheral nervous system, on which Dr. Campbell and I have published.[9] We have not looked at the central nervous system, but based on the work I have seen published by Dr. Kilburn,[10] I would hazard a very good probability that it's going on in the brains of adults and children.

LB: Is there anything people can take to protect the brain from demyelinization?

---

9   Campbell AW et al., "Neural Antibodies and Neurophysiologic Abnormalities in Patients Exposed to Molds in Water-Damaged Buildings," p. 46-56, in Kilburn, KH, editor, *Molds and Mycotoxins* (Heldref Publications, Washington, DC, 2004). Available at http://www.ncbi.nlm.nih.gov/ entrez/query.fcgi?CMD=search&DB=pubmed (accessed August 2006).

10   Kilburn KH, "Indoor Mold Exposure Associated with Neurobehavioral and Pulmonary Impairment: A Preliminary Report," p. 3-11, in Kilburn, KH, editor, *Molds and Mycotoxins* (Heldref Publications, Washington, DC, 2004). Available at http://www.ncbi.nlm.nih.gov/entrez/query. fcgi?CMD=search&DB=pubmed (accessed August 2006).

Dr. T: Again, antioxidants are the only thing I can think of.

LB: What about lecithin? Isn't it a protective brain sheath of some sort?

Dr. T: Lecithin may help, yes. You're asking some very key questions. I've thought and thought about all of these things. Let's go back to the *Gorman* case. The Gorman child was GSTM1 null.

LB: So we're talking again about the 50 percent of the population that lack the particular enzyme you talked about earlier. He falls into that 50 percent?

Dr. T: That's right. Based on the genetics that I could do, the father and the two sisters were affected. The mother was affected but didn't claim that she was, so let's say she was not affected. That means that she may have had one of the genes. There are two genes—one comes from the mother and one from the father. She may have had one gene for GSTM1 null. When you start playing around with the genetics, the biggest probability is that the father had no such genes, the two daughters may have been a mixture, and the child had no genes. That was my testimony about the genetic defect in that particular case.

LB: It seems clear to me, after reviewing many studies, that a direct cause-and-effect relationship exists between mold exposure and mold-related illnesses. However, the Institute of Medicine (IOM) states in its 2004 report that there is an insufficient correlation between mold exposure and the more serious diseases.

Dr. T: But if you look at that report very closely, the IOM stopped reviewing the literature as of December 2003, so its report doesn't include the last two to three years of scientific and medical literature that's been coming out. By the time it was published, the information was already outdated.

LB: I noticed on your resume that you've done some work with formaldehyde. We've noticed that since we were exposed to mold, we are now incredibly sensitive to formaldehyde.

Dr. T: Oh, absolutely. You're going to be sensitive to other chemicals, in addition. First of all, formaldehyde is a great conjugator or adductor. What I mean by that is, it binds to every protein in your body and changes that protein so that the immune system recognizes it as foreign. By doing this, formaldehyde is very antigenic, meaning it stimulates the production of antibodies. When you breathe it in, it starts binding with all the proteins in the nasal cavity, which changes those proteins, and then antibodies to those changed proteins are created. I have published several papers on how formaldehyde does this.

Secondly, when you inhale through the nose, there is a strong likelihood that the inhaled toxins will go directly into the brain. I don't know if you're aware of this, but the olfactory neurons are in direct communication with the brain, and there is no blood-brain barrier. Anything you have inhaled and smelled, whether you detect an odor or even if it doesn't have an odor, goes directly into the brain through the olfactory neurons.

LB: I did not know that.

Dr. T: The government doesn't want people to know these things. A recent study published in *Environmental Health Perspectives* by Dr. [James] Pestka's group at Michigan State University has demonstrated this with satratoxins in mice.[11] They instilled the mycotoxins into the olfactory neurons in the nasal cavities and demonstrated cell death and inflammatory conditions in the olfactory neurons, the olfactory tract, and the olfactory lobe of treated mice.

LB: Is there anything we haven't covered that we should?

Dr. T: Let's look at indoor versus outdoor counts. You've seen statements such as, "Well, if it's not any greater than outdoor counts, you don't have a problem." Bull! The reason I say that is you have to look at the species indoors versus the species outdoors. Invariably, what you will find in certain species of *Aspergillus* and *Penicillium* are higher levels indoors than outdoors. Also, the bacteria that we find indoors probably are not found outdoors to any great extent. In addition, one must include the fine particulate matter à la Dr. Straus and Dr. Górny.

LB: And all that adds to the toxicity level in the air.

Dr. T: Yes, I want to make sure that people understand that it's not only mold; it's all the other stuff that's going on with it that we talked about. The indoor air of water-damaged buildings contains a variety of biological matter such as spores, hyphae, fine particulates, endotoxins, 1-3 beta glucans, and bacterial fragments, among others. In addition, both bacteria and molds produce a variety of volatile organic compounds. All of these impinge upon the occupants, creating an environment that is probably additive as well as synergistic in toxic effects. Remember, we are dealing with a variety of toxins that affect all organs of the body, particularly the nervous system, the immune system, and the upper and lower respiratory tract.

LB: What about tight building syndrome?

Dr. T: Tight building situations can be resolved by opening the buildings up and

11   MSU News Release, "MSU Researchers Say Black Mold Toxins Could Affect Sense of Smell," February 28, 2006. Available at http://www.newsroom.msu.edu/site/indexer/2673/content. htm (accessed August 2006); Zahidul Islam et al., "Satratoxin G From the Black Mold *Stachybotrys chartarum* Evokes Olfactory Sensory Neuron Loss and Inflammation in the Murine Nose and Brain," *Environmental Health Perspectives*. Available at http://www.ehponline.org/docs/2006/8854/abstract. html (accessed April 2006).

quit trying to conserve so much energy. Some 30–40 percent of buildings are now suspected to have a mold problem. The original thought about sick building syndrome was that because the buildings were tight and all the internal furnishings off-gassed chemicals, this created a chemical soup that people inhaled. That was a belief back in the '80s and early '90s. As a matter of fact, I was one of the experts brought in back in August of 1988 when the EPA had its own sick building problem. I know 50 or 60 individuals in that building who are chemically sensitive and can never work in that building again. The administration, while I was sitting in the audience, tried to tell everybody that they recognized sick building syndrome, but that they didn't have one. Rufus Morrison, who was the president of the EPA union at the time, hired me as an expert to come in. I found they had a mold problem in that building.

LB: You've definitely been in the forefront in many areas. Thank you for sharing your expertise.

Update from Dr. Thrasher 7/2013: It is important to not only remove all structural mold, but also to eliminate any and all forms of water intrustion. Furthermore, a great emphasis is placed on using disinfectants to kill the mold and/or also remove or detoxify mycotoxins. Such efforts are not necessarily efficient. Dr. Erica Bloom tested the effectiveness of disinfectants on killing mold and denaturing mycotoxins. She used ten disinfectants including an ammonia-based one, peroxide, boron, and bleach (sodium hypochlorite). She tested and examined mycotoxins produced only by *Stachybotrys chartarum* and *Aspergillus versicolor*. None of the disinfectants were fully effective in killing the molds. In addition, the mycotoxins may have been altered by the disinfectants, but the toxicity of the changed chemical strucutre of the mycotoxins was not tested. Dr. Bloom concluded: None of the decontamination methods completely eliminated viable mold. No remediation treatment eliminated all toxins from the damaged materials. These results emphasize the importance to work preventatively with moisture safety throughout the construction processes and management to prevent mold growth on building materials (source: Mirko Peitzsch et al., "Remediation of mould damaged building materials—efficiency of a broad spectrum of treatments", *Journal of Environmental Monitoring*, Dec. 2011, Available at: http://www.ncbi.nlm.nih.gov/pubmed/22286589)

David C. Straus, PhD, *is a Professor of Microbiology and Immunology at Texas Tech University Health Sciences Center in Lubbock, Texas. He received his PhD in microbiology from Loyola University in Chicago and has done postdoctoral work at the University of Cincinnati and the University of Texas Health Sciences Center in San Antonio. He has been a consultant for indoor-air research for over ten years and has served as an expert in litigation involving sick building syndrome.*

He has been extensively published in the field with such literature examining the microbiology of indoor air, including Sick Building Syndrome, *of which he was editor. Dr. Straus coauthored the first findings revealing the ability to detect trichothecene mycotoxins in sera from individuals exposed to* Stachybotrys chartarum *in indoor environments.*[12] *Dr. Straus can be contacted at David.Straus@ttuhsc.edu. The interview was conducted April 2006.*

LB: After Hurricane Katrina, the NRDC took air samples in New Orleans and surrounding areas. The main molds identified were *Aspergillus*, *Penicillium*, *Cladosporium*, and *Stachybotrys*. Spores were identified by genus, not species, which would have told us if mycotoxins were being produced and, if so, which ones. In fact, *Aspergillus* and *Penicillium* were grouped together as *Aspergillus/Penicillium*.

Dr. S: Would you like me to tell you why that is and explain it? Because that's very important.

LB: Yes, it didn't make sense to me.

Dr. S: There are essentially two ways to examine spore counts. One is called *viable*, which is when we take the spores out of the air, deposit them on petri dishes, and grow them. If they're alive, they'll grow and produce a colony that's viable. The other method is the *nonviable*, which is what you're talking about with the NRDC. In the nonviable way, we just collect spores from the air and deposit them on a glass slide. Then we look at them under the microscope. We don't grow them, so we can't tell if they're dead or alive. Because *Aspergillus* and *Penicillium* spores under the microscope look almost identical, we can't tell what they are in the nonviable technique. That's why the spores are classified as *Aspergillus/Penicillium* or *Penicillium/Aspergillus*. What we're saying is that we really don't know if they are *Penicillium* or *Aspergillus,* and truthfully, they may be something else.

12   Brasel TL et al., "Detection of trichothecenes mycotoxins in sera from individuals exposed to *Stachybotrys chartarum* in indoor environments," *Arch Environ Health*, 2004; 59 (6): 317-23. Available at www.medscape.com/medline/abstract/16238166?prt=true (accessed April 2006).

LB: The report stated the NRDC collected samples using a Burkard sampler.

Dr. S: Yes, that's a nonviable technique.

LB: So live spores don't get collected with the Burkard sampler?

Dr. S: Well, they may. Again, we call it a nonviable technique because we don't try to grow them. When we look at the sample under the microscope, the spores could be dead; they could be alive. We don't know because we don't try to grow them. That's why we call it nonviable because we don't make any attempt to grow them and find out if they're dead or alive.

LB: The NRDC tested air samples for mold to find out the level of risk to human health. However, since the nonviable technique doesn't identify the species within a genus (which would have told us if a fungus was mycotoxin producing) and sometimes doesn't even identify the genus, is there any value in this type of testing to indicate the level of risk to human health?

Dr. S: Yes, there is a benefit. There are two ways in which fungi can make people sick, aside from infection: inhalation of fungal spores and inhalation of mycotoxins. With the inhalation of fungal spores, it really doesn't matter so much what the organism is because spores don't grow inside the lungs. What happens is you're inhaling high concentrations of living particles that can actually produce biological compounds inside the lungs, some of which can do damage.

LB: I thought that *Aspergillus* does colonize inside the lungs.

Dr. S: *Aspergillus* is the one organism that can colonize. There's a difference between colonization and infection.

LB: Can you clarify that?

Dr. S: A colony of something doesn't necessarily mean that it's causing pathology or causing infection. It can, but it doesn't necessarily. Colonization just means that it's there. Infection, or pathology, means that it's beginning to produce compounds, maybe even infect tissue and cause disease. So colonization does not necessarily mean disease. It just means that the organism is present and there's a colony of it, but it doesn't tell you if it's causing any disease or pathology.

LB: Can *Stachybotrys* grow inside humans?

Dr. S: No. It does not cause human infections.

LB: But it can exist inside the human body?

Dr. S: Of course. Let's say, for example, that you walk into a room and you inhale *Stachybotrys* spores. They have the potential to make you sick because when you inhale the spores, the mycotoxins are on the spores. The mycotoxins then become solubilized, get into your blood stream, and begin to kill cells. But *Stachybotrys* does not grow, does not multiply, inside the lungs, so you will never see a *Stachybotrys* lung infection.

LB: Regarding the cases in Ohio that involved *Stachybotrys* and bleeding lungs in infants, what exactly caused the infants' lungs to bleed?

Dr. S: *Stachybotrys* produces a compound that's been named stachylysin, which is a hemolysin. Hemolysin lyses [disintegrates] red blood cells. The spores of *Stachybotrys* either have on their surface or produce these hemolysins, so when these infants inhaled the *Stachybotrys* spores, the hemolysin then dissolved or diffused into the lungs of these infants and began to lyse red blood cells. What that means is red blood cells began to pop, so then the physicians began to see the disease called pulmonary hemosiderosis—the presence of heme inside of macrophages, which are the cells in the lungs that phagocytize [engulf and absorb] things that shouldn't be there. Now you have red blood cells popping; you have a lot of heme, which is the red portion and one of the most important components of red blood cells, getting into the macrophages—that's what pulmonary hemosiderosis is. So you now have a lot of pulmonary bleeding due to the fact that these red blood cells were lysing, and there was heme just going everywhere due to the compounds that these organisms were producing.

LB: Thank you for explaining that. Regarding the testing that the NRDC did in New Orleans, do we gain any indication as to the risk to human health from these mold spore counts?

Dr. S: The truth of the matter is—and this is why there are no state or federal standards saying what levels of fungal spores in the air are safe—that nobody knows what those levels are. No one will ever know what those levels are because no one will ever be able to do experiments on human beings—put them in a room, make them inhale high concentrations of fungal spores, and see when they begin to get sick. What the NRDC is trying to do in New Orleans is just get a sense of the fungal spore level in the outside air that people who are outside are going to be breathing. My guess is that in the mold-infested buildings the levels are going to be even higher than the levels outside. If the levels are higher than we normally see (10 or 20 times higher), then obviously that is a cause for concern.

LB: Did you see the total spore counts from the NRDC testing?

Dr. S: Yeah, I think I saw this all months and months ago. They certainly were elevated. The reason is because people are going back into their buildings and homes, taking all that mold-infested carpet, sheetrock, and whatever else it is that was mold-infested and throwing it out in the street, so you have this huge source of fungal-contaminated building material lying around in the streets. The way in which these fungi reproduce is to grow on wet building material, which is now lying everywhere in the streets. When fungi reproduce, they throw their spores into the air. When a spore falls on new, wet building material, you've got a new fungal colony. So, of course, the spore levels are going to be incredibly elevated because you have this tremendous source of fungal food sitting around in the streets of New Orleans. I wasn't particularly surprised at the elevated levels. That's what I would expect.

LB: Even some of the fully remediated buildings still had quite high spore counts. One had 45,000 spores/m^3 and another had 100,000 spores/m^3.

Dr. S: Once again, the probable reason is that when those buildings were remediated, they were exposed to such high spore counts in the outside air. Remember, outside air comes inside, and probably a lot of those very high spore counts in the outside air came inside. When you remediate a building and pull out all contaminated material, you're throwing spores into the air as well. So, of course, those levels in the newly remediated buildings are going to be high.

LB: Then how will they ever get the levels back down? Are those buildings going to have to be re-remediated?

Dr. S: No, just because you have high spore counts in the air does not mean that the building is now in trouble. If the building no longer has any fungal growth on its building surfaces, you can remove fungal spores from the air by using a HEPA filter or a negative air machine. We have machines that essentially filter the air and remove large particles like fungal spores. The most important thing people have to worry about in buildings that have fungal problems is fungal growth on building material. That's the whole cause of all of the problem. If you don't have any more fungal growth in a building and you still have a high spore count, we can clean the air in that building by running the air through a machine that filters out the fungal spores.

LB: Since certain levels of humidity will provide enough moisture for some mold species to grow, could humidity fuel mold growth all over again after remediation if a spore lands on a building surface?

Dr. S: Of course. The thing that people don't understand about humidity is that the amount of humidity in the air doesn't really matter. That's not what causes the trouble. What causes the trouble is when the water in the air, which of course is what humidity is, condenses out of the air, winds up on the building surface, and the building surface continually remains wet. That's why you have such large fungal problems naturally in New Orleans—because you have such tremendously high humidity. When you air condition a building, you cool down the building surfaces, and then, of course, the water condenses out of the air onto the building surface. For example, when you go out in the evening, even in Montana as dry as that is, with a glass of iced tea, what happens? You get condensation on the outside of the glass.

Water condenses out of the air onto cool surfaces. That's what happens when you go out in the morning and there's dew, water, on the grass. Dew forms because the surface of the earth has cooled down overnight, and water condenses out of the air onto the grass. That's what dew is; that's why it's called the dew point—the temperature at which dew forms. In New Orleans when you run your air conditioning system, you need to remember that you have a tremendous amount of water in the air. Of course, the water is going to condense out of the air onto the cool building surfaces. That's exactly what happens, and if those building surfaces stay wet long enough, yes, mold is going to grow.

LB: When I spoke with Dr. Thrasher, he explained that factors other than the mold and mold spores affect the toxicity in the air and must also be considered, such as mold fragments and mycotoxins.

Dr. S: Yes, that's my research.

LB: What about bacteria? We see reports of gram-negative bacteria in buildings with mold. Do molds produce bacteria?

Dr. S: No, there's no such thing as mold bacteria. Bacteria and mold are entirely different organisms. It's like the difference between a dog and a cat.

LB: But there are general bacteria already existing in the air?

Dr. S: Yes, there are general bacteria, which are in the air already, but bacteria in the air have nothing to do with mold.

LB: Why is it that gram-negative and gram-positive bacteria often accompany mold growth? There were high levels of endotoxins reported post-Katrina as well.

Dr. S: Bacteria also grow in wet buildings, as do fungi; but in most cases, bacteria require even more water than do molds. Most bacteria like to grow in liquid settings, for example, a pool of water. When this occurs, if the bacteria in question are gram-negative, they will produce a lot of endotoxins as only gram-negative bacteria produce endotoxins. Also, when sewage systems flood and get into the general water systems, this will introduce a lot of endotoxins because you normally have gram-negative bacteria, like *Escherichia coli,* in sewage.

LB: So to measure the *true* toxicity of air, we have to consider mold fragments, mycotoxins, bacteria, mold spores, and other air contaminants that are usually in the air, such as off-gassing from VOCs. If all these other components were considered in New Orleans, Dr. Thrasher's opinion is that the air would probably be 300 times more toxic than what was indicated from the tests that reported only mold spore counts. Dr. Thrasher suggested I ask your opinion as well.

Dr. S: Yeah, I would say that he's probably right. The only problem is we will never know because there's no way to measure things that you don't know are there. For example, we know we can measure bacteria; we can measure bacterial products; we can measure fungi; and we can measure fungal products. I'm sure there are toxins in the air in New Orleans that we don't even know are there, say, from the oil and chemical industries. If you don't know what's there, you don't know how to measure for it.

LB: I've read articles about measuring cytotoxicity to get a more accurate assessment of air quality. Cytotoxicity, from what I understand, has something to do with components that kill our cells.

Dr. S: Yes, cytotoxicity is really a very simple concept. "Cyto" means *cell*, and you know what toxicity means. We can grow cells in culture. For example, we could take heart cells and grow them in a petri dish. Then we could expose them to the air, say from New Orleans, and see if that air does any damage to the cells or kills them. That's cytotoxicity testing. Cytotoxicity tells you how toxic the air is against a particular cell line; it doesn't tell what the air sample does to a human being, because cells in a culture are very different from cells in human beings. Also, it doesn't tell you what you're measuring. It doesn't tell you if you're measuring mycotoxins, fungal toxins, bacterial toxins, or chemical toxins. It just tells you that there's something in the air that's killing these cells, but that's all it tells you.

LB: So, it's an indicator for further study?

Dr. S: Yes, it's an indicator that there's something toxic in the air. That's all it tells you. It doesn't tell you if it would hurt human beings because, as I told you, tissue-cultured cells are very different from cells in human beings. Cytotoxicity doesn't tell you what the toxin is or even if the stuff would be dangerous to humans.

LB: It would seem to me that if something toxic is damaging or killing tissue-cultured cells, we could draw the correlation that the toxic components could also do the same damage to the tissue cells inside our bodies. Can't we draw that correlation or make that assumption?

Dr. S: Well, no. You can't draw that correlation, but you can make the assumption that if there's something toxic in the air to tissue-cultured cells, then it's not going to be good for humans, but you can't make the correlation. There's a difference between an assumption and a correlation. The correlation is something we have to prove scientifically. Anybody can make an assumption.

LB: Tell us about the studies you've done at Texas Tech. You supervise undergraduate and graduate students who work with you on experiments, some of which have been published in peer-reviewed journals. The articles of particular interest regarding air quality have to do with *Stachybotrys chartarum* and the trichothecenes they produce.

Dr. S: In a nutshell, we believe, at least from the fungal point of view, that we understand what causes sick building syndrome. It's really pretty simple. We believe that the inhalation of high concentrations of fungal spores causes respiratory diseases in human beings, and most people who work in this field agree with that statement. There's not much controversy about that. We also believe that when people inhale the trichothecenes, it also causes human disease, and that's very controversial, as I'm sure you probably have read. We've come a long way in the last 12 years toward proving all of this. What we know at this point is that when *Stachybotrys* grows in buildings, you oftentimes find people who are sick. We know that when *Stachybotrys* grows in buildings, it produces these trichothecene mycotoxins. There's no question about that, so there's no question that you'll find sick people in buildings where there's *Stachybotrys* growing. (Sometimes you don't find sick people, but oftentimes you do.) There's no question that *Stachybotrys* produces these trichothecene mycotoxins on building materials in these buildings.

LB: Even though *Stachybotrys* does not constantly produce the trichothecene mycotoxins?

Dr. S: It may very well. According to our research, it is constantly producing them.

LB: I've read several papers in which the authors describe how mycotoxins are produced only under adverse conditions.

Dr. S: None of that's true.

LB: One article said that when spores are forming, mycotoxins are being produced.

Dr. S: We believe that when spores are produced, the mycotoxins are packaged into the spores at the same time. We don't think the spores produce the mycotoxins. We think the colony produces them and packages them into the spores when the spores are formed.

LB: To clarify the statement: When spores are forming, mycotoxins are being produced, does that mean when the spores are forming that the mycotoxins are formed and released into the environment?

Dr. S: We don't think the spores produce the mycotoxins. The mycotoxins are in the spores and are only released from the spores when they become solubilized in water. Spores don't release mycotoxins into the air. The mycotoxins get into the air when they are in the dust and picked up by air currents.

LB: So the *Stachybotrys* are constantly producing the trichothecene mycotoxins?

Dr. S: Yes.

LB: And that's been proven?

Dr. S: Yes.

LB: What source can I quote on that?

Dr. S: Well, we've not published that yet, but we've done the experiments and we've shown it. Thomas Rand (Halifax, Nov Scotia) has also looked at this in *Stachybotrys* and he also says that these mycotoxins are produced constitutively. This means all the time. In fact, he has published this. We have not published yet.

LB: I look forward to reading the report when it is published.

Dr. S: The next thing we know is that when *Stachybotrys* grows inside buildings on building surfaces, the trichothecene mycotoxins get into the air on fragments smaller than spores. I'm talking about all stuff that is true and that we've published. We also know that these trichothecene mycotoxins, because they're floating around in the air and are highly respirable, get into the human body—no question about that.

LB: Your experiments have proven that, correct?

Dr. S: Yes. Now, here's really the last question to be answered: Do the trichothecenes, which float around in the air in these buildings and actually get into human beings, get into them in concentrations sufficient to cause the human diseases that we see? We don't know the answer to that, and that really is the last question yet to be addressed.

LB: In one of your studies, you documented trichothecene mycotoxins in human blood sera. Are the trichothecenes showing up in human blood serum because people inhale them and then the trichothecenes are transported throughout people's systems, or are they being produced inside people's bodies from spores that they inhaled?

Dr. S: You can't tell. We also know that some of the compounds produced by *Stachybotrys* cause what we call immunosuppression.

LB: What exactly happens inside the body that results in the immunosuppression?

Dr. S: These compounds—for example, cyclosporin A—cause inhibition of the production of interleukin 2 by activated T cells. Interleukin 2 is necessary for the development of T cell immunologic memory. This stops expansion of the number and function of antigen-selected T cell clones, resulting in immunosuppression.

LB: Does this occur from exposure to other mycotoxins?

Dr. S: I don't know.

LB: When scientific and medical studies are published in journals and republished on Medscape and PubMed, do people still need to carefully weigh article content even though these articles have gone through a peer-review process?

Dr. S: Absolutely. Here's the perfect example. The two best biological journals in the world are *Science* and *Nature*. You know about the South Korean scandal with the stem cells?

LB: I vaguely remember reading about it.

Dr. S: Okay, what happened was a South Korean MD claimed that he could essentially grow human stem cells in tissue culture, which would be an incredible thing. I know you know all about the controversy about abortion and all that. If we could grow stem cells in tissue culture, then this idea of having to get stem cells from aborted fetuses goes away, and all of a sudden you've got this tremendous source of stem cells and no controversy—no abortions, none of that stuff. Essentially, he published in the two best journals in the world—*Science* and *Nature*—where all the big papers are published. It turned out that he had faked his data, so here you have the premier journals in the world, and it turned out that he had faked his data. While the quality of the journal is extremely important, some journals are very good and some are not. Science is what we call self-correcting. Even if somebody publishes something in *Science*, which is probably the best of the biological journals, it doesn't mean that it's true. It just means it was published in *Science*. Then, other scientists will try to reproduce the results, and if no one can reproduce them, they're totally worthless. That's self-correcting because even though it's already published, peer reviewed, and everyone says, "Oh, what a great paper. What a great journal," if no one else can reproduce the results, then the published findings are disregarded.

LB: When a paper has been published in one of these journals, doesn't that mean it's been through the peer-review process?

Dr. S: Yes, but some journals are much, much better than others, and some journals are just really bad. The layperson is not going to know the difference between a high-quality journal and one that's really bad. They just don't have that kind of knowledge.

LB: Do you have studies planned at Texas Tech to work on the last question regarding the correlation between trichothecenes and the induction of diseases?

Dr. S: I do, and I don't. Let me explain. In my opinion, the key question in this field now is the only question that hasn't been answered: Do the mycotoxins that get into human beings get there in concentrations sufficient to cause human disease? We don't know the answer to that yet. I will never be able to do experiments to answer that question because the only way you're ever going to be able to answer that is with an MD who actually can work with patients. As a PhD, I can't work with human beings. Now we are attempting to do these experiments in animals, but they're very expensive. It's going to cost $1 million. I'm asking NIH (National Institutes of Health) for the money to do that. Essentially, this entails putting animals in a *Stachybotrys*-contaminated environment, watching them get sick, measuring the trichothecenes in their systems, and correlating that with illness.

LB: There was *Stachybotrys* exposure involved in the *Gorman* case. Can you draw any correlations from data collected from human exposure to *Stachybotrys* that unwittingly occurred in these types of unfortunate situations?

Dr. S: No, I don't think so. There are correlations that you can draw, but you can't answer the final question because in order to answer the final question, you have to actually measure the trichothecenes in the air and in the environment, measure the trichothecenes in the actual person in that environment, and then have some MD say, "Yes, this level would be sufficient to cause these symptoms."

LB: They didn't do that in that particular case?

Dr. S: No, they didn't have the ability to do that. The ability to measure trichothecenes in the air and in human beings just came out in the last six months in our studies. No one else has done that yet.

LB: The EPA and many medical reports and studies state that molds and mycotoxins usually cause infection only in people who are immunocompromised with conditions such as HIV, chemotherapy, leukemia, etc. We were not immunocompromised, yet we became ill from mold exposure. Can you explain this? [A note to the reader: As you can see, at this point we did not understand that each member of our family fell into a high-risk category and that Kurt was considered immunosuppressed/immunocompromised because of his pre-hurricane treatment with prescription steroids for allergy management. This was revealed as we continued our research.]

Dr. S: You normally see fungal infections only in the immunocompromised and that's very different from what we've been talking about.

LB: So you're differentiating between fungal infections and . . . ?

Dr. S: And the respiratory diseases and the mycotoxin diseases you're referring to. You don't have to be immunocompromised to be poisoned by mycotoxins or to have a respiratory disease due to inhalation of high concentrations of fungal spores. You do have to be immunocompromised, in most cases, to have a fungal lung infection. There are just a few fungi that can cause what we call primary disease in normal people.

LB: I'm surprised that the EPA and medical reports don't explain that differentiation more clearly, as you just did. The CDC report published in the aftermaths of hurricanes Katrina and Rita states, "Invasive fungal infections can occur in individuals with normal host defenses and in certain situations can even be life threatening."[13] This statement, on the surface, appears to contradict your previous statement and the general belief reflected in scientific and medical reports that only the immunocompromised are at risk for these invasive fungal infections.

Dr. S: I understand exactly what the CDC people are saying, and they are correct, but you have to understand, you have to have knowledge about what they are saying and what they don't necessarily understand. There are some fungi that can cause primary human disease in healthy people. An example would be a *Coccidioides immitis*, which is a fungus that can cause a human pneumonia in a healthy person. That's a fungus, for example, you or I could become infected with just by inhaling the arthroconidia,[14] even if we were healthy—but that organism does not grow in water-damaged buildings. That organism can be found in the soil. So *Coccidioides immitis*, while it is a fungus and while it can cause human lung infections, is not an organism that grows in water-damaged buildings. The main organisms that grow in water-damaged buildings are *Aspergillus*, *Penicillium*, and *Stachybotrys*.

LB: Do you think we're going to see cancers and other diseases stemming from the exposure to molds, mycotoxins, chemicals, and debris that people got exposed to during and/or after Hurricane Katrina?

Dr. S: All I can say is that it's possible, but I think anybody who would predict that is what's going to happen would be foolish. The truth of the matter is nobody knows what was in the air that people inhaled; nobody knows how much of that air each person inhaled; and there would be no way to predict what's going to happen to people in the future.

LB: We have heard of epidemic levels of mold in brand new multi-unit buildings because of defects in building construction. In general, I think it's hard for people to understand that mold can be a problem even in new buildings.

---

13   The CDC Mold Work Group, "Mold Prevention Strategies," p. 25.
14   Asexual spores.

Dr. S: Yes, people don't necessarily understand that, but new buildings will have just the same problems that old buildings have if they get wet. Water is the common denominator. The age of the building does not matter at all. It's the amount of water that gets into places that it shouldn't be that causes the problem.

LB: When we were researching medical treatment options, we were cautious about being treated with antifungal medications because they are so toxic.

Dr. S: Yes, they are. I think some are too toxic to use on someone who is not infected.

LB: When you say not infected, do you mean someone who does not have an actual lung infection from fungal exposure?

Dr. S: Yes.

LB: Does this mean that you think antifungals are too toxic to use on people who are suffering only from respiratory problems and/or the effects of mycotoxin poisoning from fungal exposure?

Dr. S: Yes. Here is what I mean: Because antifungals have the potential to be toxic to patients, why would you use them on someone who is not infected? To me, it makes no sense.

LB: From a biological standpoint, regarding the effects of antifungals on the body, you stated that antifungals are very toxic. Can you explain exactly what takes place in the body that makes them toxic to the body?

Dr. S: Let me give you a good example of what is and isn't toxic. Penicillin is a wonderful antibiotic that we use against bacteria. The reason penicillin works so well against bacteria and doesn't harm humans is because penicillin attacks the cell walls of bacteria, which are entirely different from our cells. Human cells and bacteria cells are quite different. The reason that fungal antibiotics are so toxic is that fungal cells and human cells are more similar than bacterial and human cells. Some fungal antibiotics attack the cell membranes, and when they do, they also do damage to human cells. However, there are now some antifungal antibiotics that are considerably less toxic to humans. An example would be fluconazole, which is a first-generation triazole with relatively low toxicity. For example, the antifungal Sporanox has the potential to cause toxic reactions. It does inhibit the growth of fungi, but it can also do damage to our cells as well because we have very similar types of cholesterol-type compounds in our cell membranes. Some of these antifungals attack the fungal cell membrane, and because the fungal cell membranes and our cell membranes are similar (we have cholesterol in our cell membrane and fungi have ergosterol, which are both sterols) when you attack one with an antibiotic, you have the potential to attack the other as well. That's why you have to be very careful when you take antifungals and have to be closely monitored because they have the potential to damage human tissue.

LB: That is very important information to know. Thank you very much.

Andrew Puccetti, PhD, CIH, has been an Independent Consultant providing litigation support in the field of industrial hygiene for the past 25 years. Most of his experience has involved indoor environmental quality with much of his work focusing on microbial investigations. Dr. Puccetti has an MS in Environmental Health from the University of Michigan and holds a PhD in Chemistry from the University of Hawaii. His professional experience includes teaching chemistry at Honolulu Community College for five years. He is certified in the chemical aspects of industrial hygiene by the American Board of Industrial Hygiene (ABIH). The interview was conducted April 2006.

LB: What exactly is an industrial hygienist, and what kind of training and education are required?

Dr. P: An industrial hygienist is a person who is trained and has experience in the field of anticipation, recognition, evaluation, and control of environmental hazards. In order to become a CIH, one has to comply with the ABIH's requirements, which include five years of experience working under an ABIH-approved industrial hygienist—an apprenticeship-type situation. In addition, you have to have a minimum of a bachelor's degree in a scientific or engineering discipline; you have to pass a two-day, 16-hour written examination, which is administered by the ABIH. It's not an easy exam; a lot of people fail it the first time. It's a lot like the bar exam for attorneys.

LB: That's quite extensive. I don't think most people realize that.

Dr. P: They don't realize it because what has happened lately, especially in the field of indoor environmental quality, is a lot of really low-end people have come into the field, which in my opinion destroys the credibility of the more qualified people because of some of the things some of the less qualified people say and do.

LB: These people you are referring to are obviously not industrial hygienists. Do they have some sort of lower certification? Are there levels of classifications within the industrial hygienist field?

Dr. P: Well, it's become a little industry to create little professional societies that quickly produce credentials in the field of indoor environmental quality and mold investigation just by having the applicants attend a one- or two-day seminar and pass an incredibly simplistic little exam that requires minimum reading ability. As a result, the profession has been dummied down, in my opinion. A lot of the extremes in the profession are due to these poorly qualified individuals.

LB: Do you find that your company and other CIHs have to come behind these types of companies to essentially cleanup after them?

Dr. P: Yes, we often do.

LB: Do people hire these less-qualified companies, not realizing and not understanding that there can be a price to pay for hiring someone less qualified?

Dr. P: Oh, absolutely; there are a lot of people who have a tendency to hire the lowest bidder and unfortunately, or fortunately, depending on which way you look at it, they get what they pay for.

LB: That's one of the reasons we're covering this information. If people understand why it is important to compare apples to apples versus apples to oranges regarding the level of education and training when evaluating a bid, then they can make an informed decision. Are you familiar with the *IICRC S520 Standard and Reference Guide for Professional Mold Remediation*?

Dr. P: No, but it sounds like a reference intended for remediation contractors, the people who actually do the cleanup of mold.

LB: Do remediation contractors fall into a whole different category than industrial hygienists?

Dr. P: Yes, that's a totally different field. Well, not a different field, but a totally different type of work that deals with the same issue that we're talking about, namely mold. An industrial hygienist investigates and assesses mold contamination, and a remediation contractor does the actual work of removing the mold contamination. A good analogy is an architect and a home builder. The architect doesn't get out there, pound nails, erect walls, etc. He designs the building, and the contractor actually does the work that causes the building to be built. An industrial hygienist provides recommendations as to how to conduct proper remediation. There are some remediation companies that get these extremely wacky ideas about how to conduct mold remediation that aren't really scientifically based. As a result, you get a lot of methods, gadgets, chemicals, and other approaches to conducting mold remediation, which in most CIHs' opinions are inappropriate.

LB: After you assess a home or commercial building and provide recommendations for remediation, do you then give the owners a list of remediation contractors to call, or do you have one that works specifically with you?

Dr. P: It really depends on the situation. Most of the work I do is litigation-oriented, but in the cases where I do get involved on a hands-on basis in remediation activities, I usually provide clients with names of various remediation contractors that I've worked with in the past and who I think are reliable, honest, and use appropriate techniques. In most instances, people have their own lists. Then I produce the recommendations that I have, and eventually a remediation contractor is chosen. Probably the most respected organization of mold remediators is the Institute of Inspection, Cleaning, and Restoration Certification (IICRC).[15]

One thing that needs to be understood is that mold remediation is an interdisciplinary activity. Not only do you need someone such as me [a CIH] and a mold remediation contractor, but also, in many instances, you need a building expert (a contractor or an architect) who can understand or help to understand the dynamics of the water intrusion or moisture problem that caused the mold growth. You have to address the cause of the water damage before you can even hope to permanently get rid of the mold. If you haven't removed the problem that allowed the mold to grow in the first place, then once the mold is removed, it'll come back again. It takes only about 24 to 72 hours for mold growth to occur if building materials get wet enough.

LB: So people need to gather a team—an industrial hygienist, a remediation contractor, and possibly a building-related specialist. What is the best way for people to locate an industrial hygienist? Is there an industrial hygienist organization with a website where people can get referral information for an industrial hygienist in their area who is board certified, if that's the correct term?

Dr. P: Yes, that's the correct term. The American Industrial Hygiene Association (AIHA) website[16] would be a good place to start. Once you contact the AIHA, I would recommend that you request a *certified* industrial hygienist (CIH), certified by the ABIH, rather than just an industrial hygienist.

LB: If people locate an industrial hygienist through the AIHA's referral system, can they be assured that they are actually getting someone who is board certified?

Dr. P: Only if they specifically request one that is certified. The AIHA accepts any scientific professional who is interested in the field of industrial hygiene. An easy way to determine if an industrial hygienist is certified is to note if "CIH" is after his or her name. It's against the law to put "CIH" behind your name unless you are board certified.

LB: Do all CIHs handle mold-related problems?

---

15   According to the IICRC, the organization is ". . . a nonprofit certifying body for the inspection, cleaning, and restoration industry. Organized in 1972, the IICRC currently represents over 3,320 certified firms and 35,000 certified technicians in over 33 countries." The IICRC website is http://www.iicrc.org/.

16   To search the AIHA data base for a consultant by state, country, specialty, and/or location of work, see http://www.aiha.org/Content/AccessInfo/consult/consultantsearch.htm.

Dr. P: Industrial hygiene is a discipline that encompasses a wide variety of environmental issues. Most industrial hygienists, if they are in the consulting business, tend to focus on just a few areas of expertise within the discipline of industrial hygiene. The best way to determine if an industrial hygienist is proficient in any given aspect of the discipline is to look at his or her experience and the types of professional development courses he or she has taken.

LB: You mentioned that building experts might be needed as well. What types of building experts might be needed to help with a water- and mold-related building problem?

Dr. P: Well, it depends on what the problem is. Sometimes it's pretty obvious what kind of building expert you have to consult. For instance, if you have a plumbing leak, you call a plumber because he has to address that leak. If you have water intrusion through the roof, you call a roofer. You have to address the specific issue that led to the water problem in the first place. In many cases, it's very easy. More subtle cases can involve drainage problems. For example, moisture may be coming up through concrete slab flooring or accumulating under a crawl space. Another example of a situation that can be very challenging is water condensation problems due to the design of the building, the heating and air conditioning, or the ventilation system. It is important to have help from the proper building science professional. Usually, an experienced industrial hygienist will be able to recognize the type of problem, so he or she can select the right professional to help deal with the specific issues causing the water damage.

LB: Take us through a situation, an easy situation, in which people can visually see the mold, so they know it exists. At what level is there so much mold that they need to hire an industrial hygienist to come in and help? In other words, is there a quantity threshold at which a person should say, "This is too much mold. I'd better call a professional"?

Dr. P: Well, unfortunately, I don't think I can answer that. That's a question that's really site-specific. It's almost impossible to answer in a general sense and give an appropriate answer. A lot of problems in buildings involve hidden mold growth, so you don't really know how big a problem you have until you start investigating it. If you have a flood-like situation from a hurricane or flooding, it's pretty simple to determine how big a problem you have because you're more than likely going to get an extremely visible problem. You also have to understand that the mold is not just in areas where you can see it; it's also in areas that got wet and are hidden from view, like wall cavities, ceiling plenums, and other areas of the building that aren't directly observable by doing a walk-through of the occupied space of the building.

LB: Since most of the work you do is litigation-oriented, do you spend much time in the investigation phase?

Dr. P: I do a lot of fieldwork. That's part of my work as an expert witness. I conduct the investigation. In some instances, there is something to investigate, while in other instances, the evidence no longer exists, and I have to rely on the work of others.

LB: Are you usually brought into a case once an attorney is involved and it's already been established that medical problems are arising from the mold exposure?

Dr. P: Right.

LB: Are you often brought in to determine what is causing a family or workers in a building to get sick?

Dr. P: That's the situation, yes. In many instances, I'm the first one to conduct an investigation.

LB: I've read that off-gassing of VOCs can contribute to illnesses. Additionally, I have read that there can be a chemical synergy between components, for example, formaldehyde, VOCs, and mycotoxins, and that it's uncertain what level of toxicity is created from this chemical mixture and its impact on health. Can you talk about that a bit?

Dr. P: Other than what you just said, there's not a whole lot known in great detail. There's a lot of research, a lot of work that's going on, but in many instances, we don't have the luxury of having a budget to conduct an investigation that takes into account everything such as microbial VOCs and mycotoxins. Sometimes we have the luxury of having a budget that's high enough to do some mycotoxin analysis, and that's helpful, but all of these issues are still in their infancy regarding clearly identifying exposure-dose relationships and relating them to symptoms. The science is still in its infancy, but right now we do have enough anecdotal evidence to associate these exposures to deleterious health effects. There is a very simple way of looking at it: If many of these alleged health effects go away or diminish in their intensity when an occupant leaves a contaminated building, and when he returns, he gets sick again, then there is a temporal relationship between the symptoms and being inside the building. Not having symptoms while being out of the building is a pretty strong case that there is something in the building that's a problem.

LB: Which method do you use to quantify indoor spore levels, cfu/m^3 or spores/m^3?

Dr. P: Ideally, you want to use both methods because there are advantages and disadvantages to both. Often what you can see using one method, you can't see using the other method. In the viable methodology, the spores that are captured are identified very specifically as to the genus and species because they are cultured, or grown, into mold colonies that have significant differences in their morphology. This approach is analogous to cultivating lemon seeds and orange seeds and growing a lemon tree and an orange tree. Although the lemon and orange seeds cannot be differentiated, the orange tree and lemon tree can easily be.

In the nonviable method, the spores are trapped on what are normally referred to as a spore-trap device. There are several different types that are commonly used. The spores are not cultured; they are just examined microscopically. Under these examination conditions, microscopists can't often identify with very high specificity the type of spore that they see through the microscope. It's like looking at a lemon seed versus an orange seed. They're not that different and are difficult to differentiate. However, if the spores present are significantly morphologically different, then their identities can be specified. This situation is analogous to a lemon seed and an avocado seed. They are easily differentiated because of their significant morphological differences.

LB: So that's another good reason to use both methods.

Dr. P: Right. You want to be able to see as much as you can so that you get as much information from your sampling as possible. For example, a lot of times it's very difficult to culture *Stachybotrys* spores. On the other hand, you can see them quite readily and identify them microscopically on spore traps.

LB: The NRDC report stated that not all air samples collected in New Orleans were taken over a 24-hour period and that test results could be underestimates because some species produce more spores at night. Do you know which mold species produce more spores at night?

Dr. P: No, but fungi all have different preferences. Various species will not produce spores at certain times of the day. They have cycles in which they produce spores off and on. It's very species specific, and they all vary. I'm not aware of any specific type of species that only sporulates at night. I think that the concept to take away from those statements is that when you're conducting air sampling from airborne spores, you're dealing with a highly variable situation; you can have the same amount of mold growth in a building, go in there at various times of day, and get results that differ quite significantly. The reason for that is you're going to get variability in the spore-producing propensities of the different types of molds that are there. Sometimes they will be emitting spores, while other times they won't. Usually, when molds are drying out is when they emit the most spores.

LB: *Aspergillus* can grow inside someone's lungs causing aspergillosis. Do the mold spore cycles that occur in an external environment continue inside a person's body?

Dr. P: Well, I think you have a different situation. First of all, there are only a few species of *Aspergillus* that can grow in human tissue at the temperature of the human body. When they're growing, they're infecting the tissue. They're basically living off the tissue. They're not throwing spores anywhere; they're just growing. The mold is producing spores which germinate because they are using the human tissue for metabolism.

LB: I am unclear. Does the cycle continue internally or not necessarily?

Dr. P: We're talking about two different things. In the outdoor environment, we are talking about the cycles that the mold growth undergoes and how they emit spores into the air. When we're talking about mold growth in human tissue, we're talking about mold growth that's confined to tissue. It's not emitting spores anywhere because the spores are going to be confined to your body. The parameters under which the mold will grow depend on the species and on the host individual.

LB: Regarding alkalinity and acidity, can mold grow in an alkaline medium or when an alkaline chemical is used on mold?

Dr. P: Different species of mold have different likes and dislikes with regard to pH, which is a measure of alkalinity, so it really depends on the species. In my opinion, you don't worry too much about that because if the environment is not wet, mold is not going to grow. If a material is not wet enough to support mold growth, it doesn't matter if it's alkaline, acidic, or neutral. The limiting factor in mold growth is moisture content. If a material has enough moisture content, then mold will grow. The amount of moisture that's required in a given material will vary, depending on the species of mold. Some molds require a relatively small amount of moisture content within a building material, while other molds require a high moisture content. I can give you two specific examples: *Aspergillus versicolor*, a species of *Aspergillus*, can grow on building material with relatively low moisture content; *Stachybotrys chartarum*, on the other hand, requires a very high moisture content on building materials.

It's the moisture content that is the limiting factor. If you remove the moisture, you're not going to have any mold growth. The whole idea behind proper remediation is not to pick and choose what chemicals should be used; it should be to dry things out and physically remove the mold growth from the surface. This is necessary because mold retains its allergenic and toxic properties even if it is dead. Often plumbing leaks or other types of water intrusion problems are repaired without removing any of the mold that grew during the time moisture was present. If you don't eliminate the mold and you've only eliminated the source of water, you still have a mold problem. The mold, if there's enough of it, will still be in the wall cavities, and spores from that mold will come out of the wall cavities through the switch plates, power plugs, and other penetrations. These spores settle on surfaces; they collect on carpeting and other surfaces and are re-emitted into the air by occupant activity, resulting in exposure to building occupants.

LB: For example, when landlords fix the source of the water or moisture but do not take the next step to remove the mold-contaminated materials.

Dr. P: Right, and sometimes they remove part of it, and in many cases it's done improperly. Instead of hiring a remediation contractor that can do it properly, they just hire a handy man to do it. The result is more spore contamination than there was to begin with because they don't do it under proper containment conditions.

LB: So there can actually be more spores after remediation than before?

Dr. P: In the ambient space, yes.

LB: Is there any help for residents living in multiple-unit residential buildings when their landlords won't test for mold? Are there any state or federal agencies from which tenants can get assistance in facilitating mold testing?

Dr. P: Well, unfortunately, it's a pretty expensive proposition to conduct this kind of testing, and most health departments just don't have the manpower or money to do it. You're talking about a significant amount of money, and you just can't get a typical health department to do that. It's just not within their budget.

LB: The NRDC report revealed very high spore counts in fully remediated buildings. In general, what level of spore counts should we see after a building has been remediated?

Dr. P: Ideally, after remediation the spore levels inside should be much the same as those outside. In fact, in most instances they are lower because the remediation contractors have scrubbed the air with filtered negative-air machines, cleaned everything, and removed all settled dust. I would expect that if a building is properly remediated, it should have airborne spore levels that reflect what is generally found outside at the time of sampling.

LB: When the remediation work is completed, should people have someone come in to reassess the air quality, and if so, whom should they have conduct these air quality tests?

Dr. P: Sure, they should, and it should be an independent industrial hygienist who is not part of the remediation company. It should be an independent entity.

LB: Is there any value in using the do-it-yourself mold tests sold at stores as a preliminary testing measure before calling in an industrial hygienist?

Dr. P: No, in my opinion that would be sort of stupid. Most people know what mold looks like. The issue with mold growth is how it got there and how big a problem you have. Is it just a few inches of surface mold that's insignificant; is it a building problem or a home problem? The only way to determine that is to have an experienced indoor air-quality professional actually assess the building using a visual inspection, his or her experience, and proper methodology to determine whether or not there is a real mold problem or just an insignificant amount of mold growth in a few corners for reasons that are not significant to a major mold problem.

LB: A number of factors in a building could cause people to get ill. If people are getting sick in a building and they don't see mold, could a do-it-yourself mold test rule out mold as a possible cause?

Dr. P: Well, I'm not that familiar with the test you are talking about, but all I can say is that it's not enough to just take a swab of something and say, "Is this mold?" because mold is ubiquitous. You can have a little bit of mold anywhere. You will more than likely get results that are almost impossible to properly interpret.

LB: Another indication of mold being present is the moldy, musty smell that often accompanies it.

Dr. P: VOCs are produced in the metabolic process, and that is what you perceive when you perceive an odor.

LB: If people smell that moldy, musty smell but don't see mold, is that a sure indicator that there is mold present somewhere, but they just can't see it?

Dr. P: Right, it means that there's active mold growth, and there's enough moisture present to support active mold growth. It's not dormant, and it's not dead. It's actively growing because the odor that you're smelling is due to the metabolic process that the mold is undergoing.

LB: When you go into court with mold measurements, what do you compare them with since there are no indoor air-quality limits with regard to mold?

Dr. P: Well, you always compare indoor airborne concentrations of mold spores with outdoor concentrations. We pretty much know what normally is present outdoors; there are a lot of databases. In addition to that knowledge, you also sample outdoors at the same time you're sampling indoors. The assessment of the air sampling results is always based on what you find indoors versus what is normally present outdoors. In a normal building that has no sources of indoor growth, you would expect that the airborne spore concentration should mirror what is present outdoors because the air that's inside comes from the outside. In most cases, the total concentration of spores is lower indoors because there is less air movement, so some of the spores settle on surfaces, especially on carpet.

The relative types of spores that are present in the air indoors are much the same as the relative types of spores that are found outdoors. So in other words, if you find 50 percent *Cladosporium* outdoors, chances are you're going to get 50 percent of all those spores that you detect indoors to be also *Cladosporium*. That's obviously to be expected since it makes sense that the air that is inside comes from the outside. However, if you have a home where there is a significant amount of mold growth inside, then you have a situation where you have two sources of mold spores in the ambient air of that home. One source is the outdoors, but you also have another source and that's the mold growth inside the building. Usually that mold growth is of the type that is generally found to be at lower relative levels outdoors. In a building that has mold growth, you'll see the type of mold that grows inside is usually species of *Aspergillus*, *Penicillium*, *Chaetomium*, *Trichodema*, *Stachybotrys*, and other types of molds that generally produce low levels of spores outdoors. In cases where there's significant mold growth indoors, you're going to find very elevated levels of airborne spores indoors compared with what is normally found outdoors.

LB: This has been very informative. Thank you for your time.

*Jim Pearson, CMH, is President and CEO of Americlean Corporation,
one of the country's first full-service disaster restoration companies that
he helped start over 27 years ago in Billings, Montana. Mr. Pearson was
the appointed chairman of the IICRC S520 Consensus Body, which was the
committee that oversaw the completion of the second edition of the industry
publication,* IICRC S520 Standard and Reference Guide for Professional
Mold Remediation. *The IICRC is the Institute of Inspection, Cleaning, and
Restoration Certification that is an industry-respected organization of mold
remediators. Mr. Pearson is one of several coauthors of the first edition of
the standard and reference guide, which OSHA recognizes on its website as
a one of three sources that ". . . may be referenced by OSHA inspectors for
informational purposes."*[17]

*Mr. Pearson began studying indoor air quality issues and began performing
mold testing and remediation services over 20 years ago when Americlean
Corporation expanded into the field of mold remediation. He is a frequent
presenter at national technical seminars and a speaker at international
industry conventions. Mr. Pearson has served as an expert witness
in restoration and environmental legal matters. He is a contributing
consultant to industry magazines and has authored hundreds of nationally
distributed newsletters and procedural guides. Mr. Pearson can be reached
via his e-mail at jim@americleancorporation.com. The interview was
conducted April 2006.*

KB: What qualifications should a person look for in an industrial hygienist?

JP: I'm an industrial hygienist because I study the discipline. Anyone who studies
indoor air quality or works in the field (or a related field) can be called an industrial
hygienist. You have to be careful to hire a qualified industrial hygienist because in
the last several years, it seems everyone has been trying to get into the field, and
some of these people have questionable training and experience or none at all. Look
for someone who is certified because that means they have undergone a certain
level of formal training. For example, I'm not a certified industrial hygienist (CIH);
for that you must have a four-year college degree in microbiology or a related
science and pass certain exams. I am, however, a certified mechanical hygienist
(CMH). CMHs specialize in indoor air quality as it relates to the mechanical
systems of a structure.

---

17   OSHA, "Molds and Fungi Standards." Available at http://www.osha.gov/SLTC/molds/
standards.html (accessed February 2007).

KB: Many smaller communities will not have a local CIH and may not even have a local CMH, so aside from certifications, what else should people look for to help them determine if an industrial hygienist is qualified.

JP: Whatever they call themselves is not nearly as important as their credibility. A credible company will have a track record filled with experience, environmental insurance coverage, and industry connections. They will have invested in specialized training and equipment. With the potential health problems involved, it's easy to scare people into making a quick decision and into hiring the first person that comes to their rescue.

KB: Aside from the four-year college degree, what are the other differences between a CIH and a CMH in regard to training and the types of work they do?

JP: I think I should clarify this designation thing. There are thousands of CIHs. Most of them deal with health and safety issues in the industrial workplace. A few pursue a specialty such as microbiology. Some of these people may then focus their attention on bioaerosols, which include mold. Most CIHs, however, can design a sampling protocol for mold and other biohazards and provide a report and remediation instructions. CIHs almost never perform the actual remediation. It has the potential for a conflict of interest, and most are not trained to do such work.

Mold remediation contractors, on the other hand, are trained to follow protocols and perform the work to established criteria. There are many certification designations given by industry associations, and I hate to place too much emphasis on any particular combination of letters following a person's name.

Smaller communities like mine lack CIHs and even indoor environmental professionals, and this mold and allergy thing keeps getting worse. The need is there, and we have to do what's necessary to help. I consider myself to be an indoor environmental professional with the CMH focus a subspecialty because I provide services of problem identification, sampling, and remediation. When we do have an opportunity to work with CIHs, they are typically called in first. They scope the property, evaluate the problem, sample, interpret the samples, and make recommendations as to what should be done to correct the problem. This information is then given to remediation contractors to bid, contract, and accomplish. Often, CIHs are called back in at the end of projects to provide testing and issue clearance reports.

KB: Are there fly-by-night mold remediators that people need to be careful of?

JP: In the South particularly, there are many pickup-truck remediators—that's what we call them—who run around with chain saws after hurricanes, and now they're running around with masks and big bills for killing mold with disinfectants and covering it up with sealer. Because of this, people get ripped off and start suing.

In fact, there were so many lawsuits involving insurance companies not stepping up and dealing with this type of problem, that insurers simply quit covering mold losses.

All this was putting a stain on my industry, so a group of professionals got together to deal with the quality control problem. The Institute of Inspection, Cleaning, and Restoration Certification (IICRC) and the Indoor Air Quality Association (IAQA) formed a committee to write a standard for remediation that was effective, prudent, and helped the industry deliver best practices to its customers. That is why we wrote the *IICRC S520 Standard and Reference Guide for Professional Mold Remediation*. There were approximately 80 industry people in total who worked on this project with a dozen of us who were chairpersons of the various chapters. Hopefully, this book will come to be recognized by the government as *the* standard for mold remediation.

Before we wrote our book, the only thing that people were relying on was the New York City Health Department's guidelines for dealing with mold. They weren't too bad, but they weren't very specific. For example, they used size: If you have more than 10 square feet, it is not a maintenance issue anymore; you need a professional. The problem is, if you see 10 square feet, it could simply be the tip of the iceberg. You don't know if there are a hundred more square feet inside the wall. You don't know if the stuff has gotten out and contaminated the airspace or if there are settled spores on the desks, carpets, and things. You just don't have enough information to make a determination based on the amount of mold that is visible.

KB: What is the proper way to test for mold?

JP: Mold is ubiquitous. It's everywhere. If you buy bread at the store and don't even open it, in a week it will be moldy. Mold is in building materials; it's in everything. There's mold in this air that we're breathing now. There's lots of mold outside, too.
So, we come in and take a sample of the air with what we call a spore trap or sampling cassette. Five years ago, it cost $3,000–4,000 to come in with a stack of plates with agar, which is mold food. The technician would suck the air through these different growth mediums, send them to a lab that would incubate them for a couple of weeks, and see what grew. The concern was for what viable mold spores there were in the air. Viable mold spores are spores that will reproduce. It wasn't long before we discovered that it doesn't matter if the mold spores are viable or dead because even when they're dead, they're still an allergen, which affects most people. You can't "kill" a poison, so if a mold spore is toxigenic, it's still poisonous—dead or alive.

Speaking of killing mold, some people think spraying mold with bleach to kill it and then sealing it over with paint is all that's necessary. While this may inhibit mold's growth a bit, all it needs is water and it'll continue to grow. Bleach is about the worst product you can use on mold simply because it leaves so much water behind after the chlorine gas quickly flashes off. Folks think it's working because the mold seems to disappear when it has actually become colorless from the bleach.

We often use a commercial disinfectant on our projects primarily because of the bacteria that accompany mold growth, but we don't rely on it to keep the mold from coming back.

KB: Let's talk about testing or sampling. How does that work?

JP: People have a variety of test regimens. For example, I just learned of a fellow, a competitor, I suppose, that operates out of his house and claims to be a mold inspector. He goes out and takes swab samples. He puts the swab on the mold, puts it in a bottle, and sends it to a lab. They'll send it back saying, "That's mold." They'll even speciate and say, "That's *Aspergillus* mold." The problem is that you can put a swab on any mold growth, send it in, and the lab will say, "That's mold," and they will tell you what kind. However, that doesn't tell us what you're breathing; it doesn't tell us how much mold there is.

What I like to do is take at least two tests. I like to do air sampling, which is pulling a fixed quantity of air through a spore trap, which paints a trace on a sticky slide. It's all sealed, so there's no fooling with the test results. Then I send it to an accredited lab. They break it apart, affix the sticky slide under a microscope, and actually count the number and types of spores. They do a mathematical interpretation and come up with X number of spores/m^3. They use the metric system, of course, because it is scientific. So spores/m^3 is our gauge for measuring mold.

KB: Is spores/m^3 the normal gauge for the industry?

JP: Yes, except in bulk, swab, or contact-tape-lift sampling where we use spores or colony forming units per square centimeter [spores/cm^2 or cfu/cm^2]. A bulk sample is simply a piece of moldy material placed in a baggie and sent to the lab for analysis. Taking a swab sample is accomplished by dipping the swab in a solution and rolling it over a specific area, say 2 inches by 2 inches. A tape lift is clear tape pressed lightly on a hard, flat surface and then placed on a glass slide for viewing.

When the lab gives me the tests results, they'll report it in spores/cm^2 or cfu/cm^2. So the bulk collection, tape lift, and swab sampling are all measured in amount/cm^2, whereas air sampling is measured in amount/m^3.

KB: What is the second test that you like to do?

JP: Sometimes we take many samples, but one thing I always do is take a control sample of the air outside to see what molds are there. On some days, for example, there are high levels of *Aspergillus*—a potentially toxic and infectious mold coming from gray weathered wood, rotting leaves, and so forth. I take a sample of this outdoor air as a comparison or a baseline. Mold levels change outdoors all the time. If the wind is blowing, you'll probably have a fairly high count; but if it's snowing, you'll probably have none. If it's just rained, probably you'll have millions of spores because liquid water tends to disperse the mold. If I find the same amount of spores of the same species inside, I don't suspect there is a mold problem in the building.

KB: Are there any set indoor limits for mold?

JP: No standards have been set yet on permissible exposure limits because everyone's resistance level is different. That's why the government has not set standards. That's why we can't set standards. I sample the air outdoors, of course, and then I sample the living area, where the people are. Then I go to the source area or where we suspect the mold to be coming from. Normally, it's a crawl space or a basement that's been wet. If I don't see visible mold, it really is cause for sampling. If I go in and see a lot of visible mold, there is no point in taking an air sample except to see how far it actually has spread and if you're actually breathing these mold spores.

I do, however, often take a bulk sample, which is a piece of the moldy material, or a tape-lift sample to speciate to see if we're dealing with *Stachybotrys*, *Aspergillus*, or what. If it is *Cladosporium* or ascospores, it might not be as important, but I like to be aware of what I'm dealing with. I also take tape-lift samples of the upstairs and the downstairs. That tells us what types of settled spores we have and gives us an idea of how long this problem has existed. If there are no settled spores, we can assume that it is relatively fresh. If there are settle spores, we know the extent of spread and that it's probably been there awhile.

KB: Do you have to be careful of cross contamination when you are taking samples?

JP: Yes, you do. I start outdoors first because if I start at the source, I will be contaminated—I'll have spores on my clothing and on my hands by the time I get to the living room and on to the outside for what is usually my cleanest sample. There is a good chance I will contaminate the cassette. So there's a lot to be said for a good sampling methodology. You need to know what you're doing. You need to have a sampling plan and know what you are trying to accomplish. You need to have good scientific reasoning for why you sample here and not there. Some people take dozens of samples and the cost is thousands of dollars. I think that's on the ridiculous side. I like to see what's representative of what we're breathing, what mold occurs naturally outside, what's right at the source, and the species of mold. Then, when we're finished remediating, we come back in and test to find out if we've cleaned it all up. Mold fools me all the time because the places I expect to see it, I don't, and the places I would never expect to see it, there it is. The conditions have to be just right for the mold to grow.

KB: What happens if you can smell the mold and have a reaction to it, but it doesn't test out?

JP: I have occasionally had that happen. But generally, when I can smell it, I can detect it in the air with the samples. The odor is usually a result of the microbial volatile organic compounds (mVOCs), which are emitted from some molds under the right conditions. The mVOCs can contain mycotoxins as well.

KB: What are you finding regarding tight building syndrome and sick building syndrome?

JP: Surprisingly, it's more of a problem in new homes than old. Most people think, "Montana is dry, so there's no mold problem." Well, we're not talking about mold outdoors; we're talking about indoor mold from interior floods, minor leaks, drips, and condensation. If it's cold outside and you've got a cardboard box against an exterior wall, mold will form on the surface layer of condensation on that wall because there's no air flow. Mold spores are everywhere, and all they need is moisture and organics—which is dirt, wood, paper, wallpaper glue, any of that— and mold will start to grow. You're question was . . . ?

KB: Tight building . . .

JP: Yes, the mold causes me to forget. Actually, it probably could; there are some molds that I can't even be around because I have been sensitized to them. When I was growing up, I was the guy who washed his hands in gasoline to get the grease off and didn't think much about environmental issues. But we all should be concerned with environmental issues. A newer home can be just as sick as an older home, sometimes more, because they're tight. They're weather proofed; they're sealed; there's no air flow exchanged; there's no outdoor air introduced, and anything that's in the air tends to concentrate and build up.

The EPA says indoor air can be up to nine times more polluted than outside air. This is because contaminants become trapped in the building. The old house with holes in it has fewer problems because it breathes. If a little condensation forms from a small leak, it dries out right away. But now houses are so sealed up with house wrap, caulk, and all of that, contaminants from any source just build up. Sick building syndrome occurs frequently in newer, tighter construction. It also occurs in some office buildings, and yes, older buildings have their problems, too—they do have a tendency to have been moldy at one time or another from an improperly restored flood or whatever, so it could hit any building. It could be in a new house, a new apartment building, or an older home.

KB: We have seen a lot of news coverage of remediation workers, homeowners, and volunteers in Louisiana not wearing any personal protection equipment. What is the proper personal protection equipment that a person should wear when remediating a mold-contaminated building?

JP: I just talked with my brother yesterday—he just got back after working in New Orleans for many months—about how the workers and volunteers are out there, without any personal protection equipment or proper personal protection equipment, tearing this mold out and breathing this stuff. Some people wear only an N95 face mask. The N95 rating means 95 percent of particulate is filtered. The N95 is the little white mask that you put on over your mouth and nose. The one-strap masks are totally ineffective; the two-strap ones are a little better. OSHA recommends a minimum of N95, which is the two-strap dust mask with the nose-pinch thing, if you're working around mold. That is not enough. Our industry says a HEPA filter P100 is what needs to be worn. HEPA-filtered masks—full face—are best. The half-face mask is okay, but you'd better have sealed goggles because mold can enter through your eyes or through a cut in your skin. So it's latex gloves and Tyvek suits with hoods and booties, all duct taped at all the joints—between your hands and your wrists and your ankles and your feet.

KB: It sounds basically like a "chem" suit.

JP: Yes, exactly. It's an environmental suit. Ideally, a powered full-face respirator is used because it has a motor and pump that delivers filtered air. That's what we wear. We have to because we're working in it all the time. The non-powered models are more difficult to wear all day because you have to labor to draw in the air through the HEPA filters.

If I didn't supply personal protection equipment for my employees, I would be negligent and in violation of the law. The proper thing to do is to protect yourself from the mold and other contaminants. When you're working with mold, you walk into a house and see mold. If there's no one living there, it's not so bad because it isn't being disturbed, but when you start tearing that stuff out, it amplifies by hundreds and hundreds of times.

KB: The NRDC disclosed in the results of their air sampling that the dust was not disturbed while air samples were collected. They stated that this could make the test results underestimates. How much higher do you think their test results would have been if the dust had been disturbed? Astronomically higher?

JP: Yes, in fact, there's a very good study by Dr. Gene Cole, who's a friend of mine, and Jim Holland, another friend of mine who runs a restoration consultancy firm. The report is in the appendix of the *IICRC S520 Standard and Reference Guide for Professional Mold Remediation* that I spoke of earlier. It addresses exactly that issue—what you are exposed to when you start messing with mold. With activity, you will certainly see much higher levels.

After remediation, the air should be retested to ensure mold levels are back to an acceptable level. You have to be sure that no procedures are used that will temporarily lower the mold levels so that they are acceptable at clearance testing. I'll give you a little tip: In Louisiana, some remediators utilize methods that are not correct in order for a property to pass tests. The way they do it is when they are all finished, they fog some sticky disinfectant in the containment area. It forms a fog and grabs onto the airborne mold and spores and settles them to the floor. Then they shut off all the air scrubbers and air movers, seal it up, and let it sit there for a few days. Then the CIH sneaks in very carefully—so he doesn't disturb the settled mold and spores—and pulls his air sample. It passes the test.

When I go in to do sampling, I run as real a test as I can because we know it's just a snapshot in time. I don't create an abnormal condition such as trying to stir up the dust. Most mold spores stay airborne simply from air currents.

During remediation, lots of airflow is used because we want as much mold as possible airborne so the air scrubbers will remove it from the air. Of course, all surfaces are HEPA vacuumed, and moldy non-structural materials are removed, too.

When the source of water is removed—be it through normal evaporation, blowing air on it, using heat, dryers, or whatever—when it starts to dry, mold freaks out. It needs to propagate, and it starts to send out spores, so it can reproduce somewhere else where there is moisture. I like to use the example of a dandelion with the little parachute seeds. With the slightest breeze, those things fly all over the place. That's how mold works; the spores get ready and go when it starts to dry out, and any air current at all, even standard convection from heat rising, will carry these spores around. Some will float in the air for days and days until they find a wet spot, and then they'll grow just as dandelions do.

KB: When should a person bring in a professional to remediate mold? Is there a certain level of visual mold that is a red flag?

JP: I've got a few rules of thumb. One is that only healthy people should allow themselves to take care of mold. That means that people who are sick, have had chemotherapy, are very elderly, have some kind of pulmonary or respiratory illness, have had bone marrow transplants or recent surgeries, have allergies, asthma, or chemical sensitivities, or are a child should not be exposed to mold. They should call a professional.

If you are a healthy adult and have no one in the building with the above conditions and you see a patch of mold, you need to ask, "What's it doing there?" That mold could be the tip of the iceberg and could be a serious issue. I've gone to many locations where people have started to remodel their bathrooms only to discover when they open up the wall, it's full of black mold. They say, "Now what?" Well, the "Now what?" is to back up slowly, turn on the vent fan, close the door, and call a professional. That's really key because you can spread that stuff all over the place, and it will be even more expensive to cleanup.

KB: How do you as a professional dispose of the mold that you have remediated from someone's house?

JP: This is a new area, and there are no requirements for dumping moldy waste. You can take it to the landfill. The way we dispose of it—and we're very careful with it—is to put it in very thick plastic bags. Sometimes we double bag if there are big pieces of sheetrock that can poke through. We vacuum the outsides of the bags before we take them through the house; then we take them to the dump. It's not a regulated waste. It's certainly not a toxic waste. It's handled differently than asbestos, although we use a lot of the same principles for remediation of asbestos.

KB: After mold remediation is completed, what should a person look for to make sure the remediation was properly done?

JP: Certainly, it has to look visibly clean when the work is done, and I mean white- and black-glove clean. One of the best indicators of a proper job is to have clearance testing performed. The results should show fewer spore counts inside than outside or the same and you don't want to see anything unusual on the report. For example, if you have no *Stachybotrys* outside but have *Stachybotrys* inside, that's a bad deal, and it tells you there is still a source of mold inside somewhere.

I guess it all goes back to choosing a reputable, experienced contractor. Check references, ask questions, and educate yourself. It's important to be well-informed.

KB: That is a good point. Thank you.

Richard L. Lipsey, PhD, *is a Forensic Toxicologist and President of Lipsey & Associates Inc. in Jacksonville, Florida. For over 30 years, he has worked closely with Fortune 500 corporations, government agencies, nonprofit organizations, and concerned individuals worldwide to provide scientific solutions for creating healthy environments in homes, offices, and the workplace.*

Dr. Lipsey has an MS in Entomology from the University of Arkansas and a PhD in Entomology with an emphasis in Toxicology from the University of Illinois.[18] He has been an Associate Professor at the University of Florida where he taught classes in pesticide environmental hazards, especially to pesticide applicators and field workers, and trained industrial hygienists and safety officers. Dr. Lipsey served as the liaison professor with, and consulted for, the USDA, the EPA, and the U.S. State Department, both nationally and internationally. He serves as an expert witness in litigation and continues to educate professionals through multiple speaking engagements each year. In February 2006, Dr. Lipsey collected post-hurricane air samples in St. Bernard Parish, Louisiana. Dr. Lipsey can be contacted at www.richardlipsey.com. The interview was conducted April 2006.

LB: First of all, I want to say I admire your courage in going to St. Bernard Parish to evaluate the mold contamination after the hurricane, knowing the health dangers of mold exposure.

Dr. L: Both of us who went got sick. Even though I was fully protected with a HEPA respirator, a moon suit, and rubber gloves when I went into the buildings, it was too late because I got the Katrina Cough from just walking through the neighborhoods.

LB: You got sick from exposure to the air outdoors?

Dr. L: Yeah. The EPA says that the endotoxins alone were 20 times higher than normal in the outside air.

---

18   According to Leta Summers, Secretary IV in the Department Entomology at the University of Illinois, Richard L. Lipsey graduated ". . . from the University of Illinois with a degree in Entomology, because the department did not have an official toxicology program. . . . His advisor was Dr. Robert Lee Metcalf, who specialized in toxicology." This information is from a phone conversation with Leta Summers on August 3, 2006, and was confirmed in an e-mail on August 4, 2006. According to Dr. James Sternburg, a professor in the Entomology Department in 1972, toxicology was an area of specialty in the Entomology Department back in 1972. This information was provided in an e-mail from Jerald Kimble, Account Technician in the Department of Entomology, September 12, 2006.

LB: Explain what endotoxins are and the dangers they present to us.

Dr. L: An endotoxin is similar to a mycotoxin from mold. All gram-negative bacteria produce endotoxins and can become airborne. They are chemicals known to be highly toxic, similar to mycotoxins produced by certain pathogenic molds. Some examples of gram-negative bacteria are *E. coli*, *Salmonella*, and *Legionella*, which causes Legionnaires' disease.

LB: Are these bacteria infectious?

Dr. L: Of course they are. The infectious state depends on which gram-negative bacterium is in the air. *Salmonella*, you know, can be left when you cut up chicken on your drain board. Then the next time you cut anything on your drain board, you can get a serious, life-threatening infection. In fact, you can die from any of those bacteria.

LB: Can people develop diseases from just breathing air that contains these bacteria?

Dr. L: Of course. Legionnaires' disease, for example, comes from breathing bacteria from a heating and cooling system contaminated with *Legionella* bacteria.

LB: Why were there high levels of gram-negative and gram-positive bacteria present in St. Bernard Parish after Hurricane Katrina? Is this a common occurrence in buildings with extended water damage?

Dr. L: Yes. Both gram-negative and gram-positive bacteria form in conditions where carpets, couches, wallpaper, etc., have been saturated with moisture for days or longer. Additionally, the rot and decay process from a moldy environment creates an environment conducive to bacteria formation. Some bacteria can double in population every 15 minutes under ideal conditions.

LB: Why hasn't there been a mandate from federal or state officials requiring that people wear personal protective equipment in the aftermath of hurricanes, especially one the size of Katrina?

Dr. L: Well, not only that, but while I was getting a tour by the mayor's office, the Habitat for Humanity workers were allowed to join my tour of the devastated St. Bernard Parish. The parish has 45,000 homes, of which 40,000 will have to be bulldozed. When the mayor's staff asked the Habitat for Humanity workers if any of them wanted to ask the toxicologist a question, the first question asked was, "Do we need protective equipment to go into these homes and tear out these moldy walls?" I said, "Of course you do." I told them why, and their response was that the FEMA people who trained them said it was unnecessary to have any protective equipment. That really makes me angry.

LB: Many people are also getting misinformation from their doctors. They are being told that mold can't hurt them and that they can't get sick from mold exposure.

Dr. L: A medical doctor who would say that mold can't hurt you is either ignorant or *ignorant*. I don't think any medical doctor would willingly lie about mold, knowing that molds and the mycotoxins they generate have been purified and developed for germ warfare to kill people within minutes. We know that 20 percent of all molds can produce mycotoxins; the other 80 percent may be harmless, but we don't know if a high enough level of a so-called harmless mold could actually be very harmful to some people.

LB: With all of those facts known from scientific and medical research, where is the ball being dropped in regard to educating the public? We see news coverage, repeatedly, of residents, volunteers, and remediation workers cleaning and doing repairs without wearing any personal protective equipment whatsoever.

Dr. L: On all levels—from FEMA to the CDC. The CDC doesn't want to scare anybody and neither does NIOSH nor the EPA, but if you go on their websites, you'll see stories about how some molds can cause you to vomit blood, can cause you to die from pulmonary hemorrhaging, and can cause cancer. This is on the websites. Regulatory agencies are not educational institutions; people need to take the initiative to educate themselves.

Government regulatory agencies have regulators; they don't have educators. I used to work as a consultant for both the EPA and the USDA in Washington, DC. I had offices in Washington in both agencies. I have also served with NIOSH and the CDC in Atlanta on cases related to toxic issues; for example, schools in Charleston, South Carolina, and the International Marine Terminal in Portland, Maine. These agencies don't have enough time, money, or staff to educate the public properly.

Compliance with governmental regulations put forth by the EPA and OSHA is mostly voluntary. They usually do take the middle of the road in *new* toxic issues, for various reasons. They don't want to scare the public unnecessarily or get sued by chemical manufacturers, etc. By the time an agency comes forth with information for the public on the health dangers of a new chemical or mycotoxin, etc., they may have known about the dangers for many years but they could not prove it with epidemiological data. For example, in my 41 years of studying toxic materials, I have seen the hazards of pesticides documented in the book *Silent Spring*,[19] and years later the USDA, followed by the EPA, passed new regulations on diisocyanates in urethane paint requiring supplied air for auto painters. Or high doses of Tylenol with alcohol causing liver damage resulting in a warning on the label. I have been involved in lawsuits in all of the above areas of toxicology.

---

19   Rachel Carson, *Silent Spring*, Houghton Mifflin Company, 1st edition 1962.

Regulatory agencies depend to a great extent on the manufacturer to inform them of health problems with their products in the field. For example, Dursban had possible adverse effects in humans that resulted in a voluntary withdrawal from the market. Remember, no pesticide has ever been banned by the EPA. They can stop manufacturing, and after most of the pesticide has been sold, they will then stop the sales of that pesticide except for maybe the least hazardous use; for example, under a slab of concrete for termite control. Therefore, people must do their own research with search tools like Google and PubMed databases.[20]

When I taught this subject as a professor at the University of Florida and at the University of North Florida as an adjunct professor, I trained claims adjusters and managers. I trained 1,800 of them last year through speeches I gave around the country and as a professor training industrial hygienists and safety officers. That's the way I got the word out. But, my goodness, we're talking only a few thousand people when, really, there are a few million that need to know. Every time the press comes out with a story about someone being poisoned, they stand a chance of being ridiculed for fanning the flames.

LB: Maybe I've missed it, but I haven't seen any in-depth media coverage on the health effects of mold or any interviews with professionals, such as you, who could put these mold-related issues into perspective for the masses of people who watch TV.

Dr. L: Well, I've been interviewed by *CNBC Live, Good Morning America, CBS Evening News*, and a lot of magazines and newspapers. But again, that's one story on the bottom of page 19 or whatever—I'm just saying that as an example, but you know what I mean. Remember how long it took for *Silent Spring* to scare everybody when it came to pesticides—10 to 15 years, and children kept dying. Eagles kept dying, sitting on eggshells that were so thin that just their body weight squashed the eggs. It took a long time. When it came time for studying oil-based paints and asthma from isocyanates, it took federal agencies 20 years or more to even get the manufacturer to put a simple word of warning on the label. This is a slow process. We are now in the process of trying to educate people.

LB: How many years do you think we are into this education process?

Dr. L: We are probably in the preliminary stages because 10 or 15 years ago, when mold was called mildew, everybody thought that a little mold couldn't hurt you; just wipe it off with bleach. Now we know better. The first to come out with warnings was the *New York City Department of Health Guidelines* for mold

---

20   An Internet search shows that Dursban is still available in various forms. Applications include use on food and in agricultural fields. See article by Jon R. Luoma, "The Ban that Wasn't," September/October 2000 issue, *Mother Jones*. Available at http://www.motherjones.com/news/outfront/2000/09/dursban.html (accessed August 2006).

remediation. In the old days, you just ripped it out and in a few days you couldn't understand why you were so sick. Now, as I told you, only 20 percent of the molds are harmful in the first place.

LB: What molds cultured out in the samples from St. Bernard Parish?

Dr. L: The most toxic of all was *Stachybotrys*. It is ten times more toxic than *Penicillium*. Generally, *Penicillium* is ten times more toxic than *Cladosporium*. It all depends on the species, the strain, the substrate it's feeding on, and the population. More importantly, it all depends on you. How susceptible are you?

LB: Tell us about the samples you cultured from New Orleans.

Dr. L: I don't culture samples. I send them off to P & K, the top lab in the country, and they culture, identify, and count the spores. There were highly pathogenic molds in the millions per gram of dust—the highest populations I've ever seen. Five years ago, I saw *Stachybotrys* at 2 million spores per gram of dust, and I said that was the highest population I'd ever seen. Then I saw it at 10 million, and then I saw the Katrina results. I just heard from Dr. Chin Yang that he found a billion *Stachybotrys* spores generated by one square foot of mold-contaminated wallboard. Now, that's the highest I've ever heard.

LB: You were in New Orleans several months after the hurricane. What do you think the conditions are in New Orleans now, two months after you collected samples?

Dr. L: They haven't changed. I mean, how can they change until they bulldoze 90 percent of those homes or burn them? I recommended that they bulldoze the homes in one-square-block areas toward the center of each block and set them on fire. If it catches a house afire across the street, it's no big deal simply because that house is going to have to be bulldozed and burned anyway. There are very few homes in St. Bernard Parish that can be salvaged. Remember, that's where the eye of the storm went through, not through downtown New Orleans, which was primarily saved.

LB: Have you seen the test results from the NRDC's air samples?

Dr. L: I saw them about six months ago.

LB: How did the results from your mold samples compare with the results from the NRDC's air samples?

Dr. L: I never compared the two. I'm sure they were both very high.

LB: The NRDC utilized a nonviable method that measured spores/cm³.

Dr. L: I did swab samples because in unoccupied homes it's unethical to do air samples. In other words, measuring the number of spores/cm³ of air in an unoccupied home, unless you agitate the air to mimic a family living there, won't accurately reflect air quality. The federal agencies, NIOSH and the EPA, recommend a 20-inch box fan—in one case I saw even a leaf blower—to get the spores into the air to mimic a heating and cooling system running, children running, and doors opening and closing because if it's an unoccupied home, the spores are going to land on the floor, ceiling, and walls and stay there until they are disturbed.

LB: The NRDC report stated its spore counts might be underestimated because air samples were taken with no disturbance of settled dust.

Dr. L: A lot of insurance defense experts go in and leave their air scrubbers running for a week to take all the spores out of the air and then go in there quickly and do three, four, or maybe five air samples of an unoccupied home without disturbing the spores. Then they tell the residents that the home is safe, so the family goes back in and gets sick again. The insurance company will walk away saying, "We have data that says it was safe." Well, no, they found a fly-by-night company that didn't know what they were doing. I don't think they are committing fraud; it's just that they're probably ignorant of what they should have done and put the lives of the family in jeopardy. I remember the Ed McMann lawsuit where I was asked to come out twice and do sampling. It turned out that the real guilty party was the remediation company, and it was sued. I think Ed McMann got $7 million in that case.

There are a lot of unqualified so-called experts out there who a few years ago were studying asbestos, and now, since that's not as big a problem, they're moving into mold remediation; or they studied pharmaceuticals, and now they're moving into the field of mold as so-called experts. In 35 years of testifying in court, nationwide, 80 to 90 times a year, I don't think I have ever faced a mold toxicologist in a mold case, other than a pesticide toxicologist in a pesticide case, who was a qualified expert. In other words, I've always faced, in the courtroom, people who weren't real experts in my field.

LB: Going back to the NRDC report, those test results reflected 1 to 2 percent of *Stachybotrys* with the remaining total spore counts being *Penicillium/Aspergillus* and *Cladosporium*. Your report states that *Stachybotrys* was heavily present. Why didn't the NRDC's testing reveal a higher percentage of *Stachybotrys*?

Dr. L: I don't know. I'm not going to criticize them. It's the way they sampled more than anything else. They take air samples. *Stachybotrys* is a heavy, wet spore that often does not get into the air unless it's disturbed. It's not like some of the other pathogenic molds and bacteria that physically throw their spores into the air to ensure that they get somewhere to find moisture and nutrients so they too can grow. *Stachybotrys* is not like that. *Stachybotrys* will produce a mycotoxin, like trichothecene, to physically kill the other molds around it. That's the way Fleming[21] discovered penicillin—the mycotoxin from *Penicillium*—because it killed the other molds and bacteria around it. That's how *Stachybotrys* becomes dominant. If a surface stays wet for an extended period of time, it will grow very well, and then if it's disturbed, the spores will get into the air. That may be why NRDC didn't find much *Stachybotrys* in the air.

LB: Where did you take the swab samples from in the buildings you sampled?

Dr. L: The homes were all empty and ready to be destroyed. I just took the swabs off the walls.

LB: You stated that New Orleans is probably in much the same condition in terms of toxicity of air as it was when you collected samples there. Yet, we see news footage of people who fall into the high-risk group—children, elderly, and pregnant women—living in New Orleans and surrounding areas. Should these high-risk people be there?

Dr. L: That's too general a question. If a person's immune system is such that it can be severely affected by low levels of pathogenic mold and bacteria, then no; I'd find a safer place to live, but that's just common sense. I'm not going to tell people that they shouldn't live in New Orleans.

LB: But while the cleaning, rebuilding, and reconstruction process is going on, there is an increased risk for people in the high-risk groups.

Dr. L: If you're disturbing homes that are heavily contaminated, you shouldn't have anyone around there that's not fully protected with protective equipment, much less sick people and elderly people that don't have any protective equipment.

LB: Do you have any opinion as to the long-term health effects that people are going to see from exposure to these very high levels of mold?

Dr. L: You can have chronic effects from mold exposure. How long depends on you and how severely you were damaged, but that's more of a question for a medical doctor.

LB: What is the most important piece of information you would impart to readers?

Dr. L: When you're in a moldy environment and you smell it and see it, you've got an increased risk of having symptoms related to that exposure. If it's your house, you should probably hire a professional to come in and do it right, to remediate

21   Alexander Fleming accidentally discovered penicillin in 1928 and published his findings in 1929. This information is available at http://www.pbs.org/wgbh/aso/databank/entries/dm28pe/html (accessed April 2006).

properly, as opposed to a lick and a promise, such as a bucket of bleach and a gallon of water to wipe down the wall, not realizing that it's a pipe in the wall that's going to continue to leak and continue to give you upper respiratory problems. I would follow the NYC and the EPA guidelines for remediation and have it done by a qualified, licensed, insured expert who doesn't get paid until it gets done safely, properly, and final clearance testing is done by an unbiased, independent expert, not somebody related to the company doing the cleaning. My pet peeve in this field is you've got companies out there that do their own inspections, then recommend remediation, do their own remediation, do their own lab analysis, and then do their own final clearance testing. That is unethical and clearly a conflict of interest.

LB: That certainly is a good point. Thank you.

Cynthia Coulter Mulvihill, Attorney at Law, is the managing partner in the Law Offices of Cynthia Mulvihill, located in Monrovia, California, which is a suburb of Los Angeles. She specializes in mold-related legal cases and provides consultation services to contractors, insurance companies, experts, potential claimants, and litigation firms handling both plaintiff and defense cases.

Ms. Mulvihill received a bachelor's degree awarded with distinction from Chaminade University of Honolulu, Hawaii, and a JD from Western State University College of Law in Fullerton, California. She served in the US Army for four years and in the Army Reserve for five years. After working for an Orange County environmental and geotechnical engineering firm during law school, Ms. Mulvihill began her career as a construction-defect defense attorney in 1994, working for a firm in Universal City, California. She worked for Reliance Insurance Company in Glendale and then formed her own practice in 2000.

Ms. Mulvihill has had a life-long fascination with mycology. From 2004 to 2007, she published Monday Morning Mold, *an electronic digest of news reports and studies of interest to scientists, researchers, and other professionals interested in mycological issues. Ms. Mulvihill can be contacted via her website at www.cmsynergy.com. The interview was conducted April 2006.*

LB: When people realize they've been exposed to mold and that they're sick from the mold exposure, when should they seek the advice of an attorney, and what are possible avenues that may keep them out of court since court should be reserved as a last course of action?

CM: Those are really good questions. I get calls from people who have discovered that they have mold and are wondering if they possibly have a lawsuit and what should they do about it, all the way to people working on appellate cases wondering about the scientific issues associated with mold. When someone calls me and says, "Hey, I have a problem with mold. What should I do about it?", the first thing I tell people is if they think they are being made sick from mold, while it is nice to talk to me, if they want to consider a lawsuit, they need to get to a doctor. Never delay treatment. You want to get medical help and get a diagnosis.

LB: When people call you, have most of them already seen a doctor?

CM: It ranges from suspecting a problem to having a diagnosis. What I frequently run into—and I run into this everywhere, from people just making an initial call to folks working on appellate stuff—is people who realize that they have mold, realize that they are sick but have gone from "I have mold." to "I am sick. Therefore, the mold made me sick." Legally, if you are a plaintiff, you have to prove it is more likely than not that mold is making you sick.

LB: Do you get calls from homeowners who have purchased homes in which mold has developed due to tight building construction?

CM: I get calls from homeowners about mold problems. I don't know that I could say that it is a result of problems caused by tight buildings; more often I see problems with construction causing leaks rather than having the problems caused by a lack of adequate ventilation.

LB: These people call regarding suing the contractors then?

CM: Yes.

LB: Are the terms *sick building* and *tight building* used in a legal sense interchangeably, or are there distinctions in how these terms are used in court?

CM: In California, you might call something, as a catchall, a sick building or something like that, but as a legal term, there is no statute or common law cause of action for sick building syndrome. You would, for example, if you were a tenant, bring a lawsuit for breach of the implied warranty of habitability. If you were a homeowner, you would be bringing an action for a construction defect or defects. If you are someone who is working and get sick on the job, you probably would end up with a workers' compensation claim.

LB: Is there more recourse for employees who become sick from mold in the work environment versus people who become sick from mold in the home environment?

CM: In a workers' compensation claim in California, employers are strictly liable when an employee is injured on the job. It doesn't matter what causes the injury. For example, I, as an employer, carry workers' compensation for my employees. If I have an employee who trips going out the door, it doesn't matter if that employee trips because he was not looking where he was going, because I left something lying on the floor that caused him to trip, or because my landlord built an unsafe threshold that caused him to trip. It doesn't matter; workers' comp is going to cover it. That doesn't mean, for example, if my landlord has caused a construction problem, that my employee might not have a cause of action against the landlord. When a third party—not the employer, the employee, or other employees—is responsible for an injury, there might be liability for the third party. Otherwise, if it's just between the employer and employee, the "sole and exclusive" remedy is workers' compensation insurance.

LB: California has some statutes regarding mold that you have posted on your website. Is California more progressive than other states, or are all states creating statutes to address the mold issue?

CM: With respect to California, I think it's progressive in that the statutes mention the term, but there are no enforceable standards. You can't take a California statute or ordinance and say, "Okay, if there's X cubic feet of *Aspergillus* airborne, this is a problem." It's progressive in that they've recognized that there's a potential problem, but after studies on it, they haven't enacted anything. I believe Texas is probably as proactive, if not more so, as a result of the *Ballard* case—Melinda Ballard in Dripping Springs, Texas. That was a very large verdict she received relating to the failure of her insurance company to investigate adequately a mold claim. I believe the initial judgment was $32 million, and it was later reduced to $4 million. That's an interesting case because it appears to be a judgment on mold and mold issues, when in actuality, the law on the case had to do with insurance bad faith and an insurer's failure to investigate a first-party claim.

LB: Can you explain more from a legal point of view what happened in the *Ballard* case?

CM: She purchased a large, expensive house for her and her family, and it had construction defects. She made a claim to her homeowner's insurer, which was Farmers, and it was a first-party claim. Farmers failed to conduct an adequate investigation or take actions that would have prevented the mold from spreading. So the claim definitely arose from mold, but the judgment was on Farmers' failure to investigate the claim and to retain contractors for the Ballards that would take measures needed to take to stop the mold from spreading.

LB: And this was the mold *Stachybotrys*, correct?

CM: Yes, *Stachybotrys*. That's another interesting situation I run into. *Stachybotrys* is the black mold that came to the attention of the general public in 1994 when Dr. Dorr Dearborn, a doctor in Cleveland, Ohio, published a study showing a possible link between *Stachybotrys* and pulmonary hemorrhagic incidents in infants. In other words, bleeding lungs.

When people say "black mold," they are usually referring to *Stachybotrys*. When I get a call from someone who says, "Oh, I have found black mold," I always have to remind people that you cannot look at mold and say what specific genus it is, much less the species. If you look at mold and say, "This mold is black," you might be able to rule out certain species. For example, most species of *Penicillium* are green, so you might be able to look at mold and say, "Okay, this is black, so it's probably not *Penicillium*," but it could be another kind of mold that also has melanin that makes it black. You actually have to test to determine the genus and species.

Stachybotrys is exciting because of the studies about infants and also because *Stachybotrys* has been successfully cultured and grown to produce mycotoxins that are used in biological warfare. People get excited because they have *black mold*, and people think that all black mold is toxic, but when I talk to folks I say, "You have got to have it tested—just because it's that color doesn't mean it's necessarily *Stachybotrys*."

LB: Walk us through, if you don't mind, an initial call from a potential client with mold concerns.

CM: I think I have gotten only one or two calls where people have found mold and called me right away. Usually, I get the case at some later stage of the investigation. But, if someone were to call me up and say, "I have this problem," the first thing I would do is talk about the way that molds possibly could have come to grow in their home. Generally, I would talk about mold needing something to grow on (a food source), oxygen, and water. Basically, you have to have all three of those things, but some molds need less water or oxygen than others. Generally, people say, "Oh, I think this is being caused by a leak I have," maybe a leaky pipe, or what have you. People will usually have some idea of what is causing the water or the moisture source to get there.

Then, I would talk to them about the three basic ways that molds can make somebody sick. The first way, of course, is as an allergen. I explain to them mold is an allergen and, like any other allergen, may affect a certain percentage of the population. A certain percentage of that population is going to have allergies to certain species of molds or sometimes to an entire genus.

Generally, I give folks a five- to eight-minute kind of speech that I am essentially repeating to you right now. First, there are the allergies, and that is a common way that mold is going to cause problems for someone. I tell them that repeated exposure to allergens can sensitize someone, in other words make them more allergic. Second, I talk to them about mold colonization in the body, which is not very common at all, generally, in healthy people.

LB: Really?

CM: Yes. There are species that can colonize in the body. Generally, in colonization, someone is immunocompromised for some reason. Diagnosis of significant colonization can be seen on x-rays; also, if it's colonizing in the lungs, it can be diagnosed by lavage.

The third way that folks can get sick from mold is if the mold growing in the environment is capable of producing mycotoxins (if it's in its mycotoxin-producing stage) and if the individual ingests those mycotoxins. That's where things start to get very interesting. For example, as an attorney, when I am looking at molds, I actually get more excited about *Aspergillus* than *Stachybotrys*. The reason is that *Aspergillus*, I believe, has 168 separately identified species. *Aspergillus* can be an allergen, it can colonize in the body, and it can produce mycotoxins—on one of which, aflatoxins, there have been extensive studies. *Stachybotrys* does not colonize in the body. So, as an attorney, if I am looking for a problem mold, I am going to look first for *Aspergillus*.

Fusarium is also is a concern. I'm not saying *Stachybotrys* is not a concern, but as far as evidence of proof of having a problem, I am going to have an easier time proving or disproving a problem with *Aspergillus*. About half the work I do is defense work and half is plaintiff, or claimant, work.

LB: Which means that 50 percent of the time you are defending an insurance company?

CM: No, I haven't ever defended an insurance company itself for mold claims—actually, an insurance company would be defended for bad faith arising from its handling of mold claims. When I am employed by an insurance company, I defend their insureds, usually contractors. The insurance company pays my bills, but I am reporting to the insured individual or company and not to the insurance company.

LB: And with this type of work, you are not actually litigating the whole case, but acting as an expert consultant?

CM: Exactly.

LB: If people don't have their legal counsel selected when they call you, do you then refer them to a litigation attorney for whom you consult?

CM: Yes, and what I try to do is talk with them long enough so I can determine what kind of case they have. A lot of times what happens, quite frankly, is that people talk to me too late. For example, they'll call me after they have moved. They may be in a rental unit, have a problem and move out, which in a lot of cases is completely appropriate. In the meantime, the landlord completely cleans things up, destroying all the evidence.

LB: Are there any prelegal avenues of mold testing for people, maybe by state or federal agencies?

CM: As far as the actual physical testing of it, rent control boards and HUD. Obviously, if it's a work-related issue, they may want to go to OSHA or, in California, CAL-OSHA. I have seen cases, actually a lot of them, in which OSHA and CAL-OSHA have conducted indoor air quality testing. Sometimes, city, county, or state departments of health may look into the problem.

LB: OSHA has no indoor air regulations regarding mold limits, correct?

CM: There are none that I know of.

LB: Are there any air quality limitations set by any other agencies regarding mold?

CM: To the best of my knowledge, other than the ones referenced in the EPA guidelines for mold remediation, no.

LB: When you go into court, you don't have any established mold limits to use as guidelines when presenting a case to prove that certain levels of mold have made someone sick. Doesn't that make proving a case for a plaintiff more difficult?

CM: If I had a statute that said a certain amount of cfu's is too high, I could plead what's called *negligence per se*, which means that there is a statute and they are violating the statute—that's all I have to prove. I don't have to say how it happened or why, so it's easier if I have standards, but I don't. If I have to plead too many cfu's as a cause of action, I have to plead it as *negligence*. In other words, there was a duty, for example, by a landlord to have a habitable unit, and the landlord breached the duty, let's say, because a window was leaking, mold grew, that mold was the actual cause of the illness, and it damaged someone. It is a harder element of proof, but it can be proven if the facts are there. Keep in mind that the burden of proof at trial is usually *more likely than not*—in other words, a 51 percent chance of negligence. It's not *beyond a reasonable doubt*, and it's not the higher standard of *clear and convincing evidence*, which is generally between 70 percent and 80 percent.

LB: So many of the policyholders in the hurricane-hit areas aren't getting their claims paid by their insurance companies, so they don't have money with which to remediate and/or rebuild.

CM: Right. What happened, of course, is after the *Ballard* case, insurance companies started to write mold exclusions. Whether or not these exclusions are effective is a whole other issue. Some of what I do is consult on policy exclusions relating to molds. For example, I am in the middle of a dispute relating to a policy that had a toxic substances exclusion that supposedly covered mold. We are in a discussion because the insurance company is saying the cause of the loss in this particular case—it involves a single residence—was mold. In insurance litigation, there is a term called *efficient proximate cause*. Efficient proximate cause in this particular case was not mold, but the release of water. I have the insurance company trying to argue that the toxic substances exclusion excludes mold. To accept that argument, you would have to accept that the water itself was toxic, not the mold. It's an interesting dispute. But I don't know if I am going to be resolving this dispute in court. It turns out, in this particular case, another carrier is going to provide coverage. In the Louisiana cases, then, what was the official proximate cause of the loss—the rise of the water or the mold? Depending on how the policy exclusions are written, I think that some of the coverage decisions are going to be incorrect.

LB: Are there any insurance companies that offer their insureds a broader range of coverage regarding mold-related losses, or are they all cracking down?

CM: I see the bigger carriers drafting mold exclusions as tight as they possibly can, or if they do have mold coverage, it may be really broad—Allstate comes to mind, and perhaps State Farm—but coverage is limited to $5,000.

LB: Wow, that's not much.

CM: And again, it depends on the efficient approximate cause of the loss. With the mold exclusions, what the insurers are trying to do, of course, is protect themselves from losses. You will find older policies without mold exclusions or with mold exclusions that are not as effective as the newer policies. When a homeowner makes a claim under a homeowners' policy, usually only the current policy (and perhaps policies from a year or two before the loss) is going to provide coverage. So, if someone makes a claim today for a mold problem, he or she is likely to have either very limited coverage or no coverage at all.

It's different for a CGL [Commercial General Liability] policy that would insure a developer or contractor who builds a project. A current CGL policy may not have mold coverage or may have very little mold coverage. However, if the project was built much earlier—in California, the Statute of Limitations and Repose for a Construction Defect Action is ten years—there may be coverage available under earlier policies that were written because the carriers were not writing policies with mold exclusions.

LB: When cases go to trial, do they tend to settle pretrial?

CM: Oh sure.

LB: You were saying that some cases can go to court and other cases can't. Can you give us an example of a case that could go to court and one that couldn't?

CM: The first big thing that I run into is the statute of limitations. In California, the statute of limitations for personal injury is two years. It's two years from the date of discovery. What happens is someone will say, "Hey, I've lived in my apartment for ten years, and I've known for the last eight years that this mold has been making me sick."

LB: And since they didn't act within the first two years from when they realized that they had mold, the statute of limitations has expired?

CM: Right, but not from when they realized that they had mold, but rather from the time that they realized the mold itself was making them sick. It's like, "I've lived in this apartment for ten years. I always had this mold problem, and eight years ago I read about mold being a problem. I thought then [eight years ago] it was making me sick." The problem you have there is if you knew eight years ago, or six years ago, or whenever you knew, that a particular problem was the cause of a personal injury, you needed to bring it in within the statute of limitations.

LB: Okay, I follow.

CM: The next major problem that I run into with people with these problems is that they will have moved out, or whatever happens, without preserving the evidence. I tell people that if they are going to litigate, they're going to have to get a test that is going to identify something sufficiently. The other thing is medical tests; they need to have the medical tests show that they have a problem with mold. That would be, for example, an allergy test, a skin test. In the case of possible colonization, a lavage, an x-ray, or some other appropriate testing must be performed to see if there is fungal growth in the body.

LB: What exactly is a lavage?

CM: It's a throat wash. They basically wash the throat and culture it to see if you've got mold in your throat—or for that matter, it's used to test for bacterial infections as well.

LB: Are there any other types of medical tests to prove mold exposure and mold-related illnesses?

CM: Some blood tests are going to test for both repeated exposure as an allergy and exposure to mycotoxins.

LB: And all of these medical test results hold up in court?

CM: If a reputable individual does them and if the chain of custody is appropriately kept, they might. Keep in mind, though, there was a recent California Appellate Court decision (*Geffcken v. D'Andrea (2006) 137 Cal.App.4th 1298*) that held that certain serology [blood] tests for mold exposure were inadmissible. On top of the technical issues, you've also got the issues that you would have with the inadmissibility of any other kind of evidence.

LB: Can you explain the legal importance of an appropriately kept chain of custody?

CM: For example, suppose that you are going to have a blood test done. You have to have the blood drawn, and it is going to have to be sent to a lab. The person who is going to draw the blood is going to initial off, and the person who tests at the lab is going to initial off. You have to know that this is the blood of the person that was said to have been tested because if you're on the plaintiff's side, the last thing you want to have happen is if you are sitting there with a good case with someone who shows that they did, for example, have exposure to mycotoxins in their blood, but be unable to say that it was actually that person's blood that was tested. You've got not only the technical issues, but you also have to dot your i's and cross your t's on the legal evidentiary issues.

LB: What about issues of admissibility?

CM: Are you familiar with the *Daubert* test?

LB: No.

CM: *Daubert*[22] is an extremely important Supreme Court case on the admissibility of expert witness evidence. I am going to give you the website that I usually refer people to—it's www.daubertontheweb.com. If you go to "Substance" and then to "Chapter 2, *Daubert* in a Nutshell," it talks about the Federal Rules of Evidence— what's admissible and what's not admissible. It's the best explanation I've seen of this particular test.

LB: So it is a legal test.

CM: Yes, it is—in addition to the *Kelly-Frye*[23] test, which is still used in many jurisdictions and is a test for admissibility of expert witness testimony. *Daubert* is a U.S. Supreme Court decision, and it involves state court laws. Obviously, most of the time the cases that you have on mold are going to be tried in state courts. Occasionally, you could end up with a federal court on cases, but it is usually state court. However, some U.S. Supreme Court decisions are going to affect many courts—one of them is the *Daubert* test on the admissibility of evidence; another is called *Campbell*. *Campbell*[24] is a test for punitive damages, which are almost never an issue in mold cases, although punitives were an issue the *Ballard* case.

LB: That's because negligence was found?

CM: No, in *Ballard*, it was insurance bad faith, and in insurance bad faith, there is punitive damage, at least if it's brought as a state court action.

LB: Okay.

---

22   Daubert v. Merrell Dow Pharmaceuticals (1993), 509 U.S. 579.

23   People v. Kelly (1972) 17 Cal.3d 24; Frye v. United States (D.C.Cir.1923) 293 F.

24   State Farm v. Campbell (2003) 538 U.S. 408.

CM: Here, essentially, is a checklist for the admissibility of evidence under *Daubert*:

"Whether the theories and techniques employed by the scientific expert have been tested; whether they have been subjected to peer review and publication; whether the techniques employed by the experts have a known error rate; whether they are subject to standards governing their application; and whether the series and techniques employed by the experts enjoy widespread acceptance."

So, if you are going to have admissible evidence, that's what you're looking for. The first thing I look for is that not only have the theories been tested in accordance with scientific procedures, but also whether they've been peer reviewed and published. If something is peer reviewed and published, the testing procedures have usually been verified, and that's why I try to base what I can on peer-reviewed publications.

Kelly-Frye is a different test, and it is still used in California. That test is

"The proponent of evidence derived from a new scientific methodology must satisfy three prongs, by showing, first, that the reliability of the new technique has gained general acceptance in the relevant scientific community, second, that the expert testifying to that effect is qualified to do so, and, third, that correct scientific procedures were used in the particular case."

On appeal, the *general acceptance* finding under prong one of *Kelly* is a mixed question of law and fact subject to limited *de novo* review [*de novo* is a Latin term for *anew*, which means "starting over"]. The appellate court reviews the trial court's determination with deference to any and all supportable findings of historical fact or credibility and then decides as a matter of law, based on those assumptions, whether there has been general acceptance.

LB: So cutting-edge treatment protocols or new scientific testing methods could be ruled inadmissible because they do not "enjoy widespread acceptance" yet, simply because they are too new?

CM: Exactly. There are a lot of theories of what does and doesn't work.

LB: Some expert medical testimony is ruled inadmissible in court by judges based on *Daubert*, *Kelly-Frye*, and other admissibility tests, because the scientific and medical theories supporting the testimony are not viewed as generally accepted in scientific and medical communities. What do you recommend that a plaintiff's attorney do to improve the likelihood that the client's medical condition will be assessed in a manner that will be ruled admissible by a judge, especially if it is a medical condition that is still being debated in scientific and medical communities?

CM: This is a complex question. An attorney for either the plaintiff(s) or the defendant(s) needs to break down the issues. First, is the illness a new, novel illness, the very existence of which has not been established? Second, is the illness a recognized illness, but the cause for the illness has not been established? Or, are you looking at a new, previously unknown illness, the cause of which has not been established?

Those of us in the legal profession need to remember that our job is not to prove or disprove the existence of an illness or to prove or disprove the cause of an illness. Our job is to convince the trier of fact—whether a judge or a jury—that it is more likely than not that a person does—or does not—have an illness and whether it is more likely than not that the cause of the illness was—or was not—exposure to mold.

It's hard to make general recommendations, but an attorney must remember that burden of proof and what test is going to be applicable for expert testimony. It's also important to remember that in cases involving mold, an attorney is almost always going to need more than one expert—sometimes many experts. For example, a medical expert isn't going to be the right expert to opine that mold exists in a certain environment. That would be an industrial hygienist. A medical expert would give an opinion about the medical effects of mold on a person.

LB: As an attorney, is it advisable to hire a mold expert, whose prior testimony in other mold cases was ruled admissible, to review your client's medical records and testify in court—even if it isn't the exact medical condition addressed in the prior testimony?

CM: If I have a choice of experts, I'm going to go with an expert whose testimony has been ruled admissible by other courts—especially if those court decisions are published. I would especially look for an expert whose testimony has been admitted by a court in the state that I practice in.

LB: Does a history of prior admissible court testimony increase the likelihood that an expert's testimony regarding a plaintiff's medical condition would be ruled admissible?

CM: It may or may not. I would want to know what the previous testimony was and if the plaintiff's condition is similar to that of the previous person(s) that the doctor testified about. I wouldn't, for example, want an expert who testified in one case that a specific mold species definitely caused a certain illness under certain circumstances—and then have the expert testify that the same specific mold species did not cause the same illness as in the previous case, under the same circumstances as the other case. In other words, if the expert is making a 180-degree turn, there are admissibility—and credibility—problems.

LB: When you are a consultant for a plaintiff's team and you are trying to attribute a medical condition to mold exposure, how do you prove the medical condition is a result of the mold exposure and not other contaminants? Often there is a combination of things going on, such as off-gassing of formaldehyde and VOCs.

CM: You always need to consider other potential contaminants or conditions. If I'm investigating a problem I think might be caused by a particular building, I will try to determine the age of the building. Obviously, if it's an older building, one of the first things I'm going to look at is asbestos. Believe it or not, even though I am in California, I'm in a town with many, many older homes, and asbestos is an issue. They used to make floor tiles out of it, and if those are improperly removed, that is a problem. If it's newer, I will want to consider off-gassing. I will also try to rule out other sources for allergens. For example, I'll want to have an allergist test for allergies to mold, but also for other allergen sources. The last thing I want to do is walk into court and try to prove that someone is allergic to mold in their new house, when in fact, they got a new cat at the same time they moved in, and they are really allergic to only the cat.

LB: But legally, the case can still go forward; you just bring in the medical evidence with formaldehyde or whatever chemical agent is causing the trouble instead of the mold, and you proceed?

CM: Obviously, if I'm defense and their only claim is mold, I'm going to try to say, "No, it's something else," if that's the case. If I'm the plaintiff, I am going to include in the complaint anything that might reasonably have caused a problem, so I'm not precluded from introducing that as evidence.

LB: This has been very helpful and enlightening. Is there any other area that I may not be familiar enough with the legal aspect of mold-related cases to ask?

CM: No. The one thing I wish I could tell people when they think they might have a problem is, if they are really anticipating that they are going to need to put in a claim, institute litigation, or whatever they are going to do, that they talk to an attorney before they do something that allows the environment to be altered. I'm saying, if you suddenly find this mold, document it; get someone in to photograph it and test it, preferably someone your attorney recommends. If it's going to go to trial, the evidence has to be admissible.

LB: Thank you. This has been great information. I appreciate your time.

Melinda Ballard, MBA, (4/21/58-7/2/13) was founder and President of Policyholders of America (POA), a nonprofit organization designed to help policyholders receive all the benefits to which they were entitled under a variety of insurance policies. Ms. Ballard formed POA in 2001 after having fought an insurance giant over what should have been a simple claim filed because of leaks in her home. A dispute ensued over the handling of the claim, during which time mold grew in epic proportions in Ballard's large home just outside of Austin, Texas.

Ultimately, Ms. Ballard had to litigate the matter, and the case captured the attention of media worldwide including nearly all major media outlets such as USA Weekend, New York Times, New York Times Magazine, Vanity Fair, Wall Street Journal, *CBS News, 48 Hours, ABC News, 20/20, The Early Show, Court TV, and too many regional and local newspapers and TV shows to list. Ms. Ballard forewent any salary from POA and actually funded the organization herself. Membership grew to approximately 3 million members.*

Ms. Ballard earned an MBA from New York University and completed post-graduate work at Columbia University, during and after which she enjoyed a successful career in advertising and public relations in New York. In 1983 Ms. Ballard formed her own international advertising and public relations agency, Ficom International Inc. In 1989 she sold Ficom and moved to Texas, purchasing the house that was the catalyst to her career as an insurance crusader. Up until her death, Ms. Ballard resided in South Carolina, published a monthly newsletter, The Policyholders Advocate, *and was director of Capital Bank, an Arkansas-based bank. The interview was conducted April 2006.*

LB: Your story is probably one of the most widely publicized mold cases, which inadvertently has brought increased awareness to molds, the possible health repercussions associated with mold exposure, and the implications of the fine print of most insurance policies. First, share a bit about your story for those who may not be familiar with it.

MB: There is nothing inadvertent about it. The sole purpose for me to endure all of the media scrutiny and intrusion was to educate the public about water damage and the consequences of mold. Unfortunately, my story is not unique. It's symbolic of a systemic problem that has been allowed to continue at insurance companies.

Most insurance companies—at least the big ones—practice delay tactics, and such tactics, when it comes to water damage, can actually cause toxic mold problems.

For example, in 1998 we had a leak; it was immediately fixed and then reported to the insurance company. The adjuster came out and realized that it was not just a minor damage report. It was fairly significant in that the entire first floor of my home had custom hardwood floors, and they were buckling. Instead of allowing us to pull up what was clearly covered under the policy, meaning the hardwood floors, the insurance company insisted that it not be done until they finished investigating the claim. Months passed. The leak in the pipe had been fixed immediately, but the water that had escaped during the leak was still trapped between the floor and the subfloor.

LB: Since you were living in Texas, where the humidity is very high, did that contribute to the problem?

MB: Actually, the humidity level doesn't have anything to do with the indoor environment. We have tons of POA members in Arizona and Nevada who have very little humidity in their environment and yet have tremendous mold problems in their homes. What matters is the *inside* environment. Waterlogged building materials cause mold to colonize and spread.

LB: Why didn't the floors dry out once the leak was fixed?

MB: Because it was so saturated with moisture—we're not talking a gallon of water; we're talking probably 500 gallons of water trapped in between a floor and a subfloor. The floor was literally mushy when you walked on it; you could hear a sound that was similar to walking on waterlogged leaves in the garden. When you stepped on the wood floor, it had that soggy feel. The insurance company acknowledged the claim was covered, yet they would not allow us to remove that wet floor and begin repairs because they wanted to argue, I guess, about price. We got bids from the contractors who actually installed the floor initially. The insurance company got bids. These bids were tremendously apart, and I'm not talking about a couple of dollars; I'm talking about $150,000 apart. The insurance company wanted to replace a custom, patterned hardwood floor with a cheap alternative and call it a day. Well, we wanted what we had, just as anybody would, which was a custom hardwood floor, and "like kind" and "quality replacement" is what the policy—a contract—specified. During their long investigation process, we were not allowed to remove the existing wet floor and subfloor. Not only did mold spread throughout the subfloor, but the sheetrock on the walls began to soak up water from the floor. Within months, sizable mold growth had climbed 5 to 6 feet up the walls.

LB: At this point, your family—consisting of your husband Ronnie, your son Reese, and you—were still living in the home?

MB: Oh yeah, absolutely.

LB: And, how old was your son at this point?

MB: Reese was two at the time, and that means he was spending a lot of time crawling around on the floor. We started having some really bad respiratory problems, and my son started bleeding from the ears and nose, coughing up blood, and this kind of thing. We couldn't really figure out what it was—he was in day care, and oftentimes when kids are in day care, they get sick. They give one another flus, colds, and whatever else. We really didn't have any clue that this would be because of the home. On April 1, which is April Fool's Day, I got on an airplane going to a bank board meeting in Arkansas, and I sat across from a guy who is affiliated with Texas Tech University. He saw me coughing up blood and asked me, "Lady, what in the world is wrong with you?" I said, "I just don't know. Everybody that comes into my house gets sick like this. There must be a really vicious flu going around." He asked, "Have you had water damage?" I said, "Oh, yeah." He said, "Have you seen any mold growth?" I said, "Everywhere." And he asked, "Have you ever heard of a mold called *Stachybotrys*?" Now, I couldn't even pronounce it, let alone spell it, having never heard of it.

LB: Right. I've been there myself!

MB: Yeah. So I, of course, said, "No." And he said, "Well, it's a toxic mold." At this point I'm rolling my eyes, thinking he's a freak.

LB: Because you were thinking that mold really couldn't hurt you—you didn't know?

MB: Of course! Why would I think any differently? There was never any publicity on this subject back in the 1998–1999 time frame.

LB: Many people down South, living in the midst of the mold infestation from Hurricane Katrina, still don't have any idea.

MB: I know, I know, and we've tried to educate them, but you can't educate somebody who refuses to be educated. Also, many hurricane victims are without TVs, radios, and newspapers because they are out of their homes, and many just do not have easy access to the Internet. It's very difficult to reach these potential mold victims because they are displaced.

Anyway, back to the fellow on the plane. He told me he was affiliated with Texas Tech University and was doing some work at the governor's mansion for then-Governor Bush of Texas, now President Bush, because the governor's mansion had this problem. He said, "I'm going to be in Austin. I'll come and test your home for free because I think I know what your problem is." Well, all he had to say was "free"; I was delighted to have him come by my house and test it for free. Had he said I'm going to charge $8,000 to do this, I would have said, "No, thank you," and we would have never known.

A few days later—April 5, 1999, I believe—he showed up at my doorstep with test tubes and other testing gizmos in hand, and again I'm rolling my eyes because I really did not believe that mold could cause this. Then four days later I'm getting a call from Dr. David Straus at Texas Tech University Health Sciences Center, who tells me that we've got this toxic mold that can cause all kinds of property damage, potential health problems, this, that, and the other thing, and that I need to think about leaving the home.

LB: David Straus was a colleague of the person you met on the airplane?

MB: Correct. And I'm thinking this is something out of "The X Files." The gravity of the situation really did not hit me. At the same time my child—if you're a parent, you understand that that's probably the most important thing in your life—is bleeding from the ears, nose, and coughing up blood. Of course, I am too, but I'm not worried about me; I'm worried about him. I couldn't help but take that seriously because of the symptoms that Reese was having and that Ronnie, my husband, was beginning to have. Ronnie was beginning to lose his memory—he couldn't remember things that would normally be second nature; for example, what car he drove. He would go to the grocery store and have to walk every row of the parking lot trying to find the car because he couldn't remember where he had parked.

Anyhow, there was an apartment attached to that house, and we moved into it, thinking we were out of harm's way. The insurance company came out, tested the house and that apartment, and found the same results that Texas Tech had found. The mold had even spread to the attached apartment where we had been staying. At that point, we decided to evacuate the home entirely. So we moved out, and, I'm saying this sarcastically, the fun began. If the insurance company would have let us make repairs, this wouldn't have happened! You know what I mean? If you remove the wet building materials immediately, you're not going to have a mold problem. If you let it sit, fester, and stay waterlogged, you're going to have not just a mold problem, but a serious, epic mold problem, like what you see in New Orleans today.

LB: What would have happened if you had removed the floorboards, made repairs with your own money, and then asked for reimbursement?

MB: The insurance company would have denied the claim. This is what the policy—which is a legal, binding contract—says. Under the contract (or policy), a policyholder has a duty to make temporary repairs like tarping a roof or fixing a broken pipe, but the policyholder *cannot* make permanent repairs like ripping up a floor if the insurance company will not authorize it or if the insurance company is still "investigating" the claim. The reason for this is that by removing the floor,

I would be removing or tampering with "evidence" that would be or could be used by the insurer to determine coverage or replacement value. If policyholders take it upon themselves to destroy such evidence, the insurance company can and will deny coverage. We asked—no begged—our insurer to allow us to do this, and they refused to authorize the tearing out of the waterlogged walls and floors.

LB: But isn't there usually also a clause in insurance policies that states that owners have a duty to mitigate, which means to make repairs to limit further loss?

MB: Yes, there is the duty to mitigate damage. That is what I mean by "temporary repairs." If you have a leak, you should fix the leak, turn off the water, or put a tarp over your roof to stop further leaking. But if you remove your entire roof, floor, or both to stop the damage, that is going above and beyond mitigation and can be construed as destroying evidence if the insurer is still investigating the claim. Another duty spelled out in the contract (or policy) is that the policyholder must cooperate with the insurer. If not, you run the risk of losing coverage. Basically, when it comes to water damage, these policyholder duties are in conflict and are actually mutually exclusive.

Let me give you another example: Let's say you're in a car accident, and your car is totaled, but you need your car for work. You will lose your job if you cannot get to work, and your car is the only mode of transportation available. You can't wait around for your stupid insurance adjuster to give you a check, so you get your car repaired so that you can drive it. Finally, the insurance adjuster gets around to inspecting your totaled car, and he or she sees no damage because you already got it fixed. Per the policy, the adjuster can, and usually will, reject the claim, even if you produce receipts and photos, because they have not been given a chance to investigate the damage.

LB: I follow what you are saying. That is amazing.

MB: Well, it is amazing because on one hand, you have a duty to mitigate, and on the other hand, you have the duty to not tamper with evidence.

LB: Right, it contradicts itself.

MB: Of course, it contradicts itself in terms of anything that is a growing problem—like water damage. You see, the duties of a policyholder, as spelled out in the contract, are mutually exclusive: if you make repairs quickly (meaning, you are really mitigating damage), you are tampering with evidence and you lose coverage.

LB: I spoke with an industrial hygienist the other day, and he said that a lot of his cases arise because water leaks are often only partially dealt with. For example, the broken pipe, roof leak, or whatever caused the water source in the first place will get fixed, but then property owners do not take the next step to correct the water-damaged portions of the materials, which then leads to mold growth and ensuing health problems.

MB: Absolutely! Remember, insurance companies today have one goal: reduce claims payouts. It's in the best interest of the insurance company to *not* make complete repairs, as that would be more costly.

LB: After hearing your story, I have to wonder if these partial repairs are a result of lack of action on the part of the insurance companies that insure the buildings, and if these property owners aren't running into the same type of scenario that you did.

MB: Well, of course they have and will continue to experience the exact same problems! An insurance company is going to try its best to hold onto as much of its money as long as possible.

LB: Tell us what the health repercussions were for your family, and give us a rough timeline of when symptoms started to occur from the time of the leak.

MB: It was a couple of months before any health effects started taking hold; I'm going to say two, maybe three, months after the initial water leak. The ill health effects started with headaches and coughing. Then they progressed into a loss of muscle strength, dry mouth, hair loss, the coughing up of blood, blood coming through the nose and ears, and then, finally, the short-term type of memory problems—cognitive impairment. These things occurred in kind of a sequence. It took, I'm going to say, two to three months before we started experiencing the initial symptoms. It was probably at or around month four that we experienced the onset of the cognitive impairment.

LB: At that point, when you and your family were just starting to exhibit these symptoms and you didn't know what they were from, did you go to the doctor?

MB: Oh, of course!

LB: And the doctors didn't have any clue as to what it was?

MB: That is correct. I even mentioned to Reese's pediatrician the fact that we had mold growing all over the place. At that point, there had been the American Academy of Pediatrician's policy statement on *Stachybotrys*, and after the test results came back from Texas Tech, I told the pediatrician that *Stachybotrys* had been found. If you look at the American Academy of Pediatrician's policy statement in April of 1998, you will see that there were infant deaths in Cleveland from this very stuff.[25]

LB: Was that already on the record at that time?

MB: Yes, it was already on the record. Unfortunately, most pediatricians had not read it, and most patients did not know about it.

LB: Did you ever address that with the pediatrician?

MB: Well, sure, and he said, "Well, I didn't see it. I didn't know." Well, he's a pediatrician. How could he not know?

LB: Wow! Ignorance is still prevalent today in the medical field regarding the health dangers of mold exposure and identifying the symptoms of mold-related illnesses.

MB: Of course it is! What's also important and probably what you've found is that just because one person has the symptoms, it doesn't mean that everyone in the house exposed to the same exact stuff is going to have those same exact symptoms. For me, I had only some coughing up of blood, muscle weakness, headaches, and some hair loss. I did not have bleeding through the ears, short-term memory problems, or any of the more severe symptoms experienced by my son and husband.

LB: Yes, some people are genetically predisposed in such a way that their bodies aren't able to detoxify the effects of mold exposure as well as others.

MB: My husband's system does not. He was so exposed that he was unable to fight it. I was exposed to the exact same quantity, but my genetic engineering is such that I was able to fight it; for Reese, it just took time—seven years later, he has made, I will say, an almost full recovery.

LB: Oh, good! How old is Reese now?

MB: He is ten.

LB: How is your husband doing now?

MB: The same. He has seen no improvement.

LB: What type of testing was done with your family?

---

25   AAP Policy Statement: American Academy of Pediatrics, "Toxic Effects of Indoor Molds," *Pediatrics*, vol. 101 No. 4 April 1998, 712-714. Available at http://aappolicy.aappublications.org/cgi/content/full/pediatrics;101/4/712 (accessed August 2006).

MB: We did the serum study test, which showed trichothecene mycotoxins in our blood. We did that because Texas Tech had asked us immediately after evacuating the home to participate in a study they were doing. Actually, the study was finally published last year, and the results have been very helpful to those seeking proper diagnoses of their problems.

LB: I've read the report on that study. So your family was part of that study?

MB: Yes, we were one of many in the study. I mean, they did more than just three people for a more diverse sample, but we were the first three that ever had their blood serum tested. There were not any of these tests that are available now back in 1998 and 1999. There were IgA tests,[26] IgE tests,[27] and IgG tests[28] that simply show that you have an antibody that was produced after exposure. The new tests go beyond that and can determine if mycotoxins have invaded the bloodstream.

LB: The "old" tests tell if you're having an allergic reaction to the mold by checking for antibodies your body has created in response to the fungal antigens.

MB: Well, yes and no. I think you're going down a very slippery slope when you say an allergic reaction. Do you have an allergic reaction when you eat a hamburger laced with *E. coli*? Is that what you call it? Did the miners who died of carbon monoxide exposure have an allergic reaction? The answer is no. These are not allergens. They are toxins. Patients exposed to toxins should not seek the help of allergists. They should seek medical advice from doctors familiar with *toxic* exposures.

LB: That's a good point. The body would produce antibodies in those two scenarios, and yet those situations aren't referred to as allergic reactions. The CDC report published in the aftermath of Katrina and Rita to address "prevention strategies and possible health effects" of mold called it an allergic reaction.

MB: Well, they're also there primarily to protect industry interests. Don't forget what administration we're in—the net-net of it is that this administration protects industry interests. The FDA disregarded research on Vioxx—that it can cause heart attacks—to protect the goose that laid their golden egg (pharmaceutical companies fund the FDA). Builders, insurance companies, and large landlords do not want this issue front and center in medical research or have the CDC say it's toxic. This is an issue fraught with political interests, not public interests.

---

26   IgA tests reveal the presence of a class of immunoglobulins that include antibodies found in external bodily secretions, such as saliva, tears, and sweat, according to Medline Plus.

27   IgE tests reveal the presence of a class of immunoglobulins including antibodies that function especially in allergic reactions, according to Medline Plus.

28   IgG tests reveal the most prevalent type of serum antibody, which is produced after re-immunization (the memory immune response or secondary immune response). IgG protects the tissues from bacteria, viruses, and toxins, according to the Merck Manual.

LB: How many months was it, in total, from when the water problem started to when you and your family moved out of the house?

MB: I'm going to say it was about five months.

LB: Five months of exposure? Wow! I've read that the increased amount of exposure, in terms of quantity and duration, affects people's ability to recover, as well as the genetic component.

MB: Right, I'm sure that's true. In our case, we're talking about stuff that you could see literally growing up every wall, coming out, bleeding out from the interior of the wall cavity about 7 or 8 feet.

LB: *Stachybotrys* is one of the most toxic of the molds, too.

MB: Right. We also had just a "boatload" of *Chaetomium*, which is really bad. It is worse than *Stachybotrys*. David Straus can tell you a little about it. It actually can grow in the brain. Autopsies have been done where the cause of death was from *Chaetomium* growing in the brain.

LB: That is certainly a strong example that shows there can be serious health repercussions from mold exposure.

MB: Well, yeah—death is kind of one of those things that is not good.

LB: Your family's current situation is that your husband's health hasn't changed, your little boy has had almost a full recovery, and you—how are you?

MB: I was fine two months after we moved out. It was very transitory in my case— very short-lived.

LB: What ultimately happened to the physical structure of your house?

MB: It was bulldozed.

LB: You went through a trial, an appeal, and then, finally, a settlement. The original judgment in 2001 was for $32 million and that was appealed. Tell us what happened.

MB: It was appealed and then reduced by the Third Court of Appeals in Texas to actual damages only, which was basically lawyers' fees, the worth of the house and its contents, plus interest on that money. If you add all that together, it would have been about $8 million.

LB: I read that it was later reduced to $4 million, but you're saying it was $8 million?

MB: The Third Court of Appeals did not calculate attorneys' fees and interest in their opinion, and their opinion stated as much. Their opinion also did not factor in money that the insurance company paid that had been put in the court registry awaiting the final outcome of the case. Monies deposited in a court registry cannot be used or withdrawn and are in safekeeping at the court's discretion.

LB: And then what happened?

MB: The insurance company appealed that decision, and that sent us into the Texas Supreme Court. We settled right before the Texas Supreme Court rendered its decision in the case. The settlement I received is not anything to write home about; the only benefit is that I had a huge tax write-off.

There was clearly another motivating factor to settling: I had a home that sat unoccupied, uninsured, and was collapsing from the inside out. Teenagers were breaking into the house, getting sick (vomit was everywhere), and I was scared the house would collapse with someone trapped inside. Burglars were stealing everything from TVs, stereos, silver, and furniture to clothing. It was a huge liability for me, and I wanted it gone. Had the Texas Supreme Court given the insurance company another trial, the house could not have been destroyed, and I would have been stuck in litigation for another decade. I wanted to move on and focus on helping others through POA.

Most people who settle are never reimbursed for the cost of litigation (which usually runs into the millions of dollars in hotly contested battles) or for medical expenses (both past and future). Most never are made whole on the actual property loss (including personal property), and most find themselves having to start to rebuild their nest egg when they are reaching retirement age. Perhaps equally troubling is that insurance companies rarely are punished for their behavior because settlements rarely include any punitive damage figure. So, their behavior continues because there is no penalty for committing such acts.

LB: I read that Ronnie was diagnosed with toxic encephalopathy.

MB: Yes, which is brain damage.

LB: Now this is a medical condition that can be proven by MRIs, yet the cost of ongoing medical treatment was not factored into the settlement at all?

MB: No, and few receive compensation for ongoing medical treatment. Texans have it particularly bad. In the state of Texas, there is a bar that you must meet from the scientific burden standpoint, and toxic encephalopathy from mold does not meet that scientific hurdle. By the way, neither does anthrax. In Texas, if an insurance company caused you to be exposed to anthrax, a jury will never hear about it because anthrax does not meet the scientific hurdle. So Ronnie's personal injury part of the lawsuit was never even heard.

LB: Correct me if I'm wrong. Ronnie's personal injury claim was not even presented to the jury because the trial judge ruled expert testimony inadmissible. This is important for people to understand. The testimony of two expert witnesses—prepared to testify that Ronnie's brain damage was from his exposure to the mold inside the house—was ruled inadmissible because it did not meet the threshold of the Daubert rule.

MB: Actually, in Texas the burden of proof is guided by a Texas Supreme Court ruling called Havner. It says that there must be a doubling of exact symptoms with a margin of error of less than 5 percent in order for a study to meet the legal burden of proof. If it were legal to intentionally expose 1,000 people to the same species and quantity of mold for the same period of time and expose a control group of 1,000 people to nothing, then a study could be conducted comparing the exposed group to the control group. If the exposed group were 100 percent more likely with a confidence ratio of 5 percent to develop brain damage, then a jury might be able to hear the merits of a personal injury case in Texas. Such research cannot legally and ethically be conducted; therefore, it does not exist. Hence, in Texas such cases of exposure are not heard. Remember, anthrax does not meet this burden in Texas either, yet we all know it is extremely harmful.

LB: Eventually, personal injury cases relating to toxic encephalopathy and other serious effects of mold may go forward as scientific and medical research progresses, even if controlled case studies are not able to be done with humans.

MB: More and more research is coming out every day, and physicians are being armed with more advanced testing, which enables more accurate diagnoses.

LB: Is your case legally on record as a bad faith judgment even though a settlement was agreed upon?

MB: Appeals Courts' findings are what make case law. Our case was a breach of contract and bad faith case, not a mold case. The opinion of the Third Court of Appeals speaks to that.

LB: One of the articles that I read about your case quoted Dr. Straus as saying that the house couldn't just be bulldozed and that people with Hazmat equipment would have to go in and remove all the mold-infested material, which would then be buried. The rest of the house, sans mold-infested materials, could then be bulldozed. Is this what happened?

MB: Yes, the house was remediated before the bulldozers hit it. I spent about $1 million to safely destroy the property. The remains were bagged and carted off to a regular landfill. Mold is actually good in a landfill because it consumes what it is growing on. Without mold, garbage would not decompose. I went to such extremes to protect surrounding homes in the neighborhood.

LB: You've certainly been proactive in helping others. Tell us about Policyholders of America.

MB: I started it in late 2001. It was formed basically to help people get their claims paid without having to hire attorneys, public adjustors, and so on—all of whom have to be paid. Claims get paid when policyholders are armed with information and document their claims effectively. POA is kind of a do-it-yourself self-help kit for people who are having problems with their insurance company. That's not to say that some people, regardless of what they do, may indeed still have to hire an attorney. I'm going to say a good 70 percent of POA members end up being able to resolve their own claims without having to hire and pay an attorney. Anytime you are going against an insurance company, it's very costly. We spent probably $2.2 million in our case. Most people don't have that kind of cash on hand. I didn't; I had to sell stuff, steal from Peter to pay Paul. It was extremely expensive. This is the whole purpose of POA—to help people avoid the pitfalls so they are not forced to sue.

LB: How many members does POA have now?

MB: We have about 3 million members, but not all of them are mold-related claims. A lot of them are auto claims and just regular fire or homeowner stuff.

LB: POA members have access to a restricted area within the website that provides a data bank of legislation, news articles, peer-reviewed articles, etc. Additionally, POA has secured discounts for its members on certain services that members otherwise may not be able to locate. For example, POA negotiated the discounted laboratory services offered by Texas Tech for two different types of mold sample testing.

MB: We negotiate discounts on services our members want. Be they paying or free members, they get discounts.

LB: Most insurance policies now exclude mold coverage. When did these exclusions begin and why?

MB: Insurance is state regulated, and all changes to policies (premium increases, coverage caps, exclusions, etc.) must be approved by the state in which the policy is sold. Before my case, there were about 235 lawsuits that included mold damage. The costs to cure the problem began to soar, along with asbestos abatement, in the mid 1990s. Beginning in the mid-to-late '90s (which was long before my case was even filed), insurers were negotiating exclusions for mold with the state insurance commissions. In 2001, insurance departments began granting their requests. However, even if mold is excluded in the insurance policy, if an insurer has not lived up to *its* duties, the policyholder has a good chance of forcing coverage. An insurer's duties under the contract include:

- Properly investigating cause of loss
- Adjusting the claim in a timely manner
- Developing an accurate scope of work required including cost of repair and/or replacement
- Paying a claim when liability is reasonably clear

LB: Looking back at the whole experience, what insights would you hope people glean from reading your story?

MB: Well, it depends, really, because so many people don't have resources. They live paycheck to paycheck. People have to make up their own minds, and I'm not going to impose my morality on someone else. I just don't do that. So what was right for me and what I did does not mean that it is right for everybody. We tell people several things. If you test your home, go into it knowing full well that you may be destroying your own home's value. If you find heavy contamination, there's no turning back. When you sell your house, it must be disclosed by law, and when you disclose it, guess what happens to its value? So we tell people, go into testing with both eyes open, knowing that if you do find there is a problem, you may end up eating some of your value. What people do with that information, I can't tell you. Some people may opt to sell their house without ever testing. As a general rule, you don't have to disclose what you suspect; you have to disclose only what you know.

I never thought that I would be such a controversial figure. By going public with my problem, I stirred up a hornet's nest of people and companies whose liability would increase with public awareness of this problem. I did not realize that I would be so hated by the insurance industry for educating the public. And I never thought I'd be so welcomed by regular homeowners who appreciate our help. It's funny, you know. I get love letters and hate mail. Fortunately, the number of letters of appreciation are about 100 times the number of death threats and hate mail I get from the insurance industry.

Of course, hindsight is always 20/20. If I knew then (in 1998) what I know now, I would have ripped out the hardwood floors of my home regardless of whether it caused me to lose coverage. The cost to replace that floor was about $250,000. Within three months, the cost escalated to more than $1.5 million. The day of trial, the cost to bulldoze the home, rebuild it, and replace all of the contents was about $6 million. That does not include litigation costs, medical costs, etc. Looking back,

I would rather lose $250,000 than $10 million, any day of the week. Of course, I didn't know then what I know now.

LB: In addition to POA, which is a non-profit organization, there also is a POA Medical Research Foundation that contributes to scientific and medical research.

MB: That's correct. We do, but POA probably has $200 in the bank. We don't make any money because we give away memberships. The last thing that we want to do is be another expense for a person who is having financial problems because of an insurance claim, so we end up giving away a lot of memberships. We have given away 600,000 since Katrina. People have to apply for a full membership, but then we just give it to them, especially because of Katrina. Basically, I fund the organization because it doesn't have any money, and I fund some of the medical research that is done on POA's dime because I think it is important that certain medical research is done so that it furthers the knowledge in this arena.

LB: You are absolutely right, and what you are doing is certainly commendable. Thank you for your time, and thank you for the work you do on behalf of all of us policyholders.

Jia-Sheng Wang, MD, PhD, *is a Professor in and Head of the Department of Environmental Sciences, College of Public Health at the University of Georgia. Dr. Wang received his MD in the field of Preventive Medicine with specialty in Toxicology from the Shanghai First Medical College in Shanghai, P.R. China, his PhD in Pathology from Boston University in Boston, Massachusetts, and his Post-doctoral Fellow in Molecular Epidemiology from Johns Hopkins University in Baltimore, Maryland.*

Dr. Wang is one of several doctors who first identified the two major factors that cause liver cancer—hepatitis B and aflatoxin—while conducting research in Qidong and Guangxi in China nearly 30 years ago. He has continued his research in this area, focusing on ways to reduce the odds of getting human liver cancer. His research includes a long-term study on the effects of green tea extracts and green tea polyphenols and the possible related reduction in liver cancer. Dr. Wang has authored and coauthored over 80 peer-reviewed publications. The interview was conducted July 2006.

LB: How much of a problem is mycotoxin contamination of food in the United States?

Dr. W: In the U.S., actually, mycotoxin contamination is still a big problem. It's not as hot an issue as in the early 1970s and 1980s when a lot of academic institutes, companies, and government agencies in Washington were doing mycotoxin studies. In recent years, the USDA has a group whose work involves mycotoxins, and the FDA has only a few people working on mycotoxins. There is really not too much funding for mycotoxin studies, except for a few aflatoxin studies, which are mainly being done overseas, as is my research project. However, aflatoxins are still a problem in the U.S. For example, in 2000–2001, there was a problem—200 dogs died in Texas because dog food had become contaminated with aflatoxins. I heard that from Dr. Timothy Phillips. He is a professor in the Department of Veterinary Anatomy and Public Health at Texas A&M University.

LB: How is food for human consumption monitored for mycotoxin contamination? I've been told that it's done only through spot-checking. Is that correct?

Dr. W: Yeah, you are correct. It is done through spot-checking, which is mainly done at the farming level. Currently, there are several companies, like VICAM[29] in Massachusetts and Neogen Corporation[30] in Michigan, that have developed testing methods to detect mycotoxin levels in grain or feed. They sell their product to farmers, grain elevators, and the government. With these types of tests, farmers can test their product themselves to see if it's over the limit or within the limit of government regulatory action and guidance levels. Also, grain elevators can test product before purchasing. You can call this spot testing. Also, when product is exported to other countries, there is some international level of inspection, but I don't know if the U.S. government really puts a lot of effort into inspecting exported product.

LB: So the government has designated levels of mycotoxins that should not be exceeded. Are these action levels set only for aflatoxins, or are there action levels for other mycotoxins?

Dr. W: There is an action level for aflatoxins and a guidance level for fumonisin, currently, I believe.

LB: Is testing for the mycotoxin levels left primarily up to private enterprise to make sure these levels aren't exceeded, or does a government agency actively monitor and enforce compliance of these action levels?

Dr. W: The USDA and FDA do food inspection where they test levels from a certain number of sites every year. I think that, after several years, the USDA compiles and presents data of inspections and monitoring of mycotoxins in U.S. food, for example, over the past ten years, because I've seen several of those kinds of reports in the meetings. So, as you said, testing is done randomly with scant (very, very few) levels from each site. Testing is not forced; it's just like a recommendation.

LB: Since our exposure to mold, we can't eat foods that are prone to mold growth or mycotoxin contamination, such as wheat, corn, peanuts, tree nuts, etc. Is that an indication that these types of foods do, indeed, have some level of mycotoxin contamination in/on them?

Dr. W: Yeah, they are unavoidable, though. Mycotoxin is a naturally occurring toxin. According to FDA standards, it is considered unavoidable. That's why the FDA has to set certain levels of set action, because it is unavoidable. Mycotoxins are not like additives that are intended to be added to foods, like hormones for animal growth or food coloring. Those are not like mycotoxins, which are unavoidable. Mycotoxin levels in grains also depend on the environment in which they are grown. For example, in southern states like Texas, the weather is usually

---

29   Information on VIACAM is available at http://www.vicam.com/.

30   Information on Neogen Corporation is available at http://www.neogen.com.

hot and humid; there is a higher level of contamination of mold than in the northern states. That's true because the local products may have higher levels of aflatoxins and fumonisins. Northern states have different problems like the trichothecene toxins, such as deoxynivalenol (DON), which tend to grow on wheat, barley, and oats. Corn and peanuts mainly have problems with aflatoxins and fumonisins.

Fumonisin is a big issue right now due to very high levels of contamination in corn and corn products in the United States. Before '97 in the U.S., there were no regulatory levels for fumonisin. Then in '97, the FDA and the USDA set up a guidance level of 0.2 ppm for fumonisin. Then they found that levels in almost 95 percent of U.S. corn products were over the point of 0.2 ppm. There was no way to follow the guidance level. For example, if they kept that guidance level, no U.S. corn products could be exported. In 2000, the USDA and the FDA lifted the guidance level to 2 ppm, raising it ten times the original guidance level. I do not think that they were allowed a consortium that reviewed the details from toxicological studies from all of the scientific fields. You know it is also a co-contaminant issue, not only fumonisins (fumonisin B1, B2, and B3), but also fumonisin and aflatoxin at the same time.

LB: How much of a contributing factor is mycotoxin consumption even at the action and/or guidance levels in the development of human cancers?

Dr. W: Well, we are working on calculating that. Actually, in 2005 a CDC group tried to organize a new research project (I don't know if they got it government funded) to use aflatoxin-albumin adduct as a molecular biomarker to measure aflatoxin exposure levels in the U.S. populations by measuring serum aflatoxin adducts. These aflatoxin adducts are used as the long-term exposure biomarker for aflatoxin exposure in many countries other than the U.S. Presence of the biomarker reflects aflatoxin exposure. They developed an LC/MS method and published that in cooperation with Johns Hopkins School of Public Health. The proposed research project would analyze 5,000 or 6,000 people's serum in the U.S. to see what the exposure levels of aflatoxins would be. The project might not get funding because the government can still see that the aflatoxin exposure from food is much lower in the U.S. than in developing countries. As a matter of fact, after 2004, a lot of government agencies and organizations have put a lot of effort into aflatoxin research in Third World countries because an outbreak of acute aflatoxicosis was reported in Kenya in 2004.[31] Have you heard that story?

LB: No, I haven't.

Dr. W: In Kenya, Africa, people ate aflatoxin-contaminated food. There were 317 people that got acute aflatoxicosis, and 125 folks died. So the Kenya government

---

31   Eduardo Azziz-Baumgartner et al., "Case-Control Study of an Acute Aflatoxicosis Outbreak, Kenya, 2004," December 2005, vol. 113, no. 12, *Environmental Health Perspectives*. Available at http://www.ncbi.nlm.nih.gov/entrez/query.fcgi?CMD=search&DB=pubmed (accessed August 2006).

asked for international help, and the U.S.'s CDC, FDA, and USDA organized a group to work there.

LB: Why do Africa and Asia have higher levels of aflatoxin problems? Is the contaminated grain grown in and exported from other countries like the U.S.?

Dr. W: No, it is locally grown. Aflatoxins and the molds that produce aflatoxins are generally from a storage problem. For example, when they harvest from the field, their corn or peanuts are usually not dry enough. Then they take it back to their homes and store it where there is no air conditioning or don't put it in a place with good ventilation and things like that, and the food gets contaminated by mold, *Aspergillus flavus*, and produces toxins under those conditions.

LB: So they store their crops in their homes in Africa and Asia, not in silos as we do here in the U.S.?

Dr. W: That's true. How the product is stored in the home is actually more important than what happens in the field—with aflatoxins. Fumonisin is a different story. The fungi that produce fumonisin regularly can contaminate in the field products like corn, and almost always they produce the toxin.

LB: Here in the U.S., most people don't store grain in their homes, but the products they buy at the grocery store can already be contaminated with elevated levels of mycotoxins from fungi that grew on the crops or in storage silos and didn't get detected prior to manufacturing.

Dr. W: Absolutely. That's true. The oddest thing in the U.S. is that rice tends to have possible problems, too. I went to a grocery store to buy rice; then when I got home, I found that the rice was full of all this mold contamination. So I took it back to the small grocery store to see if it had expired, because there was no expiration date or any of that other data on the package. Maybe this batch of rice that was imported from Thailand came from a distribution center that had stored it for a long time. I use that bag of rice in a class I teach for detection of mycotoxins because I know mycotoxins will be found in there.

LB: I thought rice was less prone to mold growth and was a safer grain when it comes to fungal and mycotoxin contamination.

Dr. W: That's true. Unfortunately, it happens. As I mentioned earlier, I used the bag of rice for teaching purposes to detect mycotoxins. We also collected some corn samples in developing countries through the collaboration studies. As a control, we used cornmeal purchased from a national grocery store chain. I wanted students to see how food from developing countries is more contaminated with fumonisins than U.S. food. However, the measurements didn't prove that. We found higher levels in the U.S. cornmeal than in the corn samples from developing countries!

LB: Do you think higher levels of contaminants in mold-prone foods—corn, wheat, peanuts, tree nuts, etc.—are why mold-sensitized people get sick from eating those foods, because their bodies are now unable to detoxify the same levels of contaminants as they were able to before the mold exposure?

Dr. W: Well, that's true, but I think there's still a genetic susceptibility for each individual that plays a big part in that because aflatoxins are metabolized by certain enzymes, like cytochrome P450 1A2, 3A4, and 2A6. If you have a defect on expression of those enzymes, then you might be getting sicker compared to other people. Besides those enzymes, aflatoxins are also detoxified through what we call the phase 2 enzymes, glutathione S-transferases (GSTs) and epoxide hydrolases. We have already found from many studies that these are genetic polymorphic enzymes [enzymes that exist in several forms within a single species]. In fact, a certain percentage of populations, dependent on ethnicity, don't even have certain subtypes of these phase 2 enzyme genes, e.g., GSTM1 or T1, that lead to no expression of the enzyme. Compared to others with the enzyme, these people's ability to detoxify aflatoxin would be very slow. That is the major reason why, when we are equally exposed to the same levels of toxins, these people have a different reaction.

LB: Are food reactions that mold-sensitized people experience a reaction to the mycotoxin contamination or to the fungal antigens (possible fungal proteins or traces of fungal proteins in the food)?

Dr. W: No. Reactions to mold and exposure to mycotoxins are different. Some people may be hypersensitive (allergic) to mold-contaminated environments, which may be caused by inhalation of or skin contact with the mold or mold spores. Human exposure to high levels of mycotoxins through food usually leads to toxicosis, such as liver toxicity for aflatoxins and hematologic toxicity for trichothecene toxins.

LB: I have been told that the antibody tests that detect mycotoxins in grains and feeds require that the mycotoxin be attached to a protein cell in order to elicit an antibody effect. One manufacturer of these tests found that the mycotoxin alone was too small for the body to recognize as a foreign substance. Does this mean that the body will not have an allergic reaction to a mycotoxin, or is there a threshold of mycotoxin levels in the body at which the body recognizes the mycotoxins as foreign substances and will have an allergic reaction?

Dr. W: Yes and no. Mycotoxins are usually small molecules and do not cause immune reactions alone. However, mycotoxins in human and animal bodies can bind to proteins, inhibit DNA, RNA, and protein synthesis, or regulate gene expressions, which may cause immune-toxic effects, including immunostimulation, hypersensitivity, and immunosuppression. On the other hand, mycotoxin-containing mold spores may play a certain role in causing individual hypersensitivity (allergic reaction).

LB: Do mycotoxins have toxicological effects on the body, and if so, what are they?

Dr. W: Yes. It has been known for many years that mycotoxins cause various toxic effects on humans, such as ergotism, aflatoxicosis (turkey X disease), red mold toxicosis, alimentary toxic aleukia, Balkan endemic nephropathy, etc.

LB: You've written that several mycotoxins have carcinogenic potency in experimental animal models—aflatoxins, sterigmatocystin, ochratoxin, fumonisins, zearalenone, and some *Penicillium* toxins. Can we draw any sort of scientific correlation yet that these mycotoxins are also cancerous in humans?

Dr. W: Right now, only aflatoxin is listed as a human carcinogen. It is certainly a human carcinogen. In 1993 the International Agency for Research on Cancer listed aflatoxin as human category 1 (confirmed human carcinogen); then in 2002, they reconfirmed the conclusion. Ochratoxin A (causing Balkan endemic nephropathy) was listed as a 2B (possible human carcinogen) and fumonisin B1 as a 2B category (possible human carcinogen). Animal studies prove it is a carcinogen to animals, but at this point, human epidemiology studies have not been done, and there are no convincing human data to prove that it's a human carcinogen.

LB: So we can't make the correlation that since fumonisin causes cancer in animals, it would do the same in humans?

Dr. W: According to the Food Safety Act, anything causing animal cancer should not be used for humans. But practically speaking, humans are different from animals, so that's why there are different categories for classification as human carcinogens.

LB: Do all mycotoxins suppress the immune system, like aflatoxins?

Dr. W: Aflatoxins suppress the immune system. The molds that produce trichothecenes are actually a problem in the U.S., in the Western society. Trichothecenes suppress the immune system, in particular, T-2 toxin and deoxynivalenol. For a long time we have known these trichothecene toxins are all strong immunosuppressors. They can really suppress protein synthesis, which will also suppress the immune system. T-2 toxin is one of the selective agents, which is another mycotoxin. It is in a very low level in feed and food, but deoxynivalenol sometimes is in a very high level in certain foods, especially in cereal. In England, a study was done in which they collected normal people's urine. What they found was that when these people ate cereal, deoxynivalenol was found in their urine.

LB: Does this mean there is a good possibility that every time we sit down to eat a bowl of cereal in the U.S., it contains deoxynivalenol?

Dr. W: Yes. It depends how much you eat and the contamination level of deoxynivalenol in the original raw cereal.

LB: Do you think the immune-suppressing effects of these mycotoxins are a contributing cause of cancer?

Dr. W: It is not causing cancer. It might be causing overall general health status to deteriorate. That is why there are programs right now, like Dr. Pauline Jolly's group at the University of Alabama, that are trying to link the increased level of mycotoxin exposure in developing countries to increased sensitivity to infectious disease, like HIV or maybe tuberculosis. Overall, we still need more data to try to link exposure to these mycotoxins to cancers. Certainly, immunological functions play a big role in cancer causation, based on current concepts.

LB: People's genetic makeup determines strengths or weaknesses in detoxification pathways and how carcinogens will be metabolized. Does this mean that aflatoxin, for example, could give one person liver cancer, but another person a different form of cancer?

Dr. W: Yes.

LB: Since we all get exposed to a certain amount of these mycotoxins in food, from normal environmental exposures, and from natural disasters such as flooding and hurricanes, what can people do to prevent the immune suppression and the denaturing of cells that can come from this exposure?

Dr. W: Well, there are many ways to prevent that. Situations like Hurricane Katrina, in which there is no electricity, no refrigeration, no air conditioning, and in which it is very hot, can create issues with food. Some people will eat food that was not able to be properly stored during this time, food that might have become contaminated by molds, bacteria, etc., and not realize it.

Right now, we are doing general prevention of food contamination, which we view as a primary prevention step. A general prevention method is to try to avoid the contamination of those foods that are prone to mold growth. You control the temperature and humidity of your cereal storage, corn storage, peanuts, and the other foods that are stored in large quantities. Gamma radiation of these large quantities of stored food can effectively prevent growth of mold. Right now, maybe in the whole U.S., and possibly in the entire Western world, where food is stored in big storage warehouses, this kind of radiation is used to kill bacteria and molds. So, that is also a way of diffusing the extent of mold growth, which limit· the production of new mycotoxins in storage. In developing countries, becai·

they produce their own food and store it on their own to use for a half a year or a year, physical methods were introduced to reduce mold growth and mycotoxin contamination. For example, they can pick the contaminated corn or peanuts out and not eat them. By doing this, they can reduce by maybe 50 percent the levels of mold and mycotoxins in the entire batch.

LB: The radiation process will kill the mold, but it won't kill the mycotoxins, because dead mold spores still contain the toxic properties.

Dr. W: That is true. It will just kill the molds, not the mycotoxins. I don't think radiation will destroy the toxins. But if you try to kill the molds, especially *Aspergillus* that produces aflatoxin, after harvest, that is also a way of reducing the level of mycotoxins that could otherwise be produced during long-term storage. This is because the molds might continue producing toxins while the product is in the storage warehouses.

LB: So the theory is to kill the molds so that they don't keep producing mycotoxins in storage, even though the radiation process is not going to reduce the already existing mycotoxin levels.

Dr. W: True.

LB: The same theory applies in Third World countries. When people pick out the moldy grain, kernels, or peanuts from a large batch, it reduces the amount of new mycotoxins the moldy pieces could otherwise produce, but that does not affect the already existing level of mycotoxins already produced by the mold-contaminated pieces before they were picked out.

Dr. W: That's true, yeah, but don't confuse them. If there are toxins there, the product already is contaminated overall, but you can pick out only the kernels or peanuts that you can visually identify as infected with molds. Usually you can find those. But even really healthy corn or those kernels that look beautiful might still possibly be contaminated with mycotoxins.

LB: I understand. You have been involved in preventive cancer research that showed chlorophyllin reduces the amount of aflatoxin DNA damage by 55 percent. Will chlorophyllin reduce the DNA damage caused by other mycotoxins as well?

Dr. W: Well, actually, there are several agents right now. Chlorophyllin, actually, ... study. Dr. George Bailey at Oregon State is using chlorophyllin, ... it to aflatoxin. Dr. Tom Kensler and Dr. John Groopman, both ... ins, and Dr. George Bailey initiated a study in China for three ... ophyllin is a physical binding agent. It was found that the levels ... hat indicate DNA damage were found in reduced levels with ... nsumption.

Studies have also been done with a drug called oltipraz. I published a couple of papers with Dr. Groopman and Dr. Kensler where we were using oltipraz. It also really detoxifies aflatoxins. It works very well. The only problem is it has side effects. Also, I am using green tea polyphenols and am now working with Dr. Phillips at A&M University, who many years ago found a clay that can specifically bind aflatoxins.

LB: What kind of clay is it? We tried taking bentonite, which is a clay, and we did not see any improvement.

Dr. W: It is a clay called NovaSil. Dr. Phillips has been working on that for many years. I think he got a patent on it. They have been testing NovaSil on many kinds of animals for the past 20 years and still are.[32] You know, animal feed has a high contamination level of aflatoxins. So, he put in a small amount, like 1 percent, of NovaSil with the feed of big and small animals—cattle, sheep, pigs, and chickens, too. From these test results, they found that aflatoxin-caused toxicity can be reduced completely.

LB: Well, that is interesting. I had not read about that.

Dr. W: Actually, NovaSil did that in all common animals. Then, the United States Agency for International Development (USAID) supported a study to try to extend research of NovaSil to see if humans can use that kind of clay. So they are conducting research now to make sure a purified NovaSil is safe for human use. For example, a pill could be taken before or after a meal to bind those toxins out.

LB: Is this purified NovaSil available to the general public yet?

Dr. W: It is available right now, but before it goes to open market, they have to test it in several human studies. Actually, I organized a study at Texas Tech of 50 normal people whom we tested for two weeks with the purified NovaSil product, and we found there are no side effects, not anything.[33] Now, we are conducting a three-month study in Ghana, Africa, testing for human use. In fact, the study has ended, and we did not see any side effects, so we are evaluating to see the efficacy of the NovaSil, which is the next step in the process.

LB: So you have not proven yet that it is effective?

---

32   Afriyie-Gyawu, E. et al., "Chronic toxicological evaluation of dietary NovaSil clay in Sprague-Dawley rats." *Food Addit. & Contam.* 2005, 22 (3): 259-269. Available at http://www.ncbi.nlm.nih.gov/entrez/query.fcgi?CMD=search&DB=pubmed (accessed August 2006).

33   Wang, JS et al., "Short-term safety evaluation of processed calcium montmorillonite clay (NovaSil) in humans." *Food Addit. & Contam.* 2005, 22 (3): 270-279. Available at http://www.ncb nih.gov/entrez/query.fcgi?cmd=Retrieve&db=PubMed&list_uids=16019795&dopt=Abstract ' August 2006).

Dr. W: We have not finished that analysis. We already know that in animals it is highly effective.

LB: That sounds promising. But the studies involving the drug oltipraz have shown side effects?

Dr. W: There are side effects of that drug, yes.

LB: But there are natural alternatives that can be taken that would do the same type of thing as oltipraz?

Dr. W: Yes, that's true. That is what we are looking at; we are working on that. Right now, I have a grant from the National Institutes of Health (NIH) and from the National Cancer Institute (NCI) working on green tea polyphenols and green tea extracts to see if they can prevent the effects of aflatoxins. Structurally, they do not bind with the aflatoxin as NovaSil and chlorophyllin do, but if you drink green tea, the natural active components can increase your body's metabolizing enzymes. In this manner, the polyphenols and green tea extracts can detoxify aflatoxins.

LB: Wow.

Dr. W: That is a large study. We have two or three papers already.[34] We are right now doing a five-year study in southern Guangxi, China, which is a high-risk area for liver cancer. They also have a higher exposure level of aflatoxins. We gave people the green tea polyphenols in a capsule. We will see if that will eventually reduce the human liver cancer rate.

LB: How many milligrams of green tea polyphenols are in each capsule?

Dr. W: 250 mg per capsule. We are using 500 mg/day for the five-year study. We used 500 mg/day and 1,000 mg/day for the three-month study. The results from the three-month study are really very good.

LB: Regarding chlorophyllin, can you get chlorophyllin by eating broccoli?

Dr. W: You can get chlorophyllin from any green vegetable.

LB: Can you get high enough levels of chlorophyllin from normal consumption of ...s to reduce the effects of aflatoxins?

et al., "Phase IIa chemoprevention trial of green tea polyphenols in high-risk ncer: I. Design, clinical outcome, and baseline biomarker data," *J. Int. Cancer* ; Luo, H. et al., "Phase IIa chemoprevention trial of green tea polyphenols in liver cancer: II. Modulation of urinary excretion of green tea polyphenols and ine," *Carcinogenesis* 2006, 27:262-268. Available at http://www.ncbi.nlm.nih. =pubmed&cmd=Retrieve&dopt=AbstractPlus&list_uids=15930028&query_ csum (accessed August 2006).

Dr. W: I do not know that exactly. That was Dr. Bailey's work.

LB: Is there anything else that people can take to increase the detoxification pathways of the body?

Dr. W: Vitamins like A, E, and C are known antioxidants. They really can improve immune functions, so those might be good to take. Also, vitamin B1 and B2 would probably be good for overall metabolism because those vitamins surge the metabolism. They also might help to protect the liver and other organs.

LB: Can you talk about some of the scientific evidence available that proves mold DNA and mycotoxins are being found in the human body?

Dr. W: Our major research focuses on biomarkers found in the human body, which include those that are toxic metabolites formed when the human body metabolizes, for example, aflatoxins. The easiest way to identify these biomarkers in the body is by analyzing the blood serum or the urine. So we do find that there are aflatoxins, fumonisins, and those metabolites in human bodies; for example, in the urine and in the blood. But in the case of tissues, there are a couple of reports that said the identification of DNA adducts by histoimmunochemistry in cells and tissues is a sign of those DNA adducts in the human organs. That has been reported in more than one report.

That was a question I asked the CDC people that reported aflatoxicosis as the cause of death of the people that died in Kenya. In a meeting I asked them if they had any reports of autopsies to support that these people died of aflatoxicosis. They said they didn't have any because when they got there, the bodies were already buried or disposed of.

LB: Do we know of any autopsy reports here in the U.S. that document human cases of aflatoxicosis?

Dr. W: No, I don't think so. Actually, I am very active and keep updated on every report and any news in the mycotoxin research area. I have been researching and studying this area for about 30 years.

LB: So if that type of information were out there, you would know it.

Dr. W: Yes, I attend the meetings in which that type of information is shared all the time.

LB: In the future, do you think studies will be done on the health effects of aflatoxin, fumonisin, and other mycotoxin contaminants in the food here in the U.S.?

Dr. W: I think so. When you eat corn, you will get aflatoxin, fumonisin, and other mycotoxins. No one knows how those toxins together affect the human body. We only know the effects based on one toxin—aflatoxin or fumonisin alone. That is why our studies really focus on development of markers for multiple toxin exposures, for studying how corn or other food really affects human cancer risks.

LB: In New Orleans and surrounding areas, testing has shown very high levels of multiple molds present after Hurricane Katrina. Will exposure to high levels of multiple molds and mycotoxins in the environment, over an extended period of time, lead to cancers the same as ingestion of aflatoxins can lead to the development of cancers?

Dr. W: It is difficult to say, and we need to do biomonitoring and follow-up studies in the population.

LB: There have been so many studies in Third World countries regarding these issues and so few in the U.S. Do you think the reason for this is that the power of the U.S. industries that produce and manufacture products susceptible to mold and mycotoxins thwart these research efforts because they don't want this type of research proceeding in the U.S.?

Dr. W: I don't know. You might be right. I think it is the arguments within the scientific field. For example, when you submit grant proposals to NIH and the FDA to apply for more funding for aflatoxin research, those scientific review teams usually do not want to support this kind of study. It is constantly happening. In the 1970s a lot of universities had their own research branches doing all kinds of mycotoxin studies, and then they gradually phased out. So right now, only maybe five or six labs in the whole U.S. are doing these kinds of studies. It is very difficult to get funding to support these kinds of studies, especially in the U.S.

LB: But it is easier to get funding for a study in a Third World country then?

Dr. W: That is true.

LB: Thank you so much for your time. I look forward to reading the findings of your ongoing research studies.

David L. Eaton, PhD, is a Professor of Environmental and Occupational Health Sciences, Toxicology Program at the University of Washington (UW) in Seattle; Associate Vice Provost of Research, School of Public Health & Community Medicine at UW; and Director of the Center for Ecogenetics and Environmental Health at UW. Dr. Eaton received his PhD in Pharmacology from the University of Kansas Medical Center (KUMC) in 1978. Following a Post-doctoral Fellowship in Toxicology at KUMC, he joined the faculty of the University of Washington in 1979.

Dr. Eaton maintains his own active research and teaching program, focused in the area of the molecular basis for environmental causes of cancer and how human genetic differences in biotransformation enzymes may increase or decrease individual susceptibility to chemicals found in the environment. He is an Elected Fellow of the American Association for the Advancement of Science and the Academy of Toxicological Sciences. He has authored and coauthored over 130 published scientific articles and book chapters in the field of toxicology and risk assessment. Dr. Eaton can be reached at deaton@u.washington.edu. The interview was conducted July 2006.

LB: How large a problem is mycotoxin contamination in the food in the U.S.? Is this a concern in our country?

Dr. E: It is an economic concern and a potential public health concern, although all of the evidence to date suggests that the efforts in regulating aflatoxins in the food supply in the U.S. are pretty effective in keeping the levels below those which cause any measurable effects on public health. It is obviously impossible to say that the effects are zero, but the primary public health concern of aflatoxin contamination of the diet is liver cancer. Liver cancer is obviously an important cancer in the United States. It is not among the top five or so, but it is an important contributor. The evidence so far is that the vast majority of liver cancers in the United States are due to hepatitis virus, and/or alcohol, and perhaps other risk factors—probably hepatitis C more than B in this country and alcohol.

LB: Is there a link between hepatitis C and aflatoxin exposure?

Dr. E: There is not a link with exposure, but there is an open question as t[...] there might be interactions between aflatoxin and hepatitis C. Studies d[...] and West Africa have suggested that there is a fairly strong interacti[...] hepatitis B virus, aflatoxin, and liver cancer in that both indepen[...] contribute to liver cancer risk. The two together are especiall[...] in certain regions of China; that's where Dr. John Groopm[...] at Johns Hopkins have done most of their work. It is l[...]

kind of interaction occurs with hepatitis C. Hepatitis C is much more prevalent in the U.S. than hepatitis B. It is a chronic, insidious type of hepatitis and generally does not affect people for many years. So, it is really unclear whether there is a similar interaction with aflatoxin and hepatitis C. There are just not enough data to know whether that is the case or not.

LB: When I started talking with the FDA, which is in charge of setting the regulations for mycotoxins, I was surprised to learn that feed with up to 300 ppb aflatoxin content could be fed to cattle.

Dr. E: There are grain tolerance levels for aflatoxins in animal feed that are different from those set for the human diet. The human tolerance levels are 15 or 20 ppb for aflatoxin. It depends on either the food item or the specific definition of whether it is a tolerance level or an action level. I do not remember how that goes for humans. I know that the levels for animal feed are considerably higher. In some feeds it is as high as, if I remember correctly, 300 ppb. That would be legal. What happens though, and the reason that that may not be a public health problem, is that there are separate tolerance levels for aflatoxin M1 in milk. That would be the primary concern about feeding aflatoxin to, say, cattle because it is very quickly broken down, and you will not find much, if any, aflatoxin B1 (the potent carcinogenic form) in the tissues of an animal. One of the metabolites though is aflatoxin M1, which is found in milk. That is regulated at a pretty low level because obviously milk is used as a dietary staple for infants. So as long as they regulate the aflatoxin M1, the metabolite in milk products, feeding relatively higher levels of aflatoxin in contaminated food for cattle is not the primary concern.

There is another reason not to do that—and this is a big problem with hog farming—at those levels, aflatoxin will cause some liver toxicity in the animal. From a practical perspective, what happens is the aflatoxin makes the animals very inefficient at converting the food they are eating to meat, which is what the farmers and ranchers are selling. So, it is an economic problem because the way that hog farmers make their money is by being able to get a certain amount of meat on the ... very pound of feed that they give the animal, and if that ... less, they make less money. It actually is a very important

... whether ... ne in China ... on between ... dently seem t ... problematic ... an and Dr. T ... ss clear whe ...

... s are probably not great, as long as aflatoxin ... ly monitored and regulated. I believe that they are ... iously, not every sample of milk is analyzed. Where ... known to be contaminated with aflatoxin, I do not ... itoring system is, so I cannot comment on the ... c health evidence that is available, aflatoxin in ... a significant public health problem in the U.S.

... ed and have a reaction to a food, say wheat,

for example, are they reacting to the fungal antigens, the fungal proteins that could be remaining in the wheat, or possible mycotoxin contamination? Is there any way to tell?

Dr. E: That is a good question and not one that is simple to answer. I know that many people who have food allergies are reacting to antigens from proteins in the food itself rather than to microbial or mycotoxin contaminates. It is an interesting question. I do not know whether there might not be people who develop an allergic reaction to the mycotoxins themselves, rather than to proteins in the food source that is contaminated with the fungus that produces the mycotoxin. Certainly, people can develop allergic reactions to the proteins in the fungi that produce mycotoxins, but that is different from food allergies. If they develop sensitization to antigens that are secondary to fungal contaminates, then their response to the food item would be highly variable because not all wheat is going to be contaminated with that particular mycotoxin antigen, but all wheat contains the gluten and other proteins that people do develop sensitization to. So, I think that the majority of food allergies, both wheat and especially peanuts, are due to specific proteins in the plant product rather than fungal contaminates, but that is not to say that people could not develop a sensitization to fungal contaminate proteins that were present in that food supply. I do not know if there is any evidence of that.

LB: So there is a possibility that people who have been exposed to large amounts of mold and mycotoxins during and after the hurricane could develop and experience food allergies that they did not have before.

Dr. E: That is exactly right. I am not an expert in immunology, but certainly there is something referred to as atopic individuals [people who are genetically predisposed to allergies]. Once people do develop sensitizations to certain types of allergens, oftentimes they will develop a broader response to other antigens that they did not have before.

LB: That's a good point.

Dr. E: It's very logical. For example, for the *Aspergillus* mold that produces the chemical aflatoxin, the mold itself will have all kinds of protein antigens that are distinct from whether it produces aflatoxin or not. So when people develop sensitization reactions, it probably has nothing to do with aflatoxin itself. It has to do with proteins in the *Aspergillus* mold. Most *Aspergillus* molds are non-aflatoxigenic. Most of them do not make aflatoxin; only a couple of strain Lots of people do develop sensitization reactions to *Aspergillus* mold separate question from aflatoxin, which is a mycotoxin.

LB: There often seem to be no concrete answers when it co

Dr. E: That is exactly right. What I know about sensiti

to molds, as well as other antigen sources—dust, mites, pollens, all kinds of things—is that people can go for years and years with exposure to allergens with no response, and then, all of sudden, they develop an allergic reaction to it. Whether it is hay fever or something more serious, subsequent exposures then cause problems. Sometimes it goes away, but once they have developed sensitization, oftentimes it is pretty persistent. So, in a situation like Katrina, where the mold exposures were just enormous in terms of both the variety of molds and especially the spore content, I have no idea how that might relate to changes in sensitivity to foods and other types of things or how long it would last. A very interesting question that I do not know the answer to.

LB: Tell us about your work.

Dr. E: Well, I have been very interested in and started working on aflatoxins probably 20 years ago or so because of my interest in how chemicals cause cancer in humans. As a toxicologist, which is my primary area of training, I rely on laboratory animals to predict how humans respond. How well the animal models predict how humans respond is obviously an important question. I was struck by data from many years ago that showed that rats were exquisitely sensitive to aflatoxin in terms of its acting as a liver carcinogen or causing liver cancer in rats. Remarkably low levels of aflatoxin in the diet of rats will cause very high incidences of liver cancer, but mice were completely resistant. You could feed them aflatoxin until it practically killed them acutely, and they did not get liver cancer. As a toxicologist who relies on animal models, that was not very comforting in terms of using animal models to predict how humans might respond. So, I became interested in what the molecular basis is for why mice are so resistant and rats are so sensitive to the cancer-causing effects of aflatoxin, and then where humans, if I understood that mechanism, might fit on that spectrum.

LB: How can animal models be used to make a correlation to human health?

Dr. E: We have to b   careful and need to recognize that responses in animals
                            very good predictors of how humans respond, but that
                            are lots of examples, and it works both ways, where
                            sitive and develop responses that humans do not;
                            good predictors. And there are examples in which
                            domide (originally developed to treat nausea
                            in the treatment of leprosy and some types of
                            t causes birth defects in humans at extremely
                            fects in rats or mice. So, this whole issue of
                            erstand.

                            gists is that we have to rely on animal
                            toward understanding the biochemical and
                            se disease in the animal so that we can see

if those same mechanisms are relevant to humans. That is largely what my work on aflatoxins has focused on—understanding why mice are resistant but rats are sensitive to aflatoxin carcinogenesis.

We have identified, with a fairly high degree of certainty, why mice are resistant. It has to do with one particular gene they have that makes a protein that is very good at detoxifying aflatoxin. Rats have that same gene, but they do not express it in their liver. Not expressing the gene means you do not make a protein from it. It does not do you any good if you have the gene and you do not express it. But, interestingly, the gene can be turned on in the rat liver by certain things in the diet. That is where our research has been focused in the last decade or so: looking at how we can manipulate things in the diet to change the way genes are expressed with the idea that we could potentially take somebody who is likely to be sensitive, because they did not express the gene, and then turn that gene on so that they could then become resistant.

That works in rats. We can make rats resistant to aflatoxin by basically feeding them broccoli or a particular chemical in broccoli. We then were interested to know if that might work in humans. Unfortunately, from what we have learned so far, we have not been able to find a gene in humans that acts exactly the same way as the gene in mice that produces an enzyme that detoxifies aflatoxin. We have been looking and there are similar genes, but they do not have the same ability to detoxify aflatoxin as the mouse gene does.

Our data suggests that humans might be quite sensitive to aflatoxin, similar to the rat, which is largely what most of the risk estimates have been based on—either human data or rat data. But we remain hopeful that we can manipulate through diet the way that people metabolize aflatoxin with the idea that you might be able to make people relatively more resistant to the cancer-causing effects of aflatoxin by changing their diet.

LB: You have not found any foods that aid in the detoxification of aflatoxins?

Dr. E: Well, we found some mechanisms. We have studied a particular chemical in broccoli. In mice, it turns on a particular gene that detoxifies what is called aflatoxin epoxide. Aflatoxin, in order for it to be both toxic and carcinogenic, is first metabolized to a reactive molecule called aflatoxin epoxide. That can bind to DNA, which is what causes the initial steps in causing cancer and probably also greater contributes in high doses to liver toxicity.

LB: So you are talking about the altering of DNA, the altering of cells.

Dr. E: Yes, that is right. Aflatoxin is metabolized in the liver to aflatoxin epoxide, and that epoxide in mice is very quickly detoxified by another gene called glutathione transferase. That gene detoxifies the aflatoxin epoxide so it does not bind to DNA in mice. It protects the DNA from damage, and the mice do not get cancer. Rats form the epoxide very well, but they do not have the gene that detoxifies it; so instead of being detoxified, the aflatoxin epoxide damages DNA, and that changes the cells to become cancer cells ultimately.

What we found in humans is that humans can form the aflatoxin epoxide, but they are terrible at getting rid of it, so it does damage DNA in humans. We did find in our most recent studies that there is a chemical in broccoli that, while it does not turn on the gene that detoxifies aflatoxin as it does in rats, actually turns off the gene in humans that activates aflatoxin. So, this chemical in broccoli potentially has a protective effect in humans, but through a different mechanism. The problem is that the gene that it turns off is also involved in the metabolism of lots of other things, including 60 percent of the pharmaceutical agents that people take. So if you were to turn that gene off completely, you could protect against aflatoxin, but you would have all kinds of other problems created by turning it off because of other chemicals it is important in detoxifying instead of activating. It's kind of a complicated story.

LB: Does this particular chemical in broccoli have anything to do with the thyroid-suppressing effects of eating broccoli, or is that a totally different thing?

Dr. E: It is actually a different effect, although it is the same class of chemicals. There are a lot of different chemicals in broccoli, and lots of other plants, vegetables, and fruits have lots of interesting biological effects. In broccoli and particular forms of cabbage, which are both cruciferous vegetables, there is a chemical that is actually called goitrin because, in high concentrations, it does affect thyroid hormones to the point that it can actually cause goiter in people. That is not typical of normal dietary levels, but it is a similar kind of biological effect from a different chemical.

LB: I have read some articles on chlorophyllin. Is that the chemical in broccoli you are talking about?

Dr. E: No, I am talking about a plant chemical called sulforaphane. Chlorophyllin has been the focus of studies from some of my colleagues at Oregon State University. It turns out that chlorophyllin physically binds aflatoxins in the gut so they do not even get absorbed, whereas sulforaphane changes the expression of genes in the liver that then change the way that aflatoxin is metabolized.

LB: So they have studied chlorophyllin, and you have studied sulforaphane.

Dr. E: Right.

LB: Based on your studies and those of your colleagues, is eating broccoli a good preventative for cancer?

Dr. E: We know from epidemiological studies that people who have diets high in cruciferous vegetables have statistically lower incidence of a variety of types of cancer. Cruciferous vegetables are the family that broccoli is in, which includes brussels sprouts, cabbage, kale, and others. But broccoli and cabbage are the principal cruciferous vegetables. So, in general, yeah, it is good. Specifically, in terms of aflatoxin, my colleagues at Johns Hopkins, Dr. Groopman and Dr. Kensler, had been studying sulforaphane in broccoli sprouts in a study in China, where people are routinely exposed to relatively high levels of aflatoxin in their diet. They did show that it looked as if broccoli sprouts protected against some aflatoxin damage to DNA. It probably does. Those are at fairly high doses, probably a little bit higher than most people would get from eating a typical amount of mature broccoli, because broccoli sprouts have much higher amounts of sulforaphane than the mature plant. Broccoli sprouts are now available in stores as a health food supplement.

LB: Are you saying that the supplement that turns off the gene that activates aflatoxins also turns off the gene that detoxifies many pharmaceuticals? Did I understand you correctly?

Dr. E: Yes, we think that might happen. It is the same gene. One of the genes that activates aflatoxin is called cytochrome P450 3A4, a drug-metabolizing gene. It is present in people's livers for a very specific reason and that is to metabolize foreign substances. When it metabolizes aflatoxin, it is trying to get rid of it, but in that example, it actually makes it into a more toxic form. In the liver, many drugs (on the order of 60 percent of pharmaceuticals) are also metabolized, at least in part, by cytochrome P450 3A4, this same gene. So if you block this completely, 60 percent of pharmaceuticals would not be metabolized either. We have shown that sulforaphane will do that at high doses. Whether you could get that high a dose from a normal consumption of broccoli is something we are studying now to see how sensitive this gene is to these effects. We do not know for sure yet. So far we have done studies only in human liver cells in vivo. *In vivo* means experiments done in, or on, the living tissue of a whole, living organism. We have not done it in people yet.

LB: So, are you saying that people could eat one to two servings of broccoli a day, for example, and that might help turn off the gene that activates aflatoxins, but yet may not be a large enough amount to affect the detoxification of pharmaceuticals?

Dr. E: Yes, we think that is possible, but I do not know whether that small amount will actually help detoxify aflatoxin that much either. It might. That is hard to say. It gets very complicated to try to understand exactly the dose it will take to be effective in reducing aflatoxin. It does turn on some detoxification genes in humans, just not the one that detoxifies aflatoxin, at least, as near as we can tell.

LB: Okay.

Dr. E: There are other benefits and other biological effects aside from turning off, in part, the gene that activates aflatoxin. It is complicated because this particular cytochrome P450 3A4 that is turned off in high doses of sulforaphane is not the only one that activates aflatoxin. There is another one called P451 A2 that also activates aflatoxin. How effective sulforaphane would be in protecting against activation of aflatoxin depends on how much of those two different genes people express. Some express lots of one and not much of the other, etc. So there will be individual differences in that effect as well.

LB: Wow, it sounded so much simpler when I was reading the articles.

Dr. E: Unfortunately, it is a complicated topic—especially the point of view: What does it all mean? What does a person actually recommend? Our interests right now in sulforaphane are largely pharmacological. We actually think it might have some benefit as a pharmaceutical agent, not in the treatment of cancer, but actually to prevent certain adverse drug interactions that happen.

There is a whole set of other drugs that increase the expression of the CYP3A4[35] gene, such as a drug called rifampicin that is used to treat tuberculosis (TB). When people take rifampicin, it totally screws up their metabolism of other drugs because it actually turns on this P450 3A4. So people who take rifampicin all of a sudden start metabolizing other drugs really quickly. This was first noticed when rifampicin was first prescribed to women who were being treated for TB and were also taking the birth control pill; they became pregnant because the rifampicin turned on the ability of their livers to metabolize the estrogen in the birth control pills so it was not effective.

Some of our research has gone away from looking at this dietary sulforaphane as a chemopreventive agent for cancer to looking at it as a substance that might be used as a codrug to block CYP3A4 induction caused by other drugs. What we have shown is that we can completely block that effect of rifampicin with sulforaphane. We can block the induction (the turning on) of the gene. There are many adverse drug interactions that happen with drugs; for example, you take two drugs and one of them turns on the gene that metabolizes the other one. Anyhow, that is the direction our research has moved into lately.

LB: A very complex issue. With all the contaminants that people were exposed to post-hurricane—chemicals, molds, mycotoxins, etc.—what advice would you give people who want to focus on a cancer-preventive diet?

Dr. E: The best thing, and it is not much different from what is recommended for anybody, is a diet high in fruits and vegetables. This seems to be helpful. I would not be averse to recommending that people (as many people do) take vitamin

---

35   CYP3A4 is the abbreviation used for cytochrome P450 3A4.

supplements, particularly antioxidant types of vitamins. Like anything, you can overdo that, too. "If a little is good, more is better" is not the best advice. But reasonable dietary supplements, vitamins, and dietary antioxidants, which are best taken in the form of natural fruits and vegetables, would be my recommendation.

LB: Are there studies to show what forms of vitamins—whole food vitamins versus synthetic, whole food vitamins with some synthetic, or synthetic vitamins taken with food—are best absorbed? Do studies prove that whole food vitamins absorb better?

Dr. E: There are a lot of studies that have looked at the various aspects of that, and it depends, in part, on the particular vitamins you are talking about. It varies. I think that the best evidence for protective effects of antioxidants come from studies when the antioxidants were from the diet—in other words, whole foods.

LB: When I was talking with the companies that actually make the tests for mycotoxins used at grain elevators and other places, it was explained to me that mycotoxins are more concentrated in the bran of the grain, which means the more refined the product is, the fewer mycotoxins it will have in it. With that in mind, it almost sounds as if it would be better to eat white bread versus whole grain breads, which have been so promoted by the grain industry.

Dr. E: That is entirely possible. I have to say that one of the concerns that I have about aflatoxins in foods is actually homemade peanut butter or peanut butters that are all natural and organic, simply because mold can grow on peanuts. One contaminated peanut kernel infected with *Aspergillus* can produce a remarkable amount of aflatoxin. Just one bad kernel can contaminate a whole, large batch of peanut butter. Natural peanut butter, especially if people are making it themselves, is not subject to periodic analysis done by big companies, such as Jif, that have the resources to self-test and are looked at enough by the FDA that they pretty much do fairly routine monitoring of their products. That is not necessarily the case with small companies or with peanuts that are sold in health food stores that people use to make peanut butter for themselves. Now I do not know that it is a problem, but it is something I have thought about often.

LB: There is an alternative to peanut butter called "Sunbutter," which is a butter made from sunflower seeds. Would that be a safer form of "butter" for people trying to avoid fungal antigens and mycotoxin contamination?

Dr. E: You know, I do not know. My gut answer is: not necessarily. Mycotoxins are not limited to peanuts. Pistachios and other nuts can be infected with *Aspergillus*. There have been some serious outbreaks of aflatoxicosis and aflatoxin contamination of other nuts in the Middle East in the past years. It is not necessarily limited to peanuts, but I do not know about sunflower seeds. I have no idea whether sunflower seeds are susceptible to *Aspergillus* contamination or not.

LB: Well, they are on a USDA list of raw materials possible for aflatoxin contamination. I was surprised to see them on the list because I had read somewhere else that they were less likely to have mycotoxin contamination since they are not a peanut and not a tree nut.

Dr. E: And so much of that depends on the environment in which they are grown. I mean, most peanuts and corn do not have any problem, but every once in awhile we have a drought in the Midwest that will be followed by rain storms and flash floods, and you will get very high levels of mold contamination in corn. In most other years, it is not a problem. Totally, it is an environmental thing and depends on the opportunistic molds. The conditions have to be right for it to grow. Sometimes those conditions will be right, and you have a big problem, and other times it is not.

LB: Some of what I have learned from the FDA has surprised me. They test only 200 samples a year of grain feeds for animal consumption.

Dr. E: Two hundred?

LB: Yes, and they test each sample for one to five mycotoxins.

Dr. E: Well, that is surprising to me, too.

LB: What I have been told is that, similar to what you have been talking about, it is an economic issue and that a lot of companies are self-testing at the level of purchase. For example, the grain elevators are checking before they purchase, and maximum levels of certain mycotoxins are written into contracts between the suppliers, elevators, and manufacturers. To a large extent, it appears we have to rely on the free enterprise system.

Dr. E: Yes, I think that is probably true. I think in many ways that actually works well. I was surprised a few years ago when I was contacted by the California Pistachio Commission, which is a trade organization comprised of pistachio growers. They were petitioning the government to regulate aflatoxin on pistachios at a strict level. The reason they wanted that was because they are concerned about aflatoxin levels, and they want to keep their product safe, but they are mostly concerned that pistachio imports from other countries that are not regulated could cause a food scare. They wanted to be able to reassure consumers that their products were safe if there was a food scare from another country; that happened a number of years ago with Iranian pistachios, I think, that were fairly seriously contaminated. They want to be able to say: We have this program in place and do this routine monitoring. They want to be able to convince consumers with data that yes, maybe it is a problem over there, but it is not a problem with their product. That makes good sense, actually, because food scares can really have a devastating impact, and if it is a large enough industry to do a good job of self-regulating—and in this case, actually wanting to be regulated by the government, which you usually do not hear from industry folks—it makes economic sense to them to ensure the safety of their commodity because, if there were a problem, it would have long-ranging impacts on everybody.

LB: Plus, if there are government regulations, they can write the regulation levels into contracts, and that could make compliance more enforceable in a court of law.

Dr. E: Exactly.

LB: What I keep being told by government agencies, the FDA and the USDA, is that testing is limited by funds.

Dr. E: I am sure that is true.

LB: I have been told that 30 percent of the population of the U.S. population is lacking the GSTM1 gene.

Dr. E: Fifty percent.

LB: Oh, my word! It is 50 percent?

Dr. E: Yeah.

LB: I was told that people who lack that gene cannot detoxify fungal metabolites, fungal fragments, and things like that.

Dr. E: That is partially true. That is actually an area of my direct work. In fact, we have looked specifically at GSTM1 and aflatoxin detoxification. It does it a little bit, but it is really lousy at detoxifying aflatoxin epoxide, at least compared with the similar gene in mice. There is a little activity there, but the data that we have from human liver studies does not suggest that it really makes any difference. There is a little bit of epidemiology data from China that suggests that people who are GSTM1 deficient and have long histories of dietary exposure to aflatoxins are at some increased risk for liver cancer compared with those that have a functional GSTM1. So, there is a little evidence that suggests that is true. Now, the difference is not dramatic, and on an individual basis it would be hard to say that GSTM1-null people are going to be very much different in their likelihood of getting liver cancer. First of all, their chances are very rare to begin with, so it is going from vary rare to a little bit less very rare. From a personal perspective, it probably does not make much difference, although it can show up statistically if you look at large populations. Now in terms of other mycotoxins, there is not much data, although it is a possibility. I have done quite a lot of work with GSTM1, and I have never seen any convincing data that says if you don't have GSTM1, you cannot detoxify zearalenone or some other mycotoxin, but it could play a role in some of them.

LB: Would people who are GSTM1 null be considered detoxification compromised?

Dr. E.: Well, marginally. That by itself probably doesn't make much difference. We have done, and other people have done, a lot of studies on cigarette smokers and GSTM1. The data suggests that in a population basis, if you are a cigarette smoker and you are GSTM1 null or negative (meaning you do not have the gene), you are maybe 25 or 30 percent more likely to get lung cancer than if you are a smoker and you have an active GSTM1. That sounds like a lot, but it is not very much when you compare the risk of smoking versus not smoking. If you are worried about the risk of smoking, then you should not smoke, and your GSTM1 status really does not make that much difference. Your chances of getting lung cancer if you smoke are very high, and they are a little bit higher if you are smoking and you are GSTM1 null, but it is not a huge difference. The same is true with bladder cancer.

LB: Would people who do not have the GSTM1 gene have a higher risk for disease development if they were in a multiple-contaminant environment, such as the hurricane-hit areas of Louisiana and/or Mississippi, since they are unable to detoxify as well as they could with the gene?

Dr. E.: Well, that is a reasonable hypothesis. Let me tell you, though, that there are literally hundreds of genes that are involved in detoxification of things in our environment, and GSTM1 is just one of them. It is probably not a particularly important one relative to all of the genetic differences that exist in how people metabolize chemicals. It totally depends on the specific things they are exposed to. The GSTM1 is one gene from a family called glutathione transferase, of which there are 16 different genes in humans. About 50 percent of Caucasians don't have GSTM1 at all, but they have other GSTs that do similar things, so there is a lot of overlap. That is why it is hard to say. A geneticist looks at what we call *penetrance*, which is the association between the genetic difference and the likelihood of the disease. The penetrance is very low. It is not zero; there may be some association, but it is not necessarily hugely important. It depends on what they are exposed to. In a very complex environment like Louisiana, where there are all kinds of different things, it is easy to hypothesize—to say, well, if you are GSTM1 deficient, you are likely to be more sensitive to things in your environment. How important that is on an individual basis is really hard to say.

LB: I was told by one of the doctors associated with a detoxification clinic that they were giving glutathione intravenously to patients sick from mold exposure, and that it made these people feel absolutely great—but only in the short term—while receiving the IV treatments.

Dr. E: Well, that is a scary proposition. I have never heard that one—intravenous glutathione. I know you can buy glutathione supplements and take them orally, but your stomach acid just breaks them down. Then your body takes the building blocks and perhaps resynthesizes glutathione. But intravenously is an interesting thing. I would be really curious to know the doses they use and how they arrived at that.

LB: I do not have any of that information, but it made us curious as to whether taking the building blocks for glutathione in amino acid form would boost the body's own production of glutathione to get similar results?

Dr. E: Right. One amino acid is rate limiting in the synthesis of glutathione, and that is cysteine. You can take cysteine. Cysteine by itself is not very effective because it oxidizes really easily, but there is something called N-acetyl cysteine (NAC) that is actually used to treat certain diseases. It is also used to treat acetaminophen (Tylenol) poisoning.

LB: Oh.

Dr. E: It works extremely well because Tylenol overdoses basically deplete the glutathione in your liver, and the subsequent amounts of Tylenol that are in the body after the glutathione is depleted then damage your liver. You can prevent that damage by letting your liver resynthesize glutathione by giving it plenty of extra cysteine. N-acetyl cysteine crosses cell membranes much better than cysteine itself and is much more stable against oxidation.

LB: So taking NAC actually helps your liver regenerate glutathione?

Dr. E: Oh, it absolutely does. It is very well known. It is the major therapeutic approach to treating acetaminophen overdose, which happens thousands of times a year in the U.S.

LB: Wow. Would there be any reason to add L-methionine to that? Would that make the NAC more effective?

Dr. E: Well, methionine finds its way into the cysteine pool, so it can do the same basic thing. It is a methyl donor, which may be useful in other antioxidant pathways. So, there is nothing wrong with that. L-methionine is probably comparable to using N-acetyl cysteine. L-methionine may end up doing the same thing in a slightly different pathway, but it gets to the same place.

LB: Will taking just one or two amino acids in supplement form, in comparison with taking a supplement that provides multiple amino acids, harm you?

Dr. E: I do not think taking one or two amino acid supplements would harm you. Most people do not have amino acid deficiencies if they have a balanced diet. It is a problem for strict vegetarians if they do not eat certain soy products and other sources of certain proteins, because there are some amino acids that are derived predominantly from meat in the diet. If you do not have meat in your diet, you can become deficient in certain amino acids if you do not have certain other proteinaceous plant components in your diet. But overall, it is not usually a problem. So, I do not know if taking supplements of a couple would necessarily be a problem if you did not take supplements of everything.

LB: You have clarified many areas. Thank you.

Regina M. Santella, PhD, *is a Professor of Environmental Health Sciences at the Mailman School of Public Health at Columbia University, Director of the Epidemiology Program at the Herbert Irving Comprehensive Cancer Center (HICCC), Director of the Biomarkers Core Facility at HICCC, and Director of the Columbia Center for Environmental Health in Northern Manhattan.*

Dr. Santella has an MS in Organic Chemistry from the University of Massachusetts and a PhD in Biochemistry from the City University of New York. Her research involves the development of laboratory methods for the detection of human exposure to environmental and occupational carcinogens and their use in molecular epidemiology studies to identify causative factors, susceptible populations, and preventive interventions. Her work has concentrated on the measurement of carcinogens bound to DNA, such as aflatoxins, with highly specific and sensitive immunoassays using monoclonal and polyclonal antibodies that her laboratory has developed. Dr. Santella can be reached through the Columbia University website: www.columbia.edu/. The interview was conducted July 2006.

LB: Most people do not think about what is happening inside their bodies when they are exposed to high levels of environmental carcinogens or toxins in either the home or the workplace. Tell us what actually takes place in the body when it is exposed to toxic substances and/or carcinogens.

Dr. S: One of the main purposes of the metabolism in the body is to rid the body of these chemicals. There are specific enzyme systems that are designed to metabolize these compounds and excrete them. Most of these chemicals are excreted. Volatile compounds come back out in your breath; some come out in urine; some are excreted in feces. However, for chemicals that are carcinogenic, sometimes the metabolism, which is designed to generate metabolites that then can be excreted easily, actually forms metabolites that are highly reactive. For example, aflatoxin in its original form is not carcinogenic. It cannot damage DNA, which, as we know, is the critical step in cancer development.

Aflatoxin is metabolized in the body to aid in excretion. When the body is exposed to aflatoxin, the body tries to metabolize it to a water-soluble form so it can be excreted in the urine. One of those metabolites is the reactive compound that damages DNA. So, when we are exposed to chemicals, whether they are carcinogenic or not, the major process of the body is to get rid of them. Generally, that works quite efficiently, but sometimes there is DNA damage, protein damage, and RNA damage as a result of the various metabolites that are formed during the process of trying to get rid of the carcinogen.

LB: Does that damage always occur?

Dr. S: Our DNA is constantly being damaged, even if we are not exposed to exogenous chemical carcinogens. There are normal cellular processes like metabolism of various hormones that generate, for example, reactive oxygen species that damage DNA. If you go out in the sunlight, your skin forms thymine dimers, a type of damaged DNA that leads to skin cancer. DNA damage is happening constantly. Since DNA is so critical to the cell, there are a number of enzyme systems in the body that can repair DNA. We have multiple DNA repair pathways, so even if you are exposed to a carcinogen and some small portion of it binds to your DNA, your body has the ability to fix that DNA, and then you are fine.

Cancer develops if you have damaged DNA in a cell that then replicates, or divides, before the damage can be removed. That is how the genetic changes occur that are responsible for cancer development. So, if your DNA is damaged and your cell fixes it, no problem. However, if the DNA damage is there when the cell decided it was time to divide, that is when you have the problem. Studies that we, and others, have been doing are showing that DNA repair capacity—the ability to repair damage to your DNA, which we know varies from one person to another—is a major factor in understanding who gets cancer.

You have probably heard the story: I have a great uncle who smoked 100 cigarettes a day and did not get lung cancer. Only about 15–20 percent of smokers develop lung cancer. We now know that some of that risk is related to genetic susceptibility for many different genes. But DNA repair genes are a critical factor that accounts for why some people develop cancer and others do not. We are all getting our DNA damaged all the time, every day. Even as I am sitting here talking with you, my DNA is being damaged because my cells are doing what they are supposed to be doing, and most of that damage is being fixed.

LB: Aflatoxins can alter DNA structure. Do all mycotoxins have that ability to alter DNA?

Dr. S: I do not know about all the mycotoxins. It is the DNA binding, the covalent attachment to the DNA, that is responsible for the change in DNA structure and conformation that leads to the mutations and other genetic changes. I cannot answer for all mycotoxins. If they can be metabolized to these reactive intermediates, then DNA damage can happen. I know there is another mycotoxin, fusarium C, that also forms DNA adducts, but those are the only two I know of—aflatoxin and fusarium C.

LB: How great a health risk do you think there was, or will develop, from exposure to the prevalent mold growth and very high levels of airborne mold spores documented after Hurricane Katrina?

Dr. S: It depends on the individual's exposure and which health outcome you are talking about. I do not think there is any good data on risks from aflatoxin. I do not know if there is data on levels in the air, in any of the homes or areas down there, so it is really hard to say anything about concrete health effects. I mean, we know a lot about levels of dietary consumption of aflatoxin and relationship to liver cancer risks, but that is a very specific outcome and very specific information on dose. You cannot do risk assessments unless you know the dose. You know the health effects of the compound, but you also need to know the exposure of the people. I do not know the exposure, so I cannot answer that.

LB: You said that when the body metabolizes aflatoxin, it creates a more toxic substance. Is there any way to alter the biological structure of aflatoxin through a food source or other natural means so that the body could metabolize it with no risk to cell damage?

Dr. S: Well, aflatoxin is very light sensitive. When we work with it in the laboratory, if we leave it out on the bench, it is gone. So, there are ways of getting rid of aflatoxin because it is relatively unstable as it is light sensitive when it is out in the air.

Regarding dietary aflatoxins, there are studies that have been done in mainland China that show you can give various interventions and decrease DNA damage. They do not know cancer risks because these are short-term studies that use what we call biomarkers. We know that aflatoxins bind to DNA and proteins, so they looked at levels of aflatoxins bound to DNA and protein in people who are normally exposed through the diet.

Three randomized trials were conducted with some people put on a placebo and some put on the test agent. They had people who were exposed and looked at before and during the intervention, noting levels of aflatoxin DNA adducts that are actually excreted in the urine. What they found was that the interventions worked to decrease the levels of DNA damage that people had.

The first intervention was with oltipraz, which is a drug, so that is not terribly practical. What the drug does though is induce certain enzymes that conjugate these reactive intermediates that can be formed. So it increased levels of a protective enzyme. By enhancing the level of this enzyme, you protect the body from the reactive metabolites that are produced. That was great, but it is not very practical to give to people who are exposed because it has side effects, and it is a costly drug.

The second intervention was with a compound called chlorophyllin. It is a chlorophyll derivative normally present in food that you can actually buy at your health food store as pills. What the study showed is that when people took this pill of chlorophyllin with each meal, the DNA damage decreased. Why? Because the chlorophyllin physically binds the aflatoxin, so it does not get absorbed from the diet. These studies are being conducted by Dr. Tom Kensler, Dr. John Groopman, and others at Johns Hopkins.

The latest study that they have done is a broccoli sprouts intervention where they grew the broccoli sprouts, soaked them in water, and fed the broccoli sprout extract to people. In broccoli and broccoli sprouts, there is a class of compounds that also induce glutathione S-transferase, one of the productive enzymes.

These three different interventions resulted in lower levels of DNA damage. So, there are ways to protect yourself, but these are research studies and not ready for prime time, as they say.

LB: And there are no side effects?

Dr. S: Chlorophyllin? Broccoli? I mean everybody is eating broccoli, right? All the best data on cancer risks: Eat your fruits and vegetables, primarily vegetables. We already know what you need to do. Stop smoking, minimize your alcohol consumption, eat your fruits and vegetables, exercise, and don't gain a lot of weight. These are the most important things you can do.

LB: They found that the broccoli sprouts, soaked and then made into a tea, induced glutathione S-transferase?

Dr. S: Well actually, that is the mechanism proposed to explain their results, but they did not measure levels of the enzyme.

LB: I have been told that many people do not have certain enzymes like the GSTM1.

Dr. S: Hey, that's me! You are talking about me.

LB: You don't have that particular gene?

Dr. S: That's correct. The gene you are talking about is one of a family of glutathione S-transferases, the one that conjugates aflatoxins. This is what the broccoli sprouts have been proposed to induce.

LB: At how much greater risk for development of cancers or other serious diseases are people who lack the GSTM1?

Dr. S: There is a lot of good data, even what they call a meta-analysis that studied huge numbers of people in relationship to lung cancer. We know that smoking is a huge risk factor for lung cancer. There is a 20- to 30-fold increase in cancer risk if you are a smoker, compared with a nonsmoker. However, if you are GSTM1 null, there is maybe a 10 percent increase compared with someone with the gene.

Yes, it has a statistically significant impact on your cancer risk, but is very, very small compared to a 2,000 percent increase from smoking. Yes, the genotype does definitely have an impact on cancer risk. It is one the few genetic susceptibility genes that I really believe in now.

A lot of other data on genetic susceptibility is conflicting. But, it conveys a very, very small increase in risk, and that is very different from, for example, women who carry BRCA mutations where they have a lifetime risk as high as 80 percent of developing breast cancer. BRCA is one of the genes that normally works to prevent breast cancer. You see, GSTM1 is not the only enzyme that detoxifies these kinds of reactive compounds. There is a whole class of GSTs: M1, T1, P1, and so forth. So, it can be a risk, but it is not huge.

LB: So we don't want to say that it is not that big a deal, but we also do not want to say that it is a huge concern?

Dr. S: Right. I used it to give my children another reason not to smoke since two of them are also GSTM1 null. Not that if you are GSTM1 positive that you should smoke, obviously, since smoking has other health consequences.

LB: Well, you have the added advantage of the lab to back you up! I have been told that between 30–50 percent of the U.S. population is GSTM1 null.

Dr. S: Yeah, about 50 percent.

LB: How much slower is the detoxification process for people who are GSTM1 null as compared with those who have the GSTM1 gene, and is this a concern?

Dr. S: I don't think you are going to find any numbers on that because it is going to vary with what compound a person is exposed to and because of the overlap in what we call substrate specificity, which means different substances are metabolized through different pathways. There are multiple detoxification enzymes, so I think that is why it really conveys a small increase in risk. It is not the only enzyme that can do the work.

LB: So the other enzymes pick up the slack?

Dr. S: Yes. Maybe not 100 percent, but they do.

LB: How big a problem do you think aflatoxin and mycotoxin contamination is in food meant for human consumption in the U.S.?

Dr. S: In the U.S. it does not seem to be as big a problem, probably because also in the U.S. we do not have high hepatitis B viral infection rates. All of our studies in Taiwan and the studies of Dr. Groopman and other groups have shown that the people at highest risk for liver cancer are those who have both the aflatoxin exposure and are also carriers of either hepatitis B or C virus. That is probably because the viruses cause cell death in the liver that leads to cell proliferation. So, the cells in the liver keep dividing, but if you have aflatoxin DNA adducts formed in the liver in a dividing cell and the adducts are there when the cell divides, that is when you get the genetic alterations. So, the biggest risk is with the combination of hepatitis virus and the aflatoxins. That said, the U.S. population does have about a 2 percent carrier rate for hepatitis C virus. So, it is a problem. The FDA does regulate the levels of aflatoxin in the food chain. Our regulations are somewhat higher in terms of what is allowed as compared with other countries, like England.

LB: I have been talking with the FDA about some of those regulations. Several things surprised me: They do only 200 field samples a year in grain and feed for animal consumption; they do not require testing; and the regulations are not binding. A court of law ruled that they were not binding.

Dr. S: I did not know that.

LB: I didn't either. Also, when I spoke with Dr. Wang, he said that he had actually bought some cornmeal at a national grocery store chain that turned out to be contaminated with mycotoxins. He uses it to teach his class. He was actually buying it to use as a control for his study but found it was contaminated also. At this point, I am a little uncertain as to how much of a problem aflatoxin and/or mycotoxin contamination in U.S. food might be.

Dr. S: A couple of years ago *Consumer Reports* actually published an article about peanut butter and said never to buy health food store peanut butter because the quality controls were worse than those in some of the larger companies like Jif, Skippy, and so forth that did much more testing. But I do not think it is a huge problem because if you look at U.S. liver cancer incidence, yes, it is going up, but that is probably almost entirely due to an increase in hepatitis C virus infection. That is not to say that aflatoxin may not play a small role in it. A number of years ago, we did a little study with Dr. Groopman where we looked at some samples from U.S. liver cancer patients. We looked at the liver DNA. Dr. Groopman looked at the aflatoxin-albumin adducts, and we both found a couple of positive samples, but that does not say anything in terms of proving aflatoxin caused that liver cancer. The little pilot study that we did just cannot answer that question. All it can say is that yes, sometimes you do find aflatoxin on the DNA of the liver tissue or the albumin adducts of the people in the U.S.

LB: But, again, you can't make any correlation that the aflatoxin present in the DNA caused the liver cancer?

Dr. S: No, you cannot do that with one sample. In the nested case control studies that we are doing in Taiwan, our collaborator recruited 25,000 people and collected blood and urine from them while they were healthy. Then, over time, liver cancer cases have developed. They have gone back to their freezers, pulled samples, and we have done assays for aflatoxin biomarkers. We have shown that in those people who went on to develop liver cancer, there were higher levels of the aflatoxin biomarkers. But there were also controls who had some aflatoxin biomarkers. It was on a population basis, not an individual, that aflatoxin was associated with liver cancer. So, you had individuals with high levels of the aflatoxin biomarker but who were controls and cases of liver cancer that have low biomarkers. But among all the cases and controls, more cases had high aflatoxin biomarkers. If you do just one single sample, you cannot say anything about whether that caused the cancer or not.

LB: For people who are genetically predisposed to poor repair of DNA, is there anything that they can do or take to increase the repair ability in their bodies? I know you talked about the broccoli sprouts and the chlorophyllin earlier.

Dr. S: Well, there are no clear data for DNA repair. There is some evidence that when you are exposed to certain things, there is induction of DNA repair capacity; for example, cigarette smoking, but that is clearly not something you want to do to increase your DNA repair capacity.

LB: Wow.

Dr. S: Well, it makes sense because the body wants to defend itself. We certainly don't want to use that exposure as a way to increase DNA repair capacity.

LB: I understand.

Dr. S: There have been one or two studies saying that if you give antioxidants, you can induce DNA repair, but I don't think there is any hard, believable data at this point.

LB: Are there any other preventive measures for people who would like to reduce their susceptibility to cancer from living or working in a toxic, possibly carcinogenic, environment?

Dr. S: Well, I think that I said it before: Eat your vegetables, minimize your alcohol intake, keep your body weight down, and exercise. Those are the main things.

LB: What levels of cancer and disease do you think we will see from the toxic wake of Hurricane Katrina?

Dr. S: I, personally, don't think there will be an increase related to Katrina. I think the biggest problem continues to be the 25–30 percent smoking rate in this country and the dramatic increases in obesity. Those are major risk factors for cancer. I think it is really going to be hard to document increases in cancer in relationship to exposure to Hurricane Katrina. You need to look at long-term occupational exposures to be able to really detect an association between an exposure and cancer risk. It is the occupational exposures that have given the positive data. That's how we have identified most known chemical carcinogens. It is because the high exposures decrease the latency and increase the number of cancers. I do not think any of the exposures down South match the kinds of exposures that people see in the occupational setting. So it is really, really hard to identify causes of human cancers outside the occupational setting.

LB: Do you think we will see a higher incidence rate in the high-risk group—children, the elderly, and the immunocompromised, who were exposed to toxins post-Katrina?

Dr. S: Personally, I do not know a lot about the data from down there—what the exposures were and how long people were exposed—so I can't say anything too definitive. Stress has been shown to have an impact on cancer and disease risk in general. I think people have to be realistic about what their total exposures are and obviously minimize exposures, but not go into a panic mode, especially about bad things that it is too late to do anything about, and from then forward, just do the best they can in terms of a healthy lifestyle.

LB: And what would you most like people to take away from this interview?

Dr. S: Stop smoking. Lose weight. Eat your fruits and vegetables. That really is the public health message. I know it is hard to do. But these are really the risk factors that we know have a big impact.

LB: It is our hope that people who went through exposure will take this information and say, "Well, I was exposed to some of this stuff, so I am going to put extra effort into eating vegetables and keeping my current exposure level down," because that way they may be able to minimize or negate some of the potential damage.

Dr. S: Right. Yeah.

LB: And even use the possibility of an increased health risk from the hurricane exposure as a catalyst to help them move forward in a healthier lifestyle. Thank you for sharing this valuable information for health.

SECTION FOUR

RESOURCE SECTION

RESOURCE SECTION:

The purpose of this Resource Section is to direct you to resources we have found helpful as well as those we have put together for you. For more information on each resource, please visit the website provided in each reference listing.

MOLD INFORMATION:

www.MoldMentor.com – This is an informational website we created as a resource for victims of structural mold and water-damaged buildings. It features locator maps to Professional Resources, such as mold assessors, mold remediators, and medical professionals who treat mold- and mycotoxin-related illnesses.

Download the following free brochures and forms from MoldMentor.com:

I.   Brochures: Don't Become a Victim of Mold
  a) Disaster Area Brochures for Residents
  b) Disaster Area Brochures for Volunteers

II. Fillable Forms with Criteria Checklists
  a) Mold Assessor Selection Form
  b) Mold Remediator Selection Form

Additional free information is available in the form of blogs and articles. Click on "Ask Mold Mentor" to submit questions on mold, indoor air quality, and other health-related topics. Please note, the website is in a growth mode, so if you don't find the resource you need, feel free to submit a request via the website.

HEALTH SUPPLEMENTS:

www.HealingNoni.com – We have been long-term users of raw (unpasteurized) noni juice. We feel it was a vital component of the natural treatment plan that restored our health when we were sick from mold-related illnesses after Hurricane Katrina. Our naturopath prescribed raw noni juice to all of the members of our family in medicinal dose levels based on age, size, and health symptoms (see the chapter "Help from the Kahunas" in Section II). We continue to use raw noni juice to this day as a regular health supplement. For more information on the raw noni juice, see the following direct link: https://www.healingnoni.com/raw-organic-noni-fruit-juice-32oz/ .

As a fermented product, noni juice contains concentrated levels of live enzymes, compounds and nutrients that we feel help the body detox and sustain health. We share our supply source of a high-quality noni juice that ships direct from the farm as it is a specialty product not easy to source. As a courtesy to our readers, Healing Noni is offering a 10% discount on your first order—just mention "MoldMentor" when placing your order at 1-877-662-4610, or enter "MoldMentor" in the coupon code field when ordering online at HealingNoni.com. Their juice is available both raw (unpasteurized) and pasteurized. Substantial discounts are offered on multiple bottle orders, and shipping is FREE!

For more information on Noni juice, download our digital book *Why Noni?* The book is available on Amazon and through other digital retailers.

AIR PURIFIERS:

We are in the process of negotiating discounts for our readers on some high quality air purifiers that we have found to be effective in increasing the quality of indoor air. Please visit www.MoldMentor.com for more information.

VACUUMS:

www.Nilfisk.com – We have been long-term users of the Nilfisk GM 80 Vacuum, which is a powerful canister vacuum featuring a standard multi-stage filtration system with a paper dust bag, main filter, microfilter and HEPA exhaust filter. We have used this vacuum as a daily household vacuum as well as in an industrial capacity when remodeling our home. The suction and filtration levels make it possible to use this vacuum to suck up not only household dirt and debris but also superfine sawdust from cutting wood and heavy sheetrock dust from cutting drywall. The design is such that the vacuum does not clog, and the suction power does not reduce as the bag fills. It also does a phenomenal job of cleaning automobile carpeting and upholstery when using the hose without the floor attachment.

For more information on the GM 80 vacuum, enter the product name in the search bar at Nilfisk.com. While on their website, also check out the Nilfisk wet-dry vacuums that offer increased versatility with dual functionality. They are excellent tools to use to remove mold from wet building materials that are salvageable for drying. For blogs on both the Nilfisk GM 80 vacuum and the Nilfisk AERO 21 INOX wet/dry vacuum, visit MoldMentor.com.

FICTION BOOKS:

www.BooksThatGive.com – We formed the Books That Give™ program to help support nonprofit causes by raising funds through book sales. The program is supported by the sales of the books in the Super Bouv Series, which are three children's books based on the story of a real-life service dog. The stories help educate people about the health risks associated with water-damaged buildings but in a fun way. The main character, Ruby, is a pup in search of her talent. After several attempts with no success, she finally finds a family of her own, gains a home, and discovers her true talent using her nose!

Each purchase of the Super Bouv Series supports the rescue of Bouviers through the American Bouvier Rescue League (ABRL) in honor of Freida, our Bouvier, who helped reduce our sensitivities to mold and other airborne contaminants.

NOTES